Sound and Structural Vibration

Radiation, Transmission and Response

Sound and Structural Vibration
Radiation, Transmission and Response

FRANK FAHY

Institute of Sound and Vibration Research
University of Southampton
Southampton, UK

PAOLO GARDONIO

Institute of Sound and Vibration Research
University of Southampton
Southampton, UK

AMSTERDAM BOSTON HEIDELBERG LONDON NEW YORK OXFORD
PARIS SAN DIEGO SAN FRANCISCO SINGAPORE SYDNEY TOKYO

ELSEVIER

Academic Press in an imprint of Elsevier

Academic Press is an imprint of Elsevier
Linacre House, Jordan Hill, Oxford OX2 8DP, UK
The Boulevard, Langford Lane, Kidlington, Oxford OX5 1GB, UK
30 Corporate Drive, Suite 400, Burlington, MA 01803, USA

First edition 1987
Second edition 2007

British Library Cataloguing in Publication Data
A catalogue record for this book is available from the British Library

Library of Congress Cataloging-in-Publication Data
A catalog record for this book is available from the Library of Congress

For information on all Academic Press publications
visit our web site at books.elsevier.com

ISBN 13: 978-0-12-373633-8

Transferred to Digital Printing in 2008

In thanksgiving for the love and support given to us by our parents, Amalia, Meg, Riccardo and Frank.

Contents

4. Fluid Loading of Vibrating Structures

5. Transmission of Sound through Partitions

8. Introduction to Numerically Based Analyses of Fluid–Structure Interaction

Preface to the First Edition

In writing this book my aim has been to present a unified qualitative and quantitative account of the physical mechanisms and characteristics of linear interaction between audio-frequency vibrational motion in compressible fluids and structures with which they are in contact. The primary purpose is to instruct the reader in theoretical approaches to the modelling and analysis of interactions, whilst simultaneously providing physical explanations of their dependence upon the parameters of the coupled systems. It is primarily to the engineering student that the book is addressed, in the firm belief that a good engineer remains a student throughout his professional life. A preoccupation with the relevance and validity of theoretical analyses in relation to practical problems is a hallmark of the engineer. For this reason there is a strong emphasis on the relationship of results obtained from theoretical analysis of idealised models and the behaviour of the less than ideal realities from which they are abstracted.

The teacher of analysis in any sphere of applied science is faced with a central dilemma: systems which can be modelled and analysed in a manner sufficiently explicit and direct to illustrate a principle are usually gross oversimplifications of the real world and are hence, to some extent, trivial; systems which are of practical concern are usually much too complex to offer suitable examples for didactic purposes. In attempting to grasp this nettle I hope I may be forgiven by any physicists and applied mathematicians who may pick up this book for sacrificing a certain amount of mathematical rigour for the sake of qualitative clarity.

In teaching mechanical engineering and engineering acoustics over a number of years it has struck me forcibly that an appreciation of structural vibration as a form of wave motion, a concept readily grasped by the student of physics, is often lacking in those reared on a diet of lumped elements and normal modes. One unfortunate effect is that the associated wave phenomena such as interference, scattering and diffraction are often believed to be the preserve of water and air, and the link between natural modes and frequencies of structures, and the component waves intrinsic to these phenomena, is not readily perceived. The subject of this book appeared to be the ideal vehicle for persuading students of the advantage to be gained by taking a dual view of vibrational motion in distributed elastic systems. Hence I have emphasised the wave 'viewpoint' right from the start, in the hope of encouraging the reader to 'think waves'.

The three main categories of practical problems to which the material of this book is relevant are sound radiation from vibrating structures, sound transmission between adjacent regions of fluid media separated by an intervening solid partition, and the response of structures to excitation by incident sound fields. Much of the source material is only available (in English at least) in articles scattered throughout the learned journals of the world. In particular, fundamental analyses in acoustics textbooks of sound transmission through partitions tend to be restricted to highly idealised cases, and the complicating effects of finite panel size, non-homogeneous structures, cavity absorption and frames, and panel curvature are at best briefly and only qualitatively described. This is why Chapter 4 is the longest in the book.

Although the aim of the book is instructional, it is different from many other textbooks in that it is not divided into neat, self-contained sections of analysis, which can be concluded with Q.E.D.; it also contains a large amount of descriptive text. The first feature is connected with the 'dilemma' previously mentioned; the second stems from a desire to provide a text from which the reader can learn in the absence of a formal lecture course, although it is hoped that my prolixity will not deter a lecturer from using the book to complement his course.

The arrangement of questions in the book does not generally follow the conventional pattern of formalised quantitative examples at the ends of chapters. The reader is challenged at various places within the text to think about the material which he is currently reading, while it is fresh in his mind. I hope, in this way, to solicit more active cooperation in the learning process, and to stimulate a questioning approach to the material, rather than passive acceptance. The questions at the ends of chapters are linked to specific sections in the text and range from straightforward numerical evaluation of quantities, intended to encourage a 'physical feel' for their orders of magnitude, to rather open-ended questions, which can only be answered in qualitative terms. The absence of a large number of formal calculation exercises reflects both the nature of the subject and the fact that the readership is expected to have developed previously the facility for performing formal analyses of fundamental vibrational and acoustical problems.

Numerous references to other books, research publications and reports are provided in the text. The list is clearly not comprehensive, but it is hoped that it will provide the reader with jumping-off points for further and deeper study. The omission of any particular relevant reference in no way constitutes reflection of its value, any more than inclusion of a reference implies that it is considered to be uniformly meritorious and correct.

It is my hope that this book, for all its faults which will no doubt emerge, will help at least a few people to understand more fully the fascinating inter-play between sound and structural vibration, and thereby serve to increase their ability to control whatever aspect of the subject commands their attention.

Frank Fahy
Southampton
November 1983

Preface to the Second Edition

The first edition of 'Sound and Structural Vibration' was written in the early 1980s. Since then, two major developments have taken place in the field of vibroacoustics. Powerful computational methods and procedures for the numerical analysis of structural vibration, acoustical fields and acoustical interactions between fluids and structures have been developed and these are now universally employed by researchers, consultants and industrial organisations. Advances in signal processing systems and algorithms, in transducers, and in structural materials and forms of construction, have facilitated the development of practical means of applying active and adaptive control systems to structures for the purposes of reducing or modifying structural vibration and the associated sound radiation and transmission.

In this greatly expanded and extensively revised edition, we have retained most of the analytically based material that forms the pedagogical content of the first edition, and have expanded it to present the theoretical foundations of modern numerical analysis. Application of the latter is illustrated by examples that have been chosen to complement the analytical approaches to solving fairly simple problems of sound radiation, transmission and fluid–structural coupling that are presented in the first edition. The number of examples of experimental data that relate to the theoretical content, and illustrate important features of vibroacoustic interaction, has been augmented by the inclusion of a selection from the vast amount of material published during the past twenty five years. The final chapter on the active control of sound and vibration has no precursor in the first edition.

The principal additions to the material of the first edition are as follows. Chapter One has been expanded to illustrate the relation between waves and modes in a more explicit, analytical manner. The section on waves in thin-walled, circular, cylindrical structures, originally presented in the chapter on sound transmission, has been relocated in this chapter. Statistical parameters such as modal density and overlap are introduced. The brief introduction to impedances and mobilities of structures that originally closed Chapter One has been replaced by an extensive treatment of the topic in Chapter Two, which also introduces the energy based model of coupling between structural elements. Chapter Three, on sound radiation by vibrating structures, concludes with a presentation of the Kirchhoff–Helmholtz integral formulation and also introduces radiation modes. Chapter Five, on sound transmission through partitions, has been expanded to cover a wider range of non-uniform structures, amplifies the original treatment of sound transmission through thin-walled, circular cylindrical shells, and presents a theoretical analysis of sound transmission through pipe walls. The original analytical models of acoustic coupling between structures and enclosed volumes of fluid are complemented in Chapter Seven by sections that present the Kirchhoff–Helmholtz integral formulation for the interior problem, and briefly analyse the waveguide behaviour of fluid-filled pipes at low frequency. The original brief introduction to Statistical Energy Analysis has been modestly expanded. Chapter Eight presents a comprehensive introduction to numerical analyses of vibroacoustic coupling. Finite Element Analysis is illustrated by application to the problem of acoustic coupling between an elastic plate and fluid in a rigid rectangular box and Boundary Element Analysis is illustrated by the associated problem of sound radiation from the plate. Chapter Nine introduces the fundamental principles, theory, devices, and implementations of active control and illustrates the benefits and shortcomings of various strategies by theoretical examples.

Notes on notation:

1. Because there is inconsistency between the symbols conventionally used in analytical treatments of fluid dynamics, acoustics and vibroacoustics, and those conventionally used in numerical analysis of structural vibration, the notation for displacements and velocities employed in Chapter Eight,

which is defined in a footnote on p. 456, differs from that in the rest of the book.

2. The tilde over a symbol indicates 'complex number'.

3. An overbar denotes time average and angle brackets denote space average.

Frank Fahy and Paolo Gardonio
Southampton
May 2006

Acknowledgements

The authors wish to acknowledge contributions to the preparation of this book made by the following colleagues. Special thanks are due to Emanuele Bianchi for scanning the first edition, of which there existed no electronic version, and Silvia Rossetti for retyping most of the mathematical content of the parts included in this edition. Advice with regard to the presentation of some of the more challenging technical material was provided by Steve Elliott, Brian Mace, Mike Brennan and Chris Jones. The work of PhD students Cristoph Paulitsch, Cristobal Gonzalez Diaz and Damien Emo formed the basis of some of the material of Chapter Nine. Advice on matters of grammar and vocabulary was sought by the second author of Joyce Shotter and Maureen Strickland and graciously given.

Lynne Honigmann and Vicki Wetherell at Elsevier waited patiently as many deadlines came and went and were always ready to offer helpful advice. We also wish to acknowledge the contribution of CEPHA Imaging Pvt. Ltd, Bangalore, India.

The first author wishes to thank his wife Beryl for accepting the deprivation of shared social activities caused by the regular and prolonged physical and mental absences of her husband on book business. Assistance with the proof reading of the mathematical content was generously provided by Thomas Fahy.

List of Permissions

Figures 1.2, 1.7, 1.13, 5.4, 5.13, 5.18 and 5.27 Reproduced with permission from Fahy, F. J. (2001) *Foundations of Engineering Acoustics*. Academic Press, London.

Figures 1.5, 1.14, 1.16, 3.8, 5.1, 5.2, 5.5–5.8, 5.14, 5.17 and 5.26 Reproduced with permission from Fahy, F. J. (1987) *Sound and Structural Vibration*. Academic Press, London.

Figure 3.5 Reproduced with permission from Fahy, F. J. (1995) *Sound Intensity* (2nd Ed.). E & FN Spon, London.

Figure 3.37 Reproduced with permission from Mead, D. J. (1990) *Journal of the Acoustical Society of America*, 88:391 © 1990 Acoustical Society of America.

Figure 3.50 Data supplied under license by ESDU International plc, its affiliated and parent companies © 1995, all rights reserved.

Figure 3.32 Reproduced with permission from Cremer, L. and Heckl, M. (1973) *Structure-borne Sound*. Springer-Verlag, Heidelberg.

Figures 4.8 and 4.9 Reproduced with permission from Stepanishen, P. R. (1982) *Model Coupling in the Vibration of Fluid-loaded Cylindrical Shells*. © 1982 American Institute of Physics.

Figure 5.3 Reproduced with permission from Fahy, F. J. (1998) *Fundamentals of Noise and Vibration*. E & FN Spon, London.

Figure 5.4 Reproduced with permission from Norton, M. P. and Karczub, D. G. (2003) *Fundamentals of Noise and Vibration Analysis for Engineers 2nd Edn.* Cambridge University Press, Cambridge.

Figure 5.9 Reproduced with permission from Quirt, J. D. (1982) *Journal of the Acoustical Society of America,* 72:834–844 © 1982 Acoustical Society of America.

Copyright on Figures from the first edition as acknowledged therein.

Introduction

> I think I shall never undertake to write a book again. If one were
> a scamp, the work would be easy enough, but for an honest man
> it is dreadful.
>
> John Tyndall, 1859

Tyndall went on to write another eighteen books!

As you read these words you are almost certainly experiencing various manifestations of the process of vibrational interaction between fluids and solid structures. Traffic noise may be heard through the windows of the room; the plumbing system may be announcing its operation; or perhaps the radio or the iPod is providing background music for your pleasure. The first two examples represent the undesirable aspect of this phenomenon and suggest that study of its mechanism is of importance to those concerned with the control and reduction of noise. The third example shows that the process may be put to good use: vibrations of musical instruments, microphone diaphragms and loudspeakers act as intermediaries in the generation and reproduction of sound which, at least for some listeners, is the very antithesis of noise. The principal function of this book is to explain the physical process of interaction and to introduce the reader to various mathematical models and theoretical analyses of the behaviour of coupled fluid–structural systems. Many examples of experimental data are presented to illustrate the theory. This field of study is known as 'vibroacoustics' or, in the USA, 'structural acoustics'.

Acoustic vibrations in fluids and solid structures essentially involve the propagation of wave motion throughout the supporting media, although explicit recognition of this fact is not always apparent in textbooks on mechanical vibration. Indeed, an emphasis on the work–energy approach to vibration analysis, which is based upon the representation of the vibrational state of a system in terms of a finite number of degrees of freedom and is fundamental to many modern forms of computer-based numerical analysis, tends to obscure the wave nature of the processes under analysis. In dealing with audio-frequency vibrations of systems involving the coupling of compressible fluids with plate and shell structures, it is important to possess an appreciation of the 'wave view' of vibration. There are three main reasons for this requirement. The first concerns the three-dimensional nature, and often very great extent, of fluid volumes that are in contact with elastic structures, which effectively rule out analytical limitation on the number of degrees of freedom describing the vibrational state of the fluid. The second relates to the interaction of sound waves in fluids with structural boundaries of diverse geometric form and dynamic properties, which is best understood in terms of the wave phenomena of reflection, diffraction and scattering. The third reason is associated with the fact that the frequency ranges of practical concern in the field of vibroacoustics usually embrace a very large number of natural frequencies and high order modes of the components involved; discrete models are not necessarily appropriate because of uncertainties in the modelling of detail such as structural joints and the large number of degrees of freedom involved. Hence, vibrational wavefield models, analogous to those employed in the modelling of room acoustics, are often more practicable and effective.

For these reasons, Chapter 1 introduces the reader to a unified mathematical description of temporal and spatial distributions of wavefield variables and illustrates the forms of vibrational waves that travel in various ideal forms of structure such as uniform beams, flat plates and thin-walled, circular cylindrical shells. The phenomenon of wave dispersion that relates wave speed and frequency is shown to form the basis for categorising the regimes of wave interaction at interfaces between different media and different forms of structure. The phenomena of natural frequencies and modes of bounded elastic systems, and the related phenomenon of resonance, are illustrated and explained in terms of wave reflection and interference. The chapter closes with a brief introduction to probabilistic modelling of natural frequency distributions and related quantities that are employed in Statistical Energy Analysis (see Chapter 7).

Chapter 2 principally concerns the concepts of structural impedance and mobility that characterise the response of beams, flat plates and circular cylindrical shells to excitation by harmonic point forces and are also central to the modelling of networks of structural elements connected at discrete points. It is shown how mathematical expressions for the mobilities of uniform beams in bending may be derived either in terms of interference between bending waves reflected from boundaries or in terms of summations of modes that individually behave as single-degree-of-freedom systems consisting of lumped masses, springs and dampers. This reinforces the duality of wave and mode models introduced in Chapter 1. A simple model of singly connected structures is used to show how the effectiveness of a vibration isolator may be expressed in terms of mobilities. More complex networks of point-connected structural components may be represented by matrix models that incorporate the impedances or mobilities of the components at the connection points. Their application is illustrated by an analysis of the flow of vibrational energy through a beam network. The chapter closes with a brief analysis of wave energy propagation in beams and plates.

The mechanics of sound radiation by vibrating surfaces is explained in Chapter 3 in terms of the spatial distribution of vibrational acceleration normal to the interface with a fluid. The phenomenon of radiation cancellation is illustrated by a model of two simple sources. Analyses of sound radiation from planar surfaces by means of far-field evaluation of the Rayleigh integral, and in terms of travelling wave Fourier component (wavenumber) synthesis, are presented in such a way that the equivalence of these two models can be appreciated. It is shown that flexural waves travelling in an unbounded flat plate radiate sound power only if they travel supersonically. By means of wavenumber spectrum analysis of the modes of finite baffled plates, this crucial result is exploited to categorise the modes into four classes of radiator. Expressions for the radiation efficiencies of vibrating plates are derived in terms of natural modes and arrays of elementary radiators. The concept of frequency-dependent radiation modes that radiate sound power independently is explained and their derivation is presented. The utility of the Fourier approach is further illustrated by application to the evaluation of the contributions to sound power radiation by a flat plate made by concentrated applied forces and imposed velocities and from forces arising from the presence of local constraints. The radiation characteristics of stiffened, corrugated, sandwich and composite plates are illustrated. The radiation characteristics of thin-walled circular cylindrical shells are then analysed in some detail. The chapter closes with a derivation of the Kirchhoff–Helmholtz integral that expresses the sound pressure anywhere in a fluid in terms of the distributions of sound pressure and normal particle acceleration on any surface that bounds the fluid. This integral provides the basis for the numerical evaluation of sound radiated by vibrating bodies and for the influence on incident sound of bodies situated in a fluid as presented in Chapter 8.

The problem of evaluating the reaction forces (fluid loading) applied by a fluid to a vibrating body is addressed in Chapter 4. The concept of complex acoustic radiation impedance is illustrated by some simple examples. The value of the concept of wave impedance in analysing wave propagation in coupled fluid–structural systems is demonstrated in cases of bounded and unbounded flat plates. The effects of fluid loading on the natural frequencies and sound radiation by point-excited plates and circular cylindrical shells are then described and illustrated.

Chapter 5 presents detailed accounts of the sound transmission characteristics of plane partitions of various forms, including single-leaf, double-leaf and non-uniform constructions. The reader is led from elementary analyses of highly idealised systems, through a discussion of the relative importance of resonant and non-resonant transmission mechanisms, to an appreciation of the current state of the art with regard to understanding and predicting the behaviour of the more complex systems used in practice. Simple analyses of the influence of the presence of connecting frames and cavity absorption on the transmission loss of double-leaf partitions are presented and selected results from experimental studies of such partitions are presented. A simple model of the sound insulation behaviour of close-fitting enclosures is followed by a largely qualitative review of the transmission characteristics of stiffened, composite, multi-layer and non-uniform panels. The chapter closes with an extended review of sound transmission through thin-walled, circular cylindrical shells, including a theoretical model of outward sound transmission through pipe walls and a largely qualitative treatment of sound transmission into large diameter cylindrical shells.

The vibrational response of structures to incident sound fields is the subject of Chapter 6. This topic is of considerable practical importance in cases where the integrity of insonified structures and attached systems is at risk due to exposure to the intense sound fields generated by, for instance, aircraft engines, rocket motors and unsteady flows in industrial piping systems. The relation between radiation and response characteristics, as determined by the principle of vibro-acoustic reciprocity, is strongly emphasised, since this property has considerable experimental significance. It forms the basis of experimental investigations of the sources of noise in passenger vehicles and the radiation of sound by ships, among many other applications. The theory of modal reciprocity is developed and illustrated by application to the analysis of the sound absorbent properties of non-load bearing building structures.

In many cases of practical concern, structures wholly or partially enclose volumes of fluid. The vibrational behaviour of the coupled media if then of interest: examples include rocket launcher payload bays, floating floors, rooms and windows at low frequency and fluid transport ducts. Chapter 7 presents theoretical approaches to the modelling and analysis of coupled system behaviour. The chapter opens with a simple one-dimensional illustration of the effects of interaction between a vibrating piston and fluid in a closed-ended tube. The form of the Kirchhoff–Helmholtz integral particular to the general problem of fluid enclosure is derived. The coupled equations of motion of a structure enclosing a fluid are derived and conditions leading to substantial differences between the natural frequencies of the uncoupled components and of the coupled system are identified: these are later quantified by numerical analysis presented in Chapter 8. As mentioned above, many vibroacoustic problems involve hundreds or thousands of interacting modes and deterministic modelling and analysis is impracticable and economically unjustifiable. The most widely used probabilistic alternative is Statistical Energy Analysis (SEA). The fundamental assumptions and principles of SEA are briefly explained and it is applied to a simple case of vibroacoustic coupling. The chapter closes with a detailed analysis of the propagation of coupled waves in a waveguide consisting of two parallel flat plates enclosing a fluid and a brief discussion of wave propagation in a fluid-filled elastic tube.

Chapter 8 presents an introduction to the principles and practice of numerically based analysis of fluid–structure interaction. After a brief discussion of the role of numerical analysis in vibroacoustics, the basic variational principle of mechanics on which the universally employed Finite Element Analysis (FEA) (and associated computational methods) is based is introduced. Expressions for the energy and work integrals that feature in the equation of variation are derived and the basic procedures of mesh generation and selection of assumed deformation functions are explained and illustrated. The FEA is then applied to the computation of the natural frequencies and mode shapes of beams. The results are compared with those of the Rayleigh–Ritz (assumed mode) method. The advantage of FEA for application to geometrically and materially complex structures is explained. The FEA is subsequently applied to the vibration of flat plates and to the acoustic vibrations of a fluid in a tube and in a rectangular box. The FEA is finally applied to the evaluation of the natural frequencies and modes of a coupled system comprising a thin rectangular flat plate that forms one wall of an otherwise rigid rectangular box filled with air. The chapter closes with an introduction to the Boundary Element Analysis (BEA) for sound radiation by vibrating surfaces that is based upon the Kirchhoff–Helmholtz integral. Techniques for increasing the rate of convergence to a stable result are discussed. The BEA is then applied to the computation of the coupled vibration of the flat plate and the fluid external to the rectangular box and the associated radiation field.

Chapter 9 is dedicated to a new approach of actively controlling sound radiation which has emerged during the past fifteen years from previous work on active noise control. The basic idea behind this new technology is to modify or reduce the vibration of a radiating structure in such a way as to diminish the radiated sound. The structure is equipped with structural actuators which are driven by a controller in such a way as to minimise the radiated sound which is measured by microphone arrays in the radiating field or estimated by structural sensors bonded on the radiating structure itself. The chapter is structured in four parts. The first one presents a short review of feed-forward and feedback control theory. Both single channel and multiple channel control systems are analysed. The basic formulations for the derivation of the optimal feed-forward control filters are presented with reference to both deterministic tonal excitations and broadband random disturbances. The implementation of single channel feedback control system is introduced in terms of the so-called classic feedback control theory.

The formulation is then extended to multiple channel control systems. In parallel, the modern feedback control theory, which is based on the state-space formulation of the problem, is also presented for both single and multiple channel feedback control systems. The second part of the chapter presents the main features of the most common types of structural actuators and sensors. In particular, the excitation fields generated on structures by strain (piezoelectric patches and films) and inertial actuators (electrodynamic proof mass actuators) are revised. In parallel, the vibration measured at a point by inertial sensors (accelerometers) and strain transducers (piezoelectric patches and films) are discussed. Practical examples are introduced to highlight the main advantages and drawbacks of these transducers for the implementation of stable and effective control systems. The third part of the chapter presents an extensive and detailed review of the various stages that have brought the development of active noise control (ANC) systems into active structural acoustic (ASAC) systems and active vibration control (AVC) systems for the control of sound radiation. Finally, the fourth part of the chapter presents a detailed analysis of the physics of active control of sound radiation by a flat rectangular panel in a baffle. The main physical features and differences of ASAC and AVC control systems are analysed with reference to both feed-forward and feedback control architectures for the reduction of tonal and broadband random disturbances.

1   Waves in Fluids and Solid Structures

1.1   Frequency and Wavenumber

In this book we shall confine our attention largely to audio-frequency vibrations of elastic structures that take the form of thin flat plates, or thin curved shells, of which the thickness dimension is very much less than those defining the extent of the surface. Such structures tend to vibrate in a manner in which the predominant motion occurs in a direction normal to the surface. This characteristic, together with the often substantial extent of the surface in contact with a surrounding fluid, provides a mechanism for displacing and compressing the fluid; hence such structures are able effectively to radiate, transmit and respond to sound. The field of study and practice that is concerned with these phenomena is known as 'Vibroacoustics' (also known as 'Structural Acoustics' in the US). In order to understand the process of acoustic interaction between solid structures and fluids, it is essential to appreciate the wave nature of the responses of both media to time-dependent disturbances from equilibrium, whether these be transient or continuous. This chapter commences with an introduction to the general mathematical description of uni-directional harmonic wave motion. The forms and characteristics of the principal types of wave that are important in vibroacoustics are then surveyed, together with their behaviours in a number of archetypal forms of structure. Finally, we shall take a look at the phenomena of natural frequency, resonance and natural modes from a wave point of view.

A mechanical wave may be defined as a phenomenon in which a physical quantity (e.g., energy or pressure) propagates in a supporting medium, without net transport of the medium. It may be characterised kinematically by the form of relative displacements from their positions of equilibrium of the particles of the supporting medium—that is to say the form of distortion—together with the speed and direction of propagation of this distortion. Wave disturbances in nature rarely occur at a single frequency; however, it is mathematically and conceptually more convenient to study single-frequency behaviour, from which more complex time-dependent behaviour can be synthesised mathematically if required (provided the system behaves linearly).

Before we consider wave motion in particular types of physical systems we shall discuss the mathematical representation of relationships between variations in time and space which are fundamental to the nature of wave motion in general. Simple harmonic variations in time are most conveniently described mathematically by means of a complex exponential representation, of which there are two forms: in one, only positive frequencies are recognised; in the other, which is more appropriate to analytical and numerical frequency analysis techniques, both positive and negative frequencies are considered (Randall, 1977). In this book the former representation will be employed, because it avoids the confusion that can arise between signs associated with variations in time and space.

The basis of the representation is that a simple harmonic variation of a quantity with time, which may be expressed as $g(t) = A\cos(\omega t + \phi)$, where A symbolises amplitude and ϕ symbolises phase, can also be expressed as $g(t) = \text{Re}\{\tilde{B}\exp(j\omega t)\}$ where $\tilde{B}$ is a complex number, say $a + jb$, and $\text{Re}\{\}$ means 'real part of'. $\tilde{B}$ is be termed the 'complex amplitude' which can also be expressed in exponential form $\tilde{B} = A\exp(j\phi)$ where

$$A = (a^2 + b^2)^{1/2}, \quad a = A\cos\phi, \quad b = A\sin\phi, \quad \phi = \arctan(b/a)$$

Hence $g(t)$ may be represented graphically in the complex plane by a rotating vector (phasor) which rotates at angular speed ω, as illustrated in Fig. 1.1. The projection of the phasor $\tilde{B}\exp(j\omega t)$ on the Real axis is the real harmonic function $g(t)$.

In a linear wave propagating in only one space dimension x, of a type in which the speed of propagation of a disturbance is independent of its magnitude, a simple harmonic disturbance generated at one point in space will clearly propagate away in a form in which the spatial disturbance pattern, as observed at any one instance of time, is sinusoidal in space. This spatial pattern will travel at a speed c, which is determined by the kinematic form of the disturbance, the properties

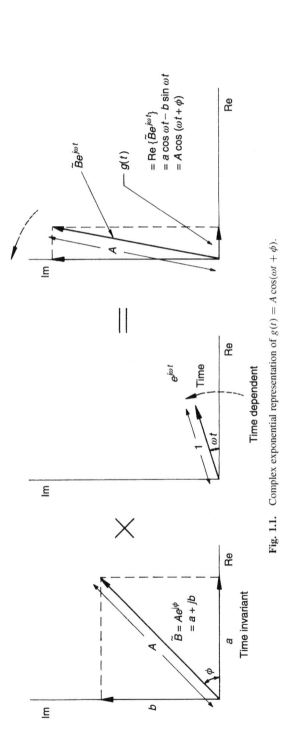

Fig. 1.1. Complex exponential representation of $g(t) = A\cos(\omega t + \phi)$.

of the medium, and any external forces on the medium. As the wave progresses, the disturbance at any point in space will vary sinusoidally in time at the same frequency as that of the generator, provided the medium responds linearly to the disturbance. Suppose we represent the disturbance at the point of generation by $g(0, t) = \mathrm{Re}\{\tilde{B}\exp(j\omega t)\}$, as represented by Fig. 1.1. The phase of the disturbance at a point distance x_1 in the direction of propagation will *lag* the phase at 0 by an angle equal to the product of the circular frequency ω (phase change per unit time) of the generator and the time taken for the disturbance to travel the distance x_1: this time is equal to x_1/c. Hence, $g(x_1,t)$ may be represented by multiplying $\tilde{B}\exp(j\omega t)$ by $\exp(-j\omega x_1/c)$ i.e., $g(x_1,t) = \mathrm{Re}\{\tilde{B}\exp[j(\omega t - \omega x_1/c)]\}$. Thus we see that the quantity $-(\omega/c)$ represents phase change per unit increase of distance, in the same way as ω represents phase change per unit increase of time. Figure 1.2 illustrates the combined effects of space and time variations. The horizontal component of the rotating phasor represents the disturbance $g(x,t)$, which may, or may not, physically correspond to displacement in the x direction, depending upon the type of wave.

The blocked-in phase angle shown in Fig. 1.2 illustrates why c is termed the 'phase velocity' of the wave, which we shall in future symbolise as c_{ph}: an observer travelling in the direction of wave propagation at this speed sees no change of phase. The spatial period of a simple harmonic wave is commonly

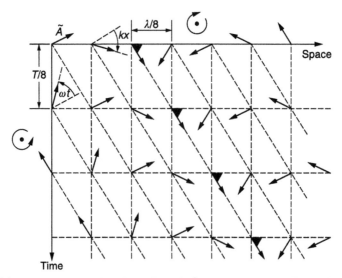

Fig. 1.2.   Phasor representation of an x-directed plane propagating wave. The blocked-in angle illustrates a contour of constant phase. $\Delta x = \lambda/8 = \pi/4k$; $\Delta t = T/8 = \pi/4\omega$ (Fahy, 2001).

described by its wavelength λ. However, the foregoing exposition of the mathematical description of a wave suggests that spatial variations are better described by an associated quantity that represents phase change per unit distance and is equal to (ω/c_{ph}). This is termed the 'wavenumber' and is generally denoted by k. One wavelength clearly corresponds to an x-dependent phase difference of 2π: hence $\omega\lambda/c_{ph} = k\lambda = 2\pi$. The following relationships should be committed to memory:

$$k = \omega/c_{ph} = 2\pi/\lambda \tag{1.1}$$

where k has the dimension of reciprocal length and unit reciprocal metre.

Wavenumber k is actually the magnitude of a vector quantity that indicates the direction of propagation as well as the spatial phase variation. This quantity $\mathbf{k}$, termed the 'wavenumber vector', is of vital importance to the mathematical representation of two- and three-dimensional wavefields. The analogy between temporal frequency ω and spatial frequency k is illustrated by Fig. 1.3. Any form of spatial variation can be Fourier analysed into a spectrum of complex wavenumber components, just as any form of temporal variation (signal) can be analysed into a spectrum of complex frequency components.

As shown in Fig. 1.4, the general complex exponential form of a simple harmonic wave travelling in the positive x direction, with complex amplitude $\widetilde{B}$ at $x = 0$, is represented by $g^+(x, t) = \mathrm{Re}\{\widetilde{B}\exp[j(\omega t - kx)]\}$. For a wave of the same complex amplitude travelling in the negative x direction, $g(x,t)$ takes the form $g^-(x, t) = \mathrm{Re}\{\widetilde{B}\exp[j(\omega t + kx)]\}$.[1]

Thus far, no assumption has been made about the dependence of phase velocity on frequency; Eq. (1.1) indicates that this relationship determines the frequency dependence of wavenumber. The form of relationship between k and ω is termed the *dispersion* relation; it is a property of the type of wave, as well as the type of wave-supporting medium. The dispersion relation indicates, among other things, whether or not a disturbance generated by a process that is not simple harmonic in time will propagate through a medium unaltered in its basic spatial form: if it does not the wave is said to be dispersive. Only if the relationship between k and ω is linear will an arbitrary spatial form of disturbance not be subject to change as it propagates: however, the amplitude of a non-dispersive wave may decrease with distance travelled on account of spreading in two or three *space* dimensions. Figure 1.5, which shows the progress of three different frequency components of a dispersive wave, illustrates the dispersion process. Any disturbance of

---

[1] In order to keep the formulations simple, in the following part of the book the real part will be assumed in the equations for wavefields and thus positive- and negative-going waves will be expressed as $g^+(x, t) = \widetilde{B}\exp[j(\omega t - kx)]$ and $g^-(x, t) = \widetilde{B}\exp[j(\omega t + kx)]$.

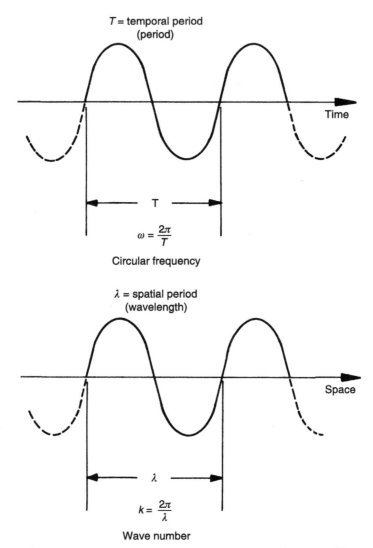

Fig. 1.3. Analogy between circular frequency and wavenumber (Fahy, 1987).

finite time duration contains an infinity of frequency components and, if transported by a dispersive wave process, will be distorted as it propagates.

We shall see later that dispersion curves can be of great help in understanding the interaction between waves in coupled media because where dispersion curves for two types of wave intersect they have common frequency and wavenumber,

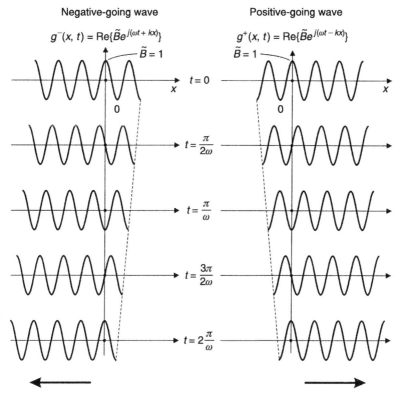

Fig. 1.4. Positive- and negative-going waves.

and therefore common wavelength and phase speed. A dispersion curve can also tell us something about the speed at which energy is transported by a wave; this speed is called the group speed c_g. It is obtained from the dispersion curve by the relation

$$c_g = \partial\omega/\partial k \qquad (1.2)$$

which is the inverse slope of the curve. It is clear from Eqs. (1.1) and (1.2) that the phase and group speeds are equal only in non-dispersive waves, for which c_{ph} is independent of ω. Knowledge of group speed is useful in vibration analyses involving the consideration of wave energy flow, such as the evaluation of vibrational energy propagation in complicated structures.

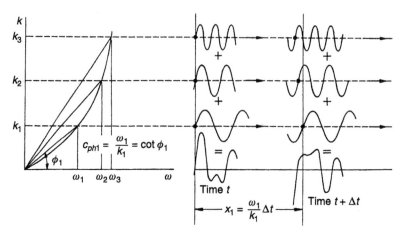

Fig. 1.5. Combination of three frequency components of a dispersive wave at successive instants of time (Fahy, 1987).

1.2   Sound Waves in Fluids

The wave equation governing the propagation of *small acoustic* disturbances through a homogeneous, inviscid, isotropic, compressible fluid may be written in rectangular Cartesian coordinates (x, y, z) in terms of the variation of pressure about the equilibrium pressure, the 'acoustic pressure' p, as

$$\frac{\partial^2 p}{\partial x^2} + \frac{\partial^2 p}{\partial y^2} + \frac{\partial^2 p}{\partial z^2} = \frac{1}{c^2}\frac{\partial^2 p}{\partial t^2} \tag{1.3}$$

The same equation governs variations in fluid density and temperature, and in particle[2] displacement, velocity and acceleration: c is the frequency-independent speed of sound given by $c^2 = (\gamma P_0 / \rho_0)$, where P_0 is the mean fluid pressure, ρ_0 is the mean density and γ is the adiabatic bulk modulus of the fluid. As shown in textbooks on acoustics, the wave equation is derived from the linearised[3] forms

---

[2] Fluids actually consist of discrete molecules moving within otherwise empty space (void). For the purpose of analysing the behaviour of fluids, fluid dynamicists adopt a voidless continuum model in which fictitious entities known as 'particles' are attributed with the average position, velocity and acceleration of the multitudes of molecules in a 'small' volume surrounding the assumed location of the particle.

[3] Products of small quantities neglected.

of the continuity equation

$$\frac{\partial \rho}{\partial t} + \rho_0 \left(\frac{\partial u}{\partial x} + \frac{\partial v}{\partial y} + \frac{\partial w}{\partial z} \right) = 0 \qquad (1.4)$$

and momentum equations

$$\frac{\partial p}{\partial x} + \rho_0 \frac{\partial u}{\partial t} = 0 \qquad (1.5a)$$

$$\frac{\partial p}{\partial y} + \rho_0 \frac{\partial v}{\partial t} = 0 \qquad (1.5b)$$

$$\frac{\partial p}{\partial z} + \rho_0 \frac{\partial w}{\partial t} = 0 \qquad (1.5c)$$

in which u, v and w are the particle velocities in the x, y and z directions, respectively.

The general solution of the wave equation may be expressed in terms of various coordinate systems: it is separable in all common forms such as rectangular Cartesian, cylindrical, spherical and elliptical. In studying the interaction between plane structures and fluids, the most useful form of the equation is a two-dimensional form involving only variations in two orthogonal directions, in association with simple harmonic time dependence:

$$\frac{\partial^2 p}{\partial x^2} + \frac{\partial^2 p}{\partial y^2} = - \left(\frac{\omega}{c} \right)^2 p = -k^2 p \qquad (1.6)$$

This is known as the two-dimensional Helmholtz equation. The propagation of sound pressure in a harmonic plane wave in a two-dimensional space may be expressed as

$$p(x, y, t) = \tilde{p} \exp(-jk_x x - jk_y y) \exp(j\omega t) \qquad (1.7)$$

Substitution of this expression into (1.6) yields the following wavenumber relation:

$$k^2 = k_x^2 + k_y^2 \qquad (1.8)$$

The interpretation of this equation is that only certain combinations of k_x and k_y will satisfy the wave equation at any specific frequency. The direction of wave propagation is indicated by the wavenumber vector **k** as shown in Fig. 1.6.

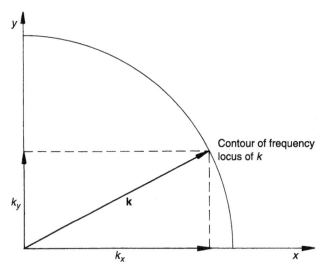

Fig. 1.6.   Two-dimensional wavenumber vector and components.

The locus of the wavenumber vector **k** for fixed frequency ω and real values of k_x and k_y is clearly seen to be a circle of radius k. It is opportune at this point to warn the reader that only wavenumber vectors, *not wavelengths*, should be directionally resolved.

The value of the linearised momentum equations (1.5) in the analysis of structure–fluid interaction is that they connect the pressure gradient in the fluid and the particle acceleration of the structure normal to the interface between the two media. In the special case of simple harmonic motion this becomes a relation between pressure gradient and particle *velocity*. Hence the normal vibration velocity at a point on a surface determines the value at that point of the pressure gradient in the direction normal to the surface; e.g., from (1.5b), $(\partial p/\partial y)_{y=0} = -j\omega\rho_0(v)_{y=0}$, where the surface is assumed to lie in the x–z plane. This relation is frequently used as a fluid boundary condition in the analysis of sound radiation, sound absorption and scattering by structures, as well as in the analysis of the influence of fluid pressure on structural vibration (see Chapter 4).

Sound radiation by a vibrating structure is commonly quantified in terms of radiated sound power. This is given by the integral over the structure's surface of the product of local surface *pressure* and normal surface velocity. The relation between these two quantities at any surface point is not uniquely determined by the fluid boundary condition expressed above in terms of the normal *pressure gradient*. In principle, the spatial distribution of vibration over the whole surface must also be known, or assumed, because acoustic disturbances emitted by all parts of a surface propagate to all other parts, thereby influencing the surface pressure at all points. The relation between local surface pressure and normal velocity at any surface point is therefore not unique either to the fluid or to the geometric form of the surface.

In the process of theoretical estimation of sound power radiated by vibrating structures, the requirement to calculate, or measure, the normal velocity distribution over the whole of the vibrating surface (in terms of both amplitude and phase at all frequencies of concern in the case of harmonic vibration) constitutes the principal obstacle to achieving accurate results. The audio-frequency vibrational fields of complicated structures such as automotive engines and aircraft fuselages are very complex, highly variable with frequency, sensitive to structural and material detail and to environmental and operating conditions. Fortunately, consideration of vibration and radiation in finite frequency bands (where appropriate) greatly eases the task, as shown in Chapter 3.

1.3 Longitudinal Waves in Solids

In pure longitudinal wave motion the direction of particle displacement is purely in the direction of wave propagation; such waves can propagate in large volumes of solids. Two parallel planes in an undisturbed solid elastic medium, which are separated by a small distance δx, may be moved by different amounts during the passage of a longitudinal wave, as illustrated in Fig. 1.7. Hence the

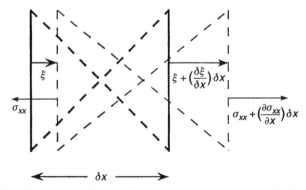

Fig. 1.7. Displacements from equilibrium and stresses in a pure longitudinal wave.

element may undergo a strain ε_{xx} given by

$$\varepsilon_{xx} = \partial\xi/\partial x \qquad (1.9)$$

where ξ is the displacement of the parallel planes in the x direction.

The longitudinal stress σ_{xx} is, according to Hooke's law, proportional to ε_{xx}. However, the constant of proportionality is not, as at first might be thought, the Young's modulus E of the material. Young's modulus is defined as the ratio of longitudinal stress σ_{xx} to longitudinal strain ε_{xx} in a *thin*, *uniform* bar under longitudinal tension. In this case, the sides of the rod are under no constraint. When longitudinal tension is applied, they move toward the axis of the rod, producing lateral strains ε_{yy} and ε_{zz}; the lateral stresses σ_{yy} and σ_{zz}, remain zero because no lateral constraints are applied. This phenomenon is termed *Poisson contraction*, Poisson's ratio being defined as the ratio of the magnitudes of the lateral strain to the longitudinal strain: $v = -\varepsilon_{yy}/\varepsilon_{xx} = -\varepsilon_{zz}/\varepsilon_{xx}$. In practice, v varies between 0.25 and 0.3 for glass and steel and nearly 0.5 for nearly incompressible materials such as rubber.

When a one-dimensional pure longitudinal wave propagates in an unbounded solid medium, there can be no lateral strain, because all material elements move purely in the direction of propagation (Fig. 1.7). This constraint, applied mutually by adjacent elements, creates lateral direct stresses in the same way a tightly fitting steel tube placed around a wooden rod would create lateral compression stresses if a compressive longitudinal force were applied to the wood along its axis. In these cases, the ratio $\sigma_{xx}/\varepsilon_{xx}$ is not equal to E. Complete analysis of the behaviour of a uniform elastic medium shows that the ratio of longitudinal stress to longitudinal strain when lateral strain is prevented, is (Cremer *et al.*, 1988)

$$\sigma_{xx} = B\partial\xi/\partial x \qquad (1.10)$$

where $B = E(1 - v)/(1 + v)(1 - 2v)$. The resulting equation of motion of an element is

$$(\rho\delta x)\partial^2\xi\partial t^2 = [\sigma_{xx} + (\partial\sigma_{xx}/\partial x)\delta x - \sigma_{xx}] = (\partial\sigma_{xx}/\partial x)\delta x \qquad (1.11)$$

where ρ is the material mean density. Equations (1.10) and (1.11) can be combined into the wave equation

$$\partial^2\xi/\partial x^2 = (\rho/B)\partial^2\xi/\partial t^2 \qquad (1.12)$$

The general solution of Eq. (1.12), which is of the same form as the one-dimensional acoustic wave equation, shows that the phase speed is

$$c_l = (B/\rho)^{1/2} \qquad (1.13)$$

which is independent of frequency; these waves are therefore non-dispersive. A simple harmonic wave can therefore be expressed as $\xi^+(x,t) = \tilde{A}\exp j(\omega t - k_l x)$ where $k_l = \omega/c_l$. Because $B > E$, $c_l \gg (E/\rho)^{1/2}$. Note that if $v = 0.5$, both B and c_l are infinite—no wonder solid rubber mats of large area are useless as vibration isolators!

1.4 Quasi-Longitudinal Waves in Solids

As we have noted, only in volumes of solid that extend in all directions to distances large compared with a longitudinal wavelength can pure longitudinal waves exist. A form of longitudinal wave can propagate along a solid bar, and another form can propagate in the plane of a flat plate. Because such structures have one or more outer surfaces free from constraints, the presence of longitudinal stress will produce associated lateral strains through the Poisson contraction phenomenon: pure longitudinal wave motion cannot therefore occur, and the term 'quasi-longitudinal' is used.

Elasticity theory shows that the relation between longitudinal stress and longitudinal strain in a thin flat plate is

$$\sigma_{xx} = [E/(1 - v^2)]\partial\xi/\partial x \tag{1.14}$$

Hence the quasi-longitudinal wave equation for a plate is

$$\partial^2\xi/\partial x^2 = [\rho(1 - v^2)/E]\partial^2\xi/\partial t^2 \tag{1.15}$$

and the corresponding frequency-independent phase speed is

$$c_l' = [E/\rho(1 - v^2)]^{1/2} \tag{1.16}$$

The ratio of longitudinal stress to longitudinal strain in a bar is defined to be E. Hence the quasi-longitudinal wave equation is

$$\partial^2\xi/\partial x^2 = (\rho/E)\partial^2\xi/\partial t^2 \tag{1.17}$$

and the phase speed is

$$c_l'' = (E/\rho)^{1/2} \tag{1.18}$$

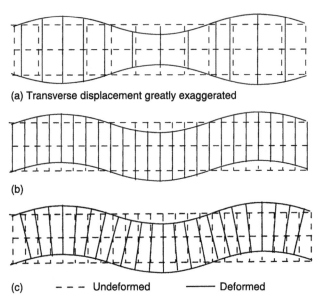

(a) Transverse displacement greatly exaggerated

(b)

(c)   – – – Undeformed   ——— Deformed

Fig. 1.8. Deformation patterns of various types of wave in straight bars and flat plates: (a) quasi-logitudinal wave; (b) transverse (shear wave); (c) bending wave.

The kinematic form of a quasi-longitudinal wave is shown in Fig. 1.8(a). Equations (1.14)–(1.18) are valid only for frequency ranges in which the quasi-longitudinal wavelength greatly exceeds the cross-sectional dimensions of the structure. This criterion will be satisfied by all practical structures and frequencies considered in this book: for example, at 10 kHz, the wavelengths in bars of steel and concrete are approximately 500 and 350 mm, respectively. Values of E, ρ, v, c_l, c_l' and c_l'' for common materials are presented in Table 1.1.

1.5 Transverse (Shear) Waves in Solids[4]

Solids, unlike fluids, can resist static shear deformation. Fluids in motion can generate shear stresses associated with flow velocity gradients, but because the agent is fluid viscosity, which generates a dissipative, frictional type of force,

---

[4] Transverse waves in solids are often described as 'shear' waves. This is not strictly correct because rotation of an element does take place. However, the latter term is widely employed in the literature and will be used in the reminder of the book to avoid confusion on the part of the reader.

TABLE 1.1

Material Properties[a] and Phase Speeds

Material	Young's modulus E (Nm^{-2})	Density ρ (kgm^{-3})	Poisson's ratio (v)	c_l (ms^{-1})	c_l' (ms^{-1})	c_l'' (ms^{-1})	c_s (ms^{-1})
Steel	2.0×10^{11}	7.8×10^3	0.28	5900	5270	5060	3160
Aluminium	7.1×10^{10}	2.7×10^3	0.33	6240	5434	5130	3145
Brass	10.0×10^{10}	8.5×10^3	0.36	4450	3677	3430	2080
Copper	12.5×10^{10}	8.9×10^3	0.35	4750	4000	3750	2280
Glass	6.0×10^{10}	2.4×10^3	0.24	5430	5151	5000	3175
Concrete							
Light	3.8×10^9	1.3×10^3				1700	
Dense	2.6×10^{10}	2.3×10^3				3360	
Porous	2.0×10^9	6.0×10^2				1820	
Rubber							
Hard (80)	5.0×10^7	1.1×10^3	0.5^d			210	125
Soft	5.0×10^6	9.5×10^2	0.5^d			70	40
Brick	1.6×10^{10}	$1.9–2.2 \times 10^3$				2800	
Sand, dry	3.0×10^7	1.5×10^3				140	
Plaster	7.0×10^9	1.2×10^3				2420	
Chipboard^c	4.6×10^9	6.5×10^2				2660	
Perspex^b	5.6×10^9	1.2×10^3	0.4	3162	2357	2160	1291
Plywood^c	5.4×10^9	6.0×10^2				3000	
Cork		$1.2–2.4 \times 10^2$				430	
Asbestos cement	2.8×10^{10}	2.0×10^3				3700	
Gypsum board^c	2.4×10^9	$7.5–8.0 \times 10^2$				1730–1790	

^{a} Mean values from various sources.
^{b} Temperature sensitive.
^{c} Varies greatly from specimen to specimen.
^{d} Poisson's ratio is only appropriate to small strains.

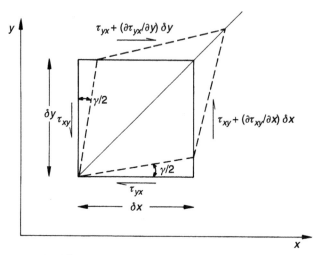

Fig. 1.9.   Pure shear strain and associated shear stresses.

shear waves generated in fluids decay very rapidly with distance and are therefore generally of little practical concern. The shear modulus G of a solid is defined as the ratio of the shear stress τ to the shear strain γ. Pure shear deformation is represented in Fig. 1.9: note that the element does not rotate and that the diagonals suffer extension and compression. The shear modulus G is related to Young's modulus E by $G = E/2(1 + \nu)$.

In *static* equilibrium under pure shear, the shear stresses acting in a particular direction on the opposite faces of a rectangular element are equal and opposite, because the direct stresses on these faces are zero, i.e., $\partial \tau_{xy}/\partial x = \partial \tau_{yx}/\partial y = 0$ and $\tau_{xy} = \tau_{yx}$ in Fig. 1.9. In the case of shear distortion associated with differential transverse dynamic displacement illustrated in Fig. 1.10, this is not necessarily so; the difference in the vertical shear stresses causes the vertical acceleration $\partial^2 \eta/\partial t^2$ of the element. Consideration of the other equilibrium equations indicates that rotation $\gamma/2$ of the element must also take place. The equation of transverse motion of an element having unit thickness in the z dimension is

$$\rho \delta x \delta y \partial^2 \eta/\partial t^2 = (\partial \tau_{xy}/\partial x)\delta x \delta y \qquad (1.19)$$

where η is the transverse displacement in y direction, and the stress–strain relationship is

$$\tau_{xy} = G\gamma = G\partial \eta/\partial x \qquad (1.20)$$

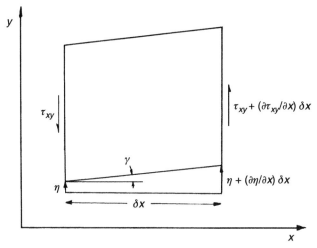

Fig. 1.10.   Displacements, shear strain and transverse shear stresses in transverse deformation.

Hence the wave equation is

$$\partial^2\eta/\partial x^2 = (\rho/G)\partial^2\eta\partial t^2 \qquad (1.21)$$

The kinematic form of the shear wave is shown in Fig. 1.8(b). Equation (1.21) takes the same form as the acoustic and longitudinal wave equations. The frequency-independent phase speed is

$$c_s = (G/\rho)^{1/2} = [E/2\rho(1+v)]^{1/2} \qquad (1.22)$$

The shear wave speed is seen to be smaller than the quasi-longitudinal bar wave speed:

$$c_s/c_l'' = [1/2(1+v)]^{1/2} \qquad (1.23)$$

This ratio is about 0.6 for many structural materials. Such shear waves can propagate in large volumes of solids and in the plane of extended flat plates of which the free surfaces have little influence, since in-plane direct stresses are negligible; hence the shear wave speed is very similar to that in a large volume. In general, in-plane shear waves are not easy to generate in plates by means of applied forces; however, they sometimes play a vital role in the process of

vibration transmission through, and reflection from, line discontinuities such as
L, T and + junctions between structures such as those of buildings and ships
(Kihlman, 1967a,b).

Torsional waves in solid bars are shear waves. The governing wave equation is

$$\partial^2\theta/\partial x^2 = (I_p/GJ)\partial^2\theta/\partial t^2 \qquad (1.24)$$

where θ is the torsional displacement, I_p is the polar moment of inertia per unit
length of the bar about its longitudinal axis and GJ is the torsional stiffness of
the bar, which is a function of the shape of the bar. The frequency-independent
phase speed is $c_t = (GJ/I_p)^{1/2}$.

Table 1.2 lists expressions for the torsional stiffness of solid bars of rectangular
and circular sections. Calculation of the torsional stiffness of structures such as I-,
L- and channel-section beams is more complex, since such phenomena such as
warping and bending-shear coupling must be taken into account (Argyris, 1954;
Muller, 1983). The effects of constraint applied to such beams by connected
structures such as plates have a substantial influence on their effective torsional
stiffness: ship and aircraft structures commonly incorporate such beam–plate construction. Shear-wave motion is of great importance in the vibration behaviour of
sandwich and honeycomb plate structures, which have thin cover plates separated
by relatively thick core layers; such cores are often designed to have low shear
stiffness. In a mid-frequency range between low frequencies, at which overall
section bending stiffness is dominant, and high frequencies, at which cover plate
bending stiffness dominates, the wave speed is controlled primarily by the core
shear stiffness. Advantage can be taken of this behaviour to design for optimum
sound radiation and transmission characteristics. Analysis of wave motion in such
structures is discussed in Chapter 5.

TABLE 1.2

Torsional Stiffness of Solid Bars

	h/b	J/b^3h
Rectangular ($b \times h$)	1	0.141
	2	0.230
	3	0.263
	4	0.283
	5	0.293
	10	0.312
Circular (radius a)	$J = \pi a^4/2$	

1.6 Bending Waves in Bars

Of the various types of wave that can propagate in bars, beams and plates, bending (or flexural) waves are of the greatest significance in the process of structure–fluid interaction at audio frequencies. The reasons are that bending waves involve substantial displacements in a direction transverse to the direction of propagation, which can effectively disturb an adjacent fluid; and that the transverse impedance (see Chapter 2) of structures carrying bending waves can be of similar magnitude to that of sound waves in the adjacent fluid, thereby facilitating energy exchange between the two media.

In spite of the fact that the transverse displacements of elements in structures carrying bending waves are far greater than the in-plane displacements of these elements, the bending stresses are essentially associated with the latter. Consequently, flexural waves can be classified neither as longitudinal nor as transverse waves. In pure bending deformation of a bar, cross sections are translated in a direction transverse to the bar axis, and are rotated relative to their equilibrium planes, as shown in Fig. 1.11. The transverse displacement η and rotation β are related approximately by

$$\beta = \partial\eta/\partial x \tag{1.25}$$

The primary assumption of pure bending theory is that plane cross sections remain plane during bending deformation of the element. Let us examine this assumption as it relates to bending wave motion. Figure 1.12 shows an element of a bar undergoing pure bending deformation under the action of purely transverse forces: note that θ is assumed to be very small. The longitudinal strain ε_r, and hence σ_r, vary linearly with r according to our assumption. Because there is no applied force component in the direction of the bar axis, we can write

$$\int_{-r_0}^{h-r_0} \sigma(r)w(r)dr = 0 \tag{1.26}$$

where $\sigma(r)$ is the longitudinal direct stress.

This equation immediately indicates that $\sigma(r)$ must vary over both positive and negative values; hence $\sigma(r)$ is zero at some unique value of r. We denote the surface on which this occurs, called the 'neutral surface', by N–S. The longitudinal strain on the deformed neutral surface is clearly zero, and the element retains its original length δx along this arc. Hence

$$\varepsilon(r) = [(R+r)\theta - \delta x]/\delta x \tag{1.27}$$

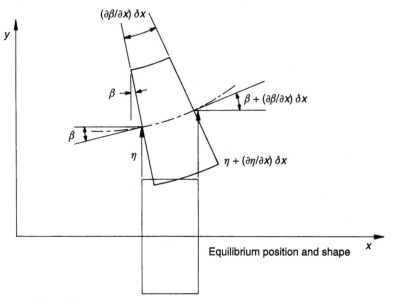

Fig. 1.11. Displacements and deformation of a beam element in bending.

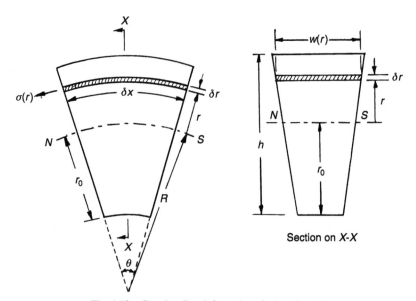

Fig. 1.12. Pure bending deformation of a bar element.

We now assume that the relationship between $\varepsilon(r)$ and $\sigma(r)$ is the same as in longitudinal loading of a thin bar, i.e., zero lateral constraint: $\sigma(r) = E\varepsilon(r)$. The local radius of curvature R is related to θ and, from Fig. 1.11, to the slope and displacement of the bar axis, by

$$1/R = \theta/\delta x = -\partial\beta/\partial x = -\partial^2\eta/\partial x^2 \tag{1.28}$$

Hence

$$\sigma(r) = E\varepsilon(r) = -Er\partial^2\eta/\partial x^2 \tag{1.29}$$

In general, the bar curvature will vary with x, and so then will $\sigma(r)$. In a static situation, the variation of $\sigma(r)$ in the axial direction must be balanced by other stresses so that the element is in longitudinal equilibrium; the corresponding force system on a section of bar above $r = r'$ is shown in Fig. 1.13.

The equilibrium equation is

$$\tau_{yx}(r')w(r') = \int_{r'}^{h-r_0} \frac{\partial\sigma(r)}{\partial x}w(r)dr \tag{1.30}$$

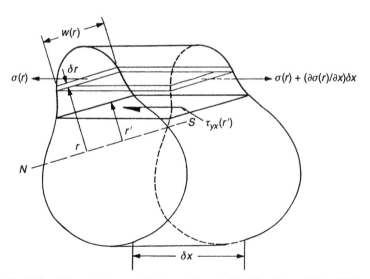

Fig. 1.13.   Balance between axial shear and longitudinal direct forces (Fahy, 2001).

From Eqs. (1.29) and (1.30)

$$\tau_{yx}(r') = -\frac{E}{w(r')}\frac{\partial^3 \eta}{\partial x^3} \int_{r'}^{h-r_0} rw(r)dr \tag{1.31}$$

The integral can only be evaluated for specific variations of the bar width $w(r)$ with r. However, for a bar of uniform width, $w(r) = w$, and $r_0 = h/2$:

$$\tau_{yx}(r') = -\frac{E}{2}\frac{\partial^3 \eta}{\partial x^3}\left[\left(\frac{h}{2}\right)^2 - (r')^2\right] \tag{1.32}$$

This parabolic relationship only holds for bars of uniform width: however, the maximum shear stress occurs on the neutral surface $r' = 0$ in all cases.

It is vital to appreciate the role of this shear stress in opposing variation in the axial direction of longitudinal direct stresses in the beam because, where its action is destroyed, as in multi-laminated beams in which the adhesive fails, or where it can create large shear distortions because of a low shear modulus, as in the core of a sandwich structure, the fundamental assumption that plane sections remain plane must be invalid. The horizontal shear stress τ_{yx} is complemented by a vertical shear stress τ_{xy} of equal magnitude which has the same distribution over the depth of the bar (Fig. 1.14). The total vertical elastic shear force acting

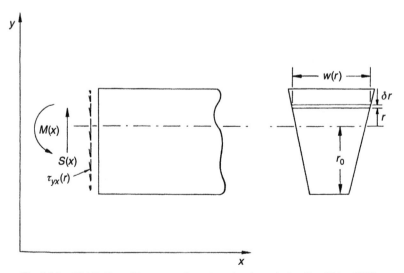

Fig. 1.14. Distribution of transverse shear stress in a beam in bending (Fahy, 1987).

on any cross section of the element is given by

$$S(x) = -\int_{-r_0}^{h-r_0} \tau_{yx} w(r) dr = E \frac{\partial^3 \eta}{\partial x^3} \int_{-r_0}^{h-r_0} \left[\frac{1}{w(r')} \int_{r'}^{h-r_0} r w(r) dr \right] w(r') dr'$$

Integration of this equation by parts yields

$$S(x) = EI \frac{\partial^3 \eta}{\partial x^3}, \quad \text{where} \quad I = \int_{-r_0}^{h-r_0} w(r) r^2 dr \qquad (1.33)$$

and I is defined to be the 'second moment of area' of the cross section about the traverse axis in the neutral plane (not the 'moment of inertia'!).

The bending moment acting on a section is reacted by the axial direct stresses $\sigma(r)$ acting about the neutral axis, and is given by

$$M(x) = -EI \frac{\partial^2 \eta}{\partial x^2} \qquad (1.34)$$

where positive M acts to produce negative curvature. (Prove.)

The presence of shear stresses within the bar naturally produces shear distortion of the cross section, and it can be seen from Eq. (1.32) that the shear strain must vary with the distance r from the neutral axis and must be zero at the upper and lower surfaces. This form of deformation is incompatible with the assumption that plane sections remain plane, but in many cases the shear distortion of homogeneous bars and beams is rather small, and the contribution to transverse displacement η of the beam is also rather small. A rough assessment of the relative contributions to the transverse shear displacement can be obtained by considering the encastré beam, Fig. 1.15. For simplicity, we assume that the vertical shear stress is uniform and equal to F/A. The vertical deflection of the

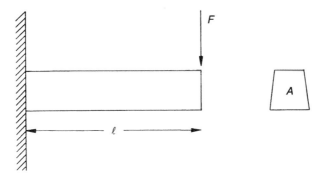

Fig. 1.15. Tip loading of a cantilever.

tip of the beam due to shear is thus approximately equal to Fl/GA, where G is the shear modulus. Simple beam theory gives the bending deflection as $Fl^3/3EI$. Hence the ratio of the shear to bending deflection is approximately equal to $3EI/GAl^2$.

For a rectangular section beam of depth h and width w, $I = wh^3/12$, and $A = wh$. Hence the ratio is approximately equal to $1/2(1 + v)(h/l)^2$. Thus, shear deflections are significant if opposing transverse forces act on the beam at separation distances that are small compared with the depth of the beam. [Why does not this criterion apply to internal shear forces?]

The equation of transverse motion of an element of bar can be derived by reference to Fig. 1.16:

$$\partial S/\partial x = -m\partial^2\eta/\partial t^2 \qquad (1.35)$$

where m is the mass per unit length of the bar.[5] The rate of change of S with x comes from Eq. (1.33); Eq. (1.35) can therefore be written

$$EI\partial^4\eta/\partial x^4 = -m\partial^2\eta\partial t^2 \qquad (1.36)$$

This is the bar wave equation, which is valid provided that the shear contribution to transverse displacement is negligible. The kinematic form is shown in Fig. 1.8(c).

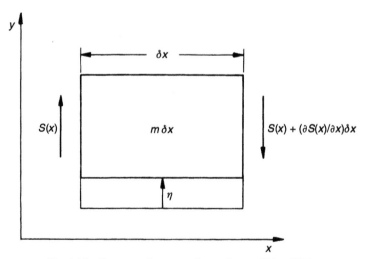

Fig. 1.16.   Transverse forces on a beam element (Fahy, 1987).

---

[5] The symbol m is used to represent mass per unit length or mass per unit area, as appropriate.

This equation differs radically from those governing all the previously considered forms of wave motion in that the spatial derivative is of fourth, and not second, order: the reason is that the bending wave is a hybrid between shear and longitudinal waves. Hence the phase speed of free bending waves cannot be deduced by inspection. Substitution of the complex exponential expression for a simple harmonic progressive wave $\eta(x,t) = \tilde{\eta}\exp[j(\omega t - kx)]$ yields

$$EIk^4 = \omega^2 m$$

Hence $k = \pm j(\omega^2 m/EI)^{1/4}$ and $\pm(\omega^2 m/EI)^{1/4}$. The complete solution is thus

$$\eta(x,t) = \left[\tilde{A}\exp(-jk_b x) + \tilde{B}\exp(jk_b x) + \tilde{C}\exp(-k_b x)\right.$$

$$\left. + \tilde{D}\exp(k_b x)\right]\exp(j\omega t) \tag{1.37}$$

where $k_b = (\omega^2 m/EI)^{1/4}$. The first two terms on the r.h.s. of Eq. (1.37) represent waves propagating in the positive and negative x directions at a phase speed $c_b = \omega/k_b = \omega^{1/2}(EI/m)^{1/4}$. The second two terms represent non-propagating fields, the amplitudes of which decay exponentially with distance; their phase velocities are imaginary and they do not individually transport energy. They cannot strictly be called 'waves' although in many references and books they are referred as evanescent waves. In this book they will be referred as 'near fields'.

The phase velocity c_b is seen to be proportional to $\omega^{1/2}$; bending waves in bars are therefore dispersive. The group speed $c_g = \partial\omega/\partial k = 2c_b$. (Prove.) The fact that the phase speed of bending waves varies with frequency has c_g a profound influence on the phenomenon of acoustic coupling between structures and fluids, as will become evident later. The dispersive nature of bending waves also produces natural bending vibration frequencies of bars that are not harmonically related, in contrast to the harmonic progression associated with non-dispersive waves. (Note: the thickness of some vibraphone bars are shaped so as to produce harmonically related overtones.)

We need to estimate the contribution of shear distortion to bending wave transverse displacement in order to evaluate the range of validity of Eq. (1.36). The free bending wavelength is equal to $2\pi/k_b$. Hence points of maximum transverse displacement and acceleration are separated from points of zero displacement by a distance $\pi/2k_b$. In the d'Alembert view of dynamic equilibrium of bending waves, 'inertia forces' may be considered to act in opposition to elastic shear forces. Maximum inertia forces act at points of maximum acceleration, and maximum shear forces act at points of zero displacement. (Prove this for yourself.) We may therefore replace the length l used in Fig.1.15 by $\pi/2k_b$. The ratio of the contributions to transverse displacement of shear and bending is therefore given approximately by $1/2(2k_b h/\pi)^2$, where h is the bar depth. An appropriate condition for negligible shear contribution is $k_b h < 1$, or $\lambda_b > 6h$. For a bar of rectangular cross section, this condition becomes $f < 4.6 \times 10^{-2}(c_l''/h)$ Hz: for I-sections, the frequency is approximately 40% lower. For example, this frequency for a 150-mm-deep rectangular section concrete bar is approximately 900 Hz. Above this frequency an extra, shear-controlled term must be included in the equation of motion. The details can be found in more advanced books on the subject (Cremer et al., 1988). The physical implication of the contribution of the shear distortion term is that at sufficiently high frequencies 'bending' waves become very like pure transverse shear waves; the phase speed asymptotes to c_s, and does not go to infinity as the expression for c_b would suggest.

1.7   Bending Waves in Thin Plates

As far as the propagation of *plane* bending waves is concerned, a uniform flat plate of infinite extent is no different from a bar, except that the relationship between the longitudinal strains and longitudinal stresses [Eq. (1.29)] must be modified to that corresponding to Eq. (1.14) to allow for the lateral constraint that is absent in a finite width bar because of its free sides. Hence, assuming that the mid-plane of the flat plate is located in the plane $y = 0$, that the plane wave travels purely in the x direction, and that the transverse vibration is defined by the displacement η in the y direction, Eq. (1.36) becomes

$$\frac{EI}{(1-v^2)}\frac{\partial^4 \eta}{\partial x^4} = -m\frac{\partial^2 \eta}{\partial t^2} \qquad (1.38)$$

where m is now the mass per unit area of the plate and I is the second moment of area per unit width: $I = h^3/12$ for a plate of thickness h. We may replace $Eh^3/12(1-v^2)$ by D, which may be termed the bending stiffness of the plate because the bending moment per unit width is given by $M = -D\partial^2 \eta/\partial x^2$. The free-wave solution is the same as Eq. (1.37) with $k_b = (\omega^2 m/D)^{1/4}$. The phase speed $c_b = \omega^{1/2}(D/m)^{1/4}$. The dispersive character of plate bending waves may be observed by listening to the sound of a stone cast onto a sheet of ice on a pond: the high frequencies arrive first.

Equation (1.38) is not, however, sufficient to describe two-dimensional bending wavefields in a plate in which propagation in the x and z directions may occur simultaneously. Derivation of the complete classical bending-wave equation, in which shear deformation and rotary inertia are neglected, is beyond the scope of this book and can be found, for example, in Cremer et al. (1988).

For a thin plate lying in the x–z plane the bending-wave equation in rectangular Cartesian coordinates is

$$D\left(\frac{\partial^4 \eta}{\partial x^4} + 2\frac{\partial^4 \eta}{\partial x^2 \partial z^2} + \frac{\partial^4 \eta}{\partial z^4}\right) = -m\frac{\partial^2 \eta}{\partial t^2} \qquad (1.39)$$

The free *plane-progressive* wave solution is the same for Eq. (1.38), and the condition for the neglect of shear deformation is rather similar to that for beams, namely, $k_b h < 1$. Derivation of the exact plate equations is extremely difficult. A rather complete approximate equation, which takes into account shear deformation and rotary inertia effects, has been published by Mindlin (1951). A monograph on Mindlin plates is also available (Liew *et al.*, 1998).

Equation (1.38) applies to plane-wave propagation in the x direction only. However, plane bending waves may propagate in a plate in any direction in its plane. Consider a simple harmonic plane wave described by $\eta(x, z, t) = \tilde{\eta}\exp[j(\omega t - k_x x - k_z z)]$. Substitution into Eq. (1.39) yields

$$[D(k_x^4 + 2k_x^2 k_z^2 + k_z^4) - m\omega^2]\tilde{\eta} = 0$$

or

$$D(k_x^2 + k_z^2)^2 - m\omega^2 = 0 \qquad (1.40)$$

If we write $k_b^2 = k_x^2 + k_z^2$, we obtain

$$Dk_b^4 - m\omega^2 = 0 \qquad (1.41)$$

which is the plane bending-wave equation for a wave travelling in the direction given by the vector sum of the wavenumber vector components $\mathbf{k}_x$ and $\mathbf{k}_z$, i.e., in a direction at angle $\phi = \arctan(k_z/k_x)$ to the x axis. Hence $\mathbf{k}_b = \mathbf{k}_x + \mathbf{k}_z$ and $k_b = (\omega^2 m/D)^{1/4}$, as before.

The effect of 'fluid loading' (the forces on a vibrating structure caused by the reaction of a contiguous fluid to imposed motion) on bending waves in a plate is treated in Chapter 4.

1.8 Dispersion Curves

In analysing the coupling between vibration waves in solids and acoustic waves in fluids it is very revealing to display the wavenumber characteristics

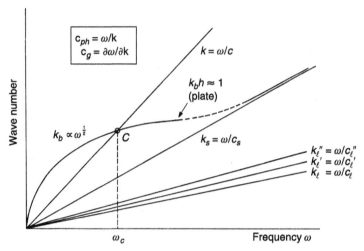

Fig. 1.17.   Dispersion curves for various forms of wave.

of the coupled systems on a common graph. This is done for the wave types described above in the form of dispersion curves in Fig.1.17. It should be recalled at this point that the phase velocity $c_{ph} = \omega/k$ and the group velocity $c_g = \partial\omega/\partial k$; hence the curves with lower slopes have higher group velocities. At the point marked C in Fig. 1.17, the phase speeds of the plate bending wave and of the acoustic wave in the fluid are equal. In wave-coupling terms this is called the *critical*, or *lowest coincidence*, frequency ω_c. (What is the ratio of the slopes of these two curves at ω_c?) The bending-wave curve is seen to approach the shear wave speed at high frequencies. The relative dispositions of the various curves depend, of course, upon the type and forms of material carrying the waves.

In cases where a wave-bearing medium is modelled as being extended in one of its dimensions, but confined within parallel boundaries in the other one (or two) dimensions, the system is known as a waveguide. The presence of the boundaries constrains the forms of wave patterns that can propagate in the medium and hence modifies the dispersion characteristics of the system. The flat plate waveguide shown in Fig. 1.18 takes the form of infinitely long strip of uniform width l located between boundaries that provide *simple supports*. Equation (1.41) is used, together with the boundary conditions, to yield the dispersion relationship for flexural modes

$$k_{zp}^2 = k_b^2 - k_x^2 = k_b^2 - (p\pi/l)^2 = \omega\,(m/D)^{1/2} - (p\pi/l)^2 \qquad (1.42)$$

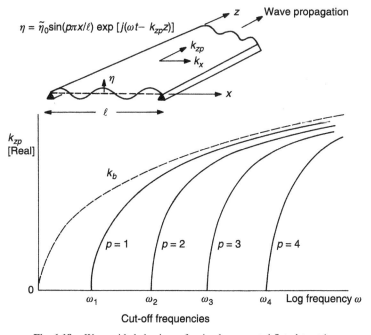

Fig. 1.18.   Waveguide behaviour of a simply supported flat-plate strip.

where k_{zp} is the wavenumber corresponding to the propagation of the waveguide 'modes': these are characteristic *spatial distributions* of the wavefield variables which take the form of standing patterns $\sin(p\pi x/l)$ across the width of the strip. This relation is represented qualitatively in Fig.1.18. Equation (1.42) shows that real (propagating) solutions exist for each value of p only at frequencies greater than that for which $k_{zp} = 0$: i.e., $\omega > (p\pi/l)^2 (D/m)^{1/2}$. The frequencies at which $k_{zp} = 0$ are the resonance frequencies of a simply supported beam of length l made from the plate material, and they are known as the cut-off frequencies of the waveguide modes of order p. Below its cut-off frequency, a mode cannot effectively propagate wave energy and its amplitude decays exponentially away from a point of excitation. At the modal cut-off frequencies the modal phase velocities $c_{ph} = \omega/k_{zp}$ are infinite and the modal group velocities $c_g = \partial\omega/\partial k_{zp}$ are zero. As will be discussed in detail in Section 1.10, a mode is a spatial pattern of the wavefield variables (characteristic of the form of waveguide) formed by interference between waves that are multiply reflected by the waveguide boundaries. Although modal phase velocity can exceed that of free waves in the unbounded medium, information and energy transported by a mode travels at the group velocity which in this case is less than that of free waves (Prove).

1.9   Flexural Waves in Thin-Walled Circular Cylindrical Shells

There are many structures that have a form approximating that of circular cylindrical shell: for example, pipes and ducts, aircraft fuselages, fluid containment tanks, submarine pressure hulls and some musical instruments. In many cases of long slender cylinders having relatively thick walls, only the transverse bending, beam-like mode of propagation, in which distortion of the cross section is negligible, is of practical interest, particularly for low-frequency vibration of pipes in industrial installations such as petrochemical plants: Section 1.6 deals with such waves. However, if the ratio of cylinder diameter to wall thickness is large, as in aircraft fuselages, wave propagation involving distortion of the cross section is of practical importance even at relatively low audio frequencies. If the wall thickness and mass per unit area are uniform, and the shell is truly circular in cross section, the allowable spatial form of distortion of a cross section must be periodic in the length of the circumference. The axial, tangential and radial displacements u, v and w of the wall must vary with axial position z and azimuthal angle θ as

$$u, v, w = [U(z), V(z), W(z)]\cos(n\theta + \phi), \quad 0 \le n \le \infty \qquad (1.43)$$

The integer n is known as the circumferential mode order. At any frequency, *three* forms of wave having a given n may propagate along an *in-vacuo* cylindrical shell: each has characteristically different ratios of U, V and W which vary with frequency (Leissa, 1993b).

Whereas small amplitude bending waves in *untensioned* plates can be considered independently of longitudinal or in-plane shear waves, it is necessary to consider all three displacements u, v and w of a cylindrical shell. The curvature of the wall of a cylinder 'allows' stresses associated with direct strains of the median surface of the wall ('membrane' stresses) to produce linearly related force components in the radial direction. It also causes radial displacement to produce membrane stresses that produce linearly related force components tangential to the median surface. The magnitude of these coupling forces decreases as the cylinder radius increases and disappear in a flat plate, of which the 'radius' is infinite. As in a flat plate, wall flexure is associated with forces transverse to the median surface. In a flat plate, the direct membrane stresses arising from transverse deflection are always tensile (Fig. 1.19) and, if significant, produce non-linear vibrational behaviour. A complete analysis of cylindrical shell vibration is beyond the scope of this book for which readers are directed to comprehensive collations of theoretical and experimental data presented by Leissa (1993b) and Soedel (2004). A thorough tutorial treatment may be found in Cremer *et al.* (1988).

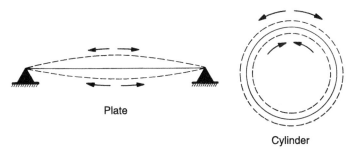

Plate

Cylinder

Fig. 1.19. Membrane stresses in a transversely deflected plate and tangential stresses in a circular cylindrical shell.

For the purposes of studying sound radiation, transmission and response behaviour of thin-walled circular cylindrical shells that do not contain, or are not submerged in, dense fluids such as water or liquid sodium, it is only necessary to consider the form of wave in which the radial displacement w is dominant. This form of wave may be classified as 'radial'; but this suggests propagation in the radial direction and, because it is significantly affected by radial forces associated with wall flexure, it is often classified as 'flexural'. This terminology is unfortunate because waveguide modes of low circumferential order formed by this class of wave are predominantly controlled by membrane forces. As explained below, the speed of propagation of a 'flexural' waveguide mode in the direction of the longitudinal axis of the cylinder is dependent upon both the circumferential mode order n and the ratio of wall thickness to shell radius, because the relative contributions to strain energy of membrane strain and wall flexure depend upon these parameters. In addition, a waveguide mode cannot propagate freely below its 'cut-off frequency', at which the corresponding axial wavenumber and group velocity are zero. A cut-off frequency corresponds to a natural frequency at which the modal pattern consists of a set of $2n$ nodal lines lying along equally spaced generators of a uniform shell of infinite length, as shown in Fig. 1.20.

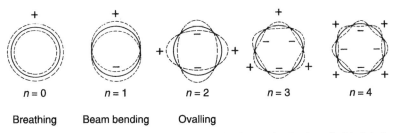

$n = 0$ $n = 1$ $n = 2$ $n = 3$ $n = 4$

Breathing   Beam bending   Ovalling

Fig. 1.20. Cross-sectional radial displacement mode shapes of a circular cylindrical shell.

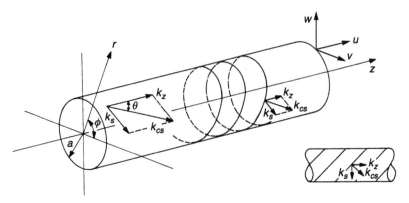

Fig. 1.21. Circular cylindrical shell coordinates, displacements, wavenumber vectors and helical wavefronts.

Flexural-type waves propagating in a uniform cylindrical shell may be characterised by axial and circumferential wavenumbers k_z and k_s, as shown in Fig. 1.21. The wavefronts of free propagating waves of wavenumber vector $\mathbf{k}_{cs}$ form a spiral pattern, the angle between the wavenumber vector and a generator of the cylinder being given by $\theta = \tan^{-1}(k_s/k_z)$. The closure of the shell in the circumferential coordinate direction requires that the wave variables be continuous around the circumference and that the characteristic circumferential patterns take the form $\genfrac{}{}{0pt}{}{\sin}{\cos}(k_s s)$, where $s = a\phi$ and $k_s = n/a$ ($n = 0, 1, 2, ..., \infty$), so that an integer number of complete circumferential wavelengths $\lambda_s = 2\pi/k_s = 2\pi a/n$ fit around the circumference. The wavenumber relation is

$$k_z^2 = k_{cs}^2 - k_s^2 = k_{cs}^2 - (n/a)^2 \qquad (1.44)$$

where k_{cs} is the magnitude of the wavenumber vector of the propagating helical wave components.

The two most important non-dimensional shell parameters are the non-dimensional frequency $\Omega = \omega/\omega_1$, where $\omega_1 = \sqrt{E/\rho a^2} = c_l''/a$ and the non-dimensional thickness parameter $\beta = h/\sqrt{12}a$, where h is the wall thickness and a is the mean radius of the shell. The shell 'ring frequency' is given by $\omega_r = c_l'/a$, so that $\omega/\omega_r = \Omega c_l''/c_l'$: we shall denote this parameter by Ω_r. The physical significance of these parameters may be explained as follows.

At the ring frequency, when $\Omega_r = 1$, the quasi-longitudinal wavelength in the shell wall equals the shell circumference and an axisymmetric, $n = 0$, 'breathing', or 'hoop', resonance occurs. Note that ω_r is dependent upon cylinder radius, but not on wall thickness. Later we shall see that the ring frequency separates frequency regions in which wall curvature effects are dominant and weak. Large-diameter shells, such as aircraft fuselages, have ring frequencies of the order of 300 Hz, whereas the ring frequencies of industrial pipes tend to be in the range 1–5 kHz. As explained in more detail below Eq. (1.49), the parameter β provides an indication of the relative contributions of strain energy associated with axial and tangential membrane strains of the median surface of the shell, and the strain energy associated with wall flexure. The greater is β, the more important is wall flexure in controlling the dispersion relation, and hence the cut-off and natural frequencies, of the flexural waveguide modes. It is generally much greater in industrial pipes than in large diameter, thin-walled cylinders.

The free flexural wavenumber in a flat plate of thickness equal to that of the shell wall is

$$k_b = \omega^{1/2}(m/D)^{1/4} = \omega^{1/2}(12)^{1/4}/h^{1/2}c_l'^{1/2} \qquad (1.45)$$

This wavenumber can be expressed in a non-dimensional form involving the cylindrical-shell parameters Ω_r and β and the radius a as

$$k_b a = \Omega_r^{1/2}\beta^{-1/2} \qquad (1.46)$$

This non-dimensional wavenumber forms a useful reference value in the subsequent analysis.

A form of bending wave that dominates the low-frequency vibration behaviour of long slender cylinders, e.g., pipelines, is the beam-bending wave. There is no cross-sectional distortion and the circumferential variation of radial displacement corresponds to the order $n = 1$. The beam-bending wavenumber is that derived for bars in Section 1.6:

$$k_{bb} = \omega^{1/2}(m'/EI)^{1/4} \qquad (1.47a)$$

in which $m' = 2\pi a h\rho$ and $I = \pi a^3 h$. The non-dimensional equivalent is

$$k_{bb}a = (2)^{1/4}\Omega_r^{1/2} \approx 1.2\Omega_r^{1/2} \qquad (1.47b)$$

which does not contain β and is therefore independent of wall thickness, unlike $k_b a$.

There are many thin-shell equations of varying degree of complexity, the differences arising largely from differences in the assumed strain–displacement relations (Leissa, 1993b). One form, due to Kennard (1953), is exploited by Heckl (1962b) to derive expressions for the natural frequencies and various forms of impedance of a thin-walled cylinder. The application of stress–strain relations in the shell, together with Newton's second law of motion, yields three coupled equations of motion in the axial, radial and tangential directions. On the assumption of non-dimensional frequency Ω, circumferential wavenumber n/a and axial wavenumber k_z, Kennard's equations become

$$\alpha \tilde{w} + n\tilde{v} + vk_z a\tilde{u} = 0 \tag{1.48a}$$

$$n\tilde{w} + \left[n^2 - \Omega^2 + \frac{1}{2}(1 - v)k_z^2 a^2 \right] \tilde{v} + \frac{1}{2}(1 + v)nk_z a\tilde{u} = 0 \tag{1.48b}$$

$$vk_z a\tilde{w} + \frac{1}{2}(1 + v)nk_z a\tilde{v} + \left[k_z^2 a^2 + \frac{1}{2}(1 - v)n^2 - \Omega^2 \right] \tilde{u} = 0 \tag{1.48c}$$

where v is Poisson's ratio and $\alpha = 1 - \Omega^2 + \left\{ (n^2 + k_z^2 a^2)^2 - \frac{1}{2}[n^2(4 - v) - 2 - v] \right.$
$\left. [1 - v]^{-1} \right\} \beta^2$.

Equations (1.48b) and (1.48c) are independent of β and are essentially membrane equations. Equations (1.48a) differs from the corresponding membrane equations by the inclusion of a term (in α) that is analogous to Eq. (1.39) for the transverse elastic force per unit area in the equation of motion of a flat plate. These yield the dispersion relation

$$\Omega^2 = (1 - v^2)\{(k_z a)^2/[(k_z a)^2 + n^2]\}^2 + \beta^2\{[(k_z a)^2 + n^2]^2$$
$$- [n^2(4 - v) - 2 - v]/2(1 - v)\} \tag{1.49}$$

This expression is accurate for thin shells ($\beta \ll 10^{-1}$). The first term in curly brackets on the right-hand side of the equation is associated with membrane strain energy and the second, which contains β^2, is associated with strain energy of wall flexure. The cross-sectional resonance, or cut-off, frequencies of an infinitely long cylinder, are given by Eq. (1.49) with $k_z = 0$ which corresponds to an infinite axial wavelength. It is important to observe that these frequencies are determined purely by strain energy of wall flexure; they correspond to Rayleigh's inextensional mode frequencies which were derived by assuming that the median surface of the shell wall does not strain. Evaluation of the resulting equation for Ω^2,

$$\Omega_n^2 = \beta^2 n^4 \left[1 - \frac{1}{2} \left(\frac{1}{1 - v} \right) \left(\frac{4 - v}{n^2} - \frac{2 + v}{n^4} \right) \right] \tag{1.50}$$

yields the values of the cut-off frequencies of the lower order modes which are presented in Table 1.3.

TABLE 1.3

Cut-Off Frequencies of Flexural Modes of
Thin-Walled Circular Cylindrical Shells

Circumferential mode order n	$\Omega_n/\beta n^2$	Ω_n/β
2	0.67	2.68
3	0.85	7.65
4	0.91	14.56
5	0.95	23.75
6	0.96	34.56
7	0.97	47.10

It is seen from Table 1.3 that, except for the lowest order modes, the cut-off frequencies are reasonably well approximated by the formula

$$\Omega_n \approx \beta n^2 \qquad (1.51)$$

Equation (1.51) indicates that the number of cut-off frequencies below the ring frequency is given approximately by

$$n_r \approx \beta^{-1/2} \qquad (1.52)$$

The dispersion relationship, Eq. (1.49) may be expressed in graphical form in three ways, depending upon which of the non-dimensional variables $k_z a$, $k_s a = n$, or Ω forms the curve parameter. The relation between the acoustic and shell wavenumbers is of crucial importance in determining the coupling of the media and therefore a two-dimensional wavenumber diagram with Ω as the parameter is found to be useful. Experimental evidence suggests that the level of excitation of long pipes by sound in a contained fluid is largely determined by an axial wavenumber coincidence phenomenon, involving internal fluid and shell waveguide modes. In this case, curves of axial wavenumber versus Ω, for given n, are revealing. Both forms of curve are presented below.

In order to illustrate the form of the shell wavenumber diagram, Eq. (1.49) is simplified by the omission of the less-important of the flexure terms in curly brackets; this omission will not significantly distort the qualitative picture of the mechanisms of sound radiation and transmission discussed in Chapters 3 and 5. In Fig. 1.22 the loci of the non-dimensional $k_z \sim k_s$ relation are presented for constant values of Ω: the radius vector represents the shell wavenumber vector $\mathbf{k}_{cs}$, the angle θ indicating the direction of component wave propagation relative

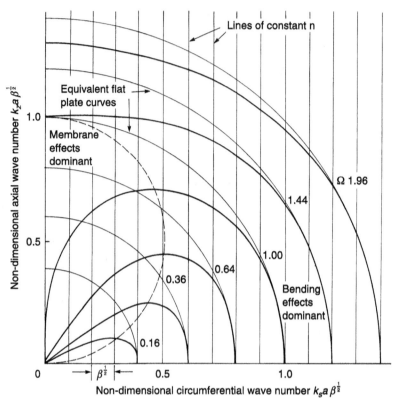

Fig. 1.22. Universal dispersion curves for flexural waves in thin-walled circular cylindrical shells $(n > 1)$.

to a generator. The particular non-dimensional form of wavenumber is chosen so that the curves are universal, as the simplified form of Eq. (1.49) shows:

$$\Omega^2 \approx \frac{(k_z a \beta^{1/2})^4}{[(k_z a \beta^{1/2})^2 + (n\beta^{1/2})^2]^2} + [(k_z a \beta^{1/2})^2 + (n\beta^{1/2})^2]^2 \qquad (1.53)$$

Also shown in Fig. 1.22 are the constant-frequency (Ω) loci for a flat plate of the same thickness and material as the cylinder walls. The broken line corresponds to equality of the first (membrane) term and second (bending) term in Eq. (1.53). It may therefore be considered to enclose the region in which membrane effects are predominant and in which the loci for $\Omega < 1$ correspond approximately to the straight-line forms appropriate to a membrane wall cylinder of vanishing wall thickness $(\beta \to 0)$, in which bending effects are negligible: consider $k_z a \sim k_s a$

for fixed Ω in Eq. (1.49) when $\beta = 0$. One striking effect of membrane stresses is to exclude the possibility of purely axial propagation of a flexural wave below the ring frequency.

Although Fig. 1.22 is universally applicable, it must be realised that the number of shell waveguide modes that can propagate in the frequency range below the ring frequency is dependent upon the shell thickness parameter, through Eq. (1.51), and that there is really not a continuum of circumferential wavenumbers, because of the requirement for continuity of the distribution of wave variables around the circumference. Hence, only the intersection points between vertical lines drawn at $n\beta^{1/2}$ and the loci are physically significant. The number of shell modes that are substantially affected by membrane effects decreases with increase in the wall thickness, as would be expected. From Fig. 1.22 we may take the formula $n_m/\beta^{1/2} = 0.5$ as an indication of this number: alternatively, we can assume that any mode with a cut-off frequency below $\Omega = 0.25$ is subject to membrane effects; for example, if $h/a = 0.003$, $n_m = 17$, and if $h/a = 0.05$, $n_m = 4$.

An alternative form of Fig. 1.22 is shown in Fig. 1.23, in which the axial wavenumbers corresponding to particular values of n are plotted against Ω: the

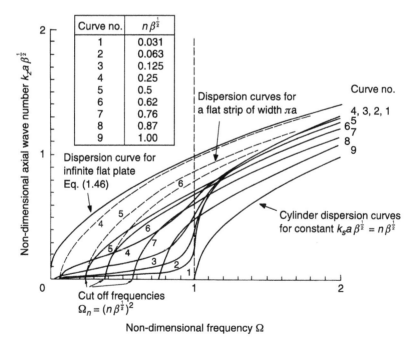

Fig. 1.23. Universal dispersion curves for flexural waves in thin-walled circular cylindrical shells.

curves correspond to cross plots along the vertical lines, corresponding to fixed values of n in Fig. 1.22. The effects of membrane stresses in increasing the axial phase speed of the low-order modes are clearly seen by comparison with the equivalent flat-strip waveguide bending wave curves plotted on the same figure.

1.10   Natural Frequencies and Modes of Vibration

Thus far, we have considered the question of what types of wave can propagate in various geometric forms of media that are uniform and unbounded in the directions of propagation. All physical systems are spatially bounded, and many incorporate non-uniformities of geometric form or material properties. Waves that are incident upon boundaries, or regions of non-uniformity, cannot propagate through them unchanged; the resulting interaction gives rise to phenomena known as refraction, diffraction, reflection and scattering. It is difficult concisely to define and to distinguish between each of these phenomena. However, broadly speaking, refraction involves veer in the direction of wave propagation due either to spatial variation of wave phase velocity or of mean flow in a fluid medium, or to wave transmission through an interface between different media; diffraction involves the distortion of wave fronts (surfaces of uniform phase) created by their encounter with one or more partial obstacles to wave motion; reflection implies a reversal of the wavenumber vector or a component thereof and scattering refers to the redirection of wave energy flux, normally into diverse directions, due to the presence of localised regions of non-uniformity in a medium or irregularity of a boundary.

Although all these wave phenomena may occur in solid structures, the one having most practical importance is reflection. The reason for its importance is that it is responsible for the existence of sets of frequencies, and associated spatial patterns of vibration, which are proper to a bounded system. An infinitely long, undamped, beam can vibrate *freely* at any frequency; a bounded, undamped, beam can vibrate *freely* only at discrete *natural* or *characteristic* frequencies, that are theoretically infinite in number. The elements of the beams that are not at boundaries satisfy the same equation of motion in both cases; they clearly only 'know about' the boundaries because of the phenomenon of wave propagation and reflection. Wave reflection at boundaries also leads to the very important phenomenon of *resonance*. Note carefully that resonance is a phenomenon associated with *forced* vibration, generated by some input, whereas natural frequencies are phenomena of *free* vibration. Resonance is of very great practical importance because it involves large amplitude response to excitation, and can lead to structural failure, system malfunction and other undesirable consequences.

In order to understand the nature of resonance, we turn to the consideration of the process of free vibration of an *undamped*, wave-bearing system. Mathematically, the questions to be answered are: At what frequencies can the equation(s) of motion, together with the physical boundary conditions, be satisfied in the absence of excitation by an external source; and what characteristic spatial distributions of vibration are associated with these frequencies? This question can be interpreted in physical terms as follows: If the system is subjected to a transient disturbance, which frequencies will be observed to be present in the subsequent vibration, and what characteristic spatial distributions of vibration are associated with these frequencies?

It is possible to answer these questions without explicit reference to wave motion at all, which is a little surprising since any vibrational disturbance is propagated throughout an elastic medium in the form of a wave. However, it must be realised that an alternative macroscopic model to that of the elastic continuum is one consisting of a network of elemental discrete mass-spring systems. In fact, some of the earliest analysis of free vibration of distributed, continuous systems was based upon such a discrete element model (Lagrange, 1788). In this case, equations of motion can be written for each element, together with the coupling condition, thereby producing n equations of motion for n elements. Alternatively, statements can be made about the kinetic and potential energies of the elements and about the work done on them by internal and external forces: the 'Hamiltonian' energy approach forms the basis of most practical methods of estimation of natural frequencies of structures. Today it is most widely implemented by the 'Finite Element Method' which is introduced in Chapter 8. However, one of the principal aims of this book is to promote the 'wave picture' of vibration in fluids and solid structures, and hence we shall initially discuss natural vibrations of bounded systems from this point of view.

In earlier sections we have seen that the wavenumbers and associated frequencies of waves propagating freely in unbounded uniform elastic systems are uniquely linked through the governing wave equation in the form of dispersion equations. Let us imagine what happens when a freely propagating wave meets an interface with a region of the system in which the dynamic properties are different from those of the uniform region previously traversed by the wave: in practice, the interface could take the form of a boundary, a change of geometry or material, or a local constraint. The relations between forces and displacements in this newly encountered region are different from those in the uniform region, and it is clear that the wave cannot progress unaltered. Compatibility of displacements and equilibrium of forces must be satisfied at the interface and yet, if the wave were wholly transmitted past the interface, the forces associated with given displacements would be different on the two sides of the interface. (Consider two

beams of different I joined together and refer to Eq. (1.33) with a progressive
bending-wave form for transverse displacement.) Consequently, a reflected wave
must be generated that, in combination with the incident wave, is compatible and
in equilibrium with the wave transmitted beyond the interface.

The amplitude and phase of a reflected wave relative to those of an incident
wave depend upon the relative dynamic properties of the medium bearing the
incident wave and those of the region at and beyond the interface, which are
manifested in the impedance at the interface. For simplicity, we consider the
case of a bending wave in an infinitely extended, undamped beam that is simply
supported at $x = 0$ (Fig. 1.24). The incident wave displacement is

$$\eta_i^+(x,t) = \tilde{A}\exp(-jk_bx)\exp(j\omega t) \tag{1.54}$$

where $\tilde{A}$ is the complex amplitude of the incident wave at $x = 0$. The presence of
the simple support completely suppresses transverse displacement and produces
shear force reaction, but does not restrain rotational displacement and hence does
not produce any moment reaction. The incident wave alone cannot satisfy the
condition of zero transverse displacement at the support at all times; therefore,
a reflected wave must be generated that in combination with the incident wave,
does satisfy this condition. From Eq. (1.37) the general form for the propagating
and non-propagating field components of negative-going, reflected and bending
wave is

$$\eta_r^-(x,t) = [\tilde{B}_1\exp(jk_bx) + \tilde{B}_2\exp(k_bx)]\exp(j\omega t) \tag{1.55}$$

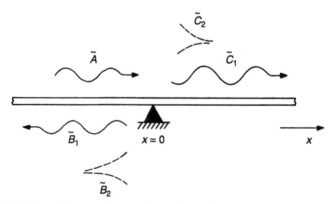

Fig. 1.24. Reflection and transmission of a bending wave at a simple support.

and for the positive-going transmitted bending wave is

$$\eta_t^+(x, t) = [\tilde{C}_1 \exp(-jk_b x) + \tilde{C}_2 \exp(-k_b x)] \exp(j\omega t) \qquad (1.56)$$

We henceforth drop the time-dependent term and the superscripts indicating wave direction.

There are four complex unknowns to be related to $\tilde{A}$ through application of conditions of compatibility and equilibrium at $x = 0$, where the terms in square brackets are evaluated:

(1) Compatibility (linear displacement)

$$[\eta_i + \eta_r] = \tilde{A} + \tilde{B}_1 + \tilde{B}_2 = [\eta_t] = \tilde{C}_1 + \tilde{C}_2 = 0 \qquad (1.57)$$

(2) Compatibility (angular displacement)

$$[\partial \eta_i/\partial x] + [\partial \eta_r/\partial x] = k_b \left(-j\tilde{A} + j\tilde{B}_1 + \tilde{B}_2\right)$$
$$= [\partial \eta_t/\partial x] = k_b \left(-j\tilde{C}_1 - \tilde{C}_2\right) \qquad (1.58)$$

(3) Equilibrium (transverse force)

$$EI[\partial^2 \eta_i/\partial x^2 + \partial^2 \eta_r/\partial x^2] = EIk_b^2 \left(-\tilde{A} - \tilde{B}_1 + \tilde{B}_2\right)$$
$$= EI \left[\partial^2 \eta_t/\partial x^2\right] \qquad (1.59)$$
$$= EIk_b^2 \left(-\tilde{C}_1 + \tilde{C}_2\right)$$

The solutions of Eqs. (1.57)–(1.59) are

$$\tilde{B}_1 = -(\tilde{A}/2)(1 + j), \quad \tilde{B}_2 = -(\tilde{A}/2)(1 - j)$$
$$\tilde{C}_1 = (\tilde{A}/2)(1 - j), \quad \tilde{C}_2 = -(\tilde{A}/2)(1 - j) \qquad (1.60)$$

It is interesting to note that, since the rate of transport of vibrational energy along a beam by a propagating bending wave is proportional to the square of the modulus of the complex amplitude, half the incident energy is reflected and half is transmitted by such a discontinuity (see Section 2.8). (Derive this result by considering the work done at a cross section by the shear forces and bending moments associated with internal stresses in the beam.)

The total displacement field on the incident side of the support $(x < 0)$ is

$$\eta(x,t) = \tilde{A}[\exp(-jk_b x) - \frac{1}{2}(1+j)\exp(jk_b x) - \frac{1}{2}(1-j)\exp(k_b x)]\exp(j\omega t)$$
(1.61)

The final term in the square brackets represents the non-propagating component of the reflected field, which decays with distance from the support. It is not, strictly speaking, a wave component because it does not possess either a real phase velocity or a real wavenumber: its phase is independent of distance and, at a sufficiently large non-dimensional distance $k_b x$ from the support, it makes a negligible contribution to the vibrational displacement; hence it is termed the 'near-field' component. The second term represents the true reflected wave, and its influence extends to $x = -\infty$. The first and second terms together represent an interference field that, following the graphical representation presented in Fig. 1.2, can be represented by Fig. 1.25. (Try to superimpose the near-field term onto the figure. Is the support boundary condition satisfied?) Unlike the phasor that represents a pure progressive wave in Fig. 1.2, the phasor that arises from the addition of the two phasors representing the incident and reflected waves does not rotate $\pi/4$ radians per unit of distance Δx, where $k_b \Delta x = \pi/4$. The interference field is not a pure standing pattern, as the absence of any nodal point except the support demonstrates. (Why is this so?) Figure 1.26 shows the displacement fields of the incident and reflected/transmitted waves, respectively

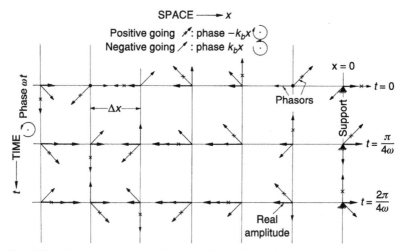

Fig. 1.25. Phasor diagram for bending wave reflection from a simple support of an infinite beam (excluding the evanescent near field).

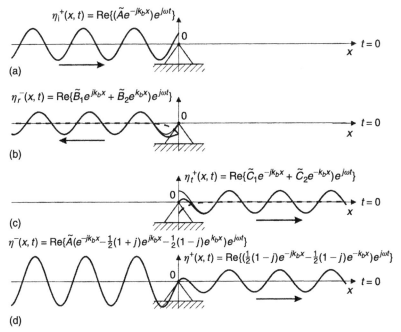

Fig. 1.26. (a) Incident, (b) reflected, (c) transmitted and (d) resultant wavefields generated by the incidence of a positive-going bending wave upon a simple support of an infinite beam. Faint solid line: propagating wave components; dashed line: near-field wave component; thick solid line: total (propagating + near-field) wavefield.

and the resultant wavefield which indeed does not show a pure standing wave pattern in the region $x < 0$ because a pure standing wave does not transport energy.

If the beam considered above had *terminated* at the simple support, the following coefficients would have been determined:

$$\tilde{B}_1 = -\tilde{A}, \quad \tilde{B}_2 = 0 \tag{1.62}$$

(Try to obtain this result.) In this case, the beam displacement is

$$\eta(x, t) = \tilde{A}[\exp(-jk_b x) - \exp(jk_b x)] \exp(j\omega t) = 2j\tilde{A}\sin(k_b x)\exp(j\omega t) \tag{1.63}$$

Of course, all the incident wave energy is reflected. The resulting phasor diagram is shown in Fig. 1.27, while Fig. 1.28 shows the displacement fields of the incident and reflected waves and the resultant wavefield.

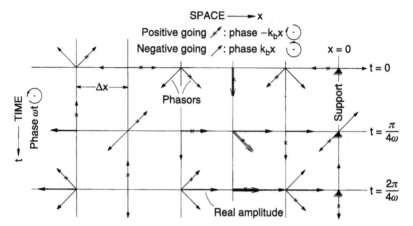

Fig. 1.27.   Phasor diagram for wave reflection from a simple support termination ($k_b \Delta x$ chosen to equal $\pi/4$ for clarity).

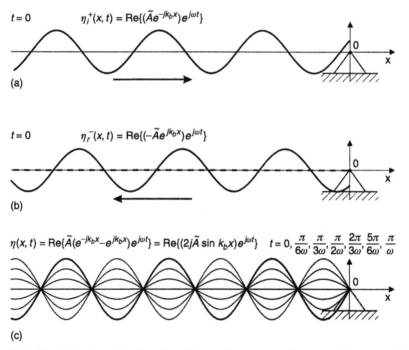

Fig. 1.28.   (a) Incident, (b) reflected, and (c) resultant pure standing-wave field generated by the incidence of a positive-going bending wave upon a simple support of a semi-infinite beam. Faint solid line: propagating wave components; dashed line: near-field wave component 0; thick solid line: total (propagating + near-field) wavefield.

   This interference field exhibits characteristics quite different from those of the pure progressive wave. The relative phase of the displacement at the various points on the beam takes only values of zero or π; the points of zero and maximum amplitude are fixed in space; the field has the form of a *pure standing wave* in which the spatial and temporal variations of displacement are independent, as indicated by Eq. (1.63). Such forms of wave can only be produced by interference between two waves of the same physical type, having equal amplitudes and frequencies, and equal and opposite wave vector components in at least one direction.

   If, instead of terminating the beam at a simple support, we chose to terminate it at a free end or a fully clamped support, or by a lumped mass or a spring-like element, none of which dissipate or transmit energy, we would find that the amplitude of the reflected wave would be of equal magnitude to that of the incident wave, although the near-field component amplitudes would be particular to the type of termination. Therefore, all such conservative terminations produce pure standing-wave interference patterns.

   Suppose now that the undamped beam is of finite length and simply supported at both ends. Consider the progress of a *single-frequency* wave after reflection from the left-hand end [Fig. 1.29(a)]. Equation (1.62) indicates that the amplitude of the reflected wave at the right-hand end [Fig. 1.29(b)] equals that of the incident wave and that its relative phase is π. This reflected wave is incident upon the left-hand end and perfectly reflected; it then returns along the path of the original incident wave [Fig. 1.29(c)]. The phase of the returning wave, relative to that of the original incident wave, depends upon the free wavenumber and upon the length of the beam. Certain combinations of these parameters will produce equality of phase, or phase coincidence,[6] of the original and returning waves. By definition, phase coincidence requires the phase change over one return journey to equal $n(2\pi)$ Since wavenumber has previously been defined as phase change per unit length, phase coincidence occurs when

$$k_b(2l) = n(2\pi) \tag{1.64}$$

or

$$k_b = \frac{n\pi}{l} \tag{1.65}$$

This condition is illustrated in Fig. 1.29(d) by the selection of the value of 5π for the non-dimensional bending wavenumber $k_b l$. The possibility of phase

---

[6] Note: the condition described as 'phase coincidence' is also termed 'phase closure' by some authors.

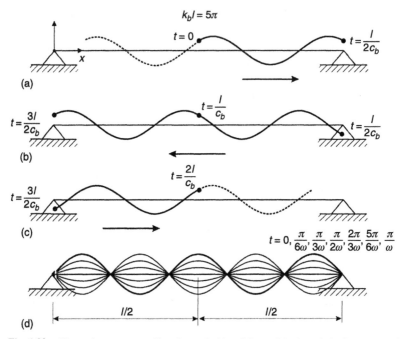

Fig. 1.29. Harmonic waves travelling forward, (a) and (c), and backward (b) from one end to the other of a simply supported beam which interfere to produce the pure standing-wave field shown in (d) (the bullet dots represent the wavefronts).

coincidence implies the possibility of existence of sustained, pure standing-wave fields in undamped structures.

Therefore, recalling that $k_b = \left(\omega^2 m/EI\right)^{1/4}$, phase coincidence occurs at those frequencies such that

$$\left(\frac{\omega^2 m}{EI}\right)^{1/4} = \frac{n\pi}{l} \tag{1.66}$$

where $n = 1, 2, 3, \dots$. The frequencies at which such pure standing-wave fields can occur in a given undamped system in free vibration are known as the 'characteristic' or 'natural' frequencies of the system. The associated spatial distributions of vibration amplitudes are known as the 'characteristic functions' or 'natural modes' of the system. According to Eq. (1.66), the natural frequencies depend on the material properties and geometric dimensions of the beam. The natural frequencies of any continuously distributed elastic system are infinite in number.

The natural modes of closed, undamped systems take the form of pure standing waves, as exemplified by Fig. 1.29. Considering a system of reference located at the left-hand end of the beam, since the relation between the complex amplitudes of the positive- and negative-going components in a simply-supported beam is given by Eq. (1.62) $\tilde{B} = -\tilde{A}$, the displacement at position x is given by

$$\eta(x, t) = \eta^+(x, t) + \eta^-(x, t) = \tilde{A}\left[e^{-jk_b x} - e^{-jk_b(2l-x)}\right]e^{j\omega t} \qquad (1.67)$$

Since the natural frequencies occur for wavenumbers such that $k_b l = n\pi$, and thus $\exp(-jk_b 2l) = 1$, the vibration fields of the natural modes take the form

$$\eta(x, t) = \tilde{A}\left[e^{-jk_b x} - e^{jk_b x}\right]e^{j\omega t} = -2j\tilde{A}\sin\frac{n\pi x}{l}e^{j\omega t} \qquad (1.68)$$

As found for the semi-infinite beam terminated by a simple support, the relative phases of the displacement at the different points on the beam take only values of 0 or π and the points of zero and maximum amplitude are fixed in space, as shown in Fig. 1.29. Conventionally, the natural modes of simply supported beams are expressed in the standard form:

$$\eta(x, t) = \phi_n(x)\exp(j\omega t) = \sqrt{2}\sin(n\pi x/l)\exp(j\omega t) \qquad (1.69)$$

where $n = 1, 2, 3, \ldots$, and the complex amplitude of the circulating wave has been taken to be $\tilde{A} = j\sqrt{2}/2$ so that the natural modes $\phi_n(x) = \sqrt{2}\sin(n\pi x/l)$ are 'normalised' in such a way that $\int_0^l [\phi_n(x)]^2\,dx = 1$ for reasons that will become apparent in Chapter 2.

A very useful device for evaluating the approximate distribution of natural frequencies of a one-dimensional bounded system of length l is to superimpose on the dispersion curves (such as those in Fig. 1.17) a set of horizontal lines representing phase coincidence, i.e., at intervals of k of $n\pi/l$. Intersections with the dispersion curves indicate the approximate natural frequencies. (What does such a construction tell you about the natural frequency distribution of a bar in flexure?) This device can be extended to geometrically regular two- and three-dimensional systems to supply estimates of the average number of natural frequencies per unit frequency; this measure is termed 'modal density' (see Section 1.12).

The question of the role of any near-field influence on phase coincidence is interesting, since a beam having different end conditions has different forms of near field at each end. In practice, this influence is only significant for the fundamental modes of uniform beam systems, because the near field decays as $\exp(-k_b x)$, and $k_b l$ is about 3 at the fundamental natural frequencies of beams of length l. However, near fields are much more important in systems incorporating multiple discontinuities, such as ribbed plates. It must be noted that the nature of a termination affects the phase of the reflected travelling wave relative to that of the incident wave, and this must be taken into account in any analysis based upon the phase coincidence criterion; for instance, both free and fully clamped beam ends produce a phase change of $-\pi/2$ rad, whereas a simply supported end produces a change of $-\pi$ rad.

Of course, all real systems possess some mechanisms of energy dissipation, or radiation, and hence waves cannot propagate to and fro forever during free vibration. Damping mechanisms are various: they include internal friction, interface friction, and radiation of energy into contiguous fluids or structures. In most practical cases of structures that are not artificially damped, the presence of damping mechanisms does not significantly change the phase velocity or the phase changes at reflection. Hence, the above arguments relating to phase coincidence are unchanged, and the damped and undamped natural frequencies are almost equal.

In a non-closed system, such as that of an *infinite* beam on two simple supports, where energy is partially transmitted (radiated) through boundaries into connected *unbounded* systems, phase changes upon reflection are affected by the presence of the sections of the beam beyond the supports. Natural frequencies exist because multiple wave reflection can occur, albeit with successive reduction of amplitude upon reflection during free vibration due to energy transmission through each support. The effect is akin to that of damping described above. Mathematically, such frequencies are complex, as explained in Section 1.12. As also explained in Section 1.12, the natural modes of systems having partially transmitting, or dissipative, boundaries are also mathematically complex. They are not pure standing waves (which cannot transport energy) and the phase does not anywhere change discontinuously by π radians because nodal points/lines do not exist. Instead, it varies continuously over the length of the beam.

Where further discontinuities occur in the beam extensions, reflections occur there also, and then considerations of phase coincidence must include all such reflections. This observation highlights the fact that there are no such entities as independent natural modes of individual parts of continuous structures, although they may exist in the minds of the theoreticians, or in partial models fed to the computer in a modelling process called 'Component Mode Synthesis'. However, in order not to completely disillusion the reader, by implying that the whole world must always be included in a vibration model, it must be said that, in many practical cases, discontinuities that divide one part of a continuous structure from the rest are often of such a form as to transmit very little energy; reflections return after a double transit of the discontinuity with so little amplitude as to alter very little the resultant phase of the motion in that part. Consequently a calculation made on the basis of an isolated subsystem with boundary conditions corresponding to an infinitely extended contiguous structure is often sufficiently accurate, at least for those models in which the vibration energy of the structure is contained largely in the particular region considered. An alternative view of ancillary structure effects is given by Bishop and Mahalingham (1981).

The condition of phase coincidence as an explanation of the existence of natural frequencies may, in principle, be extended to embrace two- and three-dimensional elastic systems. However, as we shall see, it becomes increasingly difficult to apply purely physical arguments for its action as the complexity of the geometry of the system increases. The simplest two-dimensional structural example of practical engineering interest is that of a thin, uniform, rectangular flat plate supported on all sides by simple supports. Since phase coincidence must occur anywhere on the plate at a natural frequency, it is clear that if it is satisfied independently in the two orthogonal directions parallel to the sides, then it is satisfied everywhere. It turns out in this particular case that the wavenumber vector components in these directions at the natural frequencies correspond to those for simply supported beams of lengths equal to those of the plate sides. In addition, in satisfaction of the bending-wave equation, their vector sum equals that of the free bending wavenumbers in the plate at that frequency, as explained in Section 1.7.

The difficulty of extending the phase coincidence argument to more complex geometries is simply illustrated by imagining one corner of a simply supported rectangular plate to be removed! It seems, therefore, that we must trust to the remarkable fact that mathematical solutions to mathematical models, obtained by the application of mathematical rules developed in the abstract and having no apparent empirical basis, can provide solutions that correspond to the behaviour of real physical systems. Just how pure standing-wave fields can be created in any elastic system, by reflection of waves from boundaries of arbitrary geometry, is something of a mystery [but see Langley (1999)].

We have seen that the reflection of undamped travelling waves from non-transmitting and non-dissipative boundaries, or from other such discontinuities, creates an interference field that, if composed solely of undamped waves of a frequency equal to a natural frequency of the system, takes the form of a standing-wave field in which the associated distribution of vibration amplitude is characteristic of the system and is known as a characteristic function, or natural mode, of the system. It consists of regions, or cells, of vibration of uniform phase, separated by zero-amplitude nodal lines from adjacent regions of vibration of opposite phase. A mode shape can be graphically represented most completely by contour plots of equal vibration amplitude, together with associated sections: however, it is more common simply to plot the nodal lines, especially for plates and shells of uniform thickness. It should be noted that systems in which the damping is distributed in proportion to the local mass density or stiffness exhibit *real* modes in which different points vibrate either in the same, or opposite, phase, as described above. All other distributions of damping (especially those associated with structural joints) produce *complex* modes in which the phase varies continuously with position and no pure nodal lines appear. However, in cases of light damping, modes of all physical systems approximate closely to real modes. Complex modes are discussed further in Section 1.12. Readers are referred to Newland (1989) for further reading on this subject.

For the purpose of analysing fluid–structure interactions, it is usually necessary only to plot vibration distributions normal to the interface between the media. For general formulas and numerical data, readers are referred to Chapter 2 and to general compendia of natural frequencies and mode shapes by Leissa (1993a,b) and Blevins (1995).

1.11   Forced Vibration and Resonance

So far we have considered vibration frequencies that are proper to a system in the physical sense that they predominate in the frequency spectrum of free vibration observed subsequent to a *transient* disturbance from equilibrium, and in the mathematical sense that they are solutions of the equation of free motion. The word 'predominate' is used because the spectrum of free motion of a damped system does not consist of discrete spectral lines since it is not periodic in the pure sense. The spectrum of a decaying 'sinusoid' is spread around the frequency of the undamped sinusoid: the more rapid the decay, the broader the spread. (Check this by Fourier transformation of a decaying oscillation.)

If a *linear* elastic system is subject to an unsteady disturbance that is stationary and continuous in time, the consequent vibration will contain *all and only* those frequencies present in the disturbance, not only the natural frequencies; indeed these may, in some cases, be completely absent. How can this be, when the natural frequencies have been established as those proper to the system? Perhaps a one-dimensional wave problem will help to clarify the nature of the process.

Imagine a tube with a movable piston in one end, a microphone at an arbitrary point in the tube and a rigid plug at the other end. If the piston is impulsively displaced inwards, an acoustic pulse will travel down the tube reflect off the plug, travel back to the now stationary piston, reflect down the tube and so on, theoretically endlessly. The microphone will register a succession of pulses having a basic period given by twice the length of the tube $2l$ divided by the speed of sound c. Frequency analysis of this periodic signal will produce a spectrum with discrete lines at frequencies given by $f_n = nc/2l$, where n is any positive integer. As explained in the previous section, these frequencies correspond to the acoustic natural frequencies of the tube, because the acoustic pressure phase shift at each wave reflection is zero: hence the round-trip phase shift is $k(2l) = 4\pi f_n l/c$. (What is the acoustic particle velocity phase shift?)

Imagine now that the piston is impulsively displaced periodically at intervals of time T. The pulse pattern period at the microphone will be T, and the frequencies in the received signal will be m/T, where m is any positive integer. In spite of the fact that *each* pulse does a round trip in the same time as before, namely $2l/c$, the natural frequencies will not appear in the signal spectrum unless $m/T = nc/2l$, or $T/(2l/c) = m/n$, in which case the exciting pulses will reinforce certain reflections of previously generated pulses, and strong signal components at frequencies m/T will be observed. (Check by graphical means.) Such coincidence of excitation frequency and natural frequency, leading to response reinforcement, is known as 'resonance'. This form of argument cannot be directly applied to dispersive waves. (Why not?)

The phenomenon of resonance is usually associated with single-frequency excitation, but the above example shows that it occurs whenever a continuous input possesses a frequency component that corresponds to a natural frequency of a system. The response of a system excited at resonance is clearly unlimited unless some of the energy of the continuously generated and multiply reflected waves is dissipated or otherwise lost: if not, waves pile limitlessly upon waves, as can be demonstrated by sloshing the water in a bath tub up and down the length at the right frequency. Hence resonant response is *damping* controlled: the damping may take the form of material damping, interface damping or radiation of energy into contiguous structures or fluids. (The bathroom floor may also be 'damped' in a rather different manner!)

In order further to illustrate the concept of resonance due to a single-frequency excitation, the steady state response of a simply supported beam excited in bending by a transverse harmonic force is now derived using the wave formulation introduced earlier in this chapter. As we shall see below, even for this simple problem, the derivation of the forced response using a wave approach is rather cumbersome. Indeed, most books presenting the fundamental concepts of vibrations in elastic mechanical systems derive the response to harmonic excitation with simpler mathematical formulations based on the modal summation approach which, however, do not unveil the physics of the phenomenon. The wave approach is therefore used in this chapter with the intention of providing a physical understanding of the phenomenon of resonance together with a rigorous mathematical derivation. The flexural wave response of a beam of infinite length which is subject to a transverse harmonic force $F(0, t) = \mathrm{Re}\left\{\widetilde{F}_0 \exp(j\omega t)\right\}$ at $x = 0$ is first considered. Subsequently, the multiple reflections at the two ends of the positive- and negative-going flexural waves generated by the transverse force are illustrated in order to derive the response of a bounded beam.

A force having simple harmonic time dependence generates waves in a structure that propagate away from their source at phase speeds dependent upon the wave type and, in the case of dispersive waves, upon the frequency. Small-amplitude waves in elastic structures are normally linear and, therefore the wave frequency will equal the forcing frequency. The relation between the applied force and the resulting velocity at the point of application, depends essentially upon two factors: (i) the properties, and hence wave-bearing characteristics, of the structure in the close vicinity of the driving point; (ii) the amplitude and phase at the driving point of any waves reflected back to that point from discontinuities or constraints in the surrounding structure.

The applied force is reacted by elastic shear forces in the beam. The equation of free motion of the *undamped* beam is Eq. (1.36):

$$EI\frac{\partial^4 \eta}{\partial x^4} + m\frac{\partial^2 \eta}{\partial t^2} = 0 \tag{1.70}$$

Note that the terms in this equation express the transverse elastic and inertia forces per unit length. The equation applies to all the points on the beam that are not subject to external forces such as at the driving forces or external constraints. At the driving point, the transverse shear force suffers a step change equal to the applied force, as it does in a static loading case. Mathematically such a concentrated force F applied at $x = a$ may be represented by a Dirac delta function representation of the distributed transverse loading $f(x)$ *per unit length* thus:

$$f(x) = F_0 \delta(x - a) \tag{1.71}$$

The Dirac delta function has the following properties:

(i) $\int_{-\infty}^{+\infty} \delta(x)dx = 1$;
(ii) it operates upon a continuous function $\phi(x)$ such that $\int_{-\infty}^{+\infty} \phi(x) \, \delta(x - a)dx = \phi(a)$, i.e., it 'picks out' the value of the function at $x = a$.

Note that the dimension of the one-dimensional Dirac delta function is reciprocal length, as seen from Eq. (1.71). Since the transverse loading per unit length is the derivative of the shear force with respect to x, it is seen that the delta function is the derivative of a step function. Equations (1.70) and (1.71) may be combined as

$$EI\frac{\partial^4 \eta}{\partial x^4} + m\frac{\partial^2 \eta}{\partial t^2} = \tilde{F}_0 \delta(x - a)\exp(j\omega t) \tag{1.72}$$

In this case, the force is simple harmonic, but any time history may be inserted. This form of inhomogeneous equation will be employed later in the book. Instead, we have employed solutions to the homogeneous equation (1.70) to satisfy equilibrium and boundary conditions.

Equation (1.72) is insufficient to describe the behaviour of any bounded beam to a point force; boundary conditions at the ends of the beam are also required.

Let us initially assume that the beam is infinitely extended and the external force is applied at $x = 0$. Equation (1.70) is equivalent to Eq. (1.72) at every point except that at which the force is applied. The complex solution given in Eq. (1.37), represents all the physically possible forms of free flexural vibration of the beam:

$$\eta(x, t) = \left[\tilde{A}\exp(-jk_b x) + \tilde{B}\exp(jk_b x) + \tilde{C}\exp(-k_b x)\right.$$
$$\left. + \tilde{D}\exp(k_b x)\right]\exp(j\omega t) \qquad (1.73)$$

where $k_b = (\omega^2 m/EI)^{1/4}$. Because the beam is infinite, $\tilde{A}$ and $\tilde{C}$ must be zero in the region $x < 0$, and $\tilde{B}$ and $\tilde{D}$ must be zero in the region $x > 0$. (Why?) The coefficients $\tilde{A}$, $\tilde{B}$, $\tilde{C}$ and $\tilde{D}$ may now be found by satisfying conditions of equilibrium at infinitesimally small distances to the left and right of $x = 0$, at $x = 0^-$ and $x = 0^+$, and assuming symmetry about $x = 0$. According to Fig. 1.14 and Eq. (1.33), the elastic shear stresses produce an upward shear force of magnitude $EI\left[\partial^3\eta/\partial x^3\right]$ on the left-hand end of an elemental beam section, and a downward shear force of magnitude $EI\left[\partial^3\eta/\partial x^3\right]$ on the right-hand end of a section. Since the system is symmetrical about $x = 0$, and the mass of the beam between $x = 0^+$ and $x = 0^-$ is infinitesimal, the force at $x = 0^+$ is equal in magnitude and sign to that at $x = 0^-$. If the applied force is considered to be positive in the positive η (upward) direction, each of these forces equals $-\frac{1}{2}\tilde{F}_0\exp(j\omega t)$. Thus, at $x = 0^+$,

$$\tilde{F}_0/2 - EI\left(jk_b^3\tilde{A} - k_b^3\tilde{C}\right) = 0 \qquad (1.74a)$$

and at $x = 0^-$,

$$\tilde{F}_0/2 + EI\left(-jk_b^3\tilde{B} + k_b^3\tilde{D}\right) = 0 \qquad (1.74b)$$

In addition, because of symmetry, the slope of the beam at $x = 0$ is zero. Hence

$$-jk_b\tilde{A} - k_b\tilde{C} = jk_b\tilde{B} + k_b\tilde{D} = 0 \qquad (1.74c)$$

Equations (1.74) yield

$$\tilde{A} = \tilde{B} = j\tilde{C} = j\tilde{D} \qquad (1.75)$$

and

$$\tilde{A} = -j\tilde{F}_0/4EIk_b^3 \qquad (1.76)$$

Substituting these values in Eq. (1.73), the complex transverse displacement generated by the complex point force $\widetilde{F}_0 \exp(j\omega t)$ at $x = 0$ is found to be given by

$$\eta(x > 0, t) = \frac{-j\widetilde{F}_0}{4EIk_b^3} \left[\exp(-jk_b x) - j\exp(-k_b x)\right] \exp(j\omega t) \qquad (1.77a)$$

and

$$\eta(x < 0, t) = \frac{-j\widetilde{F}_0}{4EIk_b^3} \left[\exp(jk_b x) - j\exp(k_b x)\right] \exp(j\omega t) \qquad (1.77b)$$

The time dependence of the vibration field is illustrated by Fig. 1.30, in which $\widetilde{F}_0$ is assumed to be purely real.

When the beam is bounded by simple-support terminations, the positive- and negative-going propagating wave components generated by the point force will be reflected by the right-hand and left-hand terminations of the beam, respectively. The near-field components will also be 'reflected' at the two terminations. However, they are important only at very low frequencies such that the amplitudes

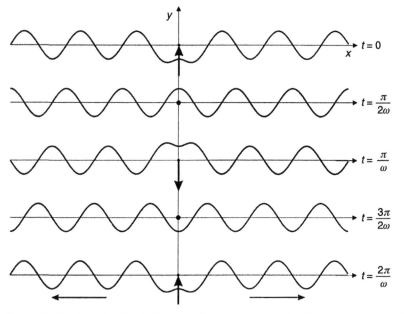

Fig. 1.30. Bending vibration field generated by a harmonic point force acting on an infinite beam.

of the decaying fields impinging on the boundaries are significant in comparison with the amplitude of the propagating wavefield. In the ideal case where no dissipative mechanisms operate on the beam, multiple reflections of positive- and negative-going propagating waves at the two terminations of the beam will continue indefinitely, giving rise to a resultant vibration field which depends on the position and frequency of the excitation, the phase speed of the bending propagating waves and the length of the beam. If the positive- and negative-going wave trains close in such a way that the phase of the doubly reflected returning waves coincides with the phases of the preceding waves then a sustained, pure standing-wave field, in which original and returning waves are indistinguishable, is produced. The amplitude of this resonant standing-wave field, being sustained by the harmonic force, will be unlimited. Alternatively, when the positive- and negative-going wave trains close in such a way that the phase of the multiple reflected returning wave does not coincide with the phase of the opening waves, the amplitude of the resultant vibration field will not be unlimited, but instead will be limited to values that could even be smaller than that of the preceding waves because of a destructive interference effect. Indeed, when the returning wave has equal amplitude but opposite phase to that of the original wave, then perfect cancellation occurs so that the response of the beam is determined only by the decaying near-field components. This state is known as 'anti-resonance'.

Let us consider the case where, as shown in Fig. 1.31, a simply supported beam of length l is excited by the transverse harmonic force $F(x_1, t) = \tilde{F}_0 \exp(j\omega t)$ at $x_1 < l/2$. The transverse displacement $\eta(x_2, t)$ generated at a point $x_2 > l/2$ can be derived by superimposing the complex amplitudes of the impinging propagating waves and decaying near fields at x_2. In order to keep the formulation simple, the time-dependent term $\exp(j\omega t)$ in the propagating and decaying components of the vibration field will be omitted. In addition, assuming that the frequency of the excitation is relatively high so that the amplitudes of the near fields at the two ends and at position x_2 are relatively small, their contributions will also be neglected. The point force produces an initial[7] positive-going propagating wave whose amplitude at x_2 is given by $(-j\tilde{F}_0/4EIk_b^3) \exp(-jk_b\Delta x)$, where $\Delta x = x_2 - x_1$. As illustrated by the wave trace scheme in Fig. 1.31(b), this wave will then propagate to the right-hand termination where its complex amplitude is given by $(-j\tilde{F}_0/4EIk_b^3) \exp\left[-jk_b(\Delta x + b)\right]$, where $b = l - x_2$. As shown in Section 1.10, the simple support reflects the incident propagating wave into a negative-going propagating wave whose complex amplitude is in antiphase with that of the incident wave. This reflected negative-going wave will travel back to position x_2

---

[7] In principle, a harmonic force exist for all time, and so does the steady-state vibrational response. Therefore the concept of 'initial' wave is, in principle, untenable. However, the following analysis obviates this conceptual difficulty by employing the sum of infinitely many wave reflections, which converges to the equivalent harmonic steady-state solution.

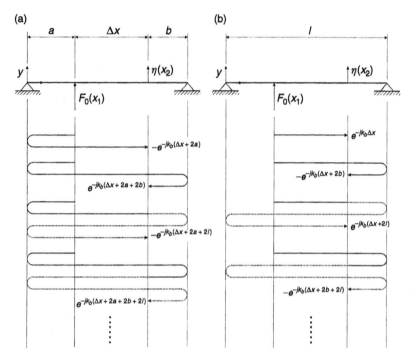

Fig. 1.31. Propagating wave trains in a simply supported beam generated by a harmonic point force acting at x_1: (a) negative-going waves; (b) positive-going waves.

where its complex amplitude is given by $(j\widetilde{F}_0/4EIk_b^3)\left[\exp(-jk_b(\Delta x + 2b))\right]$, and then further back to the left-hand support termination, where its complex amplitude is given by $(j\widetilde{F}_0/4EIk_b^3)\left[\exp(-jk_b(\Delta x + b + l))\right]$. This wave will be reflected into a positive-going wave whose complex amplitude at x_2 is given by $(-j\widetilde{F}_0/4EIk_b^3)\left[\exp(-jk_b(\Delta x + 2l))\right]$. In the ideal case where no dissipative mechanisms operate, this propagating/reflection process will continue indefinitely so that trains of positive-going and negative-going waves arrive at x_2, respectively, with complex amplitudes $(-j\widetilde{F}_0/4EIk_b^3)\left[\exp(-jk_b(\Delta x + n2l))\right]$ and $(j\widetilde{F}_0/4EIk_b^3)\left[\exp(-jk_b(\Delta x + 2b + n2l))\right]$ where $n = 1, 2, 3, \ldots$ indicates the number of return journeys completed by the waves.

As illustrated by the wave trace scheme in Fig. 1.31(a), the same multiple reflection phenomenon is produced by the negative-going propagating wave generated by the point force. In the ideal case where no dissipative effects are present in the beam, this propagation/reflection process will continue indefinitely so that trains of positive- and negative-going waves arrive at x_2 with complex amplitudes $(j\tilde{F}_0/4EIk_b^3)\left[\exp(-jk_b(\Delta x + 2a + n2l))\right]$ and $(-j\tilde{F}_0/4EIk_b^3)\left[\exp(-jk_b(\Delta x + 2a + 2b + n2l))\right]$, respectively, where $n = 1, 2, 3, \ldots$ indicates the number of return journeys completed by the wave along the beam.

As shown schematically in Fig. 1.32, the incoming waves at x_2 can be divided into positive-going (b) and negative-going waves (a). The complex displacements (η^+, η^-) produced by the positive- and negative-going waves at x_2 can therefore be expressed in terms of the following summations of exponential terms:

$$\eta^+(x_2, t) = \frac{-j\tilde{F}_0(x_1)}{4EIk_b^3} \left\{ e^{-jk_b\Delta x} - e^{-jk_b(\Delta x + 2a)} + e^{-jk_b(\Delta x + 2l)} - e^{-jk_b(\Delta x + 2a + 2l)} \right.$$

$$\left. + e^{-jk_b(\Delta x + 4l)} - e^{-jk_b(\Delta x + 2a + 4l)} + \cdots \right\} \exp(j\omega t)$$

$$(1.78a)$$

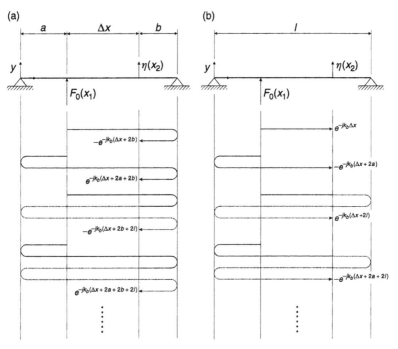

Fig. 1.32. (a) Negative-going and (b) positive-going wave trains arriving at position x_2.

$$\eta^-(x_2, t) = \frac{-j\tilde{F}_0(x_1)}{4EIk_b^3}\left\{-e^{-jk_b(\Delta x + 2b)} + e^{-jk_b(\Delta x + 2a + 2b)} - e^{-jk_b(\Delta x + 2b + 2l)}\right.$$

$$+ e^{-jk_b(\Delta x + 2a + 2b + 2l)} - e^{-jk_b(\Delta x + 2b + 4l)}$$

$$\left. + e^{-jk_b(\Delta x + 2a + 2b + 4l)} + \cdots\right\} \exp(j\omega t)$$

$$(1.78b)$$

After some manipulation, these two expressions can be rewritten in the following forms:

$$\eta^+(x_2, t) = \frac{-j\tilde{F}_0(x_1)}{4EIk_b^3}\left\{e^{-jk_b\Delta x}\left(1 + e^{-jk_b 2l} + e^{-jk_b 4l} + \cdots\right)\right.$$

$$\left. - e^{-jk_b(\Delta x + 2a)}\left(1 + e^{-jk_b 2l} + e^{-jk_b 4l} - \cdots\right)\right\} \exp(j\omega t)$$

$$(1.79a)$$

$$\eta^-(x_2, t) = \frac{-j\tilde{F}_0(x_1)}{4EIk_b^3}\left\{-e^{-jk_b(\Delta x + 2b)}\left(1 + e^{-jk_b 2l} + e^{-jk_b 4l} + \cdots\right)\right.$$

$$\left. + e^{-jk_b(\Delta x + 2a + 2b)}\left(1 + e^{-jk_b 2l} + e^{-jk_b 4l} + \cdots\right)\right\} \exp(j\omega t)$$

$$(1.79b)$$

The expression $(1 + e^{-jk_b 2l} + e^{-jk_b 4l} + \cdots)$ is a geometric series whose infinite sum is given by $1/(1 - e^{-jk_b 2l})$, so that the response $\eta(x_2, t)$ may be written as

$$\eta(x_2, t) = \eta^+(x_2, t) + \eta^-(x_2, t) \qquad (1.80)$$

$$= \frac{-j\tilde{F}_0(x_1)}{4EIk_b^3}\left\{\frac{e^{-jk_b\Delta x}\left(1 - e^{-jk_b 2a} - e^{-jk_b 2b} + e^{-jk_b(2a + 2b)}\right)}{\left(1 - e^{-jk_b 2l}\right)}\right\} e^{j\omega t}$$

The denominator of this expression goes to zero when $2k_b l$ is an integer multiple of 2π; as shown in Section 1.10, this condition corresponds to the natural frequencies $\omega_n = (n\pi/l)^2\sqrt{EI/m}$ with $n = 1, 2, 3, \ldots$ because it has been assumed that the beam is undamped. The resonance frequencies depend upon the length of the beam, the physical properties of the material and the dimensions of the cross section. The resonance phenomenon does not depend on the position of the excitation. However, considering the numerator of Eq. (1.80), we can see that the amplitude of the response depends on the positions x_1 and x_2 through the distances a and b.

In this idealised undamped case, the resonant response of the beam will grow indefinitely under continuous excitation, but in real structures it is controlled by energy loss mechanisms. These mechanisms are various; they include material damping, friction at supports and joints and transfer of vibrational energy to connected structures or fluids (radiation damping). It is found in practice that structural damping may often be reasonably represented mathematically by attributing a complex elastic modulus to the material: $E' = E(1 + j\eta)$, where η is termed the 'loss factor' and is generally much smaller than unity. In practical structures that have not been specially treated with damping material, η usually has values in the range 1×10^{-3} to 5×10^{-2}; for many structures η tends to decrease with frequency, roughly as $\omega^{-1/2}$. The inclusion of damping produces a modified form of bending-wave equation (1.36). The complex elastic modulus model is not strictly valid in the time-dependent form of equation, because it can lead to non-causal transient solutions. However it is valid if restricted to time-stationary vibration. Hence Eq. (1.36), with time dependence suppressed, may be written

$$E(1 + j\eta)I \frac{\partial^4 \tilde{\eta}(x)}{\partial x^4} = \omega^2 m \tilde{\eta}(x) \tag{1.81}$$

We now assume a one-dimensional wave solution for $\eta(x, t)$ and allow for energy dissipation by assuming a complex form for the wavenumber:

$$\eta(x, t) = \tilde{A} \exp[j(\omega t - k'x)] \tag{1.82}$$

where $k' = k(1 - j\alpha)$ and k is real. Substitution of this assumed solution into Eq. (1.81) yields

$$EI(1 + j\eta)k^4(1 - j\alpha)^4 \eta(x) = \omega^2 m \eta(x) \tag{1.83}$$

In general, α is much less than unity, and binomial expansion of the term $(1 - j\alpha)^4$, together with neglect of small terms, yields

$$EI(1 + j\eta)k^4(1 - 4j\alpha) = \omega^2 m \tag{1.84}$$

By separating real and imaginary parts, and ignoring the term $4\eta\alpha$ in comparison with unity,

$$EIk^4 = \omega^2 m \qquad (1.85a)$$

and

$$\alpha = \eta/4 \qquad (1.85b)$$

Hence $k = \pm k_b, \pm jk_b$, where $k_b = (\omega^2 m/EI)^{1/4}$, as in the case of the undamped beams. The complete solution for $\eta(x, t)$ corresponding to Eq. (1.37) is

$$\eta(x, t) = \left[\tilde{A}\exp(-jk_b'x) + \tilde{B}\exp(jk_b'x) + \tilde{C}\exp(-k_b'x) \right.$$
$$\left. + \tilde{D}\exp(k_b'x)\right]\exp(j\omega t) \qquad (1.86)$$

where $k_b' = k_b(1 - j\eta/4)$. The response of a damped beam to a transverse point force can be expressed by Eq. (1.79) by replacing the bending wavenumber k_b by the complex wavenumber k_b' and by using the complex elastic modulus $E' = E(1 + j\eta)$ in place of the Young's modulus of elasticity E.

Equation (1.80) has been derived by neglecting the effects of the near-field components generated by the point force. The near-field component generated by the force interacts with the simple support constraint to generate only a near-field component that decays in the opposite direction. Therefore the multiple-reflection formulation presented above for the propagating wave components can be used to derive the displacement field generated by the near-field components. The displacement at x_2 due to the direct and multi-reflected near field decaying components is given by

$$\tilde{\eta}(x_2, t) = \frac{-j\tilde{F}_0(x_1)}{4EIk_b^3}\left\{\frac{-je^{-k_b\Delta x}\left(1 - e^{-k_b 2a} - e^{-k_b 2b} + e^{-k_b(2a+2b)}\right)}{\left(1 - e^{-k_b 2l}\right)}\right\}e^{j\omega t} \qquad (1.87)$$

For a given frequency, the greater the distance between the points of observation and force application, the smaller the near-field effect. Moreover, as the frequency of the excitation and the bending wavenumber increase, the near-field contracts. Therefore, the contribution of the near-field components is important at low frequencies and for small distances between points of excitation and observation.

In summary, taking into account both the propagating wave and near-field components, the resultant displacement $\eta(x_2, t)$ is

$$\tilde{\eta}(x_2) = \frac{-j\tilde{F}_0(x_1)}{4EIk_b^3} \left\{ \frac{e^{-jk_b\Delta x}\left(1 - e^{-jk_b2a} - e^{-jk_b2b} + e^{-jk_b(2a+2b)}\right)}{\left(1 - e^{-jk_b2l}\right)} \right.$$

$$\left. + \frac{-je^{-k_b\Delta x}\left(1 - e^{-k_b2a} - e^{-k_b2b} + e^{-k_b(2a+2b)}\right)}{\left(1 - e^{-k_b2l}\right)} \right\} e^{j\omega t}$$

$$(1.88)$$

Figure 1.33 shows the response of the beam when a transverse point force at position $x_1 = 0.27l$ excites the beam at frequencies ω_1, $2\omega_1$, $3\omega_1$ and $4\omega_1 = \omega_2$, where ω_1 and ω_2 are the first two natural frequencies of the beam. These results have been derived assuming a loss factor $\eta = 0.01$ by using in Eq. (1.88) and a complex wavenumber $k_b' = k_b(1 - j\eta/4)$.

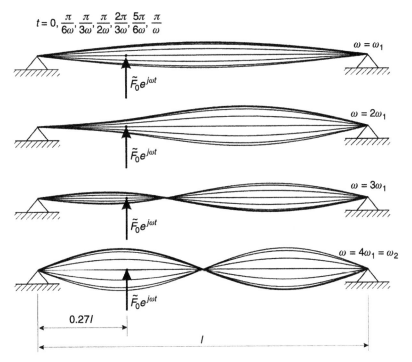

Fig. 1.33. Standing wavefields generated by a harmonic point force at position $x_1 = 0.27l$ at frequencies ω_1, $2\omega_1$, $3\omega_1$ and $4\omega_1 = \omega_2$.

In order to facilitate the comparison between the responses at the first and second resonance frequencies, ω_1 and $\omega_2 = 4\omega_1$, with the other two at off-resonance frequencies, $2\omega_1$ and $3\omega_1$, the four plots in Fig. 1.33 have been normalised to have similar amplitudes. This figure shows that the responses of the beam at the two resonance frequencies are dominated by the natural modes whose natural frequencies coincide with the resonance frequencies ω_1 and $4\omega_1 = \omega_2$. The responses of the beam at the two intermediate frequencies $2\omega_1$ and $3\omega_1$ are instead characterised by vibration fields whose shapes are summations of the first two natural modes. The response at $2\omega_1$ is principally controlled by the first natural mode such that there is no intermediate nodal point, while the response at $3\omega_1$ is principally controlled by the second natural mode which produces an intermediate nodal point.

The importance of considering the effect of the near-field component in the derivation of the response is exemplified by the two plots in Fig. 1.34 that show the modulus and phase of the ratio of complex displacement to complex force amplitude $\tilde{\eta}(x_2)/\tilde{F}_0(x_1)$ when point x_2 is located at a distance from the excitation point x_1, and when the two points are collocated. The analysis of frequency response functions such as $\tilde{\eta}/\tilde{F}_0$ is one of the principal subjects of Chapter 2 where the concepts of mechanical mobility and mechanical impedance are introduced and analysed in detail. In the first case, where x_2 is located at some distance from x_1, there is little difference when the contribution of the near field is taken into account (faint line) or neglected (dashed line). In contrast, when the response is evaluated at the excitation point ($x_2 = x_1$), the contribution of the near-field components is important, particularly at the anti-resonance frequencies where the positive- and negative-going propagating wave trains close in such a way that the returning waves interfere destructively with the original waves.

In the present and the previous sections we have seen how the phenomenon of vibrational waves underlies the existence of the natural modes and frequencies of distributed elastic structures and how wave reflection from boundaries is the origin of resonant behaviour. Closed form response solutions such as those presented in the foregoing sections can be derived only for idealised models of very simple form; modal models are usually employed in practical vibration analysis. However, the wave mechanism that generate natural modes and frequencies and resonance are common to all elastic fluid and solid systems. Most systems of practical interest comprise a number of components, each in isolation having rather different wave propagation and natural vibration characteristics. What happens when they are joined together? One approach to answering this question is based upon the concepts of the mobility (or impedance) of components, which quantify the relations between oscillating forces and velocities, and in terms of which the dynamic properties of components and assemblies of components may be quantified. The following chapter introduces and explains these concepts.

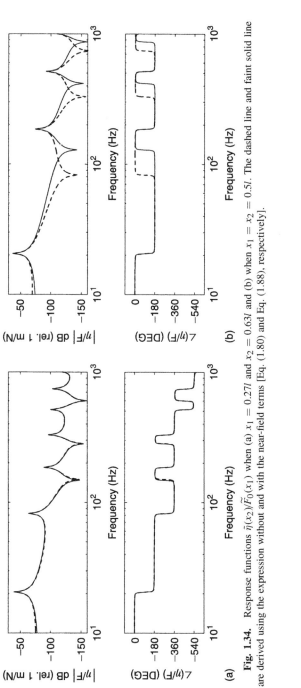

Fig. 1.34. Response functions $\tilde{\eta}(x_2)/\tilde{F}_0(x_1)$ when (a) $x_1 = 0.27l$ and $x_2 = 0.63l$ and (b) when $x_1 = x_2 = 0.5l$. The dashed line and faint solid line are derived using the expression without and with the near-field terms [Eq. (1.80) and Eq. (1.88), respectively].

1.12   Modal Density and Modal Overlap

Many engineering systems to which vibroacoustic analysis is applied are subject to forms of excitation having spectra extending to frequencies many times that of the fundamental natural frequency and consequently the responses involve very large numbers of high-order modes. Examples include aircraft, spacecraft, vehicles of many types and industrial installations such as the pipelines of process plant. The modelling of such systems presents the problem of 'high frequency' vibration (Fahy and Langley, 2004). As explained below, it is impossible to calculate the precise natural frequencies and shapes of high-order modes; consequently, probabilistic models are commonly employed. This section presents an introduction to two important statistically based parameters that are central to the most common form of such models, namely Statistical Energy Analysis (SEA).

Every linear elastic fluid and solid system that can store vibrational energy exhibits the phenomenon of natural frequencies: we do not consider non-linear systems in this book. A system is said to be 'closed' if energy cannot be transmitted through its boundaries, and 'open' if it can. The natural frequencies of an ideal closed system that does not dissipate vibrational energy into heat are mathematically real, and vibration continues indefinitely after the cessation of a transient excitation. If a closed system is dissipative, the natural frequencies are mathematically complex, the imaginary part determining the rate of decay of the associated energy with time elapsed since the cessation of the disturbance. If a system is open, but can store vibrational energy because of substantial wave reflection at the 'open' boundary, such as air in an organ pipe, its natural frequencies are also complex.

Associated with the natural frequencies there exist natural modes. As we have seen, these are spatial distributions of vibrational quantities that are characteristic of the material and geometric properties of a system and its boundaries. In the absence of dissipation, the modes take the form of pure standing waves, so that the vibrational motions at all points are relatively either in phase (0°) or in anti-phase (180°). These modes are dynamically independent and each behaves like an isolated simple damped oscillator. This quality is expressed in terms of the condition of 'orthogonality' (see Section 2.3.3). If, in a closed dissipative system, the local damping coefficient may be expressed as a linear combination of local mass and stiffness (Rayleigh damping), the modes are mathematically real, independent and orthogonal and the action of dissipative mechanisms may be represented mathematically by the attribution to each mode of a 'modal damping coefficient' or equivalent 'modal loss factor'. This allows us to express the response of a closed system as the sum of the responses of its individual modes (see Section 2.3.3). The condition of orthogonality implies that the total time-average energy of vibration in response to time-stationary excitation is equal to the sum of the individual time-average modal energies.

In nearly all closed physical systems of practical engineering interest, the dissipative mechanisms do not conform to the Rayleigh idealisation; in principle, the natural modes do not take the form of pure standing waves and are mathematically complex. Unlike real modes, the relative phase of modal vibration varies with position, and the mode 'shape' varies throughout a period (Newland, 1989). However, many engineering systems to which vibroacoustic modelling and analysis are applied exhibit pronounced resonant behaviour. This implies that the dissipative mechanisms are sufficiently weak that fraction of stored modal energy 'lost' per cycle of oscillation at the modal natural frequency is very much less than unity. This is known as a condition of 'light' damping. At low frequencies, where the natural frequencies are well separated (see below) an assumption of 'equivalent' Rayleigh damping leads to reasonably accurate estimates of response to arbitrary forms of excitation; but at higher frequencies, where the individual modal responses are not well separated, this assumption can lead to substantial errors in response estimation. Readers are warned that such idealisation can lead to non-physical results if applied to the analysis of energy *flux* (or vibrational intensity) distribution within a system. The energy of open systems continually leaks away through the open boundary and the modes are complex.

In this section, we consider a statistically based measure of the distribution of modal natural frequencies in the frequency domain, known as 'modal density'. One definition of this quantity at any frequency is the inverse of the expected (or average) interval between neighbouring natural frequencies *local* to that frequency. It may also be interpreted as the expected number of natural frequencies per unit frequency. It should be noted carefully that modal density has two alternative, but equivalent, forms. Modal density $n(f)$ is defined as 'number of natural frequencies per Hz'; $n(\omega)$ is defined as 'number of natural frequencies per radian per second': the former is 2π times the latter. The number of natural frequencies that are theoretically expected to exist below any frequency (the cumulative mode count) may be plotted against frequency in the form of a histogram. At any frequency, the local modal density may be evaluated as the local slope of the smoothed version of this histogram. The concept of modal density is not useful in the frequency range of the natural frequencies of the lowest order modes because these vary with the specific boundary conditions of the system. Theoretical estimates of the natural frequencies of physical structures are subject to increasing uncertainty as the mode order rises because they are increasingly sensitive to minor details of material properties, geometry, connections and damping distribution. Consequently, the natural frequencies of a set of nominally identical artefacts will increasingly differ as mode order rises. However, as mode order increases, the average *density* of natural frequencies along the frequency axis become increasingly less sensitive to the boundary conditions and the modal density becomes a meaningful parameter. Hence, we may consider the modal density as an expected (or ensemble-average) value that applies to a large population of grossly similar, but slightly different, systems. The modal densities of spatially uniform structures generally vary rather smoothly with frequency (except in cases of curved shells). However, the modal densities of non-uniform structures such as beam-stiffened plates and ring-stiffened shells can exhibit substantial irregularities, especially in cases of periodic structures in which modal 'clumping' occurs.

In cases where structures support a number of different types of wave, and hence exhibit sets of natural frequencies and modes formed by constructive interference between travelling waves of each type, a modal density may be attributed to each wave type. Expressions for the modal densities of spatially uniform structures may be derived by considering the form of the dispersion equation relevant to each wave type and representing the dispersion relation on a wavenumber diagram, such as that shown in Fig. 1.22 for a thin-walled, circular cylindrical shell. The modes of cylinder of length L having simply supported boundary conditions at the edges of its wall have axial wavenumbers $m\pi/L$ (m integer) corresponding to discrete values $(m\pi/L)(a\beta^{1/2})$ on the vertical axis of Fig. 1.22. The intersection of corresponding horizontal lines with the vertical lines shown represent the natural modes in wavenumber space. The non-dimensional natural frequencies may be calculated from Eq. (1.49) by replacing k_z by $m\pi/L$. The number of intersections lying between the loci of Ω and $\Omega + \delta\Omega$ can be estimated, yielding an estimate of the total modal density. A general analytical expression that mimics this graphical method may be derived: interested readers may consult Langley (1994a) for details. Expressions for the modal densities of a range of spatially uniform, homogeneous, isotropic systems are presented in Table 1.4.

As explained earlier in this chapter, the modes of thin-walled circular cylindrical shells fall into two categories: those governed by in-plane (membrane) elastic forces and those governed by the shear forces and moments associated with bending. The contributions of each mode type, and the total modal density, as proportions of the modal density of a flat plate of thickness and surface area equal to that of the shell, are shown in Fig. 1.35 (Langley, 1994a), The curve of total modal density agrees with those published by Bolotin (1984) and Szechenyi (1971a). The figure shows that the formula presented by Heckl (1962b) for 'thin-walled tubes' accounts only for the shell wall flexural modes. Note that these curves of asymptotic modal density do not exhibit the local peaks that occur just above the cut-off frequency of each circumferential order of shell waveguide modes (see Norton and Karczub, 2003). These affect the radial point force mobility, as illustrated by Fig. 2.24. Other useful publications on structural modal density include that on shallow structural elements by Wilkinson (1967), and on sandwich structures by Erickson (1969) and Renji and Nair (1996). Serati and Marshall (1989) compare a theoretical estimate of the modal density

TABLE 1.4

Asymptotic Modal Densities of Some Uniform, Isotropic Systems

Asymptotic modal densities	$n(\omega)$
Straight rod: quasi-longitudinal waves	$(L/\pi)(\rho/E)^{1/2}$
Straight beam: bending waves	$(L/2\pi)(m/EI)^{1/4}\omega^{-1/2}$
Straight bar: torsional waves	$(L/\pi)(I_p/GJ)^{1/2}$
Thin flat plate: bending waves	$(S/4\pi)(\rho h/D)^{1/4}$
Thin-walled circular cylindrical shell: flexural wave only	$\Omega \ll 1 \quad 2\omega^{1/2}a^{3/2}L/1.6h(c_l'')^{3/2}$
	$\Omega > 1 \quad (2\pi a L/4\pi)(\rho h/D)^{1/4}$
Fluid in rectangular enclosure: acoustic waves	$(V\omega^2/2\pi^2 c^3)* + (A\omega/8\pi^2 c^2) + (P/16\pi c)$

L = length, S = area (one side), V = volume, A = surface area of rectangular enclosure, P = sum of edge length of rectangular enclosure, a = cylinder radius, h = shell wall and flat plate thickness, ρ = material density, E = Young's modulus, m = mass per unit length, I = second moment of area, D = plate bending stiffness, c = speed of sound, I_p = polar moment of inertia per unit length, GJ = torsional stiffness.
*Applies to any form of proportionate enclosure at high frequency.

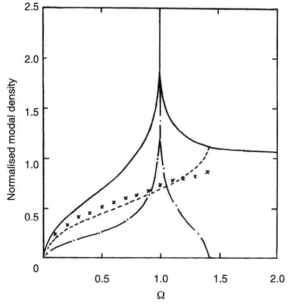

Fig. 1.35. Modal densities of a thin-walled circular cylindrical shell normalised on that of the unwrapped cylinder as flat plate. - - -, flexure-dominated modes; - . -, membrane-dominated modes; ___ , total; x, Heckl (1962b).

of a ring-stiffened cylinder with an empirical estimate based upon the relation between the space-average point force mobility of *spatially uniform* structures and the modal density, as expressed by Eq. (1.93) below. They conclude that neglect of the relatively low mobilities measured at the frame positions may be neglected for this purpose.

The 'frequency response function' (FRF) of a linear system may be broadly defined as a measure of the complex amplitude of the linear vibrational or acoustic response at one point normalised by the complex amplitude of a harmonic external source of excitation, as a function of excitation frequency. The FRF of any particular system varies with the measure of response (e.g., acceleration, displacement, pressure), with the form of excitation, and with the positions of the input and response points. The FRFs of most multimode systems of practical importance exhibit significantly different forms in the 'low' frequency range that embraces the natural frequencies of the first ten or so modes, and in the 'high' frequency range in which the natural frequencies of all the higher order modes reside. In the former range, the magnitudes of FRFs exhibit rather narrow peaks separated by broad troughs: the peaks indicate resonant modal response. As frequency increases, there is a gradual transition to a form that exhibits much broader 'hills' separated by narrow, deep, 'valleys'. The 'hills' do not represent the resonant responses of individual modes, but the magnitude of the sum of the response phasors of a number of modes having natural frequencies close to the maximum of the broad 'peak'. These features are illustrated by Fig. 1.36.

One factor that controls this transition is the modal 'half-power bandwidth'. In terms of the loss factor η, the half-power bandwidth of a mode having natural frequency $\omega_0(= 2\pi f_0)$ is $\eta\omega_0$ radians per second or ηf_0 Hz. The ratio of the half-power bandwidth to the local average interval between natural frequencies is given by $\eta\omega n(\omega)$ in terms of $n(\omega)$, or $\eta f n(f)$ in terms of $n(f)$. This ratio is termed the 'modal overlap factor', conventionally denoted by M. The modal densities of many systems of practical concern are either constant or increase with frequency, except for beams in bending vibration (see Table 1.4). Loss factors of engineering systems typically vary approximately as (frequency)$^{-1/2}$: consequently, M usually increases with frequency. As M approaches unity, the individual modal responses begin to overlap and, as it increases beyond unity, they combine to form the broad 'hills' referred to above. In between these maxima, there are frequencies at which the local modal response phasors add vectorially to produce a very small response. The precise FRF of any multimode system in the high frequency range is very sensitive to small perturbations of the individual resonance frequencies of contributing modes, as illustrated by Fig. 1.37. As mentioned above, the resonance frequencies of high-order modes themselves are very sensitive to minor details of material properties, geometry, connections and damping distribution. Consequently, the detail of the high frequency response function of any individual system is unpredictable.

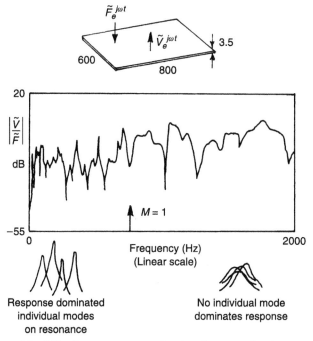

Fig. 1.36.   Frequency response function of a rectangular plate.

However, it is possible to estimate certain statistical measures of its form; these are functions of M (Langley, 1994b). The modal overlap factor plays a major role in influencing the reliability of response analyses based upon probabilistic models of systems, such as Statistical Energy Analysis, which is introduced in Chapter 7.

1.13   The Roles of Modal Density in Vibroacoustics

The response of a linear multimode system to external excitation may, in principle, be expressed as the sum of the responses of each of the theoretically infinite set of modes. In practice, it is acceptable to truncate the set because the modal contributions decrease with the difference between their natural frequencies and the excitation frequency. The same 'selective' mechanism applies also to the response to band-limited random excitation and to the equivalent band-filtered response to broadband excitation. It would therefore seem to be intuitively

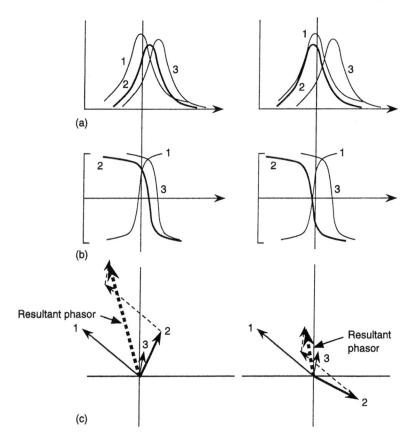

Fig. 1.37. Illustration of the sensitivity of multi-mode frequency response to a shift of one half-power bandwidth of an individual natural frequency: (a) modal amplitudes; (b) modal phases; (c) resultant complex response.

obvious that band-limited response is proportional to the local modal density. Since the modal density of a system is proportional to its length, area or volume, this would suggest that response to a given input would be proportional to the size of a system. When this is not actually the case it can be explained in two ways. As the size of a system increases, and the resonance frequencies of the modes crowd ever closer together, the modal response phasors at any point not too close to a point of localised excitation take values distributed all round the phase 'clock', and tend to cancel each other out [see Fig. 3.34(b)]. This is an indication of destructive interference between the different spatial distributions of the predominant modes. Hence, the spatial-average mean square response to such an input actually falls as the size increases. This is not the case if the excitation is distributed over the whole system, as with acoustic or boundary layer pressure fields.

The alternative view is based upon consideration of the balance between the power input by a band-limited random force in a frequency band containing at least five natural frequencies and the power dissipated by the system. It is shown by Cremer *et al.* (1988) that the frequency-, time- and position-averaged power input to a multimode system of *spatially uniform* mass per unit area by a band-limited random force of uniform spectral density is given by

$$\left\langle \overline{P}_{in} \right\rangle = \overline{F^2} \pi n(\omega_c)/2M_t \tag{1.89}$$

where $\overline{F^2}$ is the mean square force, ω_c is the band centre frequency and M_t is the total mass of the system. Since the modal density of a uniform system is proportional to the extent of the system, it is proportional to M_t. Hence, given a sufficiently large modal density, the power input is independent of size and equals the power input to an infinitely extended or boundary-damped system. By definition of the loss factor, the power dissipated is given by

$$\left\langle \overline{P}_d \right\rangle = \eta \omega_c \overline{E} = \eta \omega_c M_t \left\langle \overline{v^2} \right\rangle \tag{1.90}$$

where $\overline{E}$ is the time-average energy stored in the system and $\left\langle \overline{v^2} \right\rangle$ is space-average mean square velocity. Hence

$$\left\langle \overline{v^2} \right\rangle = \overline{F^2} \pi n(\omega_c)/2\eta \omega_c M_t^2 \tag{1.91}$$

which confirms the inference based upon modal response interference.

Chapter 2 introduces the concepts of mechanical 'impedance' and 'mobility'. The latter is defined as the quotient of the complex amplitude of velocity (or rotational velocity) response and the complex amplitude of a harmonic force (or moment) that generates the response. If the input and response points coincide, the relevant quantities are variously termed 'input impedance' or 'driving-point impedance' and 'input mobility' or 'driving-point mobility'. (Note: some publications use the term 'mechanical admittance' which is equivalent to mechanical mobility.) The time-average mechanical power input by a harmonic point force $\widetilde{F}$ to a system having input mobility $\widetilde{Y}(\tilde{v}_0/\widetilde{F})$ in the same direction is given by

$$\overline{P} = \frac{1}{2}\text{Re}\left\{\widetilde{F}\,\tilde{v}_0^*\right\} = \frac{1}{2}|\tilde{v}_0|^2\,\text{Re}\left\{1/\widetilde{Y}\right\} = \frac{1}{2}|\widetilde{F}|^2\,\text{Re}\left\{\widetilde{Y}\right\} \tag{1.92}$$

The analysis by Cremer *et al.* that leads to Eq. (1.89) yields the following expression for the space average of the frequency-band average of the real part of the input mobility of a system of *uniform mass per unit area*:

$$\langle \mathrm{Re}\{\widetilde{Y}\} \rangle = \pi n(\omega)/2M_t \qquad (1.93)$$

This equation provides one empirical means of estimating modal density from experimental estimates of the space-average input mobility.

Modal density plays a central role in SEA, which is introduced in Chapter 7. The SEA is a model in which a system that comprises an assembly of structural and/or fluid components is represented by a network of coupled subsystems. The central equation represents the balance of energies input to, transmitted by and dissipated within the subsystems under the condition of time-stationary excitation. The net flow of energy between any two subsystems in a frequency band is proportional to the difference between their 'modal energies': the factor of proportionality may be termed the 'power transfer coefficient' (Fahy, 1994). [Note: this coefficient is proportional to the product of the subsystem modal density and coupling loss factor, in terms of which it is reciprocal (see Chapter 7).] The modal energy of a subsystem is defined as the time-average energy in the subsystem divided by the expected number of subsystem modes resonant within the band: this number is equal to the modal density times the frequency bandwidth.

Although the rate of exchange of vibrational energy between subsystems may be expressed mathematically in terms of the couplings of individual pairs of modes (one of each subsystem), the modal model is employed principally for fundamental SEA research studies. A travelling wave representation of subsystem vibrational fields is generally preferred for the practical purpose of evaluating power transfer coefficients (or coupling loss factors) values. In two- and three-dimensional systems, the wavefield is generally assumed to be diffuse, although an extension to non-diffuse fields has been developed (Langley and Bercin, 1994). A wave is assumed to approach a subsystem junction where it is partially reflected and partially transmitted by the impedance discontinuity: a proportion of its energy flux is consequently transmitted. The diffuse field power transmission coefficient is obtained by integrating over wave heading. The essential link between the modal and wave models is that the both wave energy flux and modal energy are proportional to the product of energy density and group speed.

Problems

1.1 Derive an expression for the speed of longitudinal waves in a uniform circular-section bar composed of a core of material of diameter d_1, elastic modulus E_1 and density ρ_1 sheathed in a different material of outer diameter d_2, elastic modulus E_2 and density ρ_2. Neglect the Poisson effect and assume zero slip between the two materials. Check your expression by considering the extreme values of the ratio d_1/d_2.

1.2 Prove that the ring frequency of an infinitely long, uniform, circular cylindrical shell corresponds to the natural frequency of axisymmetric, 'breathing' ($n = 0$) motion by deriving the equation of radial motion of an elemental circumferential segment of shell.

1.3 By considering the dispersion curves for torsional waves in a thin flat strip of width h and thickness b, and bending waves in a plate of the same material and thickness, find an expression for the critical frequency of strip torsion with the plate, onto the edge of which the strip is welded at right angles along its centreline. What happens in the plate when the strip is excited in torsion below this frequency? [Equations (1.24) and (1.41) and Table 1.2 apply.]

1.4 Derive expressions for the complex amplitudes of reflected propagating and near fields in a semi-infinite uniform beam terminated by (a) a point mass M and (b) a transverse spring of stiffness K, in terms of the amplitude of an incident bending wave.

2 | Structural Mobility, Impedance, Vibrational Energy and Power

2.1 Mobility and Impedance Representations

It is customary in books on dynamic behaviour of structures to present analyses of response to simple harmonic excitation; this may take the form of concentrated 'point' forces and couples, or continuous spatial distributions of forces and couples. In the case of linear structures, extension to force systems having more complex variations with time can be synthesised from simple harmonic response characteristics by Fourier or Laplace transform techniques. Although this approach serves to illustrate the general frequency response characteristics of structural idealisations such as uniform beams or plates, the student is frequently uncertain how the results can be applied to a more complex system that consists of assemblies of subsystems, each of which has a different form or is constructed from a different material: this is particularly true when only the force(s) applied to one of the subsystems is known or can be estimated. As an example, we might wish to investigate the behaviour of a machine mounted on resilient isolators, which in turn are mounted upon a building floor that is connected to the supporting walls. Typically, only the forces generated within the machine by its operating mechanisms might be known. In such a case, the forces applied to the isolators, the floors and the walls are not known *a priori*. However, we can assume that the motions of connected sub-structures at their interfaces are identical and that Newton's law of action and equal and opposite reaction applies at these interfaces. In vibration analysis it is convenient to represent these physical facts in terms of equality at the interfaces of linear or angular velocities, and forces and moments. By using the complex exponential representation of simple harmonic time dependence, the ratio of the complex amplitudes of the forces and velocities at any interface for a given frequency can be represented by a complex number, which is termed the 'impedance' of the total system evaluated at that particular interface; it is sometimes more useful to use the inverse of impedance, termed as 'mobility'.[1] Hence it is convenient to characterise the individual sub-structures by their complex impedances (or mobilities) evaluated at their points, or interfaces, of connection to contiguous structures. The response of the total system to a known applied force may then be evaluated in terms of the impedances (or mobilities) of the component parts. These representations are only valid for linear systems.

In general, the mobility and impedance concepts refer to variables that can be measured. This feature helps analysts to build models that explicitly represent the physics of problems. They are very valuable in the design of experiments and in the interpretation of experimental results. This advantage comes at a cost, since mobility and impedance models of complex distributed or multi-body systems tend to be elaborate and cumbersome. Indeed, much neater formulations based on energy variational methods can be constructed which are not merely elegant mathematical expressions but, on the contrary, are also efficient approaches for the accurate prediction of the response of complex systems. Chapter 8 presents a detailed account of such approximate methods. In particular, the 'Finite Element Method' (FEM) and 'Boundary Element Method' (BEM) are introduced for the analysis of structural and acoustic problems involving complex structures and acoustic domains with complex geometry. In this chapter, the focus is instead on the mobilities and impedances of both lumped parameter and distributed one- and two-dimensional standard systems such as uniform beams, flat plates and circular cylindrical shells. It is important to emphasise that a researcher or engineer should be familiar with both the mobility-impedance and variational energy representations which are complementary.

In vibroacoustic modelling analysis, it is a common practice to employ energetic models in which vibrational state is expressed in terms of stored energy and interaction between components (subsystems) is expressed in terms of the transfer of vibrational or acoustic energy. The use of the concepts of impedance and mobility greatly facilitates the process of evaluating vibrational energy flow through a complex system. At an interface, the time-averaged power transfer

---

[1] Strictly speaking, these quantities should be termed the *mechanical impedance* and the *mechanical mobility* to distinguish them from acoustic impedances in later chapters.

by a harmonic force of complex amplitude $\widetilde{F}$ acting through a collinear particle velocity of complex amplitude $\widetilde{v}$ is given by

$$\overline{P}(\omega) = \frac{1}{T}\int_0^T P(t)dt = \frac{\omega}{2\pi}\int_0^{2\pi/\omega} \operatorname{Re}\{\widetilde{F}e^{j\omega t}\}\operatorname{Re}\{\widetilde{v}e^{j\omega t}\}dt = \frac{1}{2}\operatorname{Re}\{\widetilde{F}\widetilde{v}^*\} \quad (2.1)$$

where * indicates complex conjugate, $P(t)$ is the instantaneous power and T is the period of the harmonic vibration. The analogous expression for a harmonic moment M acting through a rotational velocity w is $\overline{P}(\omega) = \frac{1}{2}\operatorname{Re}\{\widetilde{M}\widetilde{w}^*\}$. By definition, impedance associated with a force is

$$\widetilde{Z}(\omega) = \frac{\widetilde{F}(\omega)}{\widetilde{v}(\omega)} \quad (2.2)$$

and, with a single force model only, mobility is its inverse:

$$\widetilde{Y}(\omega) = \frac{\widetilde{v}(\omega)}{\widetilde{F}(\omega)} \quad (2.3)$$

Hence,

$$\overline{P}(\omega) = \frac{1}{2}|\widetilde{F}|^2 \operatorname{Re}\{\widetilde{Y}\} = \frac{1}{2}|\widetilde{F}|^2 \operatorname{Re}\{\widetilde{Z}^{-1}\} = \frac{1}{2}|\widetilde{v}|^2 \operatorname{Re}\{\widetilde{Y}^{-1}\} = \frac{1}{2}|\widetilde{v}|^2 \operatorname{Re}\{\widetilde{Z}\}$$

$$(2.4)$$

(Why are the complex conjugate signs omitted in Eq. (2.4)?) The real part of $\widetilde{Z}$ is called the 'resistance' and the imaginary part, the 'reactance'. If we write $\widetilde{Z}(\omega) = R + jX$, then Eq. (2.4) can be written

$$\overline{P}(\omega) = \frac{1}{2}(\widetilde{F})^2|R/(R^2 + X^2)| = \frac{1}{2}|\widetilde{v}|^2 R \quad (2.5)$$

Also,

$$\widetilde{Y} = (R - jX)/(R^2 + X^2) \quad (2.6)$$

(Note: if harmonic time dependence is expressed by $e^{-j\omega t}$, all complex expressions must be replaced by their complex conjugates). The practical advantage of this formalism, which is not immediately obvious, is that in many cases it is possible to make assumptions about the magnitude of $\widetilde{F}$ or $\widetilde{v}$ at an interface from the knowledge of the impedance characteristics of the structures joined thereat, together with an observation of the force and/or velocity at the interface when a contiguous structure is disconnected or blocked. For example, consider the vibration of an instrument that is resiliently mounted upon a building floor,

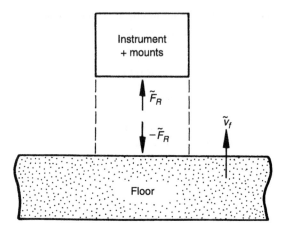

Fig. 2.1. Instrument mounted on a vibrating floor.

which itself is subject to vibration from an external source (Fig. 2.1). Suppose we have measured the vibration level of the floor and wish to estimate the vibration level of the instrument. When the instrument is placed on its mounts upon the floor, the vibration level of the floor changes—but by how much?

The reaction force on the floor created by the presence of the mounted instrument is

$$\tilde{F}_R \exp(j\omega t) = \tilde{Z}_I \tilde{v}_f \exp(j\omega t) \qquad (2.7)$$

where $\tilde{Z}_I$ is the impedance of the mounts plus instrument, and $\tilde{v}_f$ is the floor velocity in the presence of the instrument. It will be seen that the relationship between the floor velocity $\tilde{v}_o$ in the absence of the instrument, and the floor velocity $\tilde{v}_f$ with the instrument mounted, is given in terms of the floor impedance $\tilde{Z}_F$ as

$$\tilde{v}_f = \tilde{v}_o - \tilde{F}_R/\tilde{Z}_F = \tilde{v}_o - \tilde{Z}_I \tilde{v}_f/\tilde{Z}_F \qquad (2.8)$$

Hence

$$\tilde{v}_f = \tilde{v}_o[1 + \tilde{Z}_I/\tilde{Z}_F]^{-1} \qquad (2.9)$$

The floor vibration velocity is seen to be altered by the presence of the instrument to a degree characterised by the ratio $\tilde{Z}_I/\tilde{Z}_F$ of mounted instrument to floor impedance. The modulus of this ratio will normally be much smaller than unity; hence it may often be assumed that the floor velocity is unaffected by the presence of a mounted object, and an estimate of the instrument vibration may be made on this basis. (Could $\tilde{v}_f$ exceed $\tilde{v}_o$ under any circumstances?)

In this chapter, the definitions of mobility and impedance for basic elements such as lumped masses, springs and viscous dampers are first introduced and then discussed in detail. The aim is to establish a set of fundamental concepts that can be used in the interpretation of the dynamic behaviour of any mechanical system. The input force mobility and impedance functions of distributed one- and two-dimensional standard systems such as beams, flat plates and circular cylindrical shells are then derived and analysed in terms of mass-, spring- and damper-like behaviours. A brief introduction to the modelling of multi-body systems based on mobility and impedance matrices is also presented. Vibration transmission in multi-body systems is formulated in terms of energy flow and power transmission and the chapter concludes with a brief analysis of vibrational energy flux in beams.

Multi-body systems are normally interconnected at several positions where vibration transmission occurs through several degrees of freedom that can involve translations in three principal directions (axes) as well as rotations about the three axes. It is therefore difficult to analyse vibration transmission by considering only vibrational amplitude since rotations cannot be compared directly with translations. Energy-based quantities, such as the rate of energy flow along structures and power transmission from one structure to another, are better suited to the understanding and quantification of vibration transmission. In Chapter 1 we have already encountered an example which shows that suppressing the transverse vibrations of a beam vibrating in bending by means of a simple support does not stop the transmission of the incident wave but only reduces its amplitude by a factor $1/\sqrt{2}$. The energy is transmitted by the beam bending moment at the support acting through the angular velocity at the beam. Therefore, focusing the analysis solely on any one degree of freedom, for example, the transverse vibration of beams, might lead to misleading interpretations of the vibration transmission mechanisms in distributed systems. A particular advantage of energy-based models is that energy is a conserved quantity. Energy that is injected into one part of a complex system is partially dissipated therein and partly transmitted to other connected parts. Therefore vibrational response may be expressed in terms of power balance equations, as in Statistical Energy Analysis (see Section 7.8).

2.2   Concepts and General Forms of Mobility and Impedance of Lumped Mechanical Elements

The section opens with an introduction to the expressions for the mobility and impedance for mass, stiffness and viscous damper elements, together with the rules for combining these elements. It is worthwhile to examine the mobility and

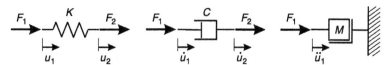

Fig. 2.2.   Linear spring, damper and mass systems.

impedance functions in some detail, because their characteristics are often sought in both measured and theoretical frequency response functions of distributed elastic systems.

In terms of the notation for the axial forces F_1 and F_2 acting *on* the elements, and axial displacements u_1 and u_2, shown in Fig. 2.2 (the dot denotes differentiation with reference to time), the constitutive equations for the assumed massless spring and viscous damper elements are

$$F_1 = -F_2 = K(u_1 - u_2), \quad F_1 = -F_2 = C(\dot{u}_1 - \dot{u}_2) \qquad (2.10\text{a,b})$$

where K and C are the stiffness and damping coefficient. Newton's second law applies to the mass element M, thus:

$$F_1 = M\ddot{u}_1 \qquad (2.10\text{c})$$

Assuming harmonic force excitation of the form $F(t) = \tilde{F}\exp(j\omega t)$ with displacement response of the form $u(t) = \tilde{u}\exp(j\omega t)$, and fixing one of each of the two points of the spring and of the damper, yields expressions for the mechanical mobilities of the elements:

$$\tilde{Y}_K = \frac{j\omega}{K}, \quad \tilde{Y}_C = \frac{1}{C}, \quad \tilde{Y}_M = \frac{-j}{\omega M} \qquad (2.11\text{a,b,c})$$

The impedances of spring, damper and mass elements are simply the reciprocals of the expressions in Eqs. (2.11a,b,c) and are thus given by

$$\tilde{Z}_K = \frac{-jK}{\omega}, \quad \tilde{Z}_C = C, \quad \tilde{Z}_M = j\omega M \qquad (2.12\text{a,b,c})$$

The moduli of the mobility and impedance functions of the three elements are plotted in Figs. 2.3(a) and 2.4(a) on a log scale as a function of log frequency, while the complex mobilities and impedances of the three elements are plotted in the complex plane in Figs. 2.3(b) and 2.4(b).

Figure 2.3(a) indicates that the mobility of a spring increases with frequency, the mobility of a mass decreases with frequency and the mobility of

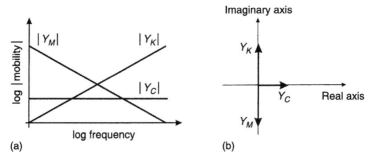

Fig. 2.3. Mobility characteristics: (a) as a function of frequency; (b) complex plane representation.

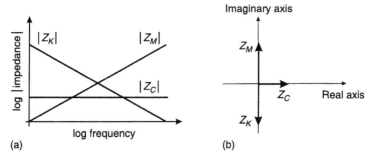

Fig. 2.4. Impedance characteristics: (a) as a function of frequency; (b) complex plane representation.

a damper is constant. According to Eqs. (2.12), the impedance functions of the three elements are characterised by a reciprocal behaviour so that, as shown in Fig. 2.4(a) the impedance of a spring decreases with frequency, the impedance of a mass increases with frequency while the impedance of a damper remains constant.

From Fig. 2.3(b) and Eqs. (2.11), it can be seen that the mobilities of a spring and mass are positive imaginary and negative imaginary respectively. For these two elements the applied harmonic force and the resultant velocity are in quadrature so that, according to Eqs. (2.1) and (2.4), they do not dissipate power and are thus referred as being 'reactive'. In contrast, the mobility of the damper element is positive real which indicates that the resultant velocity is in phase with the excitation force acting on the damper. Thus, applying Eq. (2.4), this element is seen to dissipate time-average power $\overline{P}(\omega) = (1/2C)|\widetilde{F}|^2$ where $\widetilde{F}$ is the complex amplitude of the force applied to the damper. Figure 2.4(b) and Eqs. (2.12) show the impedance of the spring to be negative imaginary and the impedance of the mass to be positive imaginary. The impedance of the damper element is also positive real and, according to Eq. (2.4), the time-average power dissipation is $\overline{P}(\omega) = \frac{1}{2}C|\tilde{v}|^2$ where $\tilde{v}$ is the complex amplitude of velocity of the vibrating terminal of the damper.

As explained in Section 1.11, structural damping is normally represented in terms of a complex stiffness given by $K' = K(1 + j\eta)$, where η is the loss factor. In this case, the mobility function is given by $Y'_K = Y_K(1 - j\eta)/(1 + \eta^2)$ which, in the case of weak dissipative mechanisms where $\eta \ll 1$, can be approximated by $Y'_K = Y_K(1 - j\eta)$. The impedance function is approximately given by $Z'_K = Z_K(1 + j\eta)$.

Mechanical systems composed of multiple spring, damper and mass elements may be arranged either in parallel or series configurations. Equivalent mobility and impedance functions can be derived for equivalent mobility and impedance elements with the following equations respectively for parallel and series connections:

Parallel
$$\frac{1}{\tilde{Y}_e} = \sum_{n=1}^{N} \frac{1}{\tilde{Y}_n}, \quad \tilde{Z}_e = \sum_{n=1}^{N} \tilde{Z}_n \qquad (2.13a,b)$$

Series
$$\tilde{Y}_e = \sum_{n=1}^{N} \tilde{Y}_n, \quad \frac{1}{\tilde{Z}_e} = \sum_{n=1}^{N} \frac{1}{\tilde{Z}_n} \qquad (2.14a,b)$$

We now consider the archetypal single-degree-of-freedom mechanical system which, as shown in Fig. 2.5, is composed of lumped mass, spring and damper elements connected in parallel (only for the mobility model). This is an important example which is used here to discuss the complex frequency response function of a mechanical system that exhibits resonant behaviour, with the aim of providing a

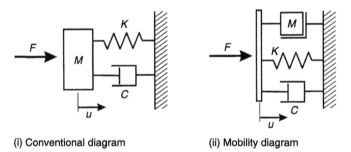

(i) Conventional diagram   (ii) Mobility diagram

Fig. 2.5. Single-degree-of-freedom mechanical system comprising lumped mass, spring and damper elements connected in parallel.

general model that forms the basis of modal analysis of distributed elastic systems whose responses are characterised by multiple resonances. Using Eq. (2.13a), the mobility function of this system is found to be given by

$$\tilde{Y}_e = \frac{1}{j\omega M + C - jK/\omega} \tag{2.15}$$

The form of this function is shown in Fig. 2.6. This plot can be interpreted by considering Fig. 2.3 which shows that the low frequency rising trend of $|\tilde{Y}_e|$ with a phase angle of $+90°$ is characteristic of a spring element having mobility $\tilde{Y}_K = j\omega/K$, whereas the decreasing trend of $|\tilde{Y}_e|$ at higher frequencies with a phase angle of $-90°$ is characteristic of a mass element having mobility $\tilde{Y}_M = -j/\omega M$.

When the damping is small, such that the damping ratio $\zeta = C/C_c$, with critical damping coefficient $C_c = 2\sqrt{KM}$, much smaller than unity, the resonance frequency is very close to the undamped natural frequency $\omega_o = \sqrt{K/M}$ of the system. By definition, at the undamped natural frequency, the spring and mass mobilities are equal and opposite since $\tilde{Y}_K(\omega_o) = j/\omega_o M$, and $\tilde{Y}_M(\omega_o) = -j/\omega_o M$. The resonant response is controlled by the dissipative effect of the damper element, the mobility $\tilde{Y}_c(\omega_o) = 1/C$ being the magnitude of the velocity response per unit excitation force at the resonance frequency.

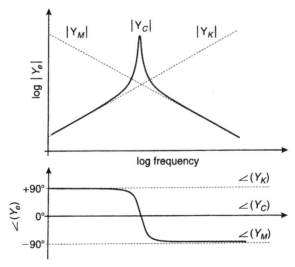

Fig. 2.6. Mobility function of the system shown in Fig. 2.5.

2.3   Mobility Functions of Uniform Beams in Bending

In this section the mobility functions of infinite and finite beams in bending are derived using the wave formulation introduced in Chapter 1, which leads to the so-called 'closed form solution' for the finite beam. Also, the so-called 'modal approach' is used to derive the mobility expressions for the finite beam. This approach is particularly suited for the study of distributed structures at low frequencies where the response is determined by the superposition of the responses of a few low frequency modes.

2.3.1   Infinite Beam

In Chapter 1 we have seen that the equation of motion in flexural vibration of an infinite, uniform, undamped beam subject to a transverse point harmonic force $F_y(t) = \mathrm{Re}\{\tilde{F}_y \exp(j\omega t)\}$ at $x = a$ is

$$EI\frac{\partial^4 \eta}{\partial x^4} + m\frac{\partial^2 \eta}{\partial t^2} = \tilde{F}_y \delta(x - a)\exp(j\omega t) \qquad (2.16)$$

The mobilities and impedances of lumped elements and at interfaces are defined in terms of collocated pairs of dynamic and kinematic quantities. In cases of distributed elastic systems, it is necessary to extend the definition to pairs of non-collocated points, in which case they are termed 'transfer' mobilities and impedances. The transfer mobility $\tilde{Y}_{v_y F_y}(x, \omega) = \tilde{v}_y(x, \omega)/\tilde{F}_y(\omega)$ between the complex velocity $\tilde{v}_y(x, \omega) = j\omega\tilde{\eta}(x, \omega)$ at position $(x \neq 0)$ and the complex force $\tilde{F}_y(\omega)$ at position $(x = 0)$ is obtained from Eqs. (1.77a,b). The general solution for the transverse displacement of the beam everywhere except at the excitation point is given by Eq. (1.77) as

$$\tilde{Y}_{v_y F_y}(x, \omega) = \frac{\omega}{4EIk_b^3}\left[e^{\mp jk_b x} - je^{\mp k_b x}\right] \qquad (2.17)$$

where the negative sign applies for $x > 0$ and the positive sign applies for $x < 0$. The mobility function evaluated at $x = 0$ is

$$\tilde{Y}_{v_y F_y}(0, \omega) = \frac{\omega(1 - j)}{4EIk_b^3} \qquad (2.18)$$

This latter mobility function is termed as 'driving point' or 'point' mobility function since it represents the response of the structure at the same point and in the same direction as that of the applied force. The real part represents the apparent 'damping' effect of energy being carried away to infinity. As we shall see below, and in the following section, other types of driving-point mobility functions can be defined. Thus, in order to avoid confusion, the form of mobility function in Eq. (2.18) will, where appropriate, be termed 'force driving-point mobility'. The driving-point mobility function in Eq. (2.18) can be expressed in the following form:

$$\widetilde{Y}_{v_y F_y}(0, \omega) = \frac{1}{C_{eq}(\omega)} - \frac{j}{\omega M_{eq}(\omega)} \tag{2.19}$$

Recalling that $k_b = (\omega^2 m / EI)^{1/4}$, the driving-point mobility corresponds to that of a frequency dependent equivalent mass $M_{eq}(\omega) = 4[EIm^3\omega^{-2}]^{1/4}$ connected in series with a viscous damper having a frequency dependent damping coefficient $C_{eq}(\omega) = 4[EIm^3\omega^2]^{1/4}$, as illustrated by Fig. 2.7.

An important practical implication of this result is that the equivalent mass and damper can interact with the stiffness of the localised supports of beam-like structures such as domestic and industrial pipes which offers the possibility of excessive vibration and possible damage due to resonance. As discussed in Chapter 1, pure bending vibrations are characterised by transverse displacement $\eta(x)$ and angular rotation of the cross section $\beta(x) = \partial\eta(x)/\partial x$. Thus, another type of mobility function can be defined as the ratio between the complex angular velocity of the cross section $\widetilde{w}_z(x, \omega) = j\omega\widetilde{\beta}(x, \omega)$ and the complex amplitude of the force $\widetilde{F}_y(\omega)$: $\widetilde{Y}_{w_z F_y}(x, \omega) = \widetilde{w}_z(x, \omega)/\widetilde{F}_y(\omega)$. This mobility function can be derived by differentiating Eq. (2.17) with respect to x, to give

$$\widetilde{Y}_{w_z F_y}(x, \omega) = \frac{\mp j\omega}{4EIk_b^2} \left[e^{\mp jk_b x} - e^{\mp k_b x} \right] \tag{2.20}$$

where also in this case the negative sign applies for $x > 0$ and the positive sign applies for $x < 0$. Note that this mobility function at $x = 0$ is zero.

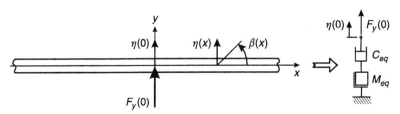

Fig. 2.7. (a) Infinite beam excited by a harmonic transverse point force at $x = 0$; (b) equivalent lumped element representation (driving point only).

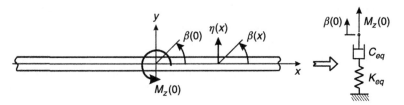

Fig. 2.8.   (a) Infinite beam excited by a harmonic couple at $x = 0$; (b) equivalent lumped element representation (driving point only).

When, as shown in Fig. 2.8(a), the beam is excited by a couple $M_z(t) = \mathrm{Re}\{\tilde{M}_z \exp(j\omega t)\}$ at $x = 0$, the coefficients $\tilde{A}, \tilde{B}, \tilde{C}$ and $\tilde{D}$ of Eq. (1.37) can be found by satisfying conditions of moment equilibrium immediately to the left and right of $x = 0$. In this case, the moments $-EI\partial^2\tilde{\eta}/\partial x^2$ generated at the left-hand and right-hand of an elemental beam section by the elastic bending stresses balance the applied moment $\tilde{M}_z$. Through symmetry, the bending moments at the terminations of the element have equal magnitude $1/2\tilde{M}_z \exp(j\omega t)$, and orientation in opposite direction to that of the moment $\tilde{M}_z$. Using Eq. (1.37), moment equilibrium condition gives at the right-hand and left-hand of the elemental beam section the following two equations: $\tilde{M}_z/2 + EI(-k_b^2\tilde{A} + k_b^2\tilde{C}) = 0$ and $\tilde{M}_z/2 + EI(-k_b^2\tilde{B} + k_b^2\tilde{D}) = 0$. Also, because the vibration field is anti-symmetric, the displacement at $x = 0$ is zero, so that Eq. (1.73) gives $\tilde{A} + \tilde{C} = \tilde{B} + \tilde{D} = 0$. Therefore the four unknowns in Eq. (1.73) are given by $\tilde{A} = -\tilde{B} = -\tilde{M}_z/4EIk_b^2$ and $\tilde{C} = -\tilde{D} = -\tilde{M}_z/4EIk_b^2$.

The transfer mobility function $\tilde{Y}_{v_yM_z}(x, \omega) = \tilde{v}_y(x, \omega)/\tilde{M}_z(\omega)$, which represents the ratio between the complex amplitude of the transverse velocity at positions $x > 0$ or $x < 0$ and the complex amplitude of the exciting couple, is

$$\tilde{Y}_{v_yM_z}(x, \omega) = \frac{\pm j\omega}{4EIk_b^2}\left[e^{\mp jk_bx} - e^{\mp k_bx}\right] \tag{2.21}$$

where the negative sign applies for $x > 0$ and the positive sign applies for $x < 0$. At $x = 0$ the mobility function is zero.

The mobility function $\tilde{Y}_{w_zM_z}(x, \omega) = \tilde{w}_z(x, \omega)/\tilde{M}_z(\omega)$ is defined as the ratio between the complex amplitude of angular velocity of a cross section and the complex amplitude of the couple. It is derived by differentiating Eq. (2.21) with reference to x so that

$$\tilde{Y}_{w_zM_z}(x, \omega) = \frac{-j\omega}{4EIk_b}\left[je^{\mp jk_bx} - e^{\mp k_bx}\right] \tag{2.22}$$

where again the negative sign applies for $x > 0$ and the positive sign applies for $x < 0$. The couple driving-point mobility function at $x = 0$ is

$$\tilde{Y}_{w_z M_z}(0, \omega) = \frac{\omega(1 + j)}{4EIk_b} \tag{2.23}$$

This expression may be written in the following form:

$$\tilde{Y}_{w_z M_z}(0, \omega) = \frac{1}{C_{eq}(\omega)} + \frac{j\omega}{K_{eq}(\omega)} \tag{2.24}$$

which suggests that, as shown in Fig. 2.8(b), this driving-point mobility function corresponds to that of an equivalent frequency dependent rotational stiffness $K_{eq}(\omega) = 4[(EI)^3 m\omega^2]^{1/4}$ in series with a viscous damper having a frequency dependent damping coefficient $C_{eq}(\omega) = 4[(EI)^3 m\omega^{-2}]^{1/4}$.

2.3.2   Finite Beam (Closed Form)

Real beams are of finite length, and therefore the mobility functions derived above are not valid if waves return from boundaries to the driving point with amplitudes significant compared to the outgoing waves. Suppose that, as shown in Fig. 2.9, the beam is of length l, is undamped, and is simply supported, i.e., the transverse displacement η and bending moment $M_z = -EI\partial^2\eta/\partial x^2$ are zero at each end. In this case all four coefficients $\tilde{A}, \tilde{B}, \tilde{C}$ and $\tilde{D}$ in Eq. (1.73) may be non-zero in both regions $x < 0$ and $x > 0$. They must take values that satisfy the boundary conditions at both ends.

Let us consider the case where the force is applied at the midpoint of the beam. This is a special case for which it is possible to derive the driving-point

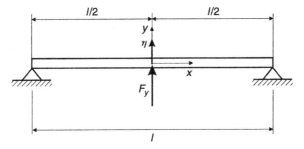

Fig. 2.9. Simply supported beam.

mobility function in closed form with a relatively simple analytical formulation. In this case at $x = -l/2$

$$
\eta(-l/2) = \tilde{A}_1 \exp(jk_b l/2) + \tilde{B}_1 \exp(-jk_b l/2)
$$
$$
+ \tilde{C}_1 \exp(k_b l/2) + \tilde{D}_1 \exp(-k_b l/2) = 0 \tag{2.25}
$$

and

$$
\frac{\partial^2 \eta(-l/2)}{\partial x^2} = -k_b^2 \tilde{A}_1 \exp(jk_b l/2) - k_b^2 \tilde{B}_1 \exp(-jk_b l/2)
$$
$$
+ k_b^2 \tilde{C}_1 \exp(k_b l/2) + k_b^2 \tilde{D}_1 \exp(-k_b l/2) = 0 \tag{2.26}
$$

Also, at $x = l/2$,

$$
\eta(l/2) = \tilde{A}_2 \exp(-jk_b l/2) + \tilde{B}_2 \exp(jk_b l/2)
$$
$$
+ \tilde{C}_2 \exp(-k_b l/2) + \tilde{D}_2 \exp(k_b l/2) = 0 \tag{2.27}
$$

and

$$
\frac{\partial^2 \eta(l/2)}{\partial x^2} = -k_b^2 \tilde{A}_2 \exp(-jk_b l/2) - k_b^2 \tilde{B}_2 \exp(jk_b l/2)
$$
$$
+ k_b^2 \tilde{C}_2 \exp(-k_b l/2) + k_b^2 \tilde{D}_2 \exp(k_b l/2) = 0 \tag{2.28}
$$

The subscripts 1 and 2 distinguish the coefficients appropriate to the two halves of the beam. In addition, shear force equilibrium equations must be applied at $x = 0^-$ and $x = 0^+$. At $x = 0^-$,

$$
\tilde{F}_y(0^-)/2 + EI\left(jk_b^3 \tilde{A}_1 - jk_b^3 \tilde{B}_1 - k_b^3 \tilde{C}_1 + k_b^3 \tilde{D}_1 \right) = 0 \tag{2.29}
$$

and at $x = 0^+$,

$$
\tilde{F}_y(0^+)/2 - EI\left(jk_b^3 \tilde{A}_2 - jk_b^3 \tilde{B}_2 - k_b^3 \tilde{C}_2 + k_b^3 \tilde{D}_2 \right) = 0 \tag{2.30}
$$

The bending moment is continuous through $x = 0$ because the applied force exerts no moment there. Hence,

$$
EI\frac{\partial^2 \eta(0^+)}{\partial x^2} = EI\frac{\partial^2 \eta(0^-)}{\partial x^2} \tag{2.31}
$$

which yields

$$-k_b^2\tilde{A}_2 - k_b^2\tilde{B}_2 + k_b^2\tilde{C}_2 + k_b^2\tilde{D}_2 = -k_b^2\tilde{A}_1 - k_b^2\tilde{B}_1 + k_b^2\tilde{C}_1 + k_b^2\tilde{D}_1 \quad (2.32)$$

Finally, the slope of the beam is continuous through $x = 0$. Hence,

$$-jk_b\tilde{A}_1 + jk_b\tilde{B}_1 - k_b\tilde{C}_1 + k_b\tilde{D}_1 = -jk_b\tilde{A}_2 + jk_b\tilde{B}_2 - k_b\tilde{C}_2 + k_b\tilde{D}_2 = 0 \quad (2.33)$$

Equations (2.25)–(2.33) contain eight complex unknowns which can be obtained as a function of frequency or wavenumber. Because we have chosen a physically symmetric configuration, it is possible by physical reasoning to obtain certain relationships between the coefficients on the two halves of the beam. (What are they?) The solutions to these equations are as follows:

$$\tilde{A}_1 = \frac{j\tilde{F}_y[1 + \exp(-jk_bl)]}{8EIk_b^3(1 + \cos k_bl)} = \tilde{B}_2 \quad (2.34a)$$

$$\tilde{B}_1 = -\tilde{A}_1 \exp(jk_bl) = \tilde{A}_2 \quad (2.34b)$$

$$\tilde{C}_1 = \frac{\tilde{F}_y}{4EIk_b^3[1 + \exp(k_bl)]} = \tilde{D}_2 \quad (2.34c)$$

$$\tilde{D}_1 = -\tilde{C}_1 \exp(k_bl) = \tilde{C}_2 \quad (2.34d)$$

The driving-point mobility function at the middle point of the beam is thus found to be

$$\tilde{Y}_{v_yF_y}(0,\omega) = \frac{\tilde{v}_y(0,\omega)}{\tilde{F}_y(0,\omega)} = \frac{j\omega}{4EIk_b^3}\left[\frac{\sin k_bl}{1 + \cos k_bl} + \frac{1 - \exp(k_bl)}{1 + \exp(k_bl)}\right]$$
$$\quad (2.35)$$
$$= \frac{j\omega}{4EIk_b^3}(\tan(k_bl/2) - \tanh(k_bl/2))$$

Comparison of this expression with that in Eq. (2.18) shows some similarity in its dependence upon the beam parameters, but it is very different in nature because it is purely imaginary and therefore no power can be transferred from the force to the beam. This makes physical sense because the beam has been assumed to possess no damping and therefore cannot dissipate power. No power can be transmitted into simple supports because they do not displace transversely, and there is no bending moment to do work through beam end rotation. However, the effect of structural damping can be accounted for by attributing a complex elastic modulus $E' = E(1 + j\eta)$ to the material. The associated complex bending wavenumber is $k_b' = k_b(1 - j\eta/4)$ where η is the loss factor.

The transverse displacement in a positive-going wave can be expressed as $\eta(x, t) = \widetilde{A} \exp[j(\omega t - k_b' x)]$. This expression can be rewritten in the following form $\eta(x, t) = \widetilde{A} \exp[j(\omega t - k_b x)] \exp(-k_b \eta x/4)$. The associated travelling-wave energy, which is proportional to $|A|^2$ [see Eq. (2.125)], suffers a fractional decrease of $\exp(-k_b \eta/2)$ per unit length. The energy of the waves generated at the point of application of force decreases by a factor $\exp(-k_b \eta l/2)$ during the passage of the wave to a beam boundary and back. If the factor $k_b \eta l/2$ is sufficiently large, the presence of the beam boundaries, as 'made known' at the driving point by the return thereto of a reflected wave, will not significantly affect the driving-point impedance: the beam will therefore 'appear to the force' to be unbounded: resonant and anti-resonant behaviour will disappear. The appropriate criterion for this to occur is $\eta \gg 2/k_b l$. Since $k_b = (\omega^2 m/EI)^{1/4}$, the value of the loss factor necessary to produce this condition decreases as the square root of frequency for a given beam and increases weakly with beam bending stiffness. If it is assumed that $\eta \gg 2/k_b l$ in Eq. (2.35), the driving-point mobility at $x = l/2$ is approximately $\widetilde{Y}_{v_y F_y} = \frac{1}{4}\omega^{-1/2}(EI)^{-1/4}m^{-3/4}(1 - j)$ which is the infinite-beam mobility [Eq. (2.18)]. (Prove.)

2.3.3   Finite Beam (Modal Summation)

Although the relatively simple case of a simply supported beam excited centrally by a point force has been considered, the mathematical analysis necessary to derive the point mobility function at any arbitrary point is quite complicated. Indeed, mobility functions in closed form are normally derived only for simple uniform structures and for particular configurations such as driving and transfer mobility functions at the terminations of the beam or driving-point mobility functions in the middle of the beam. A comprehensive account of these formulae is summarised by Bishop and Johnson (1960). Snowdon (1968) also presents an extensive set of closed form solutions with many worked examples including driving point and transfer impedance functions for arbitrary excitation points.

For more complex structures the 'modal summation' approach is normally used. This alternative approach is now presented for the simply supported beam in order to illustrate its differences and properties with respect to the closed form solution. The phenomena of natural (characteristic) frequencies and natural modes (characteristic functions) were introduced in Sections 1.10 and 1.12.

The natural frequencies of a structure are those at which it freely vibrates following the cessation of excitation. Associated with these frequencies are spatial distributions of field variables such as displacement and pressure that are characteristic of the material and geometric properties of a system and its boundaries: these are the natural modes. The distributions are normally expressed in non-dimensional forms as modal functions in which the distribution is normalised by its largest value. The response of a mode to any form of dynamic excitation is proportional to the modal (or generalised) excitation. This is given by the integral over the extent of the mode of the product of the spatial distribution of excitation agent (such as force or volume velocity) and the modal function. The response of each mode is expressed in terms of a modal coordinate (or amplitude). Thus, the harmonic transverse vibration of the beam at positon x is assumed to be given by (Timoshenko *et al.*, 1992)

$$\eta(x, t) = \text{Re} \left\{ \sum_{n=1}^{\infty} \phi_n(x) \tilde{q}_n(\omega) e^{j\omega t} \right\} \tag{2.36}$$

where $\phi_n(x)$ represents a real modal function of the n-th natural mode of the beam and $\tilde{q}_n$ represents the corresponding complex modal amplitude. In order to implement this approach it is therefore necessary to know the natural modes of the structure. (Note: mode functions may be assumed to be real in the case of undamped structures or where damping is represented in terms of a complex elastic modulus.) By definition, each mode satisfies the homogeneous bending wave equation $EI \partial^4 \eta / \partial x^4 + m \partial^2 \eta / \partial t^2 = 0$ and boundary conditions at its natural frequency ω_n. Substituting $\eta_n(x, t) = \phi_n(x) \tilde{q}_n \exp(j\omega_n t)$ into the wave equation yields

$$\left[d^4 \phi_n(x)/dx^4 - k_{bn}^4 \phi_n(x) \right] \tilde{q}_n \exp(j\omega_n t) = 0 \tag{2.37}$$

where $k_{bn} = \omega_n^{1/2} (m/EI)^{1/4}$. Assuming a solution $\phi_n(x) = A_n e^{\lambda_n x}$ yields

$$\lambda_n^4 = k_{bn}^4 \tag{2.38}$$

and

$$\lambda_n = \pm k_{bn} \text{ and } \lambda_n = \pm j k_{bn} \tag{2.39a,b}$$

giving

$$\phi_n(x) = A \cosh k_{bn} x + B \sinh k_{bn} x + C \cos k_{bn} x + D \sin k_{bn} x \tag{2.40}$$

Following from the example examined in the previous section, the case is now considered in which the beam is simply supported at both ends such that the transverse displacement η and bending moment $M = -EI\partial^2\eta/\partial x^2$ are zero at the two terminations. Assuming in this case, the origin of the system of reference to be at the left-hand termination of the beam, the following conditions apply: $\eta|_{x=0,l} = 0$ and $\partial^2\eta/\partial x^2|_{x=0,l} = 0$. If the equivalent trigonometric expression of Eq. (2.40)

$$\phi_n(x) = A_1 (\cos k_{bn}x + \cosh k_{bn}x) + A_2 (\cos k_{bn}x - \cosh k_{bn}x)$$
$$+ A_3 (\sin k_{bn}x + \sinh k_{bn}x) + A_4 (\sin k_{bn}x - \sinh k_{bn}x) \tag{2.41}$$

is considered, then the boundary conditions at $x = 0$ give $A_1 = A_2 = 0$ and the boundary conditions at $x = l$ give $A_3 = A_4$ yielding

$$\sin k_{bn}l = 0 \tag{2.42}$$

which is satisfied by $k_{bn}=n\pi/l$. Since $k_b=(m\omega^2/EI)^{1/4}$, the natural frequencies are given by

$$\omega_n = \frac{n^2\pi^2}{l^2} \left(\frac{EI}{m}\right)^{1/2} \tag{2.43}$$

where $n = 1, 2, \ldots$ as also given by Eq. (1.66) based upon the criterion of phase coincidence. The analytical expression for the modal functions can then be derived from Eq. (2.41) by substituting $A_1 = A_2 = 0$ and, for convenience, assuming $A_3 = A_4 = \sqrt{2}/2$ so that

$$\phi_n(x) = \sqrt{2}\sin\left(\frac{n\pi x}{l}\right) \tag{2.44}$$

The natural mode functions are often called 'mode shapes', since they describe the shape of the beam deformation for each natural mode as, for example, shown in Fig. 2.10 for the first four modes. A comprehensive summary of formulae for the natural frequencies and natural modes of beams in bending may be found in Gardonio and Brennan (2004).

The derivation of the forced response in terms of a modal summation is based on the condition of 'orthogonality' between natural modes. Timoshenko *et al.* (1992) shows that, considering two natural modes of order r and s, the condition of orthogonality for *uniform* beams gives

$$\int_0^l \phi_r\phi_s dx = 0 \quad (r \neq s) \tag{2.45a}$$

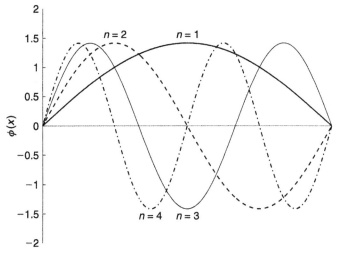

Fig. 2.10. First four natural modes of a simply supported uniform beam.

and

$$\int_0^l \phi_r'' \phi_s'' dx = 0, \quad \int_0^l \phi_r'''' \phi_s dx = 0 \quad (r \neq s) \tag{2.45b,c}$$

where $\phi_r'' = \partial^2 \phi_r / \partial x^2$ and $\phi_r'''' = \partial^4 \phi_r / \partial x^4$. If $r = s$ then $\int_0^l [\phi_r(x)]^2 \, dx$ is a constant which, for example, in the case of the natural modes of the simply supported beam given by Eq. (2.44) gives

$$\int_0^l [\phi_r(x)]^2 \, dx = l \tag{2.46a}$$

and

$$\int_0^l [\phi_r''(x)]^2 dx = k_{br}^4 l, \quad \int_0^l \phi_r''''(x)\phi_s(x)dx = k_{br}^4 l \quad (r = s) \tag{2.46b,c}$$

We now turn to the modal solution to the problem of a simply supported beam excited by a harmonic point force. Substituting the modal summation of Eq. (2.36) into Eq. (1.72) gives

$$\sum_{r=1}^{\infty} \left(EI\phi_r''''(x)\tilde{q}_r - \omega^2 m \, \phi_r(x)\tilde{q}_r \right) dx = \tilde{F}_y \delta(x-a) \tag{2.47}$$

If we multiply this equation by a natural mode $\phi_s(x)$ and integrate over the length, then we obtain

$$\sum_{r=1}^{\infty} \left[EI\tilde{q}_r \int_0^l \phi_r''''\phi_s dx - \omega^2 m \tilde{q}_r \int_0^l \phi_r\phi_s dx \right] = \int_0^l \phi_r(x)\tilde{F}_y\delta(x-a)dx$$

$$(2.48)$$

Using the orthogonality conditions given by Eqs. (2.45a–c), the three conditions in Eqs. (2.46a–c) and the properties of the Dirac delta function, it is found that for $r \neq s$ the integrals are zero and for $r = s$ Eq. (2.48) reduces to

$$-\omega^2 M_r\tilde{q}_r + K_r\tilde{q}_r = \tilde{F}_r \qquad (2.49)$$

where

$$\tilde{F}_r = \int_0^l \phi_r(x)\tilde{F}_y\delta(x-a)dx = \phi_r(a)\tilde{F}_y \qquad (2.50a)$$

represents the so-called 'modal force' and

$$M_r = m \int_0^l \phi_r^2 dx = ml \qquad (2.50b)$$

$$K_r = EI \int_0^l \phi_r''''\phi_r dx = EIk_{br}^4 l \qquad (2.50c)$$

are the so-called 'modal mass' and 'modal stiffness'. Equation (2.49) is analogous to that of a lumped mass–spring system having natural frequency $\omega_r = \sqrt{K_r/M_r} = \left(r^2\pi^2/l^2\right)\sqrt{EI/m}$ subject to a modal force $\tilde{F}_r = \phi_r(a)\tilde{F}_y \exp(j\omega t)$. The wave equation of motion is replaced by a set of second order ordinary differential equations in the modal coordinates $\tilde{q}_r$, which are normally written in the form

$$\left(\omega_r^2 - \omega^2\right)\tilde{q}_r = \frac{\tilde{F}_r}{M_r} = \frac{\phi_r(a)}{M_r}\tilde{F}_y \qquad (2.51)$$

Therefore the complex displacement response of the beam at position x is given by

$$\tilde{\eta}(x,\omega) = \sum_{r=1}^{\infty}\phi_r(x)\tilde{q}_r = \sum_{r=1}^{\infty}\frac{\phi_r(x)\tilde{F}_r}{M_r\left(\omega_r^2 - \omega^2\right)} = \sum_{r=1}^{\infty}\frac{\phi_r(x)\phi_r(a)}{M_r\left(\omega_r^2 - \omega^2\right)}\tilde{F}_y(\omega)$$

$$(2.52)$$

When the modal summation solution is employed, the effect of light damping may be taken into account by incorporating a viscous modal damping coefficient in Eq. (2.49) which therefore becomes

$$-\omega^2 M_r \tilde{q}_r + j\omega C_r \tilde{q}_r + K_r \tilde{q}_r = \tilde{F}_r \qquad (2.53)$$

where C_r is the damping coefficient of the r-th mode. However, the effect of structural damping is usually modelled by assuming a hysteretic damping model which gives a complex modal stiffness $K_r(1+j\eta)$. Equation (2.49) then becomes

$$-\omega^2 M_r \tilde{q}_r + K_r(1+j\eta)\tilde{q}_r = \tilde{F}_r \qquad (2.54)$$

These two models yield

$$\tilde{\eta}(x, \omega) = \sum_{r=1}^{\infty} \frac{\phi_r(x)\phi_r(a)}{M_r\left[\omega_r^2 - \omega^2 + j2\zeta_r\omega_r\omega\right]} \tilde{F}_y(\omega) \qquad (2.55a)$$

and

$$\tilde{\eta}(x, \omega) = \sum_{r=1}^{\infty} \frac{\phi_r(x)\phi_r(a)}{M_r\left[\omega_r^2(1+j\eta) - \omega^2\right]} \tilde{F}_y(\omega) \qquad (2.55b)$$

where $\zeta_r = C_r/2M_r\omega_r$ is the modal damping ratio that, for harmonic vibration, is related to the loss factor by $\zeta_r = \frac{1}{2}\frac{\omega_r}{\omega}\eta$ (Check). Thus, the modal summation formulation expresses the response of the beam as a sum of the responses of an infinite set of second order modal mass–spring–damper systems.

The driving point and transfer mobility functions of the simply supported beam shown in Fig. 2.11 can now be derived directly by pre-multiplying Eqs. (2.55a,b) by $j\omega$ to give

$$\tilde{Y}_{v_y F_y}(\omega) = j\omega \sum_{r=1}^{\infty} \frac{\phi_r(b)\phi_r(a)}{M_r\left[\omega_r^2 - \omega^2 + j2\zeta_r\omega_r\omega\right]} \qquad (2.56a)$$

$$\tilde{Y}_{v_y F_y}(\omega) = j\omega \sum_{r=1}^{\infty} \frac{\phi_r(b)\phi_r(a)}{M_r\left[\omega_r^2(1+j\eta) - \omega^2\right]} \qquad (2.56b)$$

An example is presented in Fig. 2.12 which shows that, unlike driving-point mobility of the infinite beam mobility, the driving-point mobility function of the bounded beam varies with frequency over a wide range of magnitude which,

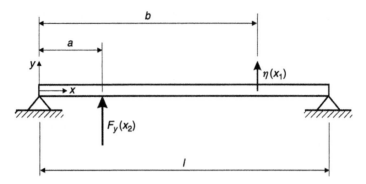

Fig. 2.11.   Simply supported uniform beam excited by a point force.

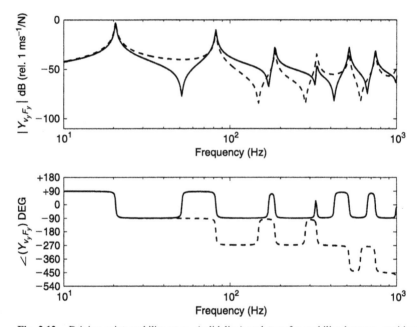

Fig. 2.12.   Driving-point mobility at x_1 (solid line) and transfer mobility between positions $x_1 = 0.27l$ and $x_2 = 0.63l$ (dashed line) of a simply supported aluminium beam of cross-section dimensions 10×20 mm, length 1.5 m and loss factor 0.01.

as discussed in Section 1.11 is determined by the phenomena of resonance and antiresonance. Comparing the magnitude and phase plots with those in Fig. 2.3 for the lumped mass–spring–damper elements, this mobility function clearly exhibits stiffness-type behaviour below resonance and mass-type behaviour above resonance with a transitional phase change of $-180°$. At the antiresonance frequency, the mobility reverts to stiffness-type behaviour with a phase change of $+180°$. As a result, the phase of the driving-point mobility function is confined between $\pm 90°$. The physics of this phenomenon is explained by considering the time average power input by the point force excitation, which, according to Eq. (2.4), can be expressed as $\overline{P} = 1/2|\widetilde{F}|^2 \operatorname{Re}\{\widetilde{Y}_{v_y F_y}\}$. Since the power input must be positive, the real part of the driving-point mobility must be positive and thus have phase confined between $\pm 90°$. It is important to emphasise that the resonances 'close up' in Fig. 2.12 because the plots are in log(frequency) scale: actually, the separation increases with frequency.

The dashed line in Fig. 2.12 indicates that the transfer mobility function is also characterised by a sequence of resonances although in this case they do not always alternate with antiresonances. For example between the first and second resonances the response does not pass through an antiresonance so that there is no $+180°$ phase recovery and thus a further $-180°$ phase lag occurs.

The alternating spring-type and mass-type behaviours of the driving-point mobility function can be interpreted by plotting the modal contributions in the summation of Eqs. (2.56). For instance the dashed, dash-dotted and dotted lines in Fig. 2.13 show the contributions from the first three modes. Below the first

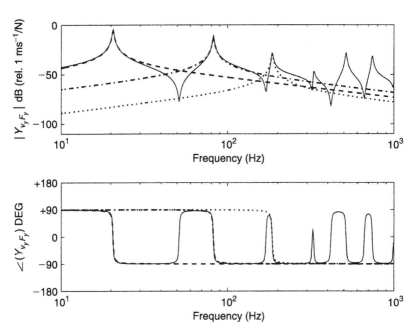

Fig. 2.13. Driving-point mobility function (solid line) of the simply supported uniform beam specified in Fig. 2.12. The dashed, dashed-dotted and dotted lines represent the contributions of the three lowest-order modes.

resonance the mobility function is controlled by the spring-type behaviour of the first mode which then converts into a mass-type behaviour above this resonance. Subsequently, the spring-type behaviour of the second mode term becomes prominent and controls the mobility function up to the second resonance where the trend is converted into a mass-type descendent behaviour up to the frequency where the spring-type behaviour of the third modal resonating term takes over. This alternating behaviour is then repeated for all the following resonances. As shown by Fig. 2.13, the transition occurs when the modal contributions have the same amplitude but opposite phase so that they cancel out. The response is therefore reduced to very low values which are determined by the spring-type contributions of the modes having higher resonance frequencies.

The expression for the mobility function in Eq. (2.56) assumes that an infinite number of terms are taken into account in the summation. It is therefore important to investigate how many terms of this summation should be considered in order to achieve convergence to an adequately precise function in a given frequency range. If, for example, the driving-point mobility function is derived in a frequency range up to 1 kHz, then, as shown in Fig. 2.14(a), the contributions of all modes with natural frequencies within this frequency range should be taken into account. Figure 2.14(b) shows that the contributions of the modes resonating at frequencies higher than 1 kHz have relatively little influence except at the antiresonance frequencies. In the frequency range of interest, these contributions all occur below the resonance frequencies so that they can be approximated as spring-type modal contributions; the summation in Eq. (2.56) can then be limited to the modes resonating in the frequency of interest plus a summation of residual spring-type terms:

$$Y_{v_y F_y}(\omega) \approx j\omega \sum_{r=1}^{R} \frac{\phi_r(b)\phi_r(a)}{M_r \left[\omega_r^2(1+j\eta) - \omega^2\right]} + j\omega \sum_{r=R+1}^{\infty} \frac{\phi_r(b)\phi_r(a)}{M_r \omega_r^2} \qquad (2.57)$$

In the case of a driving-point mobility the residual terms are all spring-like because $[\phi_r(x)]^2$ is always positive. Figure 2.14 shows that, within the frequency range of interest, the magnitude of the residual terms decreases rapidly as the modal order rises ($\omega_r^2 \propto r^4$). Thus, a relatively small number of higher order residual terms are required in the second summation of Eq. (2.57) to accurately represent the mobility function in the frequency range of interest. This factor is however magnified by the fact that the modal density of beam bending vibration decreases with the square root of frequency. Therefore, the separation between resonances rises with the mode order and thus the magnitudes of the residual contributions decrease rapidly as the modal order increases. Two-dimensional structures such as flat plates and curved shells are characterised by modal densities that are either constant or rise with frequency (see Table 1.4 in Chapter 1) so that

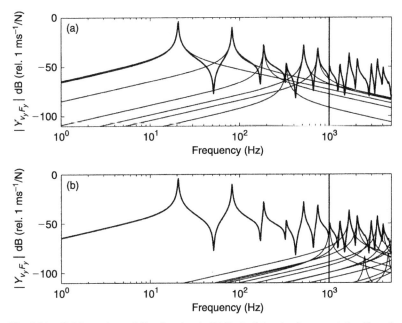

Fig. 2.14. Driving-point mobility function (solid line) of the simply supported uniform beam specified in Fig. 2.12. The faint lines represent the contributions of modes resonant (a) below 1 kHz and (b) above 1 kHz.

the average separation between resonance frequencies either remains constant or decreases with frequency. Consequently, a larger number of residual terms must be taken into account in the summation for the mobility function.

The transfer mobility function $\widetilde{Y}_{w_z F_y}(\omega) = \widetilde{w}_z(b,\omega)/\widetilde{F}_y(a,\omega)$ which represents the ratio between the complex amplitude of the angular velocity of a cross section at position $x = b$ and the complex amplitude of the force at position $x = a$ can be derived by differentiating Eq. (2.56) with reference to x, in which case the following expression is obtained:

$$\widetilde{Y}_{w_z F_y}(\omega) = j\omega \sum_{r=1}^{\infty} \frac{\psi_r(b)\phi_r(a)}{M_r\left[\omega_r^2(1 + j\eta) - \omega^2\right]} \tag{2.58}$$

where

$$\psi_r(x) = \frac{d\phi_r(x)}{dx} \tag{2.59}$$

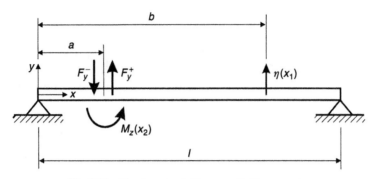

Fig. 2.15. Simply supported beam excited by a couple.

When, as shown in Fig. 2.15, the beam is excited by a harmonic couple $\tilde{M}_z \exp(j\omega t)$ at $x = a$, the response of the beam can be straightforwardly derived by modelling the couple as a pair of transverse forces of amplitudes $\pm \tilde{F}_y = \pm \tilde{M}_z/d$ where $d \to 0$. In this case the response of the beam becomes

$$
\tilde{\eta}(x) = \lim_{d \to 0} \sum_{r=1}^{\infty} \frac{\phi_r(x) \left\{ \phi_r \left(a + \frac{d}{2} \right) - \phi_r \left(a - \frac{d}{2} \right) \right\}}{M_r \left[\omega_r^2 (1 + j\eta) - \omega^2 \right]} \tilde{F}_y(\omega)
$$

$$
= \sum_{r=1}^{\infty} \frac{\phi_r(x)}{M_r \left[\omega_r^2 (1 + j\eta) - \omega^2 \right]} \lim_{d \to 0} \frac{\left\{ \phi_r \left(a + \frac{d}{2} \right) - \phi_r \left(a - \frac{d}{2} \right) \right\}}{d} \tilde{M}_z(\omega)
$$

(2.60)

where the limit corresponds to the derivative of $\phi_r(x)$ at the point where the moment excitation is acting. Thus, the mobility function $\tilde{Y}_{v_y M_z}(\omega) = j\omega \tilde{\eta}(b, \omega)/\tilde{M}_z(a, \omega)$ for the ratio between the complex amplitude of the transverse velocity of the cross section at position $x = b$ and the complex amplitude of the couple at position $x = a$ is given by

$$
Y_{v_y M_z}(\omega) = j\omega \sum_{r=1}^{\infty} \frac{\phi_r(b) \psi_r(a)}{M_r \left[\omega_r^2 (1 + j\eta) - \omega^2 \right]}
$$

(2.61)

Finally, the mobility function $\tilde{Y}_{w_z M_z}(\omega) = \tilde{w}_z(b, \omega)/\tilde{M}_z(a, \omega)$ for the ratio between the complex amplitude of the angular velocity of the cross section at position $x = b$ and the complex amplitude of the moment at position $x = a$ is derived by differentiating Eq. (2.60) with reference to x, in which case the following expression is obtained:

$$\widetilde{Y}_{w_z M_z}(\omega) = j\omega \sum_{r=1}^{\infty} \frac{\psi_r(b)\psi_r(a)}{M_r \left[\omega_r^2(1 + j\eta) - \omega^2\right]} \tag{2.62}$$

In many cases the ratios of the displacement per unit force $\tilde{\alpha} = \widetilde{\eta}/\widetilde{F}$ or acceleration per unit force $\widetilde{A} = -\omega^2\eta/\widetilde{F}$ are of interest. The former is normally known as the 'receptance function' (although it is also termed 'admittance', 'compliance' and 'dynamic flexibility') while the latter is termed 'inertance' or 'accelerance'. The inverses of receptance, mobility and accelerance are defined as 'dynamic stiffness' $\widetilde{K} = \widetilde{F}/\widetilde{\eta} = 1/\tilde{\alpha}$, 'impedance' $\widetilde{Z} = \widetilde{F}/j\omega\widetilde{\eta} = 1/\widetilde{Y}$ and 'apparent mass' $\widetilde{M} = -\widetilde{F}/\omega^2\eta = 1/\widetilde{A}$ respectively.

The plots in Fig. 2.16 shows the magnitudes of $\tilde{\alpha}$, $\widetilde{Y}$ and $\widetilde{A}$ transfer functions between points $x_1 = 0.27l$ and $x_2 = 0.63l$ of the beam. As one would expect, the three functions are similar except that the geometric mean values of the receptance $\tilde{\alpha}$ and inertance $\widetilde{A}$ functions have slopes proportional to $1/\omega$ and ω with respect to the mobility.

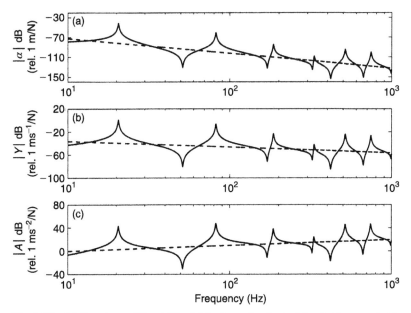

Fig. 2.16. (a) Receptance; (b) mobility; (c) inertance functions of the simply supported beam specified in Fig. 2.12.: - - - - -, geometric mean values.

2.4 Mobility and Impedance Functions of Thin Uniform Flat Plates

In this section, the mobility functions of both infinite and finite thin flat rectangular plates in bending are derived. Despite the assumption of a simple rectangular geometry and simple boundary conditions for the finite plate, the closed form solution cannot be derived and the modal summation formulation is therefore employed.

2.4.1 Infinite Plate

As discussed in Section 1.7, small amplitude bending waves in a thin flat plate are uncoupled from the in-plane longitudinal and shear waves so that out-of-plane vibration can be treated separately. Assuming 'classical plate theory' for thin plates (rotary inertia and shear strain are neglected), the equation of motion of transverse displacement of an infinite thin plate subject to a distributed transverse force per unit area $p_y(x, z, t)$ is (Reddy, 1984; Cremer *et al.*, 1988)

$$D\left(\frac{\partial^4 \eta(x, z)}{\partial x^4} + 2\frac{\partial^4 \eta(x, z)}{\partial x^2 \partial z^2} + \frac{\partial^4 \eta(x, z)}{\partial z^4}\right) + m\frac{\partial^2 \eta(x, z, t)}{\partial t^2} = p_y(x, z, t) \quad (2.63)$$

where $D = Eh^3/12(1 - v^2)$ is the bending stiffness per unit length and $m = \rho h$ is the mass per unit area. The homogeneous form ($p_y = 0$) of this equation can be solved in a similar way to that for a beam, but in cylindrical coordinates, to give the characteristics of the free wave motion in the plate.

Thus, in terms of the notation shown in Fig. 2.17(a), the complex out-of-plane displacement $\tilde{\eta}$ at any position (r, α) in response to a harmonic point force of

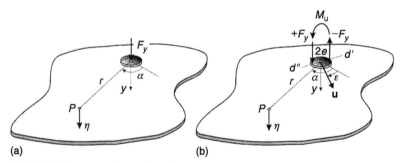

Fig. 2.17. Sign convention and coordinate systems for a thin plate excited by (a) point force; (b) a couple in direction **u**.

complex magnitude $\widetilde{F}_y$ acting on a massless, rigid indenter of diameter $2e$ is given by Cremer *et al.* (1988) and Ljunggren (1983) as

$$\widetilde{\eta}(r, \alpha) = \frac{-j\widetilde{F}_y}{8Dk_b^2}\left[H_0^{(2)}(k_b r) - j\frac{2}{\pi}K_0(k_b r) \right] \tag{2.64}$$

where $H_i^{(2)}(k_b r)$ is the *i*-th order Hankel function of the second kind, $K_i(k_b r)$ is the *i*-th order modified Bessel function of the second kind and $k_b = (\omega^2 m/D)^{1/4}$ is the bending (flexural) wavenumber.

Figure 2.18 shows the real and imaginary parts of the complex out-of-plane velocity of the plate $\tilde{v}(r, \alpha) = j\omega\widetilde{\eta}(r, \alpha)$ in the range $k_b r = 0$–5. It can be seen that for small values of $k_b r$ the real part of the velocity response is dominated by a strong near-field component, which decays rapidly as $k_b r$ increases, while the imaginary part of the displacement response tends to zero for $k_b r \to 0$.

Consider now the case shown in Fig. 2.17(b) where a couple $M_{\mathbf{u}}$ with orientation $\mathbf{u}$ acts on a small rigid indenter of radius e fixed to the plate. The resulting complex out-of-plane displacement $\widetilde{\eta}$ at position (r, α) can be calculated by adding the velocities produced by a pair of harmonic forces with opposite phase acting at points $2e$ apart on the line $d'-d''$ as shown in Fig. 2.17. These forces act on a line orthogonal to the direction $\mathbf{u}$, and in the limiting case $2e \to 0$ the

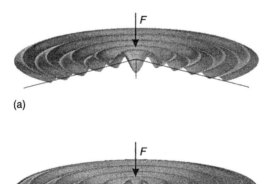

(a)

(b)

Fig. 2.18. (a) Real; (b) Imaginary parts of the transverse velocity of an infinite uniform thin flat plate excited by a harmonic transverse point force ($k_b r = 0$–5).

complex displacement is given by (Gardonio and Elliott, 1998)

$$\tilde{\eta}(r, \alpha) = \frac{-j\tilde{M}_{\mathbf{u}} \sin(\alpha - \varepsilon)}{8Dk_b} \left[H_1^{(2)}(k_b r) - j\frac{2}{\pi} K_1(k_b r) \right] \qquad (2.65)$$

where the angles α and ε are defined as in Fig. 2.17.

Figure 2.19 shows the real and imaginary parts of the complex out-of-plane velocity $\tilde{v}(r, \alpha) = j\omega\tilde{\eta}(r, \alpha)$ in the range $k_b r = 0$–5. The response to moment excitation is not characterised by the strong near-field effect generated by a point force excitation.

Expressions for the mobility functions $\tilde{Y}_{v_y F_y} = \tilde{v}_y / \tilde{F}_y$ and $\tilde{Y}_{v_y M_{\mathbf{u}}} = \tilde{v}_y / \tilde{M}_{\mathbf{u}}$ giving the ratio between the complex out-of-plane velocity $\tilde{v}_y = j\omega\tilde{\eta}$ and the complex out-of-plane force $\tilde{F}_y$ and complex couple $\tilde{M}_{\mathbf{u}}$, with orientation defined by $\mathbf{u}$, can be straightforwardly derived from Eqs. (2.64) and (2.65) respectively. Thin plates bending vibrations are characterised by angular rotations of the cross section which, as shown in Fig. 2.20, can be derived with reference to any direction $\mathbf{v}$ in the plane of the plate, so that $\beta_{\mathbf{v}}(r, \alpha) = \partial\tilde{\eta}(r, \alpha)/\partial h$, where h is orthogonal to the direction $\mathbf{v}$ (Gardonio and Elliott, 1998). Thus the mobility functions $\tilde{Y}_{w_{\mathbf{v}} F_y} = \tilde{w}_{\mathbf{v}} / \tilde{F}_y$ and $\tilde{Y}_{w_{\mathbf{v}} M_{\mathbf{u}}} = \tilde{w}_{\mathbf{v}} / \tilde{M}_{\mathbf{u}}$ giving the ratio between the complex angular velocity $\tilde{w}_{\mathbf{v}} = j\omega\beta_{\mathbf{v}}$ with orientation defined by $\mathbf{v}$, and the complex out-of-plane force $\tilde{F}_y$ and complex couple $\tilde{M}_{\mathbf{u}}$, with orientation defined by $\mathbf{u}$, can be derived by differentiating along h Eqs. (2.64) and (2.65) respectively.

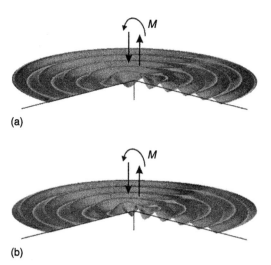

(a)

(b)

Fig. 2.19. (a) Real; (b) Imaginary parts of the transverse displacement of an infinite uniform thin flat plate excited by a couple ($k_b r = 0$–5).

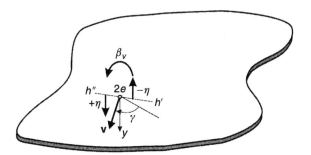

Fig. 2.20. Angular displacement in direction **v**.

The four mobility functions are:

$$\tilde{Y}_{v_y F_y} = \frac{\omega}{8 D k_b^2} T_0\,(k_b r)\,, \quad \tilde{Y}_{w_\mathbf{v} F_y} = -\frac{\omega \sin \delta_1}{8 D k_b} T_1\,(k_b r)\,, \quad \tilde{Y}_{v_y M_\mathbf{u}} = \frac{\omega \sin \delta_2}{8 D k_b} T_1\,(k_b r)$$

$$(2.66\text{a,b,c})$$

$$\tilde{Y}_{w_\mathbf{v} M_\mathbf{u}} = \frac{\omega}{8 D} \left\{ \sin \delta_1 \sin \delta_2 \left(T_0(k_b r) - \frac{1}{k_b r} T_1(k_b r) \right) + \frac{\cos \delta_1 \cos \delta_2}{k_b r} T_1(k_b r) \right\}$$

$$(2.66\text{d})$$

where $\delta_1 = \alpha - \gamma$, $\delta_2 = \alpha - \varepsilon$ and

$$T_n\,(k_b r) = H_n^{(2)}\,(k_b r) - j\frac{2}{\pi} K_n(k_b r) \tag{2.67}$$

with $n = 0, 1$.

At the excitation point, the force $\tilde{F}_y$ does not generate an angular velocity $\tilde{w}_\mathbf{v}$ and the couple $\tilde{M}_\mathbf{u}$ does not generate an out-of-plane velocity $\tilde{v}_y$. Also, the angular velocity at the excitation point generated by the couple $\tilde{M}_\mathbf{u}$ is in the same direction as that of the couple, i.e., $\mathbf{v} = \mathbf{u}$. Therefore, two driving-point mobility functions exist which are given by (Cremer *et al.*, 1988 and Ljunggren, 1984) as

$$\tilde{Y}_{v_y F_y}(0) = \frac{\tilde{v}_y}{\tilde{F}_y} = \frac{\omega}{8 D k_b^2} = \frac{1}{8 \sqrt{Dm}} \tag{2.68a}$$

$$\tilde{Y}_{w_\mathbf{u} M_\mathbf{u}}(0) = \frac{\tilde{w}_\mathbf{u}}{\tilde{M}_\mathbf{u}} = \frac{\omega}{16 D} \left\{ 1 - j\frac{4}{\pi} \ln(k_b e) \right\} \tag{2.68b}$$

where e is the radius of the indenter. Equation (2.68a) shows that the force driving-point mobility is purely real and acts as a frequency independent damper. The couple driving-point mobility in Eq. (2.68b) has both real and imaginary components, showing that it behaves as a frequency dependent damper and a frequency dependent stiffness (Why not mass?).

2.4.2   Finite Plate

Expressions for the mobilities of finite plates can be derived in terms of modal summation using the formulation as that presented for finite beams in Section 2.3.3. As an example, the case of a simply supported rectangular plate of dimensions $l_x \times l_z$ is now considered. Figure 2.21(a) shows the notation used for the force-moment excitations at position 1 (x_1, z_1) and for the linear-angular displacements at position 2 (x_2, z_2) when the Cartesian co-ordinate system of reference (O,x,y,z) is located at the corner of the plate with the y axis orthogonal to the surface of the plate.

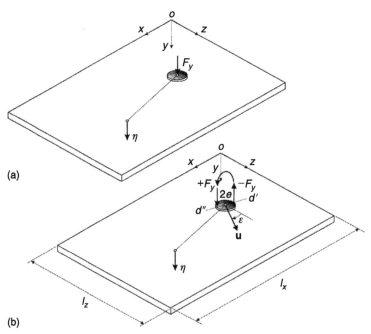

Fig. 2.21.   Sign convention and coordinate systems for a rectangular plate excited by (a) a point force; (b) a point couple in direction **u**.

The complex out-of-plane displacement $\tilde{\eta}(x, z)$ generated by a distribution of harmonic force per unit area $\tilde{p}_y(x, z)$ acting on the panel can be expressed in terms of a modal summation, which, assuming hysteretic damping, is

$$\tilde{\eta}(x, z) = \sum_{r=1}^{\infty} \frac{\phi_r(x, z)\tilde{F}_r}{M_r\left[\omega_r^2(1 + j\eta) - \omega^2\right]} \tag{2.69}$$

where $\tilde{F}_r$ is the modal force given by

$$\tilde{F}_r = \int_0^{l_x} \int_0^{l_z} \tilde{p}_y(x, z)\phi_r(x, z)dxdz \tag{2.70}$$

The natural frequencies and mode shapes of a simply supported rectangular plate having modal mass $M_r = \rho l_x l_z h$ are given by

$$\omega_r = \sqrt{\frac{D}{m}}\left[\left(\frac{r_1\pi}{l_x}\right)^2 + \left(\frac{r_2\pi}{l_z}\right)^2\right], \ \phi_r(x, z) = 2\sin\left(\frac{r_1\pi x}{l_x}\right)\sin\left(\frac{r_2\pi z}{l_z}\right) \tag{2.71a,b}$$

where r_1 and r_2 are the modal indices of the r-th mode. An extensive list of expressions for the natural modes and frequencies for rectangular plates with other types of boundary conditions was derived by Warburton (1951) and was summarised by Gardonio and Brennan (2004).

The transverse displacement generated at position (x_2, z_2) by a point transverse force F_y at position (x_1, z_1) is given by

$$\tilde{\eta}(x_2, z_2, \omega) = \sum_{r=1}^{\infty} \frac{\phi_r(x_2, z_2)\phi_r(x_1, z_1)}{M_r\left[\omega_r^2(1 + j\eta) - \omega^2\right]}\tilde{F}_y(\omega) \tag{2.72}$$

The transverse displacement due to a couple can be derived by replacing the moment with a couple of transverse forces as shown in Fig. 2.21(b). In this case the complex response generated by the couple excitation with orientation defined by **u**, is found to be given by

$$\tilde{\eta}(x_2, z_2, \omega) = \sum_{r=1}^{\infty} \frac{\phi_r(x_2, z_2)\psi_r^{\mathbf{u}}(x_1, z_1)}{M_r\left[\omega_r^2(1 + j\eta) - \omega^2\right]}\tilde{M}_{\mathbf{u}}(\omega) \tag{2.73}$$

where

$$\psi_r^{\mathbf{u}}(x, z) = -\sin\left(\varepsilon\frac{\partial\phi_r(x, z)}{\partial z}\right) + \cos\left(\varepsilon\frac{\partial\phi_r(x, z)}{\partial x}\right) \qquad (2.74)$$

As found for the infinite plate case, using the two equations for the complex displacement generated by a point force [Eq. (2.72)] and a point couple [Eq. (2.73)] the four mobility functions $\tilde{Y}_{v_y F_y} = \tilde{v}_y/\tilde{F}_y$, $\tilde{Y}_{w_{\mathbf{v}} F_y} = \tilde{w}_{\mathbf{v}}/\tilde{F}_y$, $\tilde{Y}_{v_y M_{\mathbf{u}}} = \tilde{v}_y/\tilde{M}_{\mathbf{u}}$ and $\tilde{Y}_{w_{\mathbf{v}} M_{\mathbf{u}}} = \tilde{w}_{\mathbf{v}}/\tilde{M}_{\mathbf{u}}$ may be derived in terms of the following common expression:

$$\tilde{Y}(\omega) = j\omega \sum_{r=1}^{\infty} \frac{f_r(x_2, z_2)g_r(x_1, z_1)}{M_r\left[\omega_r^2(1 + j\eta) - \omega^2\right]} \qquad (2.75)$$

where a) for $\tilde{Y}_{v_y F_y}$ $f_r = \phi_r$ and $g_r = \phi_r$,

       b) for $\tilde{Y}_{w_{\mathbf{v}} F_y}$ $f_r = \psi_r^{\mathbf{v}}$ and $g_r = \phi_r$,

       c) for $\tilde{Y}_{v_y M_{\mathbf{u}}}$ $f_r = \phi_r$ and $g_r = \psi_r^{\mathbf{u}}$,

       d) for $\tilde{Y}_{w_{\mathbf{v}} M_{\mathbf{u}}}$ $f_r = \psi_r^{\mathbf{v}}$ and $g_r = \psi_r^{\mathbf{u}}$.

Figure 2.22 shows the modulus and phase of the mobility functions $\tilde{Y}_{v_y F_y}$, $\tilde{Y}_{v_y M_x}$ and $\tilde{Y}_{w_x M_x}$ of the rectangular plate at (x_1, z_1) (thick solid lines) and the transfer mobilities between (x_1, z_1) and (x_2, z_2) (thin solid lines). The moduli of driving-point mobilities $\tilde{Y}_{v_y F_y}$ and $\tilde{Y}_{w_x M_x}$ for the infinite plate are shown by the dashed lines in the plots (a) and (c) respectively. These plots highlight a number of important features. The mobility functions $\tilde{Y}_{v_y F_y}$ and $\tilde{Y}_{w_x M_x}$ show the typical features of driving-point mobilities where the phase lies in the range $\pm 90°$ so that the real parts are positive. In contrast, the phase of the mobility function $\tilde{Y}_{v_y M_x}$ is not restricted to this range. The phase lag increases as the frequency rises so that at 1 kHz the phase lag is about 32 radians. This indicates that, although taken between two collocated positions, $\tilde{Y}_{v_y M_x}$ is not a driving-point mobility function. The functions $\tilde{Y}_{v_y F_y}$ and $\tilde{Y}_{w_x M_x}$ show the typical spectrum of driving-point mobilities, which are characterised by alternating resonances and antiresonances. The dashed lines in the two amplitudes plots (a) and (b) demonstrate that the geometric mean values of the driving-point mobility functions correspond to the equivalent mobility functions for *infinite* plates with magnitudes that, respectively, remain constant or rise in proportion to ω as given by Eqs. (2.68a) and (2.68b). This is an important observation which highlights the importance of the effects of moment excitations and angular vibrations of structures at the higher frequencies involved in vibroacoustic problems. Although, as we shall see in Chapter 3, sound radiation is determined by the transverse vibration of

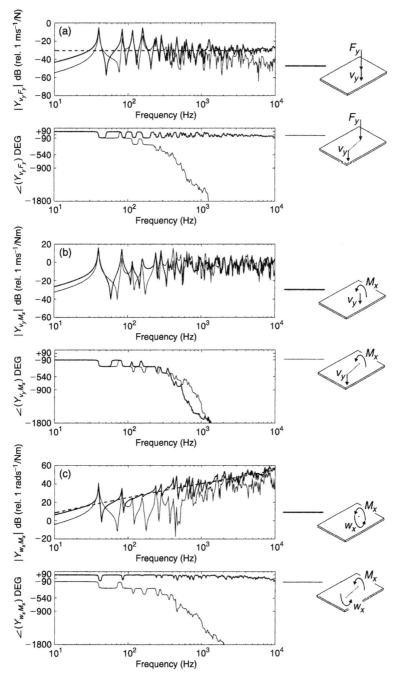

Fig. 2.22. Moduli and phases of the driving point and transfer mobilities: (a) $\widetilde{Y}_{v_y F_y}$; (b) $\widetilde{Y}_{v_y M_x}$ and (c) $\widetilde{Y}_{w_x M_x}$ between the collocated positions $(x_1, z_1) = (0.27 l_x, 0.77 l_z)$ and the non-collocated positions (x_1, y_1) and $(x_2, y_2) = (0.41 l_x, 0.12 l_z)$ of a simply supported aluminium plate with dimensions $l_x \times l_z = 0.414 \times 0.314$ mm and thickness $h = 1$ mm. The moduli of driving-point mobilities $\widetilde{Y}_{v_y F_y}$ and $\widetilde{Y}_{w_x M_x}$ for the infinite plate are given by the dashed lines in the plots (a) and (c) respectively.

the radiating structure, the effect of moment excitation and angular vibration is of great importance in controlling the transmission of structure-borne sound in complex structures made of many components. This phenomenon is further analysed in Section 2.7 where vibration transmission through linear and angular vibrations is evaluated in terms of power flow. The modal overlap of the plate considered in this simulation study becomes greater than unity above about 220 Hz, so that above this frequency the response is no more dominated by distinct resonances. On the contrary, it is characterised by a sequence of smoother and wider crests which are determined by groups of resonant modes as explained in Section 1.12.

2.5   Radial Driving-Point Mobility of Thin-Walled Circular Cylindrical Shells

Circular cylindrical shells that are widely employed in engineering systems may be placed in two general categories. One category contains structures that are characterised by very large ratios of radius to wall thickness: these are typical of aerospace structures such as aircraft fuselages, missile bodies and rocket launchers. Although shells of composite or sandwich construction are becoming more widely used, most current aircraft fuselages have thin, homogenous, aluminium skins. The thinness of the walls necessitates the incorporation of stiffening frames and stringers that substantially affect the forms and natural frequencies of the lower order modes (see ESDU, 1983). Air-conditioning, ventilation and heating ducts of circular cross section that are stiffened by flanged joints may also be considered to fall into this first category. For the sake of brevity we shall refer to shells in this category as 'thin cylinders'. The pressure hulls of submarines may also be modelled over much of their lengths as circular cylindrical shells stiffened by closely spaced frames, but the 'thin shell' assumptions that are employed in all the analyses referred to below is not adequate over the whole frequency range of acoustic interest. The assumptions of thin-shell theory are as follows: (i) the thickness of the shell wall is small compared with the smallest radius of curvature of the shell; (ii) the displacements are small compared with the shell thickness; (iii) the transverse normal stress acting on planes parallel to the shell middle surface is negligible; (iv) fibres of the shell normal to the middle surface remain so after deformation and are themselves not subject to elongation. It is also required that all structural wavelengths are much greater than the shell thickness.

Members of the second category, commonly referred to as 'pipes', form ubiquitous components of industrial plant and fluid transport systems. Although they generally have considerably smaller ratios of radius to wall thickness than 'thin cylinders', many qualify as being 'thin-walled' for the purpose of vibration modelling and analysis. They usually incorporate stiffening elements in the form of connecting flanges; but these are very much further apart in terms of the cylinder radius than those of aerospace and submarine structures. Pipe runs are, in most cases, subject to external constraints such as supports and connections to branches.

This section presents an overview of the radial force driving-point mobility characteristics of uniform, thin-walled, cylinders. The practical importance of this mobility is two-fold: the real part controls the time-average power input to a shell by a point excitation force; and both the real and imaginary parts of the mobility control the interaction between a shell and locally attached 'lumped' mechanical elements.

Ancilliary structures and systems, such as the trim panels and active noise control exciters installed in aircraft are connected to frames or other stiffening elements. Consequently, the radial force driving-point mobility of *uniform*, large diameter, thin-walled cylinders is not of great practical interest. On the other hand, the radial driving-point mobility of pipes is of practical concern for the following reasons. Ancilliary components, such as pressure and temperature transducers that are widely used to monitor the condition of fluids in process plant, are commonly mounted in small housings mounted on the outside surface of the pipes. The inertial mobility of an attachment can 'combine' with the mobility of a pipe to create a resonator, with potentially damaging results. The interaction between vibrational waves propagating in the pipes and locally connected supports is influenced by the pipe mobility. The transverse and rotational mobilities of the beam-bending mode of pipes also influences the vibration isolation effectiveness of resilient elements used to mount them on support structures such as building walls.

Most cylinders of vibroacoustic interest lying in both the above-mentioned categories may be classified as 'thin-walled' shells for which the wall thickness parameter β is very much less than unity. We initially consider the free vibration behaviour of *infinitely long*, uniform cylinders. The $n = 1$ 'beam bending mode' of an infinitely long cylinder is distinguished from all other modes that involve bending strains in that it is characterised by no cross-sectional distortion. At frequencies well below the ring frequency it is governed by the Euler-Bernouilli beam bending equation [Eq. (1.36)]: its radial mobility is given by Eq. (2.18) with $I = \pi a^3 h$, where h is the wall thickness. At frequencies higher than about $\Omega = 0.1$, shear strain becomes significant, and the pipe behaves as 'Timoshenko' beam (Brennan and Variyart, 2003) in which shear strain and rotary inertia are not negligible. This waveguide mode is exhibited by tubes and pipes for all values of β and plays a major role in the radiation of sound by hydraulic and fluid distribution systems at frequencies well below the ring frequency because it has a higher radiation efficiency than the higher order modes and is more readily excited by transverse forces such as those applied by connected machinery and internal flow disturbances.

Since the velocity response of an infinitely extended, uniform structural waveguide to a point force may be expressed as the sum of the responses of the individual waveguide modes, the radial driving-point mobility, is equal to the sum of the mobilities of the modes. Expressions for the driving-point mobility of a cylinder may be derived by introducing a radial, Dirac delta, harmonic force distribution into Eq. (1.48a). By means of spatial Fourier transformation (see Section 3.6), a radial point force can be expressed in terms of a doubly infinite set of axially travelling force wave components of all discrete circumferential orders n and all axial wavenumbers between minus and plus infinity: the (spectral) amplitudes are all equal (see Section 3.8). Heckl (1962b) introduces a force wave of circumferential order n and arbitrary axial wavenumber κ into the shell equations in order to derive an expression for the wave impedance $Z(\kappa, n, \omega)$ of the shell as a function of κ, n and frequency ω (see Section 4.3 for the case of a flat plate). Thus, the radial velocity response of a shell to a component of the point force of particular circumferential and axial wavenumber is obtained by dividing that component by the corresponding component of wave impedance $Z(\kappa, n, \omega)$. The total velocity response of each waveguide mode of order n divided by the corresponding (uniform) force amplitude (i.e., the mobility of that mode) is obtained by integrating the component responses over all axial wavenumbers. By employing certain simplifying assumptions, Heckl produces a number of simple, closed form, approximate expressions for modal mobility, but no explicit expression or approximation for the radial point-force mobility. The following expression for $\Omega \ll 1$ is presented by ESDU (2004):

$$\tilde{Y}(\omega) = 0.306 \frac{\Omega^{\frac{1}{2}}(1 + j)}{\rho h^2 c_l} \tag{2.76}$$

In a more recent, alternative form of analysis, Brennan and Variyart (2003) solved the problem by expressing the sums of the three shell displacements over circumferential order n and branch b, each trio varying with axial distance according to its corresponding free axial wavenumber $k_{n,b}$, for which analytical expressions are derived by Variyart and Brennan (2002). They introduced these into the shell equations in order to satisfy the equations of compatibility and equilibrium of a small element of shell centred on the excitation point. Simple, closed form expressions are derived for the sums of the mobilities of the waveguide modes. Comparisons between mobilities measured on a PVC pipe and calculated values are shown in Fig. 2.23. The mobility of the zero order

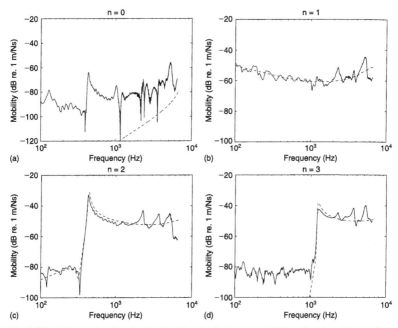

Fig. 2.23.   Measured (—) and calculated (----) radial point mobilities of modes of circumferential order $n = 1$–3 of a PVC tube (Brennan and Variyart, 2003).

(breathing) mode could not be accurately measured because it had a very small radial displacement produced only by the Poisson effect. Modes of order $n > 1$ exhibit peak responses at their respective cut-off frequencies. At any one frequency, the measured total mobility was found to be close to the sum of the modal mobilities.

In practice, all cylinders are of finite length. The terminations of cylinders reflect the incident wave components of freely propagating waveguide modes and, through constructive interference, form natural modes having associated natural frequencies. Clearly, modes of a particular circumferential order can only be formed at frequencies above the cut-off frequencies of the waveguide modes of that order. The return of the reflected waves to a point of excitation influences the driving-point mobility as described qualitatively in Section 1.11. The radial point force mobility can be evaluated in terms of the summation of modal responses as illustrated for a rectangular flat plate in Section 2.4.2. The natural frequencies of a cylindrical shell of length L that is simply supported at its ends can be evaluated from Eqs. (1.49a,b,c) by replacing k_z by $m\pi/L$. The radial mobilities of the modes controlled by flexural strain energy are greater than those of modes controlled by membrane strain energy. Figure 1.35 indicates that when $\Omega < 1$, the density of flexural modes exceeds those of the membrane controlled modes. Hence, the radial point force mobility may be rather accurately evaluated by including only the flexural modes in the model sum, except in cases where the aspect ratio L/a is of the order of unity. It should be noted that the approximate expression for Ω given by Eq. (1.53) does not account correctly for the beam bending mode of a cylinder which is a dominant contributor to the low frequency radial mobility of long pipes.

Equation (1.93) of Section 1.13 presents a relation between the space- and frequency band-averaged real part of the mobility of a uniform structure in a frequency band containing at least five natural frequencies and its local modal density. This relation is not precisely correct for pipe-like structures (Finnveden, 1997), but is adequate for the present purpose. Heckl (1962b) presents approximate formulas for the modal density of cylinders and Langley (1994a) presents a more comprehensive recent analysis. Both authors assume the number of natural frequencies below any frequency (the 'mode count') to be continuous functions of both axial and circumferential wavenumbers, whereas they actually increase discontinuously at each natural frequency. Their resulting estimates of normalised mode count are independent of the shell thickness parameter β.

Finnveden (1997) employs a highly accurate theory of thin cylinder vibration to demonstrate that the normalised mode count is a function of β and deviates significantly from these estimates, particularly at frequencies below $\Omega = 0.3$. In particular, the modal density exhibits peaks immediately above the cut-off frequencies of each waveguide mode; so, also, does the real part of the mobility and the cylinder response to a point force.

As mentioned above, thin-walled cylinders such as aircraft fuselages are stiffened by ring frames and stringers and the driving-point mobility varies with the position of the excitation point in relation to the location of the stiffening elements. However, very little open literature on this aspect of thin cylinder vibration is available. In a rare example, Eichelberger (1981) presents measurements and calculations of the force driving-point mobility of an end-damped, unstiffened, aluminium cylinder of 262 mm radius and 8 mm wall thickness. It exhibited broad peaks immediately above the cut-off frequencies of the shell waveguide modes as explained by Finnveden and illustrated by Fig. 2.24. The frequency-average mobility compared well with a prediction based upon an estimate of the modal density of the shell. Eichelberger also investigated the effect of rectangular cross-section ring stiffeners on the mobility of a damped steel cylinder of 310 mm radius and 5 mm wall thickness. The radial depth and axial width of the rings were 64 and 25 mm, respectively. The radial driving-point mobility in the plane of a ring was calculated and measured. The frequency-average mobility of the stiffened cylinder decreased slightly with frequency, unlike that of unstiffened cylinders that rises to a maximum at the ring frequency. Not surprisingly, the mobility

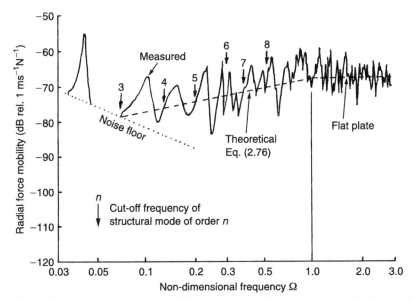

Fig. 2.24. Radial driving-point mobility of an end-damped circular cylindrical shell (Eichelberger, 1981):____ , measured value; - - - theoretical value [Eq. (2.76)];..., noise floor.

curve of the stiffened cylinder in a ring plane was dominated by the resonant responses of the ring modes. Relevant formulae are presented in ESDU (2004).

2.6 Mobility and Impedance Matrix Models

In general, the mobility representation of distributed mechanical elements involves more than one degree of freedom. For example, we have seen that the bending vibration of a beam involves both translational and angular displacements together with associated shear forces and bending moments. Such cases demand the construction of multiple-degrees-of-freedom mobility or impedance models based upon matrix formulation. If we consider a general system which, as shown in Fig. 2.25, is characterised by n degrees of freedom, then a set of $(n \times n)$ mobility functions can be defined as follows:

$$\widetilde{Y}_{ij} = \left. \frac{\tilde{v}_i}{\widetilde{F}_j} \right|_{\tilde{F}_{k \neq j} = 0} \tag{2.77}$$

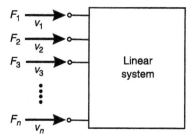

Fig. 2.25.   Linear system with n terminations.

The n velocities and n forces may be grouped into velocity and force vectors $\tilde{\mathbf{v}}$ and $\tilde{\mathbf{f}}$ and the response of the system can be expressed in terms of the mobility matrix

$$\tilde{\mathbf{v}} = \tilde{\mathbf{Y}}\tilde{\mathbf{f}} \tag{2.78}$$

where the $(n \times n)$ elements in the $\tilde{\mathbf{Y}}$ mobility matrix are given by Eq. (2.77). In parallel, a set of $(n \times n)$ impedance functions can also be defined:

$$\tilde{Z}_{ij} = \left.\frac{\tilde{F}_i}{\tilde{v}_j}\right|_{\tilde{v}_{k \neq j}=0} \tag{2.79}$$

so that the response of the system can be expressed in terms of the impedance matrix

$$\tilde{\mathbf{f}} = \tilde{\mathbf{Z}}\tilde{\mathbf{v}} \tag{2.80}$$

where the $(n \times n)$ elements in the $\tilde{\mathbf{Z}}$ impedance matrix are given by Eq. (2.79).

It is most important to note that mobility and impedance matrices are derived by applying different constraints. The mobility functions, are defined in such a way that $\tilde{F}_{k \neq j} = 0$ and thus, apart from the force $\tilde{F}_j$ in the mobility function $\tilde{Y}_{ij} = \tilde{v}_i/\tilde{F}_j\big|_{\tilde{F}_{k \neq j}=0}$, the other forces $\tilde{F}_{k \neq j}$ considered in the mobility matrix model are set to zero. On the other hand, the impedance functions are defined in such a way that $\tilde{v}_{k \neq j} = 0$ and thus, apart from the velocity $\tilde{v}_j$ in the impedance function $\tilde{Z}_{ij} = \tilde{F}_i/\tilde{v}_j\big|_{\tilde{v}_{k \neq j}=0}$, the other velocities $\tilde{v}_{k \neq j}$ considered in the impedance matrix model are set to zero. Therefore, the definitions in Eqs. (2.77) and (2.79) should not be considered merely as mathematical formalities. On the contrary, they highlight features of these functions which have very important implications for the formulation and manipulation of mobility/impedance matrix models. For example, consider the case where the model of a system has to be modified in order to take into account the effects of an additional force excitation. According to the definitions given above, the mobility model could be easily modified by adding another row and column to the existing mobility matrix. This, however, is not the case with the impedance matrix. The addition of a force does not simply imply the addition of a new row and column but also requires a redefinition of all impedance functions, since a new constraint has to be added in the derivation of the impedance function given in Eq. (2.79) (O'Hara, 1966).

Another important feature that needs to be emphasised, concerns the relation between mobility and impedance functions. If we combine Eqs. (2.78) and (2.80) then we get the following relationship

$$\widetilde{\mathbf{Z}} = \widetilde{\mathbf{Y}}^{-1} \tag{2.81}$$

which indicates that, in general, $\widetilde{Z}_{ij} \neq 1/\widetilde{Y}_{ij} : \widetilde{Z}_{ij} = 1/\widetilde{Y}_{ij}$ is only true for single input systems. There are two more important properties of mobility and impedance matrices to be mentioned. The first is that the matrices are symmetric because of the principle of reciprocity (O'Hara, 1966). The second is that the real parts of the mobility or impedance matrices are either positive definite or positive semi-definite. This condition results from passivity since the total vibrational power transmitted by external forces must be positive. As we shall see in the following section, this condition is met when the diagonal elements of the mobility or impedance matrices are positive real or zero; that is, when the matrices are either positive definite or positive semi-definite.

In some cases, mechanical systems are connected in series via single or multiple connecting points. The connection at each point can be through one or more degrees of freedom. An example of series connection is represented by double panel partitions where the structure-borne sound transmission can be modelled in terms of panel–studs–panel systems in series (see Section 5.10). In this case, the mobility and impedance matrices can be formulated in a special manner such that, as shown in Fig. 2.26, the degrees of freedom are partitioned into two groups according to the two terminals of the element they correspond to.

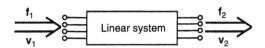

Fig. 2.26.   Linear system with n terminations partitioned according to a series connection.

In this case the classic four pole formulation (Hixson, 1976) can be implemented in which case the mobility and impedance matrix Eqs. (2.78) and (2.80) are written in the following forms:

$$\left\{ \begin{array}{c} \tilde{v}_1 \\ \tilde{v}_2 \end{array} \right\} = \left[\begin{array}{cc} \tilde{Y}_{11} & \tilde{Y}_{12} \\ \tilde{Y}_{21} & \tilde{Y}_{22} \end{array} \right] \left\{ \begin{array}{c} \tilde{f}_1 \\ \tilde{f}_2 \end{array} \right\}, \quad \left\{ \begin{array}{c} \tilde{f}_1 \\ \tilde{f}_2 \end{array} \right\} = \left[\begin{array}{cc} \tilde{Z}_{11} & \tilde{Z}_{12} \\ \tilde{Z}_{21} & \tilde{Z}_{22} \end{array} \right] \left\{ \begin{array}{c} \tilde{v}_1 \\ \tilde{v}_2 \end{array} \right\} \qquad (2.82\text{a,b})$$

where $\tilde{v}_1$, $\tilde{f}_1$ and $\tilde{v}_2$, $\tilde{f}_2$ are the vectors of the complex velocities and forces at terminals 1 and 2 respectively. Four-pole configurations can be better formulated with transfer matrices which link the vectors of the velocities and forces at the two terminals:

$$\left\{ \begin{array}{c} \tilde{v}_2 \\ \tilde{f}_2 \end{array} \right\} = \left[\begin{array}{cc} \tilde{T}_{11} & \tilde{T}_{12} \\ \tilde{T}_{21} & \tilde{T}_{22} \end{array} \right] \left\{ \begin{array}{c} \tilde{v}_1 \\ -\tilde{f}_1 \end{array} \right\} \qquad (2.83)$$

which can be written in compact form as

$$\tilde{a}_2 = \tilde{T}\tilde{a}_1 \qquad (2.84)$$

The elements in the transfer matrix can be derived in terms of the mobility matrices or impedance matrices such that $\tilde{T}_{11} = \tilde{Y}_{22}\tilde{Y}_{12}^{-1}$, $\tilde{T}_{12} = \tilde{Y}_{22}\tilde{Y}_{12}^{-1}\tilde{Y}_{11} - \tilde{Y}_{21}$, $\tilde{T}_{21} = \tilde{Y}_{12}^{-1}$ and $\tilde{T}_{22} = \tilde{Y}_{12}^{-1}\tilde{Y}_{11}$ or $\tilde{T}_{11} = -\tilde{Z}_{12}^{-1}\tilde{Z}_{11}$, $\tilde{T}_{12} = -\tilde{Z}_{12}^{-1}$, $\tilde{T}_{21} = \tilde{Z}_{21} - \tilde{Z}_{22}\tilde{Z}_{12}^{-1}\tilde{Z}_{11}$ and $\tilde{T}_{22} = -\tilde{Z}_{22}\tilde{Z}_{12}^{-1}$. The reason for the negative sign in front of the force vector $\tilde{f}_1$ in Eq. (2.83) is so that structural elements in tandem can be connected together by simple multiplication (Rubin, 1967). The mobility and impedance matrices can be written in terms of the transfer sub-matrices as

$$\left\{ \begin{array}{c} \tilde{v}_1 \\ \tilde{v}_2 \end{array} \right\} = \left[\begin{array}{cc} \tilde{T}_{21}^{-1}\tilde{T}_{22} & \tilde{T}_{21}^{-1} \\ \left(\tilde{T}_{21}^{-1}\right)^T & \tilde{T}_{11}\tilde{T}_{21}^{-1} \end{array} \right] \left\{ \begin{array}{c} \tilde{f}_1 \\ \tilde{f}_2 \end{array} \right\} \quad \text{or} \quad \left\{ \begin{array}{c} \tilde{f}_1 \\ \tilde{f}_2 \end{array} \right\} = \left[\begin{array}{cc} \tilde{T}_{12}^{-1}\tilde{T}_{11} & -\tilde{T}_{12}^{-1} \\ \left(-\tilde{T}_{12}^{-1}\right)^T & \tilde{T}_{22}\tilde{T}_{12}^{-1} \end{array} \right] \left\{ \begin{array}{c} \tilde{v}_1 \\ \tilde{v}_2 \end{array} \right\}$$
$$(2.85\text{a,b})$$

As an example of matrix formulation, we now consider the flexural response of two beams connected by an elastic element which is modelled by axial and angular lumped springs in parallel. As Fig. 2.27 shows, this type of system could be considered as a primitive idealisation of a double panel partition with a connecting frame. In reality, the vibration transmission between the two panels of a double partition occurs via all six degrees of freedom and thus a more complex model should be implemented. However, even with this simplified model, some important features of vibration transmission

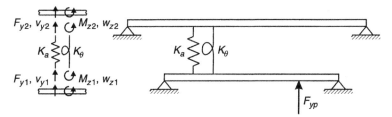

Fig. 2.27. Double beam system connected by combined axial-angular spring element. The two beams are made of aluminium and their cross-sectional dimensions and length are given by $b_1 \times h_1 = 1 \times 3$ mm, $l_1 = 0.4$ m; $b_2 \times h_2 = 1 \times 2.5$ mm, $l_2 = 0.8$ m, respectively. Their loss factor is $\eta = 0.02$. Also the complex linear and angular stiffness of the connecting element are $K_a = 785.4(1 + j0.4)$ N/m and $K_\theta = 490.9(1 + j0.1)$ Nm/rad.

between two structures connected at the point are illustrated. Indeed, the ability to formulate simple models that capture the most important features of a dynamic system is an important skill that both engineers and scientists should develop. To this end, mobility representations are of great assistance because they automatically enforce the subdivision of the system into sub-systems whose response can be represented by analytical expressions, or simulated using Finite Element codes, or measured if a test structure is available.

Returning to the problem at hand, let us assume the lower beam is excited at a point by a transverse harmonic force F_{yp}. In this section we will limit the study to the formulation of the model while in the next the multiple-degrees-of-freedom vibration transmission process will be analysed in detail in terms of structural power. The transverse and angular complex velocities $\tilde{v}_y = j\omega\tilde{\eta}$ and $\tilde{w} = j\omega\tilde{\beta}$ at the connection points 1 and 2 can be expressed in terms of the transverse force $\tilde{F}_y$ and moment $\tilde{M}_y$ at the two points and the external force excitation $\tilde{F}_p$, as

$$\left\{ \begin{array}{c} \tilde{v}_{y1} \\ \tilde{w}_{z1} \end{array} \right\} = \left[\begin{array}{cc} \tilde{Y}'_{v_{y1}F_{y1}} & \tilde{Y}'_{v_{y1}M_{z1}} \\ \tilde{Y}'_{w_{z1}F_{y1}} & \tilde{Y}'_{w_{z1}M_{z1}} \end{array} \right] \left\{ \begin{array}{c} \tilde{F}_{y1} \\ \tilde{M}_{z1} \end{array} \right\} + \left[\begin{array}{c} \tilde{Y}'_{v_{y1}F_{yp}} \\ \tilde{Y}'_{w_{z1}F_{yp}} \end{array} \right] \tilde{F}_{yp} \, ,$$

$$\left\{ \begin{array}{c} \tilde{v}_{y2} \\ \tilde{w}_{z2} \end{array} \right\} = \left[\begin{array}{cc} \tilde{Y}''_{v_{y2}F_{y2}} & \tilde{Y}''_{v_{y2}M_{z2}} \\ \tilde{Y}''_{w_{z2}F_{y2}} & \tilde{Y}''_{w_{z2}M_{z2}} \end{array} \right] \left\{ \begin{array}{c} \tilde{F}_{y2} \\ \tilde{M}_{z2} \end{array} \right\} \qquad (2.86a,b)$$

where $\tilde{Y}'$ and $\tilde{Y}''$ are the mobilities of the beams at the positions 1 and 2 which can be derived using Eqs. (2.56) and (2.62). The forces and moments at the connection points of the two beams can also be expressed by the impedance matrix

$$
\begin{Bmatrix} \tilde{F}_{y1} \\ \tilde{M}_{y1} \\ \tilde{F}_{y2} \\ \tilde{M}_{y2} \end{Bmatrix} = \begin{bmatrix} \tilde{Z}_{F_{y1}v_{y1}} & 0 & \tilde{Z}_{F_{y1}v_{y2}} & 0 \\ 0 & \tilde{Z}_{M_{y1}w_{z1}} & 0 & \tilde{Z}_{M_{y1}w_{z2}} \\ \tilde{Z}_{F_{y2}v_{y1}} & 0 & \tilde{Z}_{F_{y2}v_{y2}} & 0 \\ 0 & \tilde{Z}_{M_{y2}w_{z1}} & 0 & \tilde{Z}_{M_{y2}w_{z2}} \end{bmatrix} \begin{Bmatrix} \tilde{v}_{y1} \\ \tilde{w}_{z1} \\ \tilde{v}_{y2} \\ \tilde{w}_{z2} \end{Bmatrix} \qquad (2.87)
$$

where the impedances of the axial and angular spring elements are given respectively by

$$
\tilde{Z}_{F_{y1}v_{y2}} = \tilde{Z}_{F_{y2}v_{y1}} = -\tilde{Z}_{F_{y1}v_{y1}} = -\tilde{Z}_{F_{y2}v_{y2}} = \frac{K_a}{j\omega} + C_a \qquad (2.88a)
$$

$$
\tilde{Z}_{M_{y1}w_{z2}} = \tilde{Z}_{M_{y2}w_{z1}} = -\tilde{Z}_{M_{y1}w_{z1}} = -\tilde{Z}_{M_{y2}w_{z2}} = \frac{K_\theta}{j\omega} + C_\theta \qquad (2.88b)
$$

where K_a, K_θ are the axial and bending stiffnesses of the coupling and C_a, C_θ are the damping coefficients used to model the dissipative effect in terms of viscous dampers connected in parallel with the associated elastic elements. Equations (2.86a,b) and (2.87) can be written in compact form as follows:

$$
\tilde{v}_1 = \tilde{Y}_{11}\tilde{f}_1 + \tilde{Y}_{1p}\tilde{F}_p, \quad \tilde{v}_2 = \tilde{Y}_{22}\tilde{f}_2, \quad \tilde{f} = \tilde{Z}\tilde{v} \qquad (2.89a,b,c)
$$

Also, Eqs. (2.89a,b) can be grouped together as

$$
\tilde{v} = \tilde{Y}\tilde{f} + \tilde{Y}_p\tilde{F}_{yp} \qquad (2.90)
$$

where

$$
\tilde{Y} = \begin{bmatrix} \tilde{Y}_{11} & 0 \\ 0 & \tilde{Y}_{22} \end{bmatrix}, \quad \tilde{Y}_p = \begin{bmatrix} \tilde{Y}_{1p} \\ 0 \end{bmatrix} \qquad (2.91a,b)
$$

Therefore, imposing conditions of compatibility and dynamic equilibrium at the points where the beams are connected to the elastic joint, it is found that

$$
\tilde{v} = (I - \tilde{Y}\tilde{Z})^{-1} \tilde{Y}_p\tilde{F}_{yp}, \quad \tilde{f} = \tilde{Y}^{-1}\left[(I - \tilde{Y}\tilde{Z})^{-1} - I\right]\tilde{Y}_p\tilde{F}_{yp} \qquad (2.92a,b)
$$

These two expressions give the linear/angular velocities and forces/moments at the connections. If, for example, we wish to evaluate the vibration of the upper beam, the mobility functions for force and moment excitations at position 2, as derived in Eq. (2.92b), can be used directly.

2.7 Structural Power

Vibroacoustic problems of practical concern often involve the excitation of certain components of multi-component systems together with the transmission of the resulting vibration into the other components of the system: a very simple example is presented in Section 2.1. One way to quantify the transmission process is to evaluate the rate of transmission of vibrational energy from one component to another: this quantity is often termed 'structural power flow'. The following abstract example demonstrates how knowledge of the mobilities (or impedances) of individual components that are connected by a simple vibration isolator provides an indication of the influences of the mobilities (or impedances) on the structural power flow from one to another.

Let us assume that component 1 is excited by some external harmonic source of vibration and transmits vibration to component 2 via a rigid point connection, as shown in Fig. 2.28. The interaction between the two components generates a harmonic force of complex amplitude $\widetilde{F}$. This force transmits power into component 2 at a rate $\frac{1}{2}|\widetilde{F}|^2\mathrm{Re}\{\widetilde{Y}_2\}$, as explained in Section 2.1. The problem is how to evaluate the interaction force which depends not only upon the external excitation, but also on the dynamic properties of both components as exhibited by their mobilities (or impedances) as defined in terms of the points of interaction and direction of the interaction force. The interaction force may be determined in terms of either the vibration velocity $\tilde{v}_f$ of component 1 at the connection point with component 2 disconnected (the 'free' velocity) or in terms of the component of the reaction force $\widetilde{F}_b$ at that point when component 1 is assumed to have zero velocity at that point (the 'blocked' force). A vital simplification is that both free velocity and the blocked force are the components of these two quantities in the

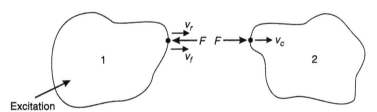

Fig. 2.28. Flexible bodies coupled by a rigid point connection.

direction of the interaction force $\tilde{F}$ in the connected configuration. For example, if the floor-mounted instrument in Fig. 2.1 actually rocked in response to the floor vibration, the interaction force would have a component in both the horizontal and vertical directions. In the following analysis we shall consider the free velocity to be a harmonic predictable or measurable quantity. When component 2 is connected to component 1, the velocity at the connection point is altered by the interaction force which generates a velocity $\tilde{v}_r$ in component 1, given by

$$\tilde{v}_r = -\tilde{F}\tilde{Y}_1 \tag{2.93}$$

where $\tilde{Y}_1$ is the mobility associated with excitation by a force applied in the direction of $\tilde{F}$. The total resulting velocity $\tilde{v}_c$ is given by

$$\tilde{v}_c = \tilde{v}_f + \tilde{v}_r = \tilde{v}_f - \tilde{F}\tilde{Y}_1 \tag{2.94}$$

Since the velocity of component 2 at the connection point is also $\tilde{v}_c$, the associated reaction force is given by

$$\tilde{F} = \tilde{v}_c \tilde{Y}_2 \tag{2.95}$$

From Eqs. (2.94) and (2.95)

$$\frac{\tilde{v}_c}{\tilde{v}_f} = \frac{\tilde{Y}_2}{\tilde{Y}_1 + \tilde{Y}_2} \tag{2.96}$$

Therefore, the time-average transmitted power is given by

$$\overline{P}_{12} = \frac{\frac{1}{2}|\tilde{v}_f|^2 \operatorname{Re}\{\tilde{Y}_2\}}{|\tilde{Y}_1 + \tilde{Y}_2|^2} = \frac{\frac{1}{2}|\tilde{v}_f|^2 \operatorname{Re}\{\tilde{Y}_2\}}{\left[\operatorname{Re}\{\tilde{Y}_1 + \tilde{Y}_2\}\right]^2 + \left[\operatorname{Im}\{\tilde{Y}_1 + \tilde{Y}_2\}\right]^2} \tag{2.97}$$

(Derive the equivalent expression in terms of the blocked force at the connection point). Since the sign of the imaginary part of the mobility of a multi-mode system fluctuates about zero as a function of frequency, the condition $\operatorname{Im}\{\tilde{Y}_1\} = -\operatorname{Im}\{\tilde{Y}_2\}$ can occur at many frequencies when two such systems are connected: the transmitted power peaks at these frequencies. It should be noted that, in a system comprising more than two connected components, the mobilities $\tilde{Y}_1$ and $\tilde{Y}_2$ are influenced by connection of components 1 and 2 to other components.

In many practical vibroacoustic problems, the source(s) of external excitation have continuous frequency spectra that extend over ranges that encompasses the natural frequencies of many high order modes of the system components. In such

cases, it is necessary to estimate this integral of Eq. (2.97) over frequency where
the mean square force velocity is replaced by its autospectrum. As explained
in Section 1.12, it is impossible to estimate the precise frequency dependencies
of mobilities (or impedances) in this 'high frequency' range. Consequently, certain approximations based upon statistical considerations must be applied to the
quantities in Eq. (2.97) in order to estimate frequency-band averages of $\overline{P}_{12}$: an
example is expressed by Eq. (1.92). The details are not presented here for lack
of space and instead readers are referred to Keane and Price (1997).

As a further example of the application of structural power flow analysis, it is
now assumed that components 1 and 2 are connected through an ideal *massless*
resilient isolator as shown in Fig. 2.29. The isolator may be represented in terms
of a differential mobility

$$\tilde{Y}_I = \left[\left(\tilde{v}_f + \tilde{v}_r \right) - \tilde{v}_t \right]/\tilde{F} = \left(\tilde{v}_f - \tilde{F}\tilde{Y}_I - \tilde{v}_t \right)/\tilde{F} \qquad (2.98)$$

where $\tilde{v}_t$ is the velocity of the component 2 at the connecting point and $\tilde{F}$ is the
reaction force caused by the deformation of the isolator. Also

$$\tilde{v}_t = \tilde{F}\tilde{Y}_2 \qquad (2.99)$$

From Eqs. (2.98) and (2.99)

$$\tilde{F} = \tilde{v}_f / \left(\tilde{Y}_I + \tilde{Y}_1 + \tilde{Y}_2 \right) \qquad (2.100)$$

The power transmitted to component 2 is given by

$$\left[\overline{P}_{12} \right]_I = \frac{1}{2} \left| \tilde{F} \right|^2 \operatorname{Re} \left\{ \tilde{Y}_2 \right\} = \frac{\frac{1}{2} \left| \tilde{v}_f \right|^2 \operatorname{Re} \left\{ \tilde{Y}_2 \right\}}{\left| \tilde{Y}_I + \tilde{Y}_1 + \tilde{Y}_2 \right|^2} \qquad (2.101)$$

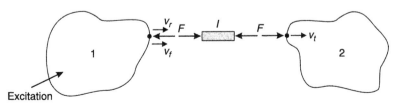

Fig. 2.29.   Flexible bodies coupled by an ideal, massless, resilient connector.

The ratio of power transmitted by the isolator to component 2 to that transmitted by a rigid connection is

$$
\frac{[\overline{P}_{12}]_I}{[\overline{P}_{12}]_R} = \frac{|\tilde{Y}_1 + \tilde{Y}_2|^2}{|\tilde{Y}_I + \tilde{Y}_1 + \tilde{Y}_2|^2} = \frac{\left[\mathrm{Re}\left\{\tilde{Y}_1 + \tilde{Y}_2\right\}\right]^2 + \left[\mathrm{Im}\left\{\tilde{Y}_1 + \tilde{Y}_2\right\}\right]^2}{\left[\mathrm{Re}\left\{\tilde{Y}_I + \tilde{Y}_1 + \tilde{Y}_2\right\}\right]^2 + \left[\mathrm{Im}\left\{\tilde{Y}_I + \tilde{Y}_1 + \tilde{Y}_2\right\}\right]^2}
$$
(2.102)

Although the quantitative value of the ratio depends upon the specific values of the six involved quantities, it is clear that, for reasonable performance, both the real and imaginary parts of $\tilde{Y}_I$ should substantially exceed the corresponding parts of $(\tilde{Y}_1 + \tilde{Y}_2)$, for the isolator to reliably be effective in reducing the transmitted power. This is the reason why it is not easy to design a resilient isolator to reduce power transmission if either component 1 or 2, or both, are lightweight structures, without compromising the isolator's static performance.

In practice, vibration transmission through systems of point-connected structural components involves the whole set of six degrees of freedom at each point junction. For example, the system shown in Fig. 2.27 involves transmission via both transverse and angular motions at the connecting points. This simple example highlights a fundamental problem where multiple degrees of freedom are involved. Although the mobility matrix model derived in the previous section allows one to calculate the fully coupled response at the connecting points of the two beams, it is not obvious how to quantify the vibration transmission from one beam to the other. For instance, there is no direct way to compare the linear and angular amplitudes of the vibration at the junction point of the upper beam: the respective velocities would be expressed in $[ms^{-1}]$ and $[rads^{-1}]$. The two vibration components cannot be simply summed to quantify the total vibration transmission. Therefore, there is a clear need for an alternative approach to the representation of vibration which may be used both to compare and sum transmissions through different degrees of freedom. Note that this problem is increased when a system includes multiple mounts or line-connecting elements such as beams or folded plates.

Both types of problem can be solved by representing vibration transmission in terms of structural power (energy flux) which is a consistent quantity whatever the form of vibration. For example, the vibration transmission to the upper beam of the system in Fig. 2.27 can be quantified in terms of the time-average power transmitted via the transverse and angular vibrations of the beam at the point where the coupling is attached. Using the definition given in Eq. (2.1), these two power contributions can be represented by the following expressions:

$$
\overline{P}_{v_{y2}F_{y2}} = \frac{1}{2}\mathrm{Re}\left\{\tilde{F}_{y2}^{*}\tilde{v}_{y2}\right\}, \quad \overline{P}_{w_{z2}M_{z2}} = \frac{1}{2}\mathrm{Re}\left\{\tilde{M}_{z2}^{*}\tilde{w}_{z2}\right\}
$$
(2.103a,b)

The total vibration transmission can then be expressed as the sum of these two terms.

According to the matrix expression in Eq. (2.86b), both the velocities $\tilde{v}_{y2}$ and $\tilde{w}_{z2}$ are related to both the force and moments at the connection point. As a result the total vibrational power transmitted to the upper beam can be expressed as a sum of four terms:

$$\overline{P}_{inp2} = \frac{1}{2}\text{Re}\left\{\tilde{F}_{y2}^{*}\tilde{Y}_{v_{y2}F_{y2}}''\tilde{F}_{y2}\right\} + \frac{1}{2}\text{Re}\left\{\tilde{F}_{y2}^{*}\tilde{Y}_{v_{y2}M_{z2}}''\tilde{M}_{z2}\right\}$$
$$+ \frac{1}{2}\text{Re}\left\{\tilde{M}_{z2}^{*}\tilde{Y}_{w_{z2}F_{y2}}''\tilde{F}_{y2}\right\} + \frac{1}{2}\text{Re}\left\{\tilde{M}_{z2}^{*}\tilde{Y}_{w_{z2}M_{z2}}''\tilde{M}_{z2}\right\} \quad (2.104)$$

The first and last terms correspond to the power transmissions generated by the transverse force and moment applied by the coupling to the upper beam as if they were acting in isolation. The remaining two terms express their concurrent power transmission. The force contributes to the power transmission partly by means of cooperation with the transverse velocity at the junction generated by the coupling moment at the junction. Similarly, the moment excitation transmits additional power via the angular displacement at the junction point generated by the coupling force at the junction. Equation (2.104) can be written in the following compact matrix form

$$\overline{P}_{inp2} = \left\{\tilde{F}_{y2}^{*}\ \tilde{M}_{z2}^{*}\right\}\text{Re}\left(\begin{bmatrix}\tilde{Y}_{v_{y2}F_{y2}}'' & \tilde{Y}_{v_{y2}M_{z2}}'' \\ \tilde{Y}_{w_{z2}F_{y2}}'' & \tilde{Y}_{w_{z2}M_{z2}}''\end{bmatrix}\right)\left\{\begin{array}{c}\tilde{F}_{y2} \\ \tilde{M}_{z2}\end{array}\right\} = \mathbf{f}_2^H\text{Re}(\tilde{\mathbf{Y}}_{22})\mathbf{f}_2 \quad (2.105)$$

where ^{H} denotes the conjugate transpose. As anticipated in the first part of this chapter, driving-point mobility functions are constrained to be positive real so that the power transmission generated by the coupling force and moment acting in isolation is positive. On the other hand we have also seen that the transfer mobility functions are not bound to be positive real.

These features are illustrated in Fig. 2.30 where the real parts of the two driving-point mobility functions $\tilde{Y}_{v_{y2}F_{y2}}''$ and $\tilde{Y}_{w_{z2}M_{z2}}''$, and the two transfer mobility functions $\tilde{Y}_{v_{y2}M_{z2}}''$ and $\tilde{Y}_{w_{z2}F_{y2}}''$ are plotted with a continuous or dotted line depending whether they are positive or negative.

Similar formulations to that presented above for the upper beam can be used to express the power input to, and output from, the lower beam. For instance, the power input $\overline{P}_{inp1} = \frac{1}{2}\text{Re}\left\{\tilde{F}_{yp}^{*}\tilde{v}_{yp}\right\}$ to the lower beam may be determined by expressing the transverse velocity of the beam $\tilde{v}_{yp}$ in terms of the primary excitation itself and the force and moment generated by the coupling, as derived

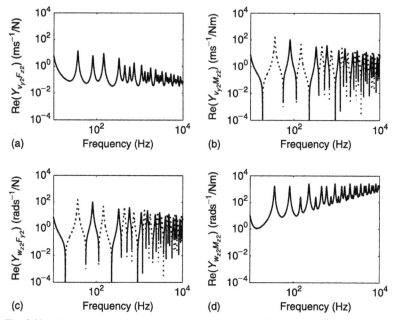

Fig. 2.30. Real part of the driving-point and transfer mobility functions at the connections between the spring and the upper beam. The solid and dotted lines indicate positive and negative values, respectively.

by the fully coupled matrix formulation in Eq. (2.91b), thus

$$\tilde{v}_{yp} = \tilde{Y}'_{v_{yp}F_{yp}}\tilde{F}_{yp} + \tilde{\mathbf{Y}}_{p1}\mathbf{f} = \tilde{Y}'_{v_{yp}F_{yp}}\tilde{F}_p + \tilde{\mathbf{Y}}_{p1}\tilde{\mathbf{Y}}^{-1}\left[\left(\mathbf{I} - \tilde{\mathbf{Y}}\tilde{\mathbf{Z}}\right)^{-1} - \mathbf{I}\right]\tilde{\mathbf{Y}}_p\tilde{F}_p$$

$$(2.106)$$

where $\tilde{Y}'_{v_{yp}F_{yp}}$ is the driving-point mobility function at the excitation position and $\tilde{\mathbf{Y}}_{p1} = \lfloor \tilde{Y}'_{v_{yp}F_{y1}} \; Y'_{v_{yp}M_{z1}} \; 0 \; 0 \rfloor$ is the transfer mobility matrix between the excitation position and the position of the coupling on the lower beam. Therefore, taking into account the fully coupled response of the two beam system, the time-average power input by the excitation force $\tilde{F}_{yp}$ is found to be given by

$$\overline{P}_{inp1} = \tfrac{1}{2}\left|\tilde{F}_p\right|^2 \mathrm{Re}\left\{\tilde{Y}'_{v_{yp}F_{yp}} + \tilde{\mathbf{Y}}_{p1}\tilde{\mathbf{Y}}^{-1}\left[\left(\mathbf{I} - \tilde{\mathbf{Y}}\tilde{\mathbf{Z}}\right)^{-1} - \mathbf{I}\right]\tilde{\mathbf{Y}}_p\right\} \qquad (2.107)$$

This equation illustrates how the vibration generated by the point force $\tilde{F}_{yp}$ does not solely depend on the local mobility of the beam at the excitation; on the contrary it also depends on the coupled response of the two-beam systems via the term $\widetilde{\mathbf{Y}}_{p1}\widetilde{\mathbf{Y}}^{-1}\big[(\mathbf{I} - \widetilde{\mathbf{Y}}\widetilde{\mathbf{Z}})^{-1} - \mathbf{I}\big]\widetilde{\mathbf{Y}}_p$. A fraction of the power input in the lower beam is dissipated by damping in the lower beam while the remaining part is transferred to the coupling element by which a further fraction is dissipated; the remainder is transmitted to the upper beam. The power transmitted from the lower beam into the coupling is given by

$$\overline{P}_{out1} = \left\{ \widetilde{F}^*_{y1} \ \widetilde{M}^*_{z1} \right\} \mathrm{Re}\left(\begin{bmatrix} \widetilde{Y}'_{v_{y1}F_{y1}} & \widetilde{Y}'_{v_{y1}M_{z1}} \\ \widetilde{Y}'_{w_{z1}F_{y1}} & \widetilde{Y}'_{w_{z1}M_{z1}} \end{bmatrix} \right) \left\{ \begin{array}{c} \widetilde{F}_{y1} \\ \widetilde{M}_{z1} \end{array} \right\} = \mathbf{f}^H_1 \mathrm{Re}(\widetilde{\mathbf{Y}}_{11})\mathbf{f}_1$$

(2.108)

The power dissipated by the damping of the coupling element is given by the difference between the power transmitted from the lower beam and the power input to the upper beam:

$$\overline{P}_{diss12} = \overline{P}_{out1} - \overline{P}_{inp2}$$

(2.109)

Figure 2.31 shows the spectra of the power input to the lower beam, the power transmitted by the lower beam to the coupling and the power transmitted to the upper beam. These three curves give considerable insight into the response of the system. For instance, they clearly indicate how the response is controlled by the resonance frequencies of the coupled system. Also, up to about 200 Hz, nearly all the power input to the lower beam is transferred to the upper one: the

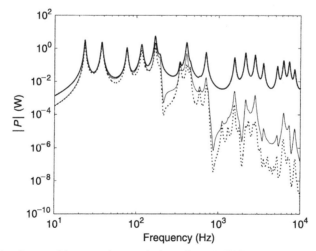

Fig. 2.31. Spectra of the power input to the lower beam (solid line), power transmitted by the lower beam (faint line) and power transmitted to the upper beam (dotted line).

insertion of the coupling is negligible. Above 200 Hz, mobility of the flexible element connecting the two beams is sufficiently high for it to act as a vibration isolator so that the power transmission to the top beam progressively decreases as the frequency rises. The power transmitted to the upper beam is almost equal to the power transmitted by the lower beam to the coupling below 1 KHz, indicating that the power dissipated by the coupling is negligible. At higher frequencies the two curves separate, indicating increased power dissipation by the coupling.

In this simple example, where a single external force is exciting the system, it is relatively straightforward to deduce that the force injects power into the lower beam which is then partially transmitted to the other beam via the coupling. If a more complicated multi-body system which is excited by several external sources is analysed, it is not so easy to intuit in which directions energy will be transmitted between the elements of the system. Indeed, in some cases, some of the external 'excitation' mechanisms could absorb rather than inject power, as in some active control systems. It is therefore important to adopt a consistent sign convention over all the elements of the system for both the linear/angular velocities and force/moment excitations. The injection or absorption of power will then be automatically identified from the sign of the power function. Power is often plotted on logarithmic or dB scales which are specifically suited to represent the responses of resonant systems exhibiting large ranges of response amplitude. Such scales cannot represent negative quantities. A simple solution to this problem is to plot the power function with different types of lines for positive or negative quantities. For example in Fig. 2.32 a solid line has been used to indicate the positive sign while the dotted line indicates the negative sign.

Returning to the two-beam system, a more detailed analysis of the vibration transmission mechanisms through the coupling can be carried out by considering the four power transmission terms as derived in Eq. (2.104). For example, comparing the magnitudes of the four plots in Fig. 2.32, it is clear that the transverse force exerted by the coupling transmits a larger amount of power than the coupling moment. Also, the power transmission due to the concurrent action of the coupling force and moment exhibit both positive and negative values (Why?). It is interesting to note that reciprocity requires that the two power terms related to the concurrent effects of the force and moment excitations are related thus: $\tilde{Y}''_{w_{z2}F_{y2}} = \left(\tilde{Y}''_{v_{y2}M_{z2}}\right)^T$.

The value of the power flow approach to the analysis of the vibration transmission can be better appreciated by comparing the four plots for the mobility functions in Fig. 2.30 and the four plots of the power terms in Fig. 2.32. For example, the plot for the mobility function $\tilde{Y}''_{w_{z2}M_{z2}}$ indicates that the angular vibration generated by unit moment excitation rises monotonically with frequency. However, no further conclusion can be drawn concerning the relative effectiveness of the moment and force excitations. Comparing the plots for $\tilde{Y}''_{v_{y2}F_{y2}}$ and $\tilde{Y}''_{w_{z2}M_{z2}}$, one might be tempted to say that at the higher frequencies the moment excitation

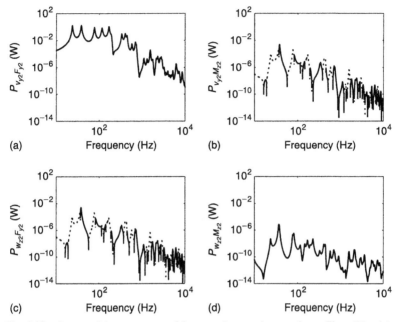

Fig. 2.32.   Spectra of the components of the power input to the upper beam. The solid and dotted lines indicate positive and negative contributions respectively.

generates relatively larger vibration than the force. That this is not the case can be seen by comparing plots (a) and (d) in Fig. 2.32.

Although the analysis of power flow has obvious advantages, its distribution within a network of linked multi-degrees-of-freedom systems is usually very complicated and highly frequency dependent, making it difficult to interpret in terms of the design of vibration control measures. This is even more the case for structural energy flow (so-called 'structural intensity') within a network of distributed elastic systems which is exquisitely sensitive to small perturbations of the system. However, integration of energy flux over frequency bands usually produces much simpler and revealing distributions.

2.8   Energy Density and Energy Flux of Vibrational Waves

In analyses of the vibrational and associated acoustic behaviour of built-up structures such as buildings and vehicles, it is often required to write expressions for the energy densities and energy fluxes of vibrational waves present in the various structural and fluid components. In the formulation of numerically based analysis of vibration and sound, expressions for the kinetic and potential energies of the elements of the media, together with the application of certain fundamental principles of mechanics, yield the governing equations of motion (see Sections 8.2, 8.3 and 8.4). In Statistical Energy Analysis, the total time-average stored energies of the subsystems into which a system is conceptually divided constitute the degrees of freedom of the subsystem (see Section 7.8). The analysis of wave energy flux and energy transfer between coupled elements forms the basis of studies of the behaviour of many types of vibroacoustic system. Chapter 5 treats the subject of the transmission of airborne sound through various forms of partition. Similar forms of analysis may also be applied to vibrational energy transfer across junctions between structural elements (Cremer *et al.*, 1988). This section introduces this subject in terms of analyses of the energy density and energy flux in quasi-longitudinal waves in uniform rods, bending waves in uniform beams and plane bending waves in uniform flat plates. The expressions for arbitrary bending wavefields in plates and for the various types of wave that propagate in fluid-filled circular cylindrical shells are more complex. The latter may be found, for example, in Pavic (1992).

The convention used in the analysis of quasi-longitudinal waves in rods is presented in Fig.1.6. The kinetic energy per unit length of a quasi-longitudinal wave travelling in a rod of cross-sectional area S and density ρ is

$$e'_k = \frac{1}{2}\rho S \, (\partial \xi/\partial t)^2 \qquad (2.110)$$

where ξ is the axial particle displacement. The potential energy per unit length is equal to the work done per unit length by forces applied by contiguous elements in straining the element. According to the convention defined by Fig. 1.7, the work done by the direct stresses on an element of unstrained length δx is

$$\frac{1}{2}S \left\{-\sigma\xi + [\sigma + (\partial\sigma/\partial x)\delta x]\,[\xi + (\partial\xi/\partial x)\delta x]\right\} \qquad (2.111)$$

To second order, this is $\frac{1}{2}S\xi \, (\partial\sigma/\partial x)\,\delta x + \frac{1}{2}S\sigma \, (\partial\xi/\partial x)\,\delta x$. The first term represents the work done in element displacement without strain and does not contribute to elastic potential energy. Therefore

$$e'_p = \frac{1}{2}S\sigma_{xx}\,(\partial\xi/\partial x) = \frac{1}{2}SE(\partial\xi/\partial x)^2 \qquad (2.112)$$

where σ_{xx} is the axial direct stress. The relation between stress gradient and element displacement is given by Newton's second law of motion as

$$S(\partial\sigma_{xx}/\partial x)\,\delta x = -\rho S\delta x\left(\partial^2\xi/\partial t^2\right) \qquad (2.113)$$

The solution of the wave equation for the particle displacement in a wave propagating in the positive-x direction takes the functional form $\xi(x, t) = f(ct - x)$, as with a plane sound wave. The strain is given by $\varepsilon_x = \partial\xi/\partial x$ and the stress therefore is given by $\sigma_{xx} = E\left(\partial\xi/\partial x\right)$. The particle velocity is given by $\partial\xi/\partial t$. Therefore the stress and velocity are related by

$$\sigma_{xx} = E\left(\partial\xi/\partial t\right)/c_l'' = (E\rho)^{1/2}\left(\partial\xi/\partial t\right) \qquad (2.114)$$

The quantity $(E\rho)^{1/2}$ is the characteristic specific mechanical impedance of the wave. In fluids, the bulk modulus $\rho_o c^2$ is equivalent to E and the characteristic specific acoustic impedance equals $(\rho_o^2 c^2)^{1/2}$.

Using Eq. (2.114) to express both forms of energy in terms of particle velocity or stress, we find that they are equal. Therefore the total energy per unit length is given by

$$e' = SE\left(\partial\xi/\partial x\right)^2 = \rho S\left(\partial\xi/\partial t\right)^2 = S\sigma_{xx}^2/E \qquad (2.115)$$

The energy flux per unit cross-section area (intensity) is given by the product of the particle velocity and associated stress. Using Eq. (2.114), this becomes

$$I = \sigma_{xx}^2\left(E\rho\right)^{-1/2} = \sigma_{xx}^2/\rho c_l'' \qquad (2.116)$$

analogous to $p^2/\rho_o c$ for sound waves. The group speed, defined as the ratio of intensity to specific energy density e'/S, equals the phase speed c_l'' because the wave is non-dispersive.

Because bending waves are dispersive, it is very awkward to develop expressions for energy and intensity in the time domain. Consequently, we shall base the following analysis on the general form of the expression for transverse displacement in a harmonic wave travelling in the positive-x direction

$$\eta(x, t) = \tilde{A}\exp\left[j\left(\omega t - k_b x\right)\right] \qquad (2.117)$$

The kinetic energy of transverse motion per unit length is

$$e'_k = \frac{1}{2} m \left(\frac{\partial \eta}{\partial t} \right)^2 = \frac{1}{2} m \left[\mathrm{Re} \left\{ j\omega \tilde{A} \exp \left[j(\omega t - k_b x) \right] \right\} \right]^2$$

$$= \frac{1}{2} m \omega^2 \left[a \sin(\omega t - k_b x) + b \cos(\omega t - k_b x) \right]^2$$

(2.118)

where $\tilde{A} = a + jb$. Note carefully that $(\partial \eta / \partial t)^2 \neq \mathrm{Re} \left\{ \left[j\omega \tilde{A} \exp \left[j(\omega t - k_b x) \right] \right]^2 \right\}$. The kinetic energy density associated with axial particle velocities (or rotational motion) is comparatively negligible in the frequency range where the assumptions of the simple bending theory are valid.

The potential energy has two components. One is associated with moments and rotational displacements of section planes, which involves axial stresses and strains; the other is associated with shear forces and associated shear deformation of elements. The former far outweighs the latter in the frequency range over which simple bending theory applies. The potential energy of a length of bar δx associated with a small rotational displacement under the action bending moments is given by

$$E_p = \frac{1}{2} M \frac{\partial \eta}{\partial x} - \frac{1}{2} \left[M + (\partial M / \partial x) \, \delta x \right] \left[\partial \eta / \partial x + \left(\partial^2 \eta / \partial x^2 \right) \delta x \right] \quad (2.119)$$

To second order $E_p = -\frac{1}{2} \left[M \partial^2 \eta / \partial x^2 + (\partial M / \partial x)(\partial \eta / \partial x) \right] \delta x$. The second term means rotational without strain. Thus the elastic potential energy per unit length is

$$e'_p = -\frac{1}{2} M \frac{\partial^2 \eta}{\partial x^2} = \frac{1}{2} EI \left(\frac{\partial^2 \eta}{\partial x^2} \right)^2 \quad (2.120)$$

Substitution of the expression for the time-harmonic displacement from Eq. (2.117) gives

$$e'_p = EIk_b^4 \left[a \cos(\omega t - k_b x) - b \sin(\omega t - k_b x) \right]^2 \quad (2.121)$$

In freely propagating bending waves, for which $k_b^4 = \omega^2 m / EI$, the sum of the two energies per unit length is independent of space and time and the time-average elastic potential and kinetic energies are equal. The total time-average energy per unit length is

$$\overline{e'} = \frac{1}{2} EIk_b^4 \left(a^2 + b^2 \right) = \frac{1}{2} EIk_b^4 |\tilde{A}|^2 \quad (2.122)$$

The energy flux has two contributions: one from the shear force acting through transverse displacement, and other from the bending moment acting through section rotation. Recalling that the shear force is given by $S = EI \partial^3 \eta / \partial x^3$, the rate at which work is done by the shear force is given by

$$W_s = EI \frac{\partial^3 \eta}{\partial x^3} \frac{\partial \eta}{\partial t} = EI \omega k_b^3 \left[a \sin(\omega t - k_b x) + b \cos(\omega t - k_b x) \right]^2 \quad (2.123)$$

and the rate at which the work is done by the moment is

$$W_m = -EI \left(\partial^2 \eta / \partial x^2 \right) \left[\partial / \partial t \, (-\partial \eta / \partial x) \right]$$

$$= EI \omega k_b^3 \left[a \cos(\omega t - k_b x) + b \sin(\omega t - k_b x) \right]^2 \quad (2.124)$$

The sum of these powers is independent of space and time. The total time-average energy flux of freely propagating bending waves is

$$\overline{W} = EI \omega k_b^3 \left| \widetilde{A} \right|^2 \quad (2.125)$$

The group speed is given by the ratio of time-average energy flux to time-average energy per unit length as $c_{gb} = 2\omega / k_b = 2c_b$.

The general expression of the potential energy and intensity of a bending wavefield in a plate are complicated by the contribution of the twisting moments. However, the expressions for the energies per unit area and energy flux per unit width of *plane* bending waves in a plate are the same as those for a beam with the mass per unit length replaced by the mass per unit area, and the bending stiffness EI replaced by the bending stiffness per unit width $Eh^3 / 12 \left(1 - \nu^2 \right)$. The group speed is again twice the phase speed.

The fact that the time-average kinetic and potential energies per unit area are equal may be exploited in experimental estimates of total energy per unit area. In the cases of travelling bending waves, or reverberant, quasi-diffuse, bending wavefields, the spatial distribution of time-average kinetic energy density may be estimated from measurements of surface vibration using accelerometers or laser systems, and the result is simply doubled. As with sound fields in enclosures, the estimates are not correct near boundaries where evanescent bending fields predominate. This method may also be applied to fields dominated by *resonant* bending wave modes, which behave like simple oscillators in that their time-average kinetic and potential energies are equal.

Direct experimental methods of estimating bending wave energy flux have been developed, but these require measurements of surface motion at multiple points. They are subject to significant errors, especially near boundaries, localised excitation points and other discontinuities, where evanescent fields are present.

Problems

2.1 An object is supported off a floor of uniform thickness by a vibration isolator consisting of an elastic spring and a viscous damper in parallel. A bending wave in the floor is incident upon the base of the isolator. Determine the ratio of the displacement of the mass to that of the bending wave in terms of the point force impedance of the floor and the mass, stiffness and damping coefficient of the isolator, and bending wave frequency.

2.2 A simple harmonic transverse force of frequency independent amplitude F is applied to an effectively infinite uniform beam of bending stiffness EI that is supported at the driving point by a spring of stiffness K. At what frequency will maximum time-average power be transmitted to the beam?

2.3 A semi-infinite uniform steel beam of rectangular cross section (20 mm × 30 mm) is welded end-on to a large, 3 mm thick, heavily damped, flat steel plate. The beam is excited into bending vibration by a harmonic transverse force acting on the wider face at a large distance from the junction. Derive an expression for the ratio of time-average transmission of vibrational energy to the plate to the time-average vibrational power flow in the incident bending at 100 and 1000 Hz. Assume zero in-plane displacement of the plate at the connection. (Sections 2.7 and 2.8 apply.)

2.4 A copper pipe has a mean diameter of 20 mm and a wall thickness of 2 mm. Turbulent water flow generates bending waves in the pipe. It is connected to a 200 mm thick porous concrete wall by a resilient isolator which has a differential mobility $\tilde{Y} = j(\omega/K)(1 - j\eta)$, where $K = 10\mathrm{e}5$ Nm^{-1} and $\eta = 0.1$. Assuming that both pipe and wall point mobilities are equal to those of their infinitely extended equivalents, determine the ratio of vibrational power transmitted to the wall by the isolator to that transmitted by a rigid connection at 500 Hz.

3 | Sound Radiation by Vibrating Structures

3.1 The Importance and Mechanism of Sound Radiation by Vibrating Structures

A large proportion of sources of sound radiate energy through the action of vibrating solid surfaces upon surrounding fluid. Some of these sources, such as pianos, radio loudspeakers and church bells, are generally desirable; many, such as internal combustion engine blocks, punch presses and train wheels, are not. There also exist many sources of sound that do not radiate through the action of vibrating solid surfaces. [Can you think of any?]

The subject of sound radiation from vibrating structures is of great practical importance. It is imperative that designers of loudspeakers understand the mechanism of sound radiation so that they can improve the quality of the product. Designers of industrial machinery must take into account the widespread operation of industrial and community noise limitation regulations and therefore must understand the mechanisms of sound generation that operate in their machines, and also how most effectively and economically to eliminate or suppress them. Acoustic comfort of the passengers, together with minimum noise interference with spoken and audio signals, are major considerations in the design of road, rail, marine and airborne vehicles. Much of the noise in vehicles is due to vibration of the structural envelope. Noise radiated externally by ground-borne vehicles is of major environmental concern. The commercial competitiveness of domestic machinery products such as washing machines, vacuum cleaners and dishwashers is strongly affected by the noise levels that they generate. The designer of military ships and fishing vessels needs to reduce the radiation of sound from hull structures in order to minimise, respectively, the chances of detection or the disturbance of fish. Engineers need to understand the physical mechanisms of sound radiation by vibrating structures in order to specify and implement suitably the cost-effective measures for controlling it.

Structures that radiate sound through vibration are extremely diverse in their geometric forms, material properties and forms of construction: contrast a violin with a marine diesel engine. The process of theoretical estimation of the detailed spectrum and directivity of the sound fields radiated by practical structures is far from simple, for two main reasons: first, it is impossible to calculate the precise distribution of phase and amplitude of the vibration field of a complex structure over the entire audio-frequency range of subjective significance because specification, and therefore mathematical modelling, of the dynamic properties of structures is subject to considerable uncertainty, particularly in respect of structural joint properties, damping distribution and operational and environmental influences; second, even with the availability of sophisticated computational software and powerful hardware, vibroacoustic models demand uneconomically large amounts of preparation and CPU time for analysis at frequencies above a few hundred Hertz. However, in many cases, it is only required to estimate the total sound power radiated by a structure, together with some broad measure of its frequency distribution, such as one-third octave band levels. If so, then analytical methods of evaluation are more easily applied.

Although the mechanism of generation of sound by surface vibration, namely the acceleration of fluid in contact with the surface, is common to all such sources, the effectiveness of radiation in relation to the amplitude of vibration varies widely from source to source. In order for a vibrating surface to radiate sound effectively, it must not only be capable of changing the density of the fluid with which it is in contact, but must do so in such a manner so as to produce significant density changes in the region of fluid remote from the surface.

It is appropriate here to consider the process of sound radiation by vibrating solid surfaces, together with the associated fluid loading, in terms of the molecular nature of fluids. We shall concentrate on the gaseous fluid media whose bulk mechanical properties are determined by the interaction between molecules that are, on average, 'widely' spaced on the scale of their individual size, unlike liquids, in which the molecules are closely spaced and bulk behaviour is partly governed by molecular attraction. However, the conclusions to be drawn from the following discussion are common to both gaseous and liquid fluids. Each gas molecule moves freely in space between 'close encounters' with others. Mutually repulsive forces cause changes of direction of movement, and of molecular momentum, in a manner similar to collisions between balls on a snooker table (but in three dimensions). In volumes of gas that are stationary from a macroscopic point of view, the direction of molecular motion is random; but in the bulk flow of gas, whether steady or oscillatory, there must obviously also exist a mean motion in the direction of flow. The pressure acting on any elementary plane surface lying within, or bounding, a gas is a manifestation of the average rate of change of the component of momentum directed normal to unit area of the surface that is experienced by the molecules in the process of 'collision'. Gas pressure is proportional to the *density* of the kinetic energy of translational motion of the molecules. The gas temperature is proportional to the average translational kinetic energy *per molecule*. Herein lies the physical basis of the Universal Gas Law: $p/\rho = RT$.

Given that local disturbances to the state of a gas can only be propagated at a finite speed via the process of molecular interaction, it is not surprising that the speed of sound propagation in a gas is very close to the average molecular speed in any one direction, as expressed by the aphorism 'the molecule is the messenger'. Hence, the bulk-fluid response to any imposed action, such as the vibration of a solid bounding surface, depends upon frequency. However, as we shall see, it also depends upon the spatial distribution of that vibration.

Consider two small adjacent regions of a plane surface that undergo equal and opposite normal displacements and then halt. It is 'easier' for the molecules in the compressed region of contiguous fluid to move toward the rarified region than toward the, as yet, unaffected fluid a little away from the surface. This bulk movement tends to equalise the pressures and densities local to the surface, producing a much weaker *propagating* disturbance in the surrounding fluid than if the two regions had displaced the fluid in unison. This phenomenon is known as 'radiation cancellation', although it is never complete. If the displacements are now reversed, the molecules will move to re-establish equilibrium.

The more rapidly the reversal takes place, and the greater the distance between the two regions, the less chance there is for the molecules to effect the cancellation process, and the more effectively sound will be radiated. The critical reversal time is given by the distance between the centres of the oppositely displaced regions divided by the speed of sound. Hence, for a given spatial separation of the two regions, such a process that takes place at a high frequency will radiate more effectively than that at a lower frequency. In terms of harmonic vibration and spatially sinusoidal wave motion of a surface, the critical half period is given by half the surface wavelength divided by the speed of sound in the fluid. If the vibrational wave speed is less than that of sound in the fluid, radiation cancellation operates.

In the first part of this chapter, sound radiation by the vibrational modes of baffled, flat plates is analysed using two different formulations: one is based upon a representation in terms of an array of elementary sources of fluid volume acceleration; the other is based upon Fourier decomposition of the surface vibration field in terms of spatially harmonic travelling waves. This is followed by an analysis of radiation by flat plates exerted by localised forces and the influence of non-uniformities such as stiffeners, corrugations and multi-layer composition. The radiation characteristics of circular cylindrical shells are then described and their differences from those of flat plates are explained. The fundamental integral equation relating the surface boundary condition of an arbitrarily shaped body to the associated radiation field is then developed, followed by some practical explanations.

3.2   The Simple Volume Source

Surfaces vibrating in contact with fluids displace fluid volume at the interface. Consequently, it is sensible initially to investigate the sound field generated by the fluid volume displacement produced by a small element of a vibrating surface. By the principle of superposition, one would expect to be able to construct the field by summation of the fields generated by surface elements distributed over the entire surface. Although such an exercise seems simple at first, it is generally not so, because the field generated by any one surface element depends upon the geometry of the whole surface of which it is a part, and upon the presence of any other bodies in the fluid. However, there are many cases of practical importance to which a simple theoretical expression applies with reasonable accuracy.

It is shown in fundamental textbooks on acoustics that the pressure field generated in free field (no reflections) by the uniform, radial, harmonic pulsation of a sphere of equilibrium radius a at frequency ω is

$$p(r,t) = \frac{1}{1+jka} \frac{j\omega\rho_0 \widetilde{Q}}{4\pi r} \exp\{j[\omega t - k(r-a)]\} \qquad (3.1)$$

where r is the radial distance from the centre of the sphere, $\widetilde{Q}$ is the complex amplitude of volume velocity of the source and $j\omega\widetilde{Q}$ is the complex amplitude of volume acceleration. The volume acceleration equals the rate of change of the rate of displacement of fluid volume: if the normal displacement of the surface of the sphere is $\xi = \widetilde{\xi}\exp(j\omega t)$ then $j\omega\widetilde{Q} = -\omega^2 4\pi a^2\widetilde{\xi}$. As indicated previously, the surface acceleration is the agent of sound radiation. The significance of the rate of change of volume flow dQ/dt may be demonstrated by bringing the cupped hands together at different rates: the sound is created when the rate of volume displacement suddenly changes as the hands meet. [Clap your hands close to your lips. You will sense the initial outflow, which produces no audible sound. The sound follows the rapid cessation of flow.]

Figure 3.1(a) illustrates the spherical source. It is clear, on account of symmetry, that any particle velocity vectors must be oriented radially; therefore, assuming an inviscid fluid, the presence of a rigid plane AB in Fig. 3.1(a) in no way alters the sound field. If the source dimension is made very small in comparison with a wavelength, $ka \ll 1$, it is known as a 'point monopole', and Eq. (3.1), reduces to

$$p(r, t) = j\omega\rho_0 \frac{\widetilde{Q}}{4\pi r} \exp[j(\omega t - kr)] \qquad (3.2)$$

in which $\rho_0 d\widetilde{Q}/dt$ (or $j\omega\rho_0\widetilde{Q}$) is known as the 'monopole source strength'. The term $\exp(-jkr)/4\pi r$ is known as the 'free space Green's function': it relates the sound pressure in free field to the harmonic monopole source strength.

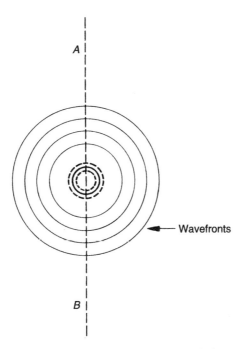

Fig. 3.1. (a) Sound field radiated by a pulsating spherical source.

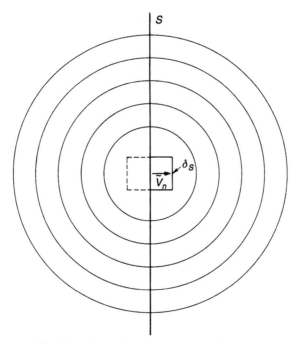

Fig. 3.1. Cont'd (b) Equivalent small baffled piston.

Other forms of Green's function appropriate to particular non-free field conditions will be encountered later in this book. The half-source on one side of the plane AB may represent an elementary volume source operating on an otherwise *infinite, rigid, plane* surface. To an observer confined to one side of the plane, the volume velocity source strength appears to equal $\tilde{Q}/2$; the pressure, however, is still given by Eq. (3.2).

Let the normal surface velocity of a small piston representing an elemental surface area δS of a plane vibrating surface be $v_n(t) = \tilde{v}_n \exp(j\omega t)$ as shown in Fig. 3.1(b); then $\tilde{Q}/2 = \tilde{v}_n \delta S$, and we can rewrite Eq. (3.2) as

$$p(r, t) = j\omega\rho_0 \frac{2\tilde{v}_n \delta S}{4\pi r} \exp[j(\omega t - kr)] \tag{3.3}$$

In writing $\tilde{Q} = 2\tilde{v}_n \delta S$, we have tacitly assumed that the field produced by a small volume velocity source is independent of the detailed form of distribution of velocity over the source surface. This is, in fact, true in the limit of vanishingly small sources, and is true except very close to the source region even for sources of finite extent, provided the typical dimension d of the source region satisfies the condition $kd \ll 1$. Equation (3.3) may be applied to infinitely extended *plane* surfaces by summation or integration over the elementary sources. The resulting integral formulation was derived by Lord Rayleigh (1896):

$$p(\mathbf{r}, t) = \frac{j\omega\rho_0}{2\pi} e^{j\omega t} \int_S \frac{\tilde{v}_n(\mathbf{r}_s)e^{-jkR}}{R} dS \qquad (3.4)$$

where $\mathbf{r}$ is the position vector of the observation point, $\mathbf{r}_s$ is the position vector of the elemental surface δS having normal velocity amplitude $\tilde{v}_n(\mathbf{r}_s)$ and R is the magnitude of the vector $\mathbf{r} - \mathbf{r}_s$: $R = |\mathbf{r} - \mathbf{r}_s|$. In the following sections, Eq. (3.4) is applied to the analysis of radiation from baffled pistons and vibrating plates.

3.3 Sound Radiation by a Pair of Elementary Surface Sources

The time-average sound power radiated per unit area of a vibrating surface is equal to the time average of the product of surface pressure and normal particle velocity. The pressure at any point on a surface receives contributions from acoustic disturbances generated by the motion of all other points on the surface. This phenomenon of 'mutual acoustic interaction' is one of the most important mechanisms governing the effectiveness of sound radiation by vibrating surfaces. The following model and analysis illustrates the geometric and frequency parameters that influence the strength of interaction.

The natural modes of vibration of lightly damped plate and shell structures exhibit spatial distributions of normal displacement in which adjacent regions that are separated by nodal lines vibrate in anti-phase. The simple volumetric source representation of a surface element may be extended to model a pair of small, non-coincident, elements separated by distance d and undergoing harmonic vibration. In the special cases of baffled flat plates and equivalent monopole sources of volumetric velocities of equal magnitude, the complex amplitude of sound pressure induced by one monopole source on the other is given by $\tilde{p}(d) = \pm j\omega\rho_0\tilde{Q}_0 \exp(-jkd)/4\pi d$, where the plus and minus signs relate, respectively, to in-phase (Fig. 3.2) and anti-phase sources.

The sound power generated by each source in working against the induced pressure is given by

$$W_i = \pm\frac{1}{2}\mathrm{Re}\{\tilde{Q}_0\tilde{p}^*(d)\} = \pm\frac{1}{2}\left|\tilde{Q}_0\right|^2 (\omega\rho_0/4\pi d)\sin(kd) \qquad (3.5)$$

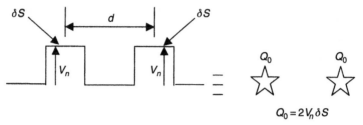

Fig. 3.2. Vibrating surface elements and equivalent monopoles.

The total power radiated by each source is the sum of this term and that produced by the source acting in isolation, which is (Fahy, 2000)

$$W_m = (\rho_0 \omega^2 / 8\pi c) |\widetilde{Q}_0|^2 \tag{3.6}$$

The total power radiated by each source is

$$W = W_m + W_i = W_m[1 \pm \sin(kd)/kd] \tag{3.7}$$

The modifying factors $1 \pm \sin(kd)/kd$ are plotted against non-dimensional frequency (or distance) $kd = 2\pi fd/c = 2\pi d/\lambda$ in Fig. 3.3. The distance dependence is caused by the effect of hemi-spherical spreading on the pressure imposed by one source on another. The frequency dependence arises from the fact that the in-phase (real) component of 'self-pressure' generated by a volume source on itself varies as the square of frequency [see Eq. (4.5)], whereas the pressure imposed on it by another non-coincident source is linearly dependent upon frequency [see Eq. (3.3)]. Hence, the 'self-power' increasingly dominates as frequency rises and each source ultimately radiates independently.

The implication of this result is that the mutual acoustic interaction between regions of surface vibrating in *anti-phase* that are close together in terms of a wavelength ($kd \ll 1$) strongly suppresses the generation of sound power by each region. As mentioned before this phenomenon is known as 'radiation cancellation', although it is never complete. It will feature prominently in subsequent sections concerning sound radiation by vibrating plates.

Although *linear* acoustic variables such as pressure and particle velocity generated by simultaneously operating sources may, according to the principle of linear superposition, be added, the principle does not apply to sound energy and power which have quadratic dependence upon source strength. In addition to affecting sound power, the combined effect of elementary anti-phase sources is to influence the directivity of the resultant sound field via the phenomenon of wave 'interference' which is quite distinct from source 'interaction'.

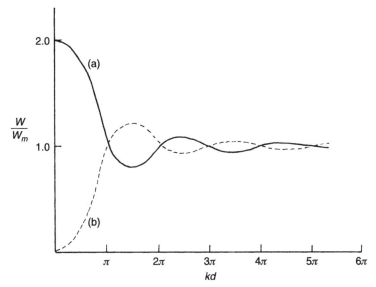

Fig. 3.3. Normalised sound power radiated by two surface elements vibrating: (a) in-phase; (b) in anti-phase.

3.4 The Baffled Piston

The model of a rigid circular disc vibrating transversely to its plane in a coplanar rigid baffle is amenable to mathematical analysis and also constitutes a reasonable representation of a loudspeaker cone in a baffle or cabinet. In addition, this model has application to the design of cylindrical sonar transducers, which are usually mounted on baffles. The model is shown in Fig. 3.4. As with most sound radiation problems, it is difficult to evaluate the field at a distance from the source surface comparable with, or much less than, a typical source dimension. The problem of evaluating the field on the vibrating surface is particularly difficult and will be treated in Chapter 4 on fluid loading of vibrating structures.

The difficulty in evaluating the integral in Eq. (3.4) for distances not greater when compared to the source dimensions is associated with the fact that the distance R between elementary source and the source point is generally a rather complicated function of the coordinates of the two points. However, if the observation point is at a distance greater compared to the source dimensions, R in the denominator of Eq. (3.4) may be approximated by a constant that is an average distance. On the other hand, the phase variation with $\mathbf{r}_s$, given by $\exp(-jkr)$, is a rapidly varying function of R: hence this term must be left within the integral.

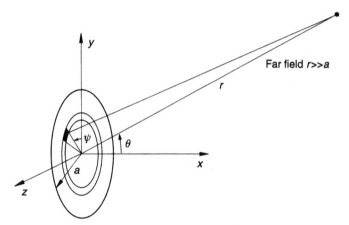

Fig. 3.4.   Piston element and field coordinate system.

In terms of the coordinate system and variables shown in Fig. 3.4, the integral becomes

$$p(r, \theta, \omega) \approx \frac{j\omega\rho_0 \tilde{v}_n e^{j\omega t} e^{-jkr}}{2\pi r} \int_0^a y\,dy \int_0^{2\pi} \exp(-jky \sin\theta \cos\psi)d\psi$$

$$= \frac{j\omega\rho_0 \tilde{v}_n e^{j\omega t} e^{-jkr}}{r} \int_0^a J_0(ky \sin\theta)y\,dy \qquad (3.8)$$

$$= j\rho_0 cka^2 \tilde{v}_n e^{j\omega t} \left[\frac{2J_1(ka \sin\theta)}{ka \sin\theta} \right] \frac{e^{-jkr}}{2r}$$

The functions J_0 and J_1 are Bessel functions of the first kind of the zeroth and first order and are plotted in Fig. 7.15 (Watson, 1966). In the limit $ka \rightarrow 0$, $J_1(ka \sin\theta)/(ka \sin\theta) \rightarrow 1/2$ and the result corresponds to Eq. (3.2), with $\tilde{Q} = 2\pi a^2 \tilde{v}_n$. The term containing the Bessel function is a far-field directivity term, which can be explained physically by interference between the fields radiated by the distributed elementary sources. Briefly, when $ka \ll 1$, $p(r)$ is nearly independent of θ, and the radiation is omnidirectional; when $ka \gg 1$ the sound field is much stronger on, and near, the polar axis, and there is very little radiation in the lateral direction. Details are given in various acoustics books such as Kinsler *et al.* (1999). The theoretical sound intensity fields of a baffled circular piston at low ($ka = 2$) and high ($ka = 25$) frequency are shown in Fig. 3.5 (Fahy, 1995b).

Before the vibrational characteristics of loudspeaker cones had been studied in detail, cone radiation was frequently modelled as piston radiation. At frequencies

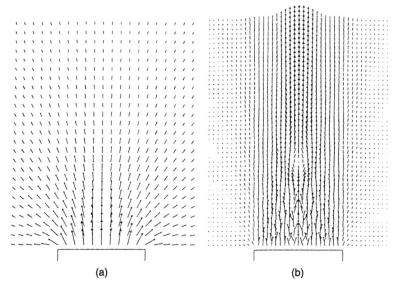

Fig. 3.5. Mean sound intensity field of a vibrating baffled piston of radius a: (a) $ka = 2$; (b) $ka = 25$ (Fahy, 1995b).

well within the audio frequency range, loudspeaker cones exhibit radial wave patterns in which annular rings of alternate phase appear in the outer region of the cone (Fig. 3.6); the number of the rings increases with frequency and consequently, the effective radiating radius decreases with frequency. Hence, loudspeaker directivity does not increase as rapidly with frequency as piston theory would suggest. The fluid loading on a piston due to its motion is studied in Chapter 4.

3.5 Sound Radiation by Flexural Modes of Plates

For the purpose of estimating their sound radiation characteristics, many structures of practical interest may be modelled sufficiently accurately as rectangular, uniform, flat plates. Examples include walls and floors of buildings, factory machinery casings, parts of vehicle shells and hulls and bulkheads of ships. The natural modes of vibration of such plates vary in shape and frequency with their edge conditions and it is not strictly correct to consider the modes of isolated

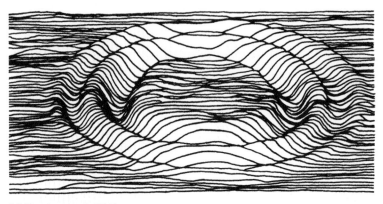

(a) New bass unit: 4 kHz

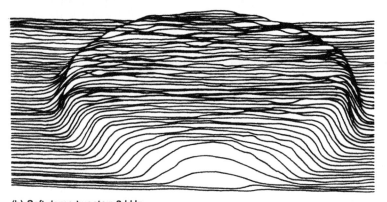

(b) Soft dome tweeter: 8 kHz

Fig. 3.6. Vibrational displacement distribution on loudspeakers (Bank and Hatthaway, 1981). (a) New bass unit, 4 kHz; (b) soft dome tweeter, 8 kHz. Courtesy of Celestion International Ltd.

panels when they are dynamically coupled to contiguous structures, except in a purely mathematical sense as component modes of larger systems. However, the isolated rectangular panel forms a useful starting point for the illustration of flexural-mode radiation behaviour, and the effects of structural complications can often be estimated reasonably accurately once an understanding of the basic physics of modal radiation has been achieved.

Flexural-mode patterns of uniform rectangular panels take the general form of contiguous regions of roughly equal area and shape, which vary alternately in vibrational phase and are separated by nodal lines of zero vibration. For simply supported edges, the normal vibration velocity distribution is

$$\tilde{v}_n(x, z) = \tilde{v}_{pq} \sin(p\pi x/a) \sin(q\pi z/b) \quad \begin{cases} 0 \leq x \leq a \\ 0 \leq z \leq b \end{cases} \quad (3.9)$$

Radiation from lightly damped (damping ratio < 0.01), mechanically excited structures is usually associated with resonant vibration of the modes, i.e., around their natural frequencies. However, we can evaluate the radiation from a modal vibration distribution at *any arbitrary* frequency ω, whether it is vibrating at resonance or not. Equation (3.4) becomes

$$p(x', y', z', t) = \frac{j\omega\rho_0\tilde{v}_{pq}e^{j\omega t}}{2\pi} \int_0^a \int_0^b \frac{\sin(p\pi x/a)\sin(q\pi z/b)e^{-jkR}}{R} dx\,dz$$
$$(3.10)$$

As already stated, this integral does not admit an analytic solution for arbitrary observer points x', y', z'. However, in the far field, where R is much greater than the source size as defined by the larger edge of the two panel dimensions a and b, Wallace (1972) has produced an analytic solution, using the coordinate system shown in Fig. 3.7, and Eq. (3.10) in the form[1]

$$p(r, \theta, \phi, t) = \frac{j\omega\rho_0\tilde{v}_{pq}e^{-jkr}e^{j\omega t}}{2\pi r} \int_0^a \int_0^b \sin\frac{p\pi x}{a} \sin\frac{q\pi z}{b}$$
$$\times \exp\left[j\left(\frac{\alpha x}{a}\right) + j\left(\frac{\beta z}{b}\right) \right] dx\,dz$$
$$(3.11)$$

where

$$\alpha = ka \sin\theta \cos\phi, \quad \beta = kb \sin\theta \sin\phi \qquad (3.12)$$

Comparison of Eqs. (3.10) and (3.11) shows that R, the distance between a surface element at (x, z) and the observation point, is related to r, the distance of the observation point from the coordinate origin, by the approximate relation

$$R \approx r - x \sin\theta \cos\phi - z \sin\theta \sin\phi \qquad (3.13)$$

---

[1] Wallace has exp(jkr) instead of exp($-jkr$) because he uses the time exponent $-j\omega t$.

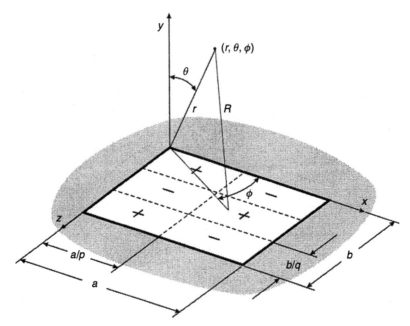

Fig. 3.7. Coordinate system, nodal lines and phases of a vibrating rectangular panel.

Provided that $R \gg a$ and $R \gg b$, this relationship is sufficiently accurate for the purpose of evaluating the integral in Eq. (3.10). The solution of Eq. (3.11) is

$$\tilde{p}(r, \theta, \phi) = j\tilde{v}_{pq}k\rho_0 c \frac{e^{-jkr}}{2\pi r} \frac{ab}{pq\pi^2} \left[\frac{(-1)^p e^{-j\alpha} - 1}{(\alpha/p\pi)^2 - 1} \right] \left[\frac{(-1)^q e^{-j\beta} - 1}{(\beta/q\pi)^2 - 1} \right]$$

$$(3.14)$$

Note that the distance dependence of $\tilde{p}$ is characteristic of the far field: $\tilde{p} \propto (1/r) \exp(-jkr)$.

We are not normally interested in the phase relationships in the far field, but more in the sound intensity and sound pressure level, both of which are functions of mean square pressure $p^2 = |\tilde{p}|^2/2$. The sound intensity is a vector quantity given by the time-averaged product of the sound pressure and the particle velocity vector $\mathbf{v}(r, \theta, \phi)$:

$$\bar{\mathbf{I}}(r, \theta, \phi) = \frac{1}{T} \int_0^T p(r, \theta, \phi, t) \, \mathbf{v}(r, \theta, \phi, t) \, dt \qquad (3.15)$$

where T is a suitable length of time over which to estimate the mean intensity value. In the far field, the component of acoustic particle velocity in phase with the pressure is radially directed. As a result, the sound intensity vector is also radially directed and given by the product of the sound pressure and the radial component of the particle velocity. In the far field, the latter is equal to $\tilde{p}(r, \theta, \phi, \omega)/\rho_0 c$ so that for harmonic vibration

$$\bar{I}(r, \theta, \phi, \omega) = \frac{1}{2}\mathrm{Re}\{\tilde{p}^*(r, \theta, \phi, \omega)\tilde{v}_r(r, \theta, \phi, \omega)\} = \frac{|\tilde{p}(r, 0, \phi, \omega)|^2}{2\rho_0 c} \quad (3.16)$$

Therefore, using Eq. (3.14), the far-field sound intensity generated by the sound radiation of a baffled rectangular panel is given by

$$\bar{I}(r, \theta, \phi) = 2\rho_0 c \left|\tilde{v}_{pq}\right|^2 \left(\frac{kab}{\pi^3 rpq}\right)^2 \left\{\frac{\cos\left(\frac{\alpha}{2}\right)\cos\left(\frac{\beta}{2}\right)}{\sin\left(\frac{\alpha}{2}\right)\sin\left(\frac{\beta}{2}\right)}\left[(\alpha/p\pi)^2-1\right]\left[(\beta/q\pi)^2-1\right]\right\}^2 \quad (3.17)$$

where $\cos(\alpha/2)$ is used when p is an odd integer, and $\sin(\alpha/2)$ is used when p is an even integer; $\cos(\beta/2)$ is used for odd q and $\sin(\beta/2)$ for even q. Equation (3.17) reveals a great deal about the characteristics of panel radiation. The form of the denominator suggests infinite intensity when $\alpha = p\pi$ or $\beta = q\pi$. However, this is not so. [Why?] The equation for the maximum intensity at a given frequency $\omega = kc$ is

$$\bar{I}_{max} = \frac{|\tilde{p}(r, \theta, \phi)|^2_{max}}{2\rho_0 c} = \rho_0 c\frac{|\tilde{v}_{pq}|^2}{2}\left(\frac{kab}{8\pi r}\right)^2 \quad (3.18)$$

where $\alpha = p\pi$ and $\beta = q\pi$, the ϕ direction of maximum radiation being given by $\tan\phi = (\beta/b)(\alpha/a) = (q/b)(p/a)$. This condition cannot occur at all frequencies because the substitution of $\alpha = p\pi$ and $\beta = q\pi$ in Eq. (3.12) gives

$$p\pi/ka = \sin\theta\cos\phi \quad (3.19a)$$

$$q\pi/kb = \sin\theta\sin\phi \quad (3.19b)$$

Since the right-hand sides of Eqs. (3.19) cannot exceed unity, the condition corresponds to $k > p\pi/a$ and $k > qn/b$; hence the frequency must be sufficiently high to make the acoustic wavelength $\lambda = 2\pi/k$ smaller than the traces of the structural wavelength in both the x and y directions, which are $2a/p$ and $2b/q$, respectively.

When the frequency is such that the acoustic wavelength greatly exceeds both the structural trace wavelengths, i.e., $ka \ll p\pi$ and $kb \ll q\pi$, the terms $\alpha/p\pi$

and $\beta/q\pi$ in the denominator of Eqs. (3.14) and (3.17) are small compared with unity, and can be neglected. The maximum value of intensity is then

$$\bar{I}_{max} = \frac{|\tilde{p}(r, \theta, \phi)|^2_{max}}{2\rho_0 c} = 2\rho_0 c \, |\tilde{v}_{pq}|^2 \left(\frac{kab}{\pi^3 rpq}\right)^2 \qquad (3.20)$$

This value is of the order of $(pq)^{-2}$ times that given by Eq. (3.18). It is instructive to compare this intensity with that which would be radiated by a single modal cell of area ab/pq in the absence of any other surface motion. Since the dimensions of the cell have been assumed to be far smaller than an acoustic wavelength, the single cell may be modelled by a point source of volume velocity given by

$$\tilde{Q} = 2\tilde{v}_{pq} \int_0^{a/p} \int_0^{b/q} \sin(p\pi x/a) \sin(q\pi z/b) \, dx dz = \tilde{v}_{pq}(8ab/\pi^2 pq) \quad (3.21)$$

Equation (3.2) applies, and the far-field intensity at all points at a distance r from the source is

$$\bar{I}(r, \theta, \phi) = \frac{|\tilde{p}(r)|^2}{2\rho_0 c} = 2\rho_0 c \, |\tilde{v}_{pq}|^2 \left(\frac{kab}{\pi^3 rpq}\right)^2 \qquad (3.22)$$

By comparing Eqs. (3.22) and (3.20), we see that the intensity that would be produced at low frequencies by one cell equals the maximum intensity generated by all the pq cells acting together. Because the intensity produced by one cell is independent of θ and ϕ, whereas the intensity produced by the complete panel is less in all other directions than the maximum given by Eq. (3.20), the total sound power radiated by a single isolated cell would exceed that radiated by the whole panel.

A more useful measure of the effectiveness of sound radiation by vibrating surfaces than that of the maximum intensity is the total radiated sound power normalised with respect to the specific acoustic impedance of the fluid medium, the panel area and the velocity of surface vibration, which is defined as the 'radiation efficiency', 'radiation ratio' or 'radiation index'.[2] The radiated power can be obtained by integrating the far-field intensity over a hemispherical surface centred on the panel. Since the far-field intensity is directed radially, the time-average power may be written as

$$\overline{P}(\omega) = \int_S \bar{I}(\theta, \phi, \omega) \, dS = \int_0^{\pi/2} \int_0^{2\pi} \bar{I}(\theta, \phi, \omega) r^2 \sin\theta \, d\theta d\phi \qquad (3.23)$$

---

[2] Logarithmic measure (radiation index) $= 10 \log_{10} \sigma$.

A commonly used measure of the surface vibration is the space-average value of the time-average squared vibration velocity defined by

$$\langle \overline{v_n^2} \rangle = \frac{1}{S} \int_S \left[\frac{1}{T} \int_0^T v_n^2(x, z) \, dt \right] dS \tag{3.24}$$

where T is a suitable period of time over which to estimate the mean square velocity $\overline{v_n^2}(x, z)$ at a point (x, z), and S extends over the total vibrating surface: $\langle \overline{v_n^2} \rangle$ is sometimes known as the 'average mean-square velocity'. For the modal distribution of velocity given by Eq. (3.9),

$$\langle \overline{v_n^2} \rangle_{pq} = \frac{|\tilde{v}_{pq}|^2}{8} \tag{3.25}$$

The radiation efficiency ratio is defined by reference to the acoustic power radiated by a uniformly vibrating baffled piston at a frequency for which the piston circumference greatly exceeds the acoustic wavelength: $ka \gg 1$. Use of Eq. (3.8) to evaluate the far-field intensity of a piston, together with integration over a hemisphere, yields the following expression for radiated power in this case:

$$\overline{P} = \frac{1}{2} \rho_0 c \pi a^2 |\tilde{v}_n|^2 = \rho_0 c S \langle \overline{v_n^2} \rangle \tag{3.26}$$

The definition of radiation efficiency σ is thus

$$\sigma = \overline{P} / \rho_0 c S \langle \overline{v_n^2} \rangle \tag{3.27}$$

The radiation efficiency of the p, q-th mode of a panel of area S is given by

$$\sigma_{pq} = \frac{\overline{P}_{pq}}{\rho_0 c S \langle \overline{v_n^2} \rangle_{pq}} \tag{3.28}$$

There is no physical reason why the radiation efficiency of a vibrating surface should not exceed unity, and therefore the term 'radiation efficiency' is rather misleading; although, in most practical cases, the radiation efficiency is either below or very close to unity. For this reason, it is also a common practice to express sound radiation in terms of the 'radiation resistance' which for the p, q-th vibration mode is defined as

$$R_{pq} = \frac{\overline{P}_{pq}}{\langle \overline{v_n^2} \rangle_{pq}} = \rho_0 c S \sigma_{pq} \tag{3.29}$$

The time-average total vibrational energy of a mode or set of modes in temporally stationary resonant or broadband vibration is equal to twice the time-average kinetic energy. For a panel, this gives

$$\overline{E} = \int_S m(x, z)\overline{v_n^2(x, z)}\, dxdz \tag{3.30}$$

where m is the mass per unit area. A radiation loss factor η_{rad} can be defined as

$$\eta_{rad} = \overline{P}/\omega\overline{E} \tag{3.31}$$

For a panel of uniform density ρ_s and thickness h executing such vibration, $\overline{E} = \rho_s h S \langle \overline{v_n^2} \rangle$. Hence, in such cases,

$$\eta_{rad} = (\rho_0/\rho_s)(1/kh)\sigma \tag{3.32}$$

The radiation loss factor indicates the magnitude of acoustic radiation damping; it rarely exceeds 10^{-3} for engineering structures vibrating in air, except in the cases of lightweight, stiff constructions such as the sandwich panels used in aerospace structures. In liquids, it can exceed the loss factors associated with structural energy dissipation mechanisms and can control the radiation process (see Section 4.8).

Wallace presents the following approximate expressions for the radiation efficiency of rectangular panels at arbitrary frequencies for which the acoustic wavelength is very much greater than either of the panel trace wavelength components $\lambda_x = 2a/p \ll 2\pi/k$, $\lambda_z = 2b/q \ll 2\pi/k$:

(a) p and q both odd integers,

$$\sigma_{pq} \approx \frac{32(ka)(kb)}{p^2 q^2 \pi^5} \left\{ 1 - \frac{k^2 ab}{12} \left[\left(1 - \frac{8}{(p\pi)^2} \right) \frac{a}{b} + \left(1 - \frac{8}{(q\pi)^2} \right) \frac{b}{a} \right] \right\} \tag{3.33a}$$

(b) p odd and q even,

$$\sigma_{pq} \approx \frac{8(ka)(kb)^3}{3p^2 q^2 \pi^5} \left\{ 1 - \frac{k^2 ab}{20} \left[\left(1 - \frac{8}{(p\pi)^2} \right) \frac{a}{b} + \left(1 - \frac{24}{(q\pi)^2} \right) \frac{b}{a} \right] \right\} \tag{3.33b}$$

(Interchanging p and q gives the result for p even and q odd.)

(c) p and q both even integers,

$$\sigma_{pq} \approx \frac{2(ka)^3(kb)^3}{15p^2q^2\pi^5} \left\{ 1 - \frac{5k^2ab}{64} \left[\left(1 - \frac{24}{(p\pi)^2} \right) \frac{a}{b} + \left(1 - \frac{24}{(q\pi)^2} \right) \frac{b}{a} \right] \right\}$$

(3.33c)

Clearly the odd–odd modes are the most effective radiators when ka, $kb \ll 1$; a good approximation to Eq. (3.33a) is then

$$\sigma_{pq} \approx \frac{2k^2}{\pi^5 ab} \left(\frac{2a}{p} \right)^2 \left(\frac{2b}{q} \right)^2 = \frac{2k^2 \lambda_x^2 \lambda_z^2}{\pi^5 ab}$$

(3.34)

where λ_x and λ_z are the structural trace wavelengths in x and z directions. A radiation efficiency based upon the total panel area, but upon the power radiated by only one cell vibrating in isolation, is obtained by integrating Eq. (3.22) over the surface of a large hemisphere of radius r.

$$\sigma_1 \approx \frac{32(ka)(kb)}{\pi^5 p^2 q^2} = \frac{2k^2 \lambda_x^2 \lambda_z^2}{\pi^5 ab}$$

(3.35)

We see that the radiation efficiency of all panel modes at low frequencies is at most equal to, and mostly less than, that corresponding to isolated cell vibration and that, for the most efficient odd–odd modes with the same wavelengths, it is inversely proportional to the area of the panel.

Consider two rectangular panels of different size but of the same aspect ratio, material and thickness. Modes of the two having equal structural trace wavelengths will have the same natural frequencies. Hence, according to Eq. (3.34), the radiation efficiencies of these two panels *at this frequency* will be different, in inverse proportion to their areas: the total power radiated by vibration of the same level in both is the same. We may ask why it is that the panel of larger area does not radiate more power. One answer lies in the far-field form of the Rayleigh integral [Eq. (3.4)] in which R in the denominator may be taken outside the integral:

$$p(\mathbf{r}, t) \approx \frac{j\omega\rho_0 e^{j\omega t}}{2\pi R_0} \int_S \tilde{v}_n(\mathbf{r}_s) e^{-jkR} dS$$

(3.36)

where R_0 is the average distance from the observation point to the panel. For a mode of order (p, q), the maximum distance between the centres of adjacent regions of volume velocity of opposite sign is $l = a/p$ or b/q, whichever is the greater. At the observation point the contributions from these regions differ in phase by $\delta\phi = kl\sin\theta$, where θ is measured from the normal to the baffle plane; the maximum value of $\delta\phi$ is kl at positions lying on the plane. If $kl \ll \pi$ or $l \ll \lambda/2$, the phase difference due to distance is negligible compared with the difference of vibrational phase between adjacent cells: hence the contributions to the field from adjacent cells almost cancel.

We may visualise the cancellation process by reference to Fig. 3.8, which illustrates the one-dimensional case. Since the radiation fields must be either phase-symmetric or phase-antisymmetric about the panel centre, for odd and even modes, respectively, cancellation must involve half-cell pairs, and not whole-cell pairs; it is seen that the half cells at the edges of the panel remain. If, in addition to kl being small, ka is also small, the out-of-phase end half cells of the even mode will also largely cancel, whereas those of the odd mode will add. This reasoning can be extended to the real two-dimensional case to provide a physical understanding of the form of the low-frequency radiation expressions in Eqs. (3.33a–c). In particular, we can see that the four uncancelled corner quarter cells may add their contributions (p, q odd; $ka, kb \ll 1$) or almost cancel (p, q even; $ka, kb \ll 1$). As frequency increases, the reinforcement or cancellation decreases in effect.

An alternative physical picture of the behaviour can be obtained by utilising the principle of acoustic reciprocity for point monopole sources. Under all time-invariant states of a non-flowing fluid and of the dynamic properties of its boundaries, the positions of a point monopole source and an observer can be reversed without altering the observed sound pressure (Fahy, 1995a). This is true for a point source on the surface of an infinite rigid plane that can be deduced from the expression for the field of a point source in Eq. (3.2), by considering

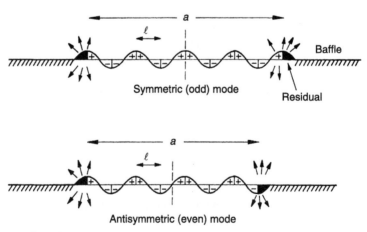

Fig. 3.8.   Intercell cancellation on vibrating plates: $kl \ll 1$ (Fahy, 1987).

a point source at any arbitrary position, together with its image in an infinite rigid plane: the introduction of the image allows the plane to be removed. The total pressure at the position of the plane is clearly equal to twice that in the absence of the image; similarly the pressure produced by an elemental volume source operating at the surface of a rigid plane is double that produced by the same source in the absence of the plane. Now, as we have seen, each element $dxdz$ of the integrand of Eq. (3.10) behaves like a point source on a rigid plane. Hence the variation of the exponential term $\exp(-jkR)$ in the integrand with the variation of source point can as well be envisaged as the variation with surface position, of the pressure phase on the surface of the panel produced by a point source operating at the observation point. When the observation point is very far from the panel surface, the curvature of the wavefronts produced by such a source at their intersection with the panel surface is negligible, and they can be considered to be plane (Fig. 3.9). The degree of matching of the vibration velocity and the exponential term can thus be visualised. The reason for the directional maximum radiation characteristics indicated below Eq. (3.18) can now be seen.

It would appear from the construction of Fig. 3.9 that matching to a given mode would not be poor only at low frequencies, where the acoustic half-wavelength becomes large compared with nodal line separation distances in both the x and

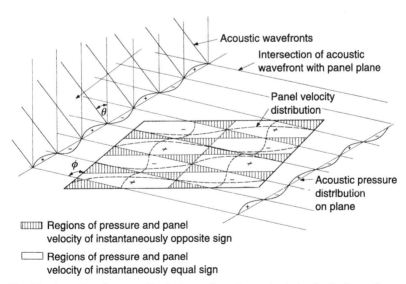

IIIIIII Regions of pressure and panel
velocity of instantaneously opposite sign

⬚ Regions of pressure and panel
velocity of instantaneously equal sign

Fig. 3.9.   Matching of pressure distributions and panel normal velocity distribution on the panel plane (panel instantaneously in its equilibrium position).

z directions, but also at high frequencies, where the inverse is true. However, it must be remembered that the distance between wavefront–plane intersections corresponding to intervals of half an acoustic wavelength increases with elevation of the observer above the source plane; some elevation angle giving optimum matching can therefore always be found below infinite frequency. This fact explains why high-frequency modal radiation is directional not only in azimuth (ϕ) but also in elevation (θ), the direction of maximum radiation moving toward the normal with increasing frequency.

Graphical presentations of radiation efficiency data for rectangular simply supported panels have been published by Wallace (1972): some representative results are shown in Fig. 3.10. It should be carefully noted that the abscissa expresses the value of $\gamma = k/k_b$, the ratio of acoustic to free bending wavenumber. For a simply supported, uniform, rectangular panel the bending wavenumber

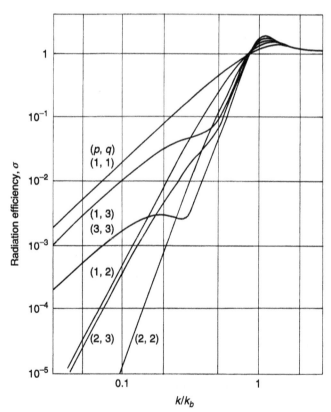

Fig. 3.10. Radiation efficiency of the low order modes of a baffle square panel (Wallace, 1972).

is given by $k_b^2 = (p\pi/a)^2 + (q\pi/b)^2$ [see Eqs. (1.40) and (1.41)]. *It is not explicitly a frequency axis, and the curves do not assume vibration of modes at their natural frequencies.* The radiation efficiency of all modes becomes asymptotic to unity when γ exceeds unity. In general, for $\gamma < 1$, the radiation efficiencies of the modes for which both p and q are odd are the highest, that of the fundamental mode (1, 1) being dominant. [Why?] Those of the even–even modes are the lowest, as we have already seen. For a given mode order (p, q) and γ, modes having square internodal areas radiate less well, and modes having internodal areas of aspect ratio aq/bp very different from unity are the most effective radiators.

Although simply supported boundary conditions are idealised and not representative of practical structures, analysis by Berry *et al.* (1990) shows that clamping the edges of a panel makes little difference to the modal radiation efficiency, slightly decreasing that of the low order modes and increasing that of higher order modes by a maximum factor of 2.5 (4 dB). This conclusion agrees with that of Lomas and Hayek (1977).

The forementioned results in no way apply to unbaffled or perforated panels. In the former case, at low frequencies, the surface velocity sources on the 'underside' of the panel, especially those near the edge, largely cancel the field produced by the 'upper side' surface sources of opposite phase, because the distances to the far field are insignificantly different in terms of a wavelength, especially at positions near the plane of the panel. In the language of idealised acoustic sources, the corresponding regions of opposite phase on the two sides of the panel constitute dipoles that are very much less efficient than pure volume velocity (monopole) sources; hence the low-frequency radiation efficiency is far less than the baffled panel equivalent. This cancellation effect decreases as frequency increases and the acoustic wavenumber approaches the structural wavenumber. An iterative, FFT-based, analysis of unbaffled plate radiation is presented by Williams (1983b), but the method of Laulagnet (1998) is more effective. Oppenheimer and Dubowsky (1997) compare the results of their semi-empirical model with measured data.

[Try holding a thin panel of plywood, or similar material, about 600 mm × 600 mm in size, parallel to a wall or table top, and tap it while moving it closer to the surface. Can you think of any explanation for the observed effect, perhaps in terms of sources and images?]

In the case of a perforated panel, the effect of pressure differences across the panel, due to opposite phase surface sources, is to drive fluid through the holes, thereby partially cancelling the surface volume velocities and hence the radiation. However, the radiation efficiency of a perforated plate is not zero, and it depends upon the geometry of the perforations and the thickness of the plate. The relevant perforation/thickness parameter for regularly spaced, circular holes is $h = (kS/2a)[1 + (2t/\pi a)]$, where S is the plate area per hole, a is the hole radius and t is the plate thickness (Fahy and Thompson, 2004).

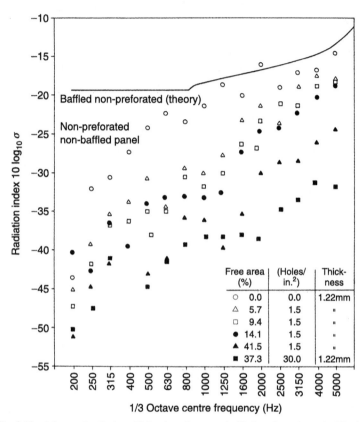

Fig. 3.11. Measured radiation efficiencies of some unbaffled perforated panels (Pierri, 1977).

The factor in square brackets is only applicable in cases where $t \gg a$. As a general guide, if $h > 100$, the perforation has a negligible effect on radiation; if $h < 1$, radiation efficiency is reduced by a factor of at least five. Some measured data are presented in Fig. 3.11. [Why is volume velocity cancellation not complete? Consider the inertia of the fluid in and near the holes.]

So far we have only briefly mentioned the *natural* frequencies of the baffled rectangular panel studied above. This is because, it is possible to analyse radiation from a given mode at *arbitrary* frequencies to produce data in the form presented by Wallace. In most practical cases of vibration of lightly damped structures, the kinetic energy of vibration is associated mainly with resonant modal vibration, which occurs at frequencies close to the *in-vacuo* modal natural frequencies. However, this condition is not sufficient for us to assume that it is only the radiation of a mode at its natural frequency that is important, because the radiation efficiency of thin panel modes is often rather low and rises with frequency, as Eqs. (3.33a–c) show. Hence, there can be a trade-off between reduction of vibration response above the resonance frequency, and increase of efficiency. In most cases of *mechanical* excitation, the reduction in response wins, and only resonant radiation is important. As we shall see in Chapter 5, this is usually not so in the case *of airborne* sound transmission through panels. Further examples of radiation efficiency of plates, based upon resonant mode vibration only, will be presented later in this chapter.

3.6 Sound Radiation by Plates in Multi-Mode Flexural Vibration

There are two methods of computing the vibration field of a structure subject to specified inputs: (i) modal summation and (ii) direct algorithms. This section presents analysis of sound radiation by baffled plates both in terms of individual modal contributions and their interactive effects, and in terms of a non-modal, discretised surface approach to the evaluation of the Rayleigh integral.

3.6.1 Formulation in Terms of Structural Modes

In Chapter 2, we have seen that the response of structures to a time-harmonic excitation can be modelled in terms of a modal expansion such that at each frequency the response is determined by a summation of modal terms. The transverse velocity of a plate may be expressed by

$$\tilde{v}_n(x, z, \omega) = \sum_{p=1}^{P} \sum_{q=1}^{Q} \phi_{pq}(x, z) \tilde{v}_{pq}(\omega) \tag{3.37}$$

where $\tilde{v}_{pq}$ represents complex modal velocity amplitude and $\phi_{pq}(x, z)$ represents normalised mode shape. Equation (3.37) can be conveniently written in matrix form as the scalar product of a line vector of the mode shapes at position (x, z), $\lfloor \varphi(x, z) \rfloor = \lfloor \phi_{11}(x, z) \; \phi_{12}(x, z) \; ... \; \phi_{PQ}(x, z) \rfloor$, and a column vector with the complex modal velocities $\{\tilde{v}\} = \lfloor \tilde{v}_{11} \; \tilde{v}_{12} \; \cdots \; \tilde{v}_{PQ} \rfloor^T$, thus:

$$\tilde{v}_n(x, z) = \lfloor \varphi(x, z) \rfloor \{\tilde{v}\} \tag{3.38}$$

In the previous section, the time-averaged total sound power radiation of a baffled plate was formulated in terms of the far-field sound intensity over a hemisphere surface positioned with reference to the centre of the panel. In the alternative approach, the time-averaged sound power radiation can be evaluated by integrating the product of the surface sound pressure, $p(x, 0, z, t)$, and the transverse velocity of the panel, $v(x, z, t)$ over the surface of a panel. For harmonic vibration,

$$\overline{P}(\omega) = \frac{1}{2} \int_0^a \int_0^b \text{Re} \left\{ \tilde{v}_n(x, z, \omega)^* \tilde{p}(x, 0, z, \omega) \right\} dx\,dz \qquad (3.39)$$

where $*$ denotes the complex conjugate. As we have seen in Section 3.2, the surface acoustic pressure $\tilde{p}(x, 0, z, \omega)$ can be written in terms of the normal surface velocity through the Rayleigh integral as

$$\tilde{p}(x, 0, z, \omega) = \frac{j\omega\rho_0}{2\pi} \int_0^a \int_0^b \tilde{v}_n(x', z', \omega) \frac{e^{-jkR}}{R} dx'\,dz' \qquad (3.40)$$

where, $R = \sqrt{(x - x')^2 + (z - z')^2}$ is the distance between the point (x, z) where the sound pressure is estimated and the vibrating surface element at (x', z'). Substituting Eq. (3.40) into Eq. (3.39), the time-average total sound power is given by the quadruple integral

$$\overline{P}(\omega) = \frac{1}{2} \text{Re} \left\{ \frac{j\omega\rho_0}{2\pi} \int_0^a \int_0^b \int_0^a \int_0^b \tilde{v}_n(x, z, \omega)^* \tilde{v}_n(x', z', \omega) \frac{e^{-jkR}}{R} dx'\,dz'\,dx\,dz \right\}$$
$$(3.41)$$

which, using the vector expression for $\tilde{v}_n(x, z)$ given in Eq. (3.38), becomes

$$\overline{P}(\omega) = \frac{1}{2} \text{Re} \left\{ \frac{j\omega\rho_0}{2\pi} \int_0^a \int_0^b \int_0^a \int_0^b \{\tilde{v}\}^H \lfloor \varphi(x, z) \rfloor^T \right.$$
$$\left. \times \frac{e^{-jkR}}{R} \lfloor \varphi(x', z') \rfloor \{\tilde{v}\}\,dx'\,dz'\,dx\,dz \right\} \qquad (3.42)$$

where H denotes the Hermitian transpose (transpose and conjugate). Since $je^{-jkR}/R = j(\cos kR - j\sin kR)/R$, and because $\{\tilde{v}\}^H \{\tilde{v}\}$ is bound to be real positive, this expression can be rewritten as

$$\overline{P}(\omega) = \frac{\omega \rho_0}{4\pi} \{\tilde{\mathbf{v}}\}^H \left(\int_0^a \int_0^b \int_0^a \int_0^b \lfloor \varphi(x,z) \rfloor^T \frac{\sin kR}{R} \lfloor \varphi(x',z') \rfloor dx' dz' dx dz \right) \{\tilde{\mathbf{v}}\}$$

(3.43)

or, alternatively, in the matrix form

$$\overline{P}(\omega) = \{\tilde{\mathbf{v}}(\omega)\}^H [\mathbf{A}(\omega)] \{\tilde{\mathbf{v}}(\omega)\}$$

(3.44)

where $[\mathbf{A}]$ is an $(n \times n)$ matrix with $n = p \cdot q$, which is normally referred to as the 'power transfer matrix' (Snyder and Tanaka, 1995), given by

$$[\mathbf{A}(\omega)] = \frac{\omega \rho_0}{4\pi} \int_0^a \int_0^b \int_0^a \int_0^b \lfloor \varphi(x,z) \rfloor^T \frac{\sin kR}{R} \lfloor \varphi(x',z') \rfloor dx' dz' dx dz$$

(3.45)

The term $A_{\alpha\beta}$, where $\alpha = p, q$, and $\beta = p', q'$ can be interpreted as the contribution to sound power radiated by mode α due to the vibration of mode β. Alternatively, since the matrix $[\mathbf{A}(\omega)]$ is symmetric, it can also be considered as the contribution to the sound power radiated by mode β due to the vibration of mode α. According to Eq. (3.45), the total sound power radiated by a baffled plate is governed by the self-radiation effects of the natural modes of the plate, which are determined by the diagonal terms in the power transfer matrix $[\mathbf{A}(\omega)]$, together with the modifications of the sound radiation of each mode by acoustic interaction with other modes, which are determined by the off-diagonal terms in the matrix $[\mathbf{A}(\omega)]$.

The specific problem of radiation by a uniform, rectangular, baffled plate, has been analysed by Snyder and Tanaka (1995) who have also derived a simple formula for the off-diagonal terms $A_{\alpha\beta}$ in terms of the radiation efficiencies σ_α and σ_β of the corresponding natural modes:

$$A_{\alpha\beta}(\omega) = \frac{\rho_0 c S}{64} \left(1 + (-1)^{p_\alpha + p_\beta}\right) \left(1 + (-1)^{q_\alpha + q_\beta}\right) \left\{ \frac{p_\alpha q_\alpha}{p_\beta q_\beta} \sigma_{p_\alpha q_\alpha} + \frac{p_\beta q_\beta}{p_\alpha q_\alpha} \sigma_{p_\beta q_\beta} \right\}$$

(3.46)

In order to obtain an expression equivalent to Eq. (3.44), the vector expression for the transverse velocity of the plate given in Eq. (3.38) is substituted into the Rayleigh integral for the far-field sound pressure given by Eq. (3.4). As a result, the far-field sound pressure is given by the scalar product of a line vector with modal radiation functions $\lfloor \mathbf{z}(r, \theta, \phi) \rfloor = \lfloor z_{11}(r, \theta, \phi) \quad z_{12}(r, \theta, \phi) \dots z_{PQ}(r, \theta, \phi) \rfloor$ and the column vector with the complex modal velocities $\{\tilde{\mathbf{v}}\}$ that is

$$\tilde{p}(r, \theta, \phi) = \lfloor \mathbf{z}(r, \theta, \phi) \rfloor \{\tilde{\mathbf{v}}\} \tag{3.47}$$

where, according to Eqs. (3.10) and (3.14), the modal radiation functions are given by

$$
\begin{aligned}
z_{pq}(r, \theta, \phi) &= \frac{j\omega\rho_0}{2\pi} \int_0^a \int_0^b \phi_{pq}(x, z) \frac{e^{-jkR}}{R} dx dz \\
&= jk\rho_0 c \frac{e^{-jkr}}{2\pi r} \frac{ab}{pq\pi^2} \left[\frac{(-1)^p e^{-j\alpha} - 1}{(\alpha/p\pi)^2 - 1} \right] \left[\frac{(-1)^q e^{-j\beta} - 1}{(\beta/q\pi)^2 - 1} \right]
\end{aligned} \tag{3.48}
$$

where from Eq. (3.12), $\alpha = ka \sin\theta \cos\phi$ and $\beta = kb \sin\theta \sin\phi$. Applying Eqs. (3.16) and (3.23), the radiated sound power is given by

$$\bar{P}(\omega) = \{\tilde{\mathbf{v}}\}^H \int_0^{\pi/2} \int_0^{2\pi} \frac{\lfloor \mathbf{z}(r, \theta, \phi) \rfloor^H \lfloor \mathbf{z}(r, \theta, \phi) \rfloor}{2\rho_0 c} r^2 \sin\theta \, d\theta d\phi \{\tilde{\mathbf{v}}\} \tag{3.49}$$

which can be expressed in compact matrix form as in Eq. (3.44) with the elements in the power transfer matrix $[\mathbf{A}(\omega)]$ given by

$$A_{\alpha\beta}(\omega) = \int_0^{\pi/2} \int_0^{2\pi} \frac{z_{pq}(r, \theta, \phi)^* z_{rs}(r, \theta, \phi)}{2\rho_0 c} r^2 \sin\theta \, d\theta d\phi \tag{3.50}$$

Recent research has produced accurate expressions for the elements in the power transfer matrix $[\mathbf{A}(\omega)]$ that can be considered valid even for frequencies where the acoustic wavelength is much smaller than either of the plate modal wavelengths. For example, Figs. 3.12 and 3.13, respectively show the diagonal and off-diagonal elements of the 'power transfer matrix' for a simply supported, rectangular, aluminium plate of dimensions 0.414 m × 0.314 m and thickness 2 mm as calculated with the analytical expressions derived by Li (2001). Both logarithmic and linear scales have been used to better highlight their main features and differences between the diagonal and off-diagonal elements of the matrix. Since the diagonal terms of $[\mathbf{A}(\omega)]$ are proportional to the radiation efficiencies of the modes of the plate, the plot in logarithmic scale in Fig. 3.12 is similar to that shown in Fig. 3.10 for the modal

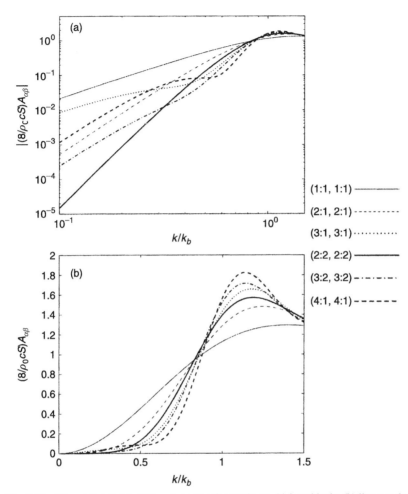

Fig. 3.12. Diagonal elements of the power transfer matrix on: (a) logarithmic; (b) linear scales.

radiation efficiencies. The plot in linear scale in Fig. 3.12 emphasises the relatively small amplitudes of the diagonal terms for $k/k_b < 0.7$ which, as seen above, are due to the modal intercell cancellation effects. The plot in logarithmic scale in Fig. 3.13 shows that the amplitudes of the off-diagonal terms of $[\mathbf{A}(\omega)]$ are much smaller than those of the diagonal terms. More importantly, the plot in linear scale in Fig. 3.13 shows that the off-diagonal terms oscillate between the positive and negative values which indicates that the mutual radiation effect

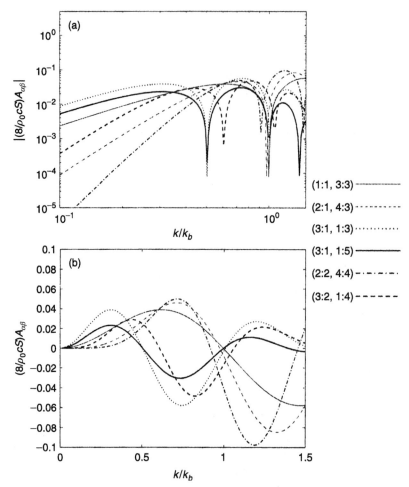

Fig. 3.13. Off-diagonal elements of the power transfer matrix on: (a) logarithmic; (b) linear scales.

between the pairs of modes may contribute either constructively or destructively to the radiated sound power.

Since the panel is the sole acoustic source, the total sound power radiated is constrained to be positive. Therefore, on the basis of Eq. (3.42), the power transfer matrix $[\mathbf{A}(\omega)]$ is constrained to be positive definite. This is an important property that we shall see in the following section that enables the definition of a special set of modes of the panel which radiate sound power independently.

3.6.2   *Formulation in Terms of Elementary Radiators*

Although the formulation presented above seems to be quite simple and neat, the derivation of the quadruple integrals for the elements in the power transfer matrix [**A**] is relatively complex and involved. An alternative non-modal approach based on the elementary radiators is therefore presented as introduced by Mollo and Bernhard (1989), Vitiello *et al.* (1989) and Cunefare and Koopmann (1991).

As shown in Fig. 3.14, the baffled panel is divided into a grid of R rectangular elements whose transverse vibrations are specified in terms of the velocities v_{er} at their centre positions so that, assuming time-harmonic motion, the overall vibration of the panel can be described by the following column vector of complex amplitudes:

$$\{\tilde{\mathbf{v}}_e\} = \lfloor \tilde{v}_{e1} \ \tilde{v}_{e2} \ \cdots \ \tilde{v}_{eR} \rfloor^T \tag{3.51}$$

If the amplitudes of the sound pressures acting on each element are also grouped into a column vector, as

$$\{\tilde{\mathbf{p}}_e\} = \lfloor \tilde{p}_{e1} \ \tilde{p}_{e2} \ \cdots \ \tilde{p}_{eR} \rfloor^T \tag{3.52}$$

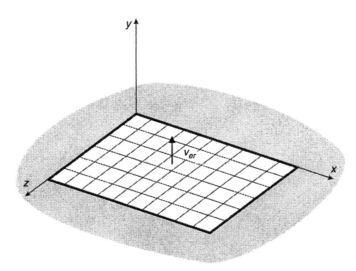

Fig. 3.14.   Subdivision of a panel into elemental radiators.

Assuming that the dimensions of the element are small compared with both the structural wavelength and the acoustic wavelength, the total radiated sound power can then be expressed as the summation of the powers radiated by each element expressed as $\overline{P}_{er} = 1/2A_e \mathrm{Re}\{\tilde{v}_{er}^* \tilde{p}_{er}\}$, so that

$$\overline{P}(\omega) = \sum_{r=1}^{R} \frac{1}{2} A_e \mathrm{Re}\{\tilde{v}_{er}^* \tilde{p}_{er}\} = \frac{S}{2R} \mathrm{Re}\{\{\tilde{\mathbf{v}}_e\}^H \{\tilde{\mathbf{p}}_e\}\} \qquad (3.53)$$

where A_e and S are respectively the areas of each element and of the whole panel. The pressure on each element is generated by the vibrations of all elements of the panel. Assuming that $\sqrt{A_e} \ll \lambda$, where λ is the acoustic wavelength, Eq. (3.4) gives

$$\tilde{p}_{ei}(x_i, z_i) = \frac{j\omega\rho_0 A_e e^{-jkR_{ij}}}{2\pi R_{ij}} \tilde{v}_{ej}(x_j, z_j) \qquad (3.54)$$

with R_{ij} the distance between the centres of the i-th and j-th elements. The vector of sound pressures can therefore be expressed by the impedance matrix relation

$$\{\tilde{\mathbf{p}}_e\} = [\tilde{\mathbf{Z}}] \{\tilde{\mathbf{v}}_e\} \qquad (3.55)$$

where $[\tilde{\mathbf{Z}}]$ is the matrix incorporating the point and transfer acoustic impedance terms over the grid of elements into which the panel has been subdivided: $\tilde{Z}_{ij}(\omega) = (j\omega\rho_0 A_e/2\pi R_{ij}) e^{-jkR_{ij}}$. Note that, because of reciprocity, the impedance matrix $[\tilde{\mathbf{Z}}]$ is symmetric, in which case $[\tilde{\mathbf{Z}}] = [\tilde{\mathbf{Z}}]^T$. Substituting Eq. (3.55) into the expression for the total radiated sound power given in Eq. (3.53), we obtain

$$\overline{P}(\omega) = \frac{S}{2R} \mathrm{Re}\{\{\tilde{\mathbf{v}}_e\}^H [\tilde{\mathbf{Z}}] \{\tilde{\mathbf{v}}_e\}\} \qquad (3.56)$$

which, because $[\tilde{\mathbf{Z}}]$ is symmetric, can be rewritten as

$$\overline{P}(\omega) = \frac{S}{4R} \{\tilde{\mathbf{v}}_e\}^H \left([\tilde{\mathbf{Z}}] + [\tilde{\mathbf{Z}}]^H\right) \{\tilde{\mathbf{v}}_e\} = \{\tilde{\mathbf{v}}_e\}^H [\mathbf{R}] \{\tilde{\mathbf{v}}_e\} \qquad (3.57)$$

The matrix $[\mathbf{R}]$ is defined as the 'radiation resistance matrix' for the elementary radiators which, for the baffled panel, is given by

$$[\mathbf{R}] = \frac{S}{2R}\,\mathrm{Re}[\widetilde{\mathbf{Z}}] = \frac{S}{4R}\left([\widetilde{\mathbf{Z}}] + [\widetilde{\mathbf{Z}}]^{H}\right)$$

$$= \frac{\omega^2 \rho_0 A_e^2}{4\pi c}
\begin{bmatrix}
1 & \frac{\sin(kR_{12})}{kR_{12}} & \cdots & \frac{\sin(kR_{1R})}{kR_{1R}} \\
\frac{\sin(kR_{21})}{kR_{21}} & 1 & & \\
\cdots & & \cdots & \cdots \\
\frac{\sin(kR_{R1})}{kR_{R1}} & & & 1
\end{bmatrix} \tag{3.58}$$

As found for the power transfer matrix $[\mathbf{A}]$, the radiation resistance matrix is also positive definite since the quadratic expression for the power radiation in Eq. (3.57) is bound to be positive. Vitiello *et al.* (1989) have compared modal power radiation predicted using the modal and the elementary radiator models. They found that the agreement between the two models is generally good provided that the dimensions of the discrete elements are both much less than the acoustic wavelength and the modal wavelengths in the panel, i.e., much greater than the Cartesian components of the plate modal wavenumber vectors. In general, the modal approach can be used for regular shapes and common boundary conditions such that analytical expressions can be derived for the mode shapes. In the elementary radiator model, the velocities of the elementary radiators can be derived using Finite Element analysis (see Chapter 8) which enables the study of non-uniform, baffled plates of complex geometry having non-classical boundary conditions which can result from interfaces with other flexible structures. With this approach, the shape of the panel is therefore not a major issue provided, it can be mapped into a mesh of regular quadrilateral elements with small distortion. An even more accurate formulation has been proposed by Naghshineh *et al.* (1992) in which the vibratory effects of both linear and angular velocities of the elements are considered in order to better model the sound radiation contributions by 'weakly radiating' velocity distributions.

According to Eq. (3.38), the complex velocities in Eq. (3.51) can be expressed by vector products between the row vectors of the modal amplitudes at the centres of the surface elements $\lfloor \varphi_j(x_j, z_j) \rfloor = \lfloor \phi_{11}(x_j, z_j)\ \phi_{12}(x_j, z_j) \ldots \phi_{PQ}(x_j, z_j) \rfloor$ and the column vector of the modal velocity terms $\{\tilde{\mathbf{v}}\} = \lfloor \tilde{v}_{11}\ \tilde{v}_{12}\ \ldots\ \tilde{v}_{PQ} \rfloor^{T}$. By grouping the row vectors $\lfloor \varphi_j(x_j, z_j) \rfloor$ into a matrix

$$[\boldsymbol{\psi}] =
\begin{bmatrix}
\varphi_1(x_1, z_1) \\
\varphi_2(x_2, z_2) \\
\vdots \\
\varphi_R(x_R, z_R)
\end{bmatrix} \tag{3.59}$$

the vector of velocities of the elements defined in Eq. (3.51) can be expressed as

$$\{\tilde{v}_e\} = [\psi]\{\tilde{v}\} \tag{3.60}$$

The expression for the total radiated sound power derived assuming elementary radiators can therefore be converted into the modal formulation by substituting Eq. (3.60) into Eq. (3.57) so that

$$\overline{P}(\omega) = \{\tilde{v}\}^H [\psi]^H [R] [\psi] \{\tilde{v}\} \tag{3.61}$$

Equation (3.61) reveals a direct relation between the power transfer matrix [A] and the radiation resistance matrix [R]; that is

$$[A] = [\psi]^H [R] [\psi] \tag{3.62}$$

We have seen that the elements of the radiation resistance matrix [R] can be calculated more easily than those of the power transfer matrix [A], particularly when complex shapes or non-ideal boundary conditions are involved. Equation (3.62) then provides a convenient means for calculation of the diagonal and off-diagonal terms in [A] for any form of baffled flat plate.

3.7 Independent Radiation Modes

The two formulations presented in the previous section for the derivation of the total radiated sound power are based on the vibrational field of a plate which is represented either by an array of modal velocities or by an array of velocities specified at the centres of a grid of elements. As indicated by Eqs. (3.44) and (3.57), both formulations involve fully populated square matrices which, as previously discussed, represent the mutual radiation effects between pairs of modes or between pairs of elementary radiators. Except for the cases where a very large frequency band or very large thin plates are considered, the dimensions of both the power transfer matrix [A] and the radiation resistance matrix [R] are relatively small and the numerical derivation of the total sound power radiation is not a problem with the power of digital computation currently available. However, these two formulations do not expose the radiation mechanisms in a simple and intuitive manner that can help engineers or scientists to understand and control the radiation of a given system. Therefore, it is desirable to express the sound radiation in terms of an alternative set of modes that radiate independently. This section shows how this set can be derived on the basis of the modal and the element radiator formulations as explained by Elliott and Johnson (1993).

3.7.1 Formulation in Terms of Structural Modes

In previous section, we have seen that the power transfer matrix $[\mathbf{A}]$ is normal: that is, it is real, symmetric and positive definite. Therefore it has an eigenvalue/eigenvector decomposition which can be written as

$$[\mathbf{A}] = [\mathbf{P}]^T [\mathbf{\Omega}] [\mathbf{P}] \qquad (3.63)$$

where $[\mathbf{P}]$ is the matrix of orthogonal eigenvectors and $[\mathbf{\Omega}]$ is the diagonal matrix whose elements Ω_n are the eigenvalues of $[\mathbf{A}]$. Since the matrix $[\mathbf{A}]$ is positive definite, its eigenvalues are all real and positive. Using the eigenvalue/eigenvector decomposition, the expression for the total radiated sound power given in Eq. (3.44) becomes

$$\overline{P}(\omega) = \{\tilde{\mathbf{v}}\}^H [\mathbf{A}] \{\tilde{\mathbf{v}}\} = \{\tilde{\mathbf{v}}\}^H [\mathbf{P}]^T [\mathbf{\Omega}] [\mathbf{P}] \{\tilde{\mathbf{v}}\} \qquad (3.64)$$

The product between the matrix $[\mathbf{P}]$ of the eigenvectors and the vector of the modal velocities $\{\tilde{\mathbf{v}}\}$ gives the vector

$$\{\tilde{\mathbf{b}}\} = [\mathbf{P}] \{\tilde{\mathbf{v}}\} \qquad (3.65)$$

of velocities of transformed 'radiation modes' which radiate sound power independently since

$$\overline{P}(\omega) = \{\tilde{\mathbf{b}}\}^H [\mathbf{\Omega}] \{\tilde{\mathbf{b}}\} = \sum_{n=1}^{N} \Omega_n |b_n|^2 \qquad (3.66)$$

The independent sound powers radiated by this new set of modes are characterised by self-radiation efficiencies which are proportional to the eigenvalues Ω_n. The spatial distribution of the n-th radiation mode is given by the sum of the natural modes of the panel weighted by the elements in the corresponding row of the matrix $[\mathbf{P}]$:

$$\eta_n(x, z) = \lfloor \phi_{11}(x, z) \ \phi_{12}(x, z) \ \dots \ \phi_{PQ}(x, z) \rfloor \begin{bmatrix} P_{n1} \\ P_{n2} \\ \vdots \\ P_{nN} \end{bmatrix} \qquad (3.67)$$

where $N = P \times Q$.

It is important to emphasise that, in general, both matrices $[\mathbf{P}]$ and $[\mathbf{\Omega}]$ depend on the frequency of vibration of the plate and on the shapes of its natural modes: therefore, the radiation modes are *frequency dependent*.

3.7.2   Formulation in Terms of Elementary Radiators

An formulation equivalent to that presented above for the radiation modes can also be derived starting from the radiation resistance matrix. We have seen that the matrix $[\mathbf{R}]$ is real and symmetric and thus it also has an eigenvalue/eigenvector decomposition such that

$$[\mathbf{R}] = [\mathbf{Q}]^T [\mathbf{\Lambda}] [\mathbf{Q}] \tag{3.68}$$

where $[\mathbf{Q}]$ is a matrix of orthogonal eigenvectors and $[\mathbf{\Lambda}]$ is a diagonal matrix of real and positive eigenvalues λ_i, since, according to Eq. (3.57), $[\mathbf{R}]$ is positive definite. In this case, substitution of the eigenvalue/eigenvector decomposition into Eq. (3.57) gives the following expression for the total radiated sound power:

$$\overline{P}(\omega) = \{\tilde{\mathbf{v}}_e\}^H [\mathbf{R}] \{\tilde{\mathbf{v}}_e\} = \{\tilde{\mathbf{v}}_e\}^H [\mathbf{Q}]^T [\mathbf{\Lambda}] [\mathbf{Q}] \{\tilde{\mathbf{v}}_e\} \tag{3.69}$$

Isolating the product between the eigenvector matrix $[\mathbf{Q}]$ and elemental radiator velocity vector $\{\tilde{\mathbf{v}}_e\}$ gives the vector

$$\{\tilde{\mathbf{y}}\} = [\mathbf{Q}] \{\tilde{\mathbf{v}}_e\} \tag{3.70}$$

of 'radiation modes' in terms of the velocities of the individual radiators. These radiate sound independently since Eq. (3.69) becomes

$$\overline{P}(\omega) = \{\tilde{\mathbf{y}}\}^H [\mathbf{\Lambda}] \{\tilde{\mathbf{y}}\} = \sum_{r=1}^{R} \lambda_r |\tilde{y}_r|^2 \tag{3.71}$$

where R represents the number of elements. The independent sound radiation generated by this new set of modes is characterised by self-radiation efficiencies, which in this case are proportional to the eigenvalues λ_r, and spatial distributions of the radiation mode which are given directly by the rows of the matrix $[\mathbf{Q}]$. Thus, the amplitudes of the radiation modes are defined directly by the velocity distributions over the mesh of elements into which the panel has been subdivided, without reference to the structural modes of the panel.

It is relatively simple to demonstrate the equivalence between the modal and element formulations for the radiation modes. In fact, according to Eq. (3.60), the velocities of the elements can be expressed in terms of the modal velocities via the matrix $[\psi]$ with the modal amplitudes at the centres of the elements, i.e., $\{\tilde{v}_e\} = [\psi]\{\tilde{v}\}$, and thus, as shown by Eq. (3.62), the power transfer matrix is linked to the radiation resistance matrix by the relation $[A] = [\psi]^H [R][\psi]$. Thus, using the eigenvalue/eigenvector decomposition in Eq. (3.68) for the radiation resistance matrix, $[R] = [Q]^T [\Lambda][Q]$, the power transfer matrix can be expressed as

$$[A] = [\psi]^H [Q]^T [\Lambda] [Q] [\psi] \qquad (3.72)$$

which is equal to the expansion given in Eq. (3.63) for the power transfer matrix, $[A] = [P]^T [\Omega][P]$. It should be noted that the two expansions refer to the modal and element bases so that the matrices $[\Omega]$ and $[\Lambda]$ have dimensions $R \times R$ and $N \times N$. Thus only a subset of the elements in $[Q][\psi]$ and $[\Lambda]$ correspond to the elements of $[P]$ and $[\Omega]$.

Works presented by Borgiotti (1990), Baumann *et al.* (1991), Cunefare (1991), Naghshineh *et al.* (1992) and Naghshineh and Koopmann (1993) were instrumental in the development of the radiation mode model.

3.7.3 *Radiation Modes and Efficiencies*

In order to better illustrate the physics behind this alternative formulation for the sound radiation, the shapes of the first six radiation modes and their corresponding eigenvalues λ_r, which are proportional to the radiation efficiencies, have been calculated and plotted in Figs. 3.15 and 3.16 for the simply supported rectangular plate considered in Section 3.6.1. To facilitate their interpretation, the first six natural modes of the panel have also been plotted in Fig. 3.16.

Figure 3.15 indicates that the eigenvalues λ_r vary rather smoothly with frequency and that their magnitude is smaller than unity for $k/k_b \ll 1$, while it gradually approaches and then exceeds unity as k approaches and then exceeds k_b. The radiation modes are frequency dependent and are plotted at two frequencies of $f_1 = 250\,\text{Hz}$ and $f_2 = 800\,\text{Hz}$. Figure 3.16 clearly shows that the first, and most efficient mode, corresponds to a piston-like mode. At low frequency the velocity distribution is nearly uniform over the surface of the panel; as frequency increases the piston-like shape is gradually distorted toward a dome-shape. The following two modes are rocking-type modes oriented along the two axes of the panel. At low frequencies the first three mode shapes are therefore described by linear functions. The next three modes are instead described by quadratic functions which, for example, assume a saddle or double saddle type shape as shown in the bottom two plots in Fig. 3.16. The following sequence of radiation modes is characterised by higher order functions. The formulation

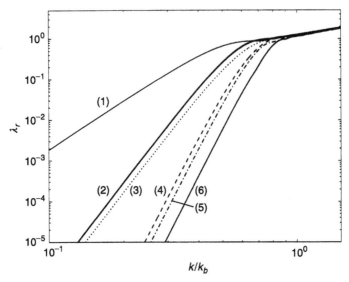

Fig. 3.15. Eigenvalues of the six lowest order modes of a simply supported rectangular plate.

presented in this section may be extended to any type of two-dimensional structure. In particular, the element approach can be combined with a Finite Element model to derive the radiation modes shapes and thus a direct interpretation of the radiation properties of the structure of concern.

At first sight, this formulation of sound radiation might appear to be merely a mathematical exercise with no practical consequences. This is not the case, since the radiation mode representation is extensively used for the design of active structural acoustic control systems where structural sensors and actuators are embedded in panel and shell structures in order to implement active control systems that reduce their sound radiation. An introduction to this type of control system is presented in Chapter 9. The radiation mode representation may also be useful in the design stage of 'quiet structures'. Indeed, the representation in terms of radiation modes give direct indications about the way a structure should be designed in order to minimise its sound radiation. Also, distributed damping treatments can be shaped in such a way to maximise their effects on targeted radiation modes. The subject of the design of quiet structures is presented in the book by Koopmann and Fahnline (1997).

3.7.4 A Comparison of Self- and Mutual Radiation by Plate Modes

The relative magnitudes of self and mutual modal sound powers are now illustrated by a calculation of the sound power radiated by a point-excited,

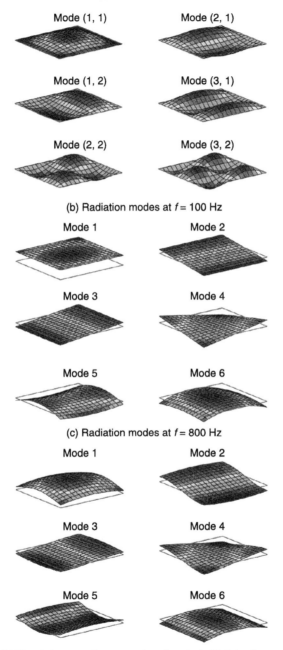

Fig. 3.16. (a) The six lowest order structural modes of a baffled simply supported rectangular plate. Corresponding radiation modes at (b) 100 Hz and (c) 800 Hz.

uniform, simply supported rectangular plate. The plate is made of aluminium, and measures 0.414 m × 0.314 m, is 2 mm thick and has a damping ratio of 1%. For harmonic vibrations, the time-average kinetic energy of the plate is computed from

$$\overline{T}(\omega) = \frac{1}{2} \int_S \frac{1}{T} \int_0^T \rho h v^2(x, z, t) \, dt \, dS = \frac{\rho h}{4} \int_0^a \int_0^b |\tilde{v}(x, z, \omega)|^2 \, dx \, dz \quad (3.73)$$

where T is a suitable period of time over which to estimate the mean energy value. Substituting the modal summation expression for the response of the plate given in matrix form in Eq. (3.38), Eq. (3.73) becomes

$$\overline{T}(\omega) = \frac{\rho h}{4} \int_0^a \int_0^b \{\tilde{\mathbf{v}}\}^H \lfloor \boldsymbol{\varphi}(x, z) \rfloor^T \lfloor \boldsymbol{\varphi}(x', z') \rfloor \{\tilde{\mathbf{v}}\} \, dx \, dz \quad (3.74)$$

Modal orthogonality dictates that the integral over the surface of the plate of the off-diagonal terms is zero. Assuming the natural modes of the simply supported plate take the form $\phi_{pq}(x, z) = 2\sin(p\pi x/a)\sin(q\pi y/b)$, the integrals of the diagonal terms in Eq. (3.74) give $\int_0^a \int_0^b [\phi_{pq}(x, z)]^2 \, dx \, dz = ab$ and the diagonal terms correspond to the modal masses terms, $M_{pq} = \rho h \int_0^a \int_0^b [\phi_{pq}(x, z)]^2 \, dx \, dz = \rho h a b = M$, which is the total mass of the plate. The total kinetic energy of the panel becomes

$$\overline{T}(\omega) = \frac{M}{4} \{\tilde{\mathbf{v}}\}^H \{\tilde{\mathbf{v}}\} \quad (3.75)$$

Alternatively, the elemental approach could also be used in which case the time-averaged total kinetic energy is given by

$$\overline{T}(\omega) = \frac{M_e}{4} \{\tilde{\mathbf{v}}_e\}^H \{\tilde{\mathbf{v}}_e\} \quad (3.76)$$

where M_e is the mass of each element into which the plate has been subdivided. Figure 3.17 shows the total sound powers radiated by the plate calculated using Eq. (3.44) with and without mutual modal sound radiation. In order to compare the total sound power and the space-averaged vibration of the plate, the kinetic energy scaled by a factor of −20 dB has also been plotted in Fig. 3.17.

Comparing the spectra of the kinetic energy and total radiated power calculated with mutual modal sound radiation, we note that both spectra are characterised by a sequence of peaks corresponding to the resonant responses of the panel modes. However, the sound powers corresponding to the resonances of weakly radiating modes, such as the second and third resonances, which are associated

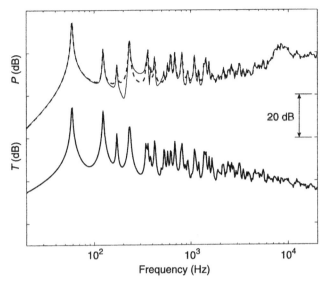

Fig. 3.17. Kinetic energy (thick solid line) and radiated sound powers of a baffled simply supported rectangular plate calculated with (solid line) and without (dotted line) taking into account the mutual radiation.

with the (1,2) and (2,1) modes are relatively lower than those corresponding to the strongly radiating modes, such as the first, which is associated with the (1,1) mode. The difference between the two curves give a quantitative indication of the radiation index of the plate since $\sigma = \overline{P}/\rho_0 c S \langle \bar{v}^2 \rangle$ and $S \langle \bar{v}^2 \rangle = 2\overline{T}/m$, where m is the plate mass per unit area. Comparing the two spectra of radiated sound power, it is clear that mutual radiation is significant only between resonance frequencies of the low order modes: it can both increase and reduce the sound power.

To conclude, it must be emphasised that this discussion is based on a specific system and a narrow band frequency analysis. Mutual modal sound power will increase with modal overlap factor and if we consider band limited sound power it is likely that the mutual radiation mechanism will affect the total sound power to a negligible extent, since the frequency-integrated power is controlled by the high values at resonance frequencies.

3.8 Sound Radiation by Flexural Waves in Plates

Before proceeding to the question of frequency distributions of natural frequencies of plates and thereby to their frequency-average resonant radiation characteristics, we pause to reconsider radiation by panel modes from a different viewpoint. Modal vibration of lightly damped structures arises from the constructive interference between travelling waves which are multiply reflected from structural boundaries or discontinuities to form standing patterns of vibration. It is therefore natural to enquire into the nature of sound radiation by transverse waves travelling along a structure–fluid interface.

We start with the two-dimensional problem illustrated in Fig. 3.18. An infinite, uniform plate is in contact with a fluid that exists in the semi-infinite space $y > 0$. A plane transverse wave of *arbitrary* frequency ω and arbitrary wavenumber κ is forced to travel in the plate; the transverse plate displacement is

$$\eta(x,t) = \tilde{\eta} \exp[j(\omega t - \kappa x)] \tag{3.77}$$

The acoustic field in the fluid is found by using the solution of the two-dimensional wave Eq. (1.7), together with the fluid momentum Eq. (1.5) at the plate–fluid interface. The x-wise variation of the acoustic field variable must follow that of the plate wave, i.e., k_x in the fluid equals κ. From Eq. (1.8)

$$k_y = \pm(k^2 - \kappa^2)^{1/2} \tag{3.78}$$

where $k = \omega/c$.

The appropriate sign of the square root is determined by the physics of the model. Three distinct conditions are possible:

(a) Plate wave phase speed is greater than the speed of sound in the fluid: $\kappa < k$. Plane sound waves travel away from the surface of the plate at an angle to the normal given by $\cos\phi = k_y/k = [1 - (\kappa/k)^2]^{1/2}$, as seen from Eq. (1.7): no wave can propagate toward the plate surface, and therefore the negative sign is disallowed.

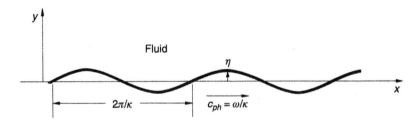

Fig. 3.18.   Transverse wave in a plate in contact with a fluid.

(b) Plate wave phase speed is less than the speed of sound in the fluid: $\kappa > k$,
When $\kappa > k$, k_y is imaginary; the disturbance of the fluid decays exponentially with distance normal to the plate and only a surface wave exists in the fluid. In this case, the negative sign of the square root must be selected so that

$$k_y = -j(\kappa^2 - k^2)^{1/2} \tag{3.79}$$

$$\tilde{p}(x, y) = \tilde{p} \exp(-j\kappa x) \exp\left\{-k[(\kappa/k)^2 - 1]^{1/2} y\right\} \tag{3.80}$$

(c) Plate wave phase speed equals the speed of sound in the fluid: $\kappa = k$. In this case $k_y = 0$; however, the boundary conditions cannot be satisfied in practice, because finite vibration produces infinite pressure, and this condition cannot occur physically.

The amplitude $\tilde{p}$ of the pressure field is determined by the application of the condition of compatibility of normal velocities, or displacements, at the plate–fluid interface. The specific acoustic impedance of the fluid at the interface is

$$(\tilde{p}/\tilde{v})_{y=0} = \omega\rho_0/k_y = \pm\omega\rho_0/(k^2 - \kappa^2)^{1/2} \tag{3.81}$$

Note that this is a particular form of fluid impedance, which can be termed as *wave impedance* because of the wave-like form of the excitation.

In case (a), $\kappa < k$, and

$$(\tilde{p}/\tilde{v})_{y=0} = \rho_0 c/[1 - (\kappa/k)^2]^{1/2} \tag{3.82}$$

The fluid wave impedance is real and positive; the plate does not work upon the fluid, and radiation damping results.

In case (b), $\kappa > k$, and

$$(\tilde{p}/\tilde{v})_{y=0} = j\omega\rho_0/(k^2 - \kappa^2)^{1/2} = j\rho_0 c/[(\kappa/k)^2 - 1]^{1/2} \tag{3.83}$$

The fluid wave impedance is imaginary and positive; no net work is done upon the fluid, but inertial fluid loading is experienced, as discussed more fully in Chapter 4.

In case (c), $\kappa = k$, and

$$(\tilde{p}/\tilde{v})_{y=0} \to \infty \tag{3.84}$$

The fluid loading is infinitely great; in practice, plate wave components satisfying this condition cannot exist.

This analysis shows that only plate wave components with phase speeds greater than the speed of sound create disturbances in the far field and radiate energy into the fluid. In the case of an infinitely extended vibrating plate, the far-field condition necessarily implies an infinitely great distance normal to the plate, whereas in the case of a vibrating plate of finite area, the far-field condition may be satisfied at great distances from the centre of the plate, even by points on or near the plane of the plate. Hence, we can associate sound energy radiation from any planar structure (strictly, in an infinite baffle) with the presence in its normal vibration velocity distribution of supersonic wavenumber components k_s satisfying the condition $k_s < k$ at the frequency concerned. This is the vital conclusion from the foregoing analysis.

Returning now to the problem of sound radiation from the modes of a simply supported panel, we observe that a mode corresponds to a particular pattern of interference between travelling waves. This suggests that the radiation from the component travelling waves could be estimated individually and then summed to give the total field. However, unlike the waves travelling in an infinite plate, those in a panel exist only within a finite interval of space, between panel boundaries. The corresponding cases in time-dependent signals are those of the pure sinewave of single frequency and the transient signal consisting of a finite duration sample of that signal. The frequency spectrum of the latter is obtained by applying the Fourier integral transform (Randall, 1977)

$$\tilde{F}(\omega) = \int_{-\infty}^{\infty} f(t) \exp(-j\omega t)\, dt \qquad (3.85)$$

The spatial equivalent of Eq. (3.85) is the wavenumber transform

$$\tilde{F}(k) = \int_{-\infty}^{\infty} f(x) \exp(-jkx)\, dx \qquad (3.86)$$

The field $f(x)$ can be considered to be constructed from an infinity of *infinitely extended* pure sinusoidal travelling waves of the form

$$f_k(x) = \tilde{F}(k) \exp(jkx) \qquad (3.87)$$

just as the transient time signal can be considered to be made up of an infinity of pure tones, in the form

$$f_\omega(t) = \tilde{F}(\omega) \exp(j\omega t) \qquad (3.88)$$

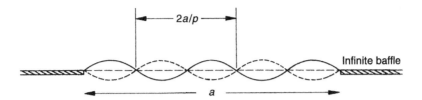

Fig. 3.19.   Modal vibration of an infinitely long strip in a rigid baffle.

The formal expression of the synthesis of $f(x)$ from $\tilde{F}(k)$ is[3]

$$f(x) = \frac{1}{2\pi} \int_{-\infty}^{\infty} \tilde{F}(k) \exp(jkx)\, dk \qquad (3.89)$$

The interval of integration is seen to extend from $-\infty$ to $+\infty$, so that waves travelling in both directions are included.

The analytical and conceptual advantage of decomposing a modal pattern into infinitely extended travelling waves is that the plane-wave impedance relationship (3.81) can be applied exactly to each component to allow a radiated pressure field to be related to the surface normal velocity distribution. This process is first illustrated by application to a one-dimensional surface velocity distribution.

Consider a simply supported panel of width a and of infinite length vibrating harmonically in an infinite rigid baffle (Fig. 3.19). The normal velocity distribution is

$$v_n(x,t) = \begin{cases} \tilde{v}_p \sin(p\pi x/a) \exp(j\omega t); & 0 < x < a \\ 0; & 0 \geq x \geq a \end{cases} \qquad (3.90)$$

The wavenumber transform of v_n is

$$\tilde{V}(k_x) = \tilde{v}_p \int_0^a \sin(p\pi x/a) \exp(-jk_x x)\, dx \qquad (3.91)$$

where k_x is used to distinguish it from the acoustic wavenumber k. The solution of Eq. (3.91) is

$$\tilde{V}(k_x) = \tilde{v}_p \frac{(p\pi/a)[(-1)^p \exp(-jk_x a) - 1]}{[k_x^2 - (p\pi/a)^2]} \qquad (3.92)$$

---

[3] Alternative definitions incorporate the factor $1/2\pi$ differently.

Hence

$$v_n(x, t) = \frac{\exp(j\omega t)}{2\pi} \int_{-\infty}^{\infty} \tilde{V}(k_x) \exp(jk_x x) \, dk_x \qquad (3.93)$$

Associated with each wavenumber component $\tilde{V}(k_x)$ is a surface pressure field wavenumber component given by Eq. (3.90) as

$$[\tilde{P}(k_x)]_{y=0} = \frac{\pm \omega \rho_0}{(k^2 - k_x^2)^{1/2}} \tilde{V}(k_x) \qquad (3.94)$$

so that the surface pressure field can be expressed as

$$[p(x, t)]_{y=0} = \frac{\exp(j\omega t)}{2\pi} \int_{-\infty}^{\infty} [\tilde{P}(k_x)]_{y=0} \exp(jk_x x) \, dk_x \qquad (3.95)$$

The power radiated per unit length of the plate is found by evaluating the expression

$$\overline{P} = \frac{1}{T} \int_0^T \int_0^a [p(x, t)]_{y=0} \, v_n(x, t) \, dx dt$$

$$= \frac{1}{2} \mathrm{Re} \left\{ \int_0^a \tilde{p}(x) \tilde{v}_n^*(x) dx \right\} \qquad (3.96)$$

Substitution for v_n and p from Eqs. (3.93) and (3.95) gives

$$\overline{P} = \frac{1}{8\pi^2} \mathrm{Re} \left\{ \int_{-\infty}^{+\infty} \left[[\tilde{P}(k_x)]_{y=0} \exp(jk_x x) \, dk_x \int_{-\infty}^{\infty} \tilde{V}^*(k_x') \exp(-jk_x' x) \, dk_x' \right] dx \right\} \qquad (3.97)$$

where k_x' has been used solely to distinguish between the integrations over k_x associated with $\tilde{P}$ and $\tilde{V}$. Notice that the range of integration over x has changed from 0 to a to $-\infty$ to $+\infty$, because the form of $\tilde{V}(k_x)$ ensures that v_n is actually zero outside the range $0 \leq x \leq a$. We can replace $\tilde{P}$ in Eq. (3.97) by substituting from Eq. (3.94) to give

$$\overline{P} = \frac{1}{8\pi^2} \mathrm{Re} \left\{ \int_{-\infty}^{+\infty} \left[\int_{-\infty}^{\infty} \frac{\pm \omega \rho_0}{(k^2 - k_x^2)^{1/2}} \tilde{V}(k_x) \exp(jk_x x) \, dk_x \right. \right.$$

$$\left. \left. \times \int_{-\infty}^{\infty} \tilde{V}^*(k_x') \exp(-jk_x' x) dk_x' \right] dx \right\} \qquad (3.98)$$

Integration is first performed over x: the only functions of x are $\exp(jk_x x)$ and $\exp(-jk'_x x)$, which together form the integrand $\exp[j(k_x - k'_x)x]$. The integral of this term over the doubly infinite range is zero if $k_x \neq k'_x$, and is clearly infinite if $k_x = k'_x$: in fact, it is equivalent to the Dirac delta function $2\pi\delta(k_x - k'_x)$. The nature of this function is such that the subsequent integration over k'_x sets k'_x equal to k_x. Hence the integral of Eq. (3.98) becomes

$$\overline{P} = \frac{1}{4\pi} \text{Re} \left\{ \int_{-\infty}^{\infty} \frac{\pm \omega \rho_0 \left|\tilde{V}(k_x)\right|^2 dk_x}{(k^2 - k_x^2)^{1/2}} \right\} \tag{3.99}$$

Now, only wavenumber components satisfying the condition $k_x \leq k$ contribute to sound power radiation and to the real part of the integral. Hence, the range of integration can be reduced to $-k \leq k_x \leq k$ to give

$$\overline{P} = \frac{\rho_0 c k}{4\pi} \int_{-k}^{k} \frac{\left|\tilde{V}(k_x)\right|^2 dk_x}{(k^2 - k_x^2)^{1/2}} \tag{3.100}$$

Therefore, in order to evaluate the radiated power, we need to consider the form of the spectrum of the modulus squared of $\tilde{V}(k_x)$, which is also known as the energy spectrum of $\tilde{V}(k_x)$ (Randall, 1977). From Eq. (3.92)

$$\left|\tilde{V}(k_x)\right|^2 = \left|v_p\right|^2 \left[\frac{2\pi p/a}{k_x^2 - (p\pi/a)^2}\right]^2 \sin^2\left(\frac{k_x a - p\pi}{2}\right) \tag{3.101}$$

which is plotted as a spectrum in Fig. 3.20. As expected, the spectrum peaks when $k_x = p\pi/a = 2\pi/\lambda$, where $\lambda = 2a/p$ is the structural 'wavelength': the value of the peak is up $|\tilde{v}_p|^2 a^2/4$, independent of p. The exception to this form is the modulus spectrum of the fundamental mode ($p = 1$), which has its maximum value at $k_x = 0$. The 'bandwidth' of the major peak in the modulus-squared spectrum is related to the panel width a by $\Delta k_x \approx 2\pi/a$, which is independent of p. For values of $k_x \ll p\pi/a$, the mean value of the spectrum has a value of $\frac{1}{2}|\tilde{v}_p|^2 (2a/p\pi)^2$, which is proportional to the square of the structural wavelength. For the condition $ka \ll p\pi$, Cremer et al. (1988) evaluated Eq. (3.100) as

$$\overline{P} = \frac{\rho_0 c k |\tilde{v}_p|^2}{2} \left(\frac{a}{p\pi}\right)^2 \tag{3.102}$$

[Try it yourself.] The corresponding radiation efficiency is

$$\sigma = 2ka/p^2\pi^2 \tag{3.103}$$

For $ka \gg p\pi$, the radiation efficiency is unity.

The significance of the shape of the modulus-squared spectrum relates to the condition necessary for sound energy radiation from travelling surface velocity waves, namely, $|k_x| < k$ [Eq. (3.82)]. At any particular frequency, $k = \omega/c$ can be marked on the wavenumber axis in Fig. 3.21. Only those wavenumber components satisfying the above condition can radiate sound energy; the others simply create near-field disturbance of the fluid close to the plate surface. As the frequency of vibration in a *given mode shape* is increased, the proportion of radiating wavenumbers increases, as shown in Fig. 3.21, in which the negative wavenumber spectrum is included. Clearly, the radiation increases rapidly as the acoustic wavenumber approaches the structural wavenumber. The fundamental mode is an exception shown in Fig. 3.22.

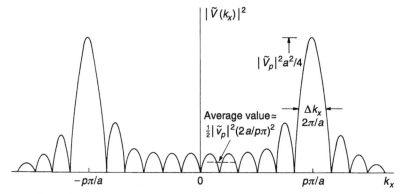

Fig. 3.20. Wavenumber modulus spectrum of plate velocity (diagrammatic).

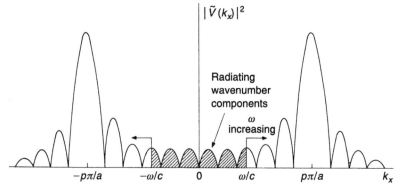

Fig. 3.21. Identification of radiating wavenumber components.

Constructions of the form shown in Figs. 3.20–3.22 can help one to form a qualitative picture of the influence of plate parameters on sound energy radiation. For instance in Fig. 3.23, two panels of the same material and thickness, but of different width, are shown vibrating in modes having the same structural wavelength $\lambda_s = 2a/p$ and which have the same natural frequency ω_n: it is assumed that $2\pi/\lambda \gg k_n$, where $k_n = \omega_n/c$.

Although the spectra are different in the region of the main peaks, because of the different widths of the panels, the mean value of the spectra at low wavenumbers are almost the same, being approximately equal to $(2a/p\pi)^2$. Hence it may

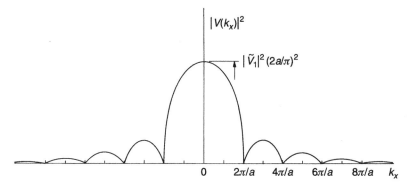

Fig. 3.22. Wavenumber modulus spectrum of the fundamental mode.

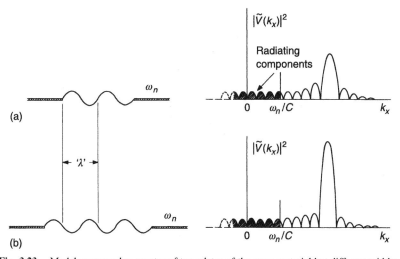

Fig. 3.23. Modal wavenumber spectra of two plates of the same material but different widths.

be concluded qualitatively that for the same vibration velocity amplitude, panel
(a) mode and panel (b) mode radiate the same amounts of sound energy at their
natural frequencies; therefore, the panel (a) mode has a higher radiation efficiency than the panel (b) mode. The conclusion is in agreement with that based
upon the Rayleigh integral approach.

Alternatively, consider two panels of the same width and material, but of different thickness. The panel (a) mode will have a lower natural frequency than the
panel (b) mode; however, they have the *same* wavenumber spectra. Figure 3.24
indicates that the panel (b) mode will have a higher radiation efficiency *at resonance* than the panel (a) mode *at resonance*, because a greater proportion of its
wave-number components are capable of radiating energy.

For modes having natural frequencies ω_n such that their wavenumber spectral
peaks occur at wavenumbers less than the acoustic wavenumber corresponding
to the natural frequency, i.e., $(k_x)_{peak} = p\pi/a < \omega_n/c$, the great majority
of their wavenumber components are supersonic, and their resonant radiation
efficiencies are close to unity. This condition corresponds to the condition
$ph/a > (12)^{1/2}(c/\pi c'_l)$, which for steel, aluminium and glass in air gives
$p > 0.075\, a/h$ where h is the plate thickness. One might enquire about the radiation from the wavenumber component that satisfies $k_x = k$, the impedance
of which Eq. (3.84) suggests is infinite. Because the impedance is infinite,
such a structural wavenumber component cannot physically exist, and those

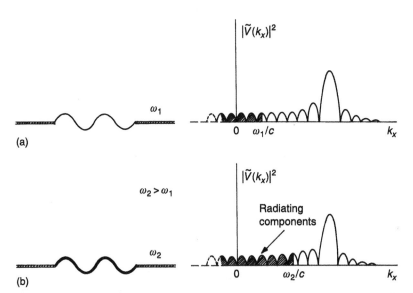

Fig. 3.24.   Modal wavenumber spectra of two plates of the same width and material but different
thickness.

Translational
constraints

Fig. 3.25.   Two of the lowest band modes of a periodically supported panel.

immediately below it are strongly damped by radiation [see Eq. (3.82) with κ/k slightly less than unity]. Consequently, the fluid loading will alter the structural mode shape from the *in-vacuo* form assumed and a fully coupled model must be assumed (see Chapters 4 and 8).

The mode shapes assumed in the analysis so far are highly idealised in their sinusoidal form: this means that adjacent internodal cells have volume velocities of equal magnitude and opposite sign, so that they largely cancel each other's ability to radiate sound of a wavelength that is substantially greater than the structural wavelength. As with time-dependent signals, deviation from a sinusoidal form spreads the frequency spectrum. In the case of modal vibration at frequencies where the structural wavelength is much less than the acoustic wavelength, any distortion from the sinusoidal form that enhances magnitudes of the low-wavenumber (energy-radiating) components increases the radiation efficiency. Consider the idealised case illustrated in Fig. 3.25. The attachment of supports to the panel allows modes to occur that have a non-zero component at zero wavenumber. Active control can have the same (adverse) effect (see Section 9.5).

Consequently, the presence of the internal supports or constraints increases the radiation efficiency. The radiation efficiency of *large, thin*, beam-stiffened panels is consequently higher than that of *large, thin*, unstiffened panels (Maidanik, 1962). This is not to say that stiffening a large panel will necessarily increase the total sound power radiated due to the action of a particular form of excitation, because the vibration level $\langle \overline{v_n^2} \rangle$ *may* be decreased by stiffening to a degree sufficient to offset the increase in radiation efficiency σ: remember, $\overline{P} = \sigma \rho_0 c S \langle \overline{v_n^2} \rangle$. Stiffening can also increase airborne sound transmission through plates, as we shall see in Chapter 5. In practice, adverse effects of stiffening can be minimised by increasing the damping of the structure. Section 3.10 discusses the effects of localised constraints on plate radiation in more detail.

3.9   The Frequency-Average Radiation Efficiency of Plates

So far we have largely concentrated upon modal radiation at *arbitrary* frequency. In practice, the vibrational response of structures to mechanical excitation forces tends to be dominated by the resonant responses of modes having natural frequencies within the frequency range of the excitation. Therefore, we must concentrate our attention on the radiation behaviour of modes at their *natural* frequencies. Again we employ the model of a simply supported, baffled panel for reasons of analytical simplicity; however, the radiation characteristics of all plate structures conform to a similar pattern.

In the application of the wavenumber spectral approach to the qualitative understanding of the influence of plate parameters on radiation, we restricted our attention to one-dimensional panel modes. Of course, most panel modes are two-dimensional in form, and we must therefore consider what effect this has on radiation. In Section 1.7, we saw that the two-dimensional thin-plate equation yields free-wave solutions of the form

$$\eta(x, z, t) = \tilde{\eta}(x, z) \exp[j(\omega t - k_x x - k_z z)] \tag{3.104}$$

in which

$$k_x^2 + k_z^2 = k_b^2 = (\omega^2 m/D)^{1/2} \tag{3.105}$$

and k_b is the free structural wavenumber at frequency ω. We have also seen that the modes of a simply supported rectangular plate take the form

$$v_n(x, z) = \tilde{v}_{pq} \sin(p\pi x/a) \sin(q\pi z/b) \quad \begin{cases} 0 \leq x \leq a \\ 0 \leq z \leq b \end{cases} \tag{3.106}$$

This standing wave can be considered to be made up of component travelling waves in the form

$$v_n(x, z) = -(j\tilde{v}_{pq}/4)\{\exp[j(p\pi x/a)] - \exp[-j(p\pi x/a)]\}$$
$$\times \{\exp[j(q\pi z/b)] - \exp[-j(q\pi z/b)]\} \tag{3.107}$$

which consists of the sum of four travelling waves, existing only within the region $0 \leq x \leq a$ and $0 \leq z \leq b$. The corresponding *primary* wavenumber components are

$$k_x = \pm p\pi/a \tag{3.108a}$$

$$k_z = \pm q\pi/b \tag{3.108b}$$

It must not be forgotten, however, that the finite extent of the wave-bearing region introduces other wavenumber components, as in the case of the one-dimensional mode. Substitution of these expressions into Eq. (3.107) yields the natural frequency equation

$$\omega_{pq} = (D/m)^{1/2}[(p\pi/a)^2 + (q\pi/b)^2] \tag{3.109}$$

In considering radiation from modes at their natural frequencies, we have to consider the relative magnitudes of the primary structural wavenumber components and the acoustic wavenumber ω/c; this is most conveniently done by means of a wavenumber vector diagram or grid.

In Fig. 3.26, the constant-frequency locus of k_b, given by Eq. (3.105), is displayed. A grid of lines corresponding to the allowed primary wavenumber components given in Eqs. (3.108a,b) is superimposed. The intersections of the grid lines represent panel modes. We can also superimpose the locus of the acoustic wavenumber $k = \omega/c$ for any arbitrary frequency. The relative radii of the two loci of k and k_b can be determined from Eq. (3.105):

$$k_b^2 = (k^2c^2m/D)^{1/2} \quad \text{or} \quad (k_b/k) = c(m/D)^{1/4}(1/\omega)^{1/2} \tag{3.110}$$

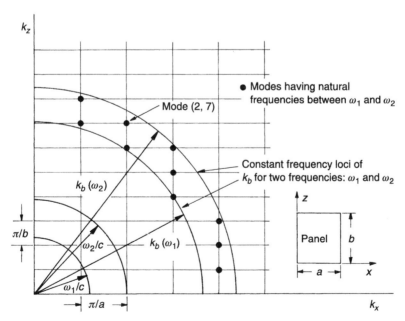

Fig. 3.26.   Constant-frequency loci of acoustic and bending wavenumbers on a wavenumber vector diagram.

This ratio is unity when

$$\omega = c^2(m/D)^{1/2} = \omega_c \tag{3.111}$$

Hence Eq. (3.110) can be written

$$k_b/k = (\omega_c/\omega)^{1/2} \tag{3.112}$$

Frequency ω_c is known as the critical frequency of the panel because it separates the two regions (see Fig. 1.17):

$$\omega < \omega_c, \qquad k_b > k \tag{3.113a}$$

$$\omega > \omega_c, \qquad k_b < k \tag{3.113b}$$

Now, if waves of wavenumber k_b existed in an *infinite* plate, regime (a) would correspond to zero energy radiation, and regime (b) to energy radiation. The finite extent of the plate makes the situation far more complicated, as we have already seen in the case of one-dimensional modes. Below the critical frequency, modes may satisfy one of the following four conditions:

(1) $k > p\pi/a,$ $k < q\pi/b,$

(2) $k < p\pi/a,$ $k > q\pi/b,$

(3) $k < p\pi/a,$ $k < q\pi/b,$

(4) $k > p\pi/a,$ $k > q\pi/b.$

Figures 3.27(a–c) illustrate the wavenumber spectra of the modal velocity distribution along the x and z axes in relation to the acoustic wavenumber. Figure 3.27 suggests that conditions (1) and (2) produce more effective radiation than condition (3). Modes satisfying the latter condition are sometimes termed *corner modes* for a reason that, we have already met in the previous section, only the corner quarter-cells contribute significantly to the radiation. By analogy, modes satisfying the first two conditions are sometimes termed *edge modes*, because strips of half-cell width along the edges normal to the axis for which the primary wavenumber is less than the acoustic wavenumber remain largely uncancelled. Modes satisfying condition (4) are the most efficient subcritical radiators. Their natural frequencies lie fairly close to the critical frequency. Each of the primary modal wavenumber components is less than the acoustic wavenumber corresponding to the modal natural frequency, but the sum of the squares of the two modal components (k_b^2) exceeds the square of the

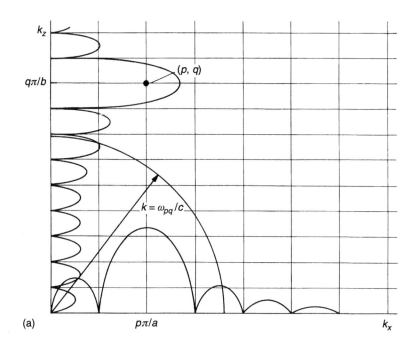

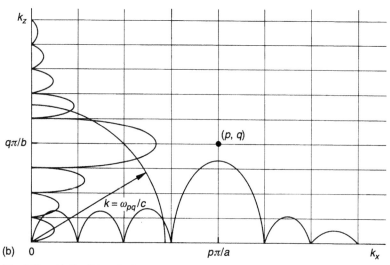

Fig. 3.27.   (a) Condition (1): $k > p\pi/a$; $k < q\pi/b$; (b) condition (2): $k < p\pi/a$; $k > q\pi/b$.

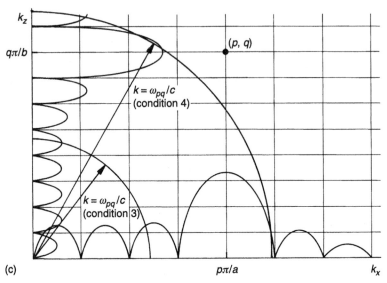

Fig. 3.27. Cont'd (c) conditions (3): $k < p\pi/a$; $k < q\pi/b$ and (4): $k > p\pi/a$; $k > q\pi/b$, $(p\pi/a)^2 + (q\pi/b)^2 > k^2$.

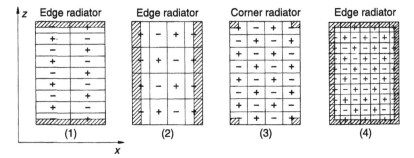

Fig. 3.28. Regions of uncancelled volume velocity at subcritical frequencies.

acoustic wavenumber. As shown by Fig. 3.27(c), modes that satisfy this condition also satisfy the condition $p/q \approx a/b$. Figure 3.28 illustrates the uncancelled areas corresponding to conditions (1)–(4). Under condition (4), the uncancelled edge strip extends along the whole perimeter. It is these modes that are responsible for the rapid rise of radiation efficiency with frequency just below the critical frequency. Figure 3.29 presents an example of the contributions of the various categories of mode to the radiation efficiency of an orthotropic rectangular plate (Anderson and Bratos-Anderson, 2005) in which $\Gamma = [k^2/(k_{bx}^2 + k_{bz}^2)]$.

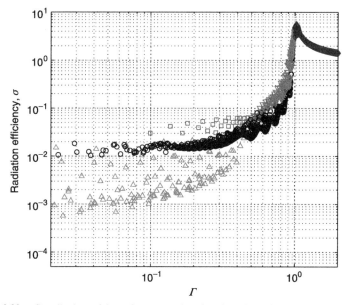

Fig. 3.29. Contributions of the various categories of mode to the radiation efficiency of a baffled simply supported rectangular orthotropic plate. $D_x = 333.2\,\text{Nm}$; $D_y = 21.3\,\text{Nm}$; $D_{xy} = 27.8\,\text{Nm}$; $\rho = 1590\,\text{kgm}^{-3}$; $a = 1.4\,\text{m}$; $b = 0.9\,\text{m}$; $h = 3.06\,\text{mm}$. Modes: $\triangle$ = corner; $\bigcirc$ = x-edge; $\square$ = y-edge; ∇ = x–y edge; $\diamond$ = surface (Anderson and Bratos-Anderson, 2005).

It is not normally possible in practice to find a simple analytical expression for the radiation efficiency corresponding to arbitrary single-frequency excitation, because a number of modes will respond simultaneously, each vibrating with a different amplitude and phase, depending upon its damping and the proximity of its natural frequency to the excitation frequency. Therefore, it is more usual to try to estimate the average radiation efficiency of the modes having natural frequencies in a frequency band; this is called the frequency-average, or modal-average, radiation efficiency. For this purpose it is necessary to assume a distribution of vibration amplitude, or energy, among the modes: a common assumption is that of equipartition of modal energy (since we do not usually know any better). On this basis, Maidanik (see Beranek and Vér, 1992) has produced a modal-average radiation curve shown in Fig. 3.30. Non-dimensional parameters of Maidanik's curves are the ratio of the centre frequency of the band to the critical frequency, the ratio the length of the plate perimeter to the wavelength at the critical frequency, and the ratio of the square of the perimetric length to the area. These parameters feature in the expression given by Cremer *et al.* (1988) for the radiation efficiency of point-excited, lightly damped, plates at frequencies

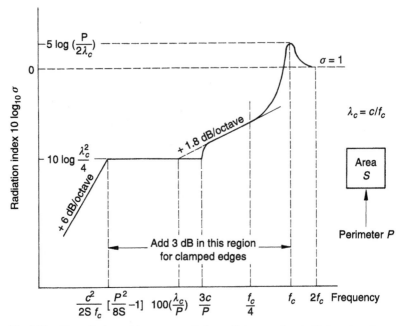

Fig. 3.30. Theoretical modal-average radiation efficiency of a baffled rectangular panel (Beranek and Vér, 1992).

far below the critical frequency: $\sigma \approx (1/\pi^2)(\lambda_c/P)(P^2/S)(f/f_c)^{1/2}$ where P is the perimetric length and S is the area of the plate. For a plate of given thickness and material, the term λ_c/P decreases with increase in plate size and P^2/S is more-or-less independent of plate size. As discussed in Section 3.5, small is not 'beautiful' in this respect.

More recent estimates have been published by Leppington *et al.* (1986) using various methods of analysis: the results of asymptotic analysis and 'exact' numerical integration are compared. An alternative model of multi-mode sound radiation by baffled flat panels has been developed by Bonilha and Fahy (1999) in which it is assumed that a modal density of a panel is sufficiently large for a considerable number of resonant modes to contribute substantially to the vibration field within a frequency band of analysis. A general expression for radiated sound power is developed in terms of the spatial correlation field formed by the superposition of many modes: Bolotin's edge effect analysis is employed to represent a range of boundary conditions (Bolotin, 1984). Over the interior region of a plate, the field exhibits spatial correlation properties that approximate closely to that of an ideally diffuse two-dimensional wave field; the associated radiation efficiency is very small. In the vicinity of boundaries, the correlation field properties are different and depend upon the boundary conditions (see Bonilha and Fahy, 1998). These regions are predicted to radiate the major part of the sound power, in accordance with Maidanik's analysis. The result of applying the spatial correlation model to a baffled, 10 mm thick, simply supported, rectangular aluminium plate, measuring 1.0 m × 0.8 m and having a critical frequency of 1250 Hz, is compared with those of Leppington in Fig. 3.31. Agreement with Leppington's numerical estimate is good except at very low frequencies. The correlation model predicts that the subcritical radiation efficiencies of rectangular plates with clamped boundaries exceed those of plates with simply supported boundaries by a maximum factor of two, except at frequencies below the natural frequency of the (2,2) mode, where the factor becomes less than unity: this is in agreement with Berry *et al.* (1990).

The measured frequency-average radiation efficiencies of some common forms of structure are shown in Fig. 3.32 [after Cremer *et al.* (1988)]. It should be carefully noticed that the radiation efficiency curves are relevant to *multi-mode resonant* vibration in response to broadband mechanical excitation forces; they are not appropriate to acoustically excited structures in which forced wave motion is dominant (see Fig. 3.33).

The concept of a frequency-average radiation efficiency based upon the space-average mean square vibration velocity of a structure is *not tenable* in cases of

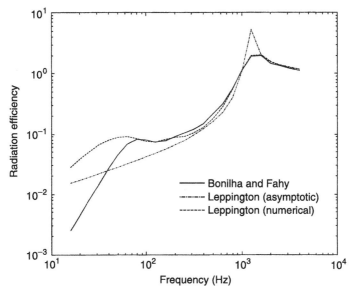

Fig. 3.31.   Comparison of various theoretical estimates of the modal-average radiation efficiency of a baffled, rectangular plate (Leppington *et al.*, 1986; Bonilha and Fahy, 1999).

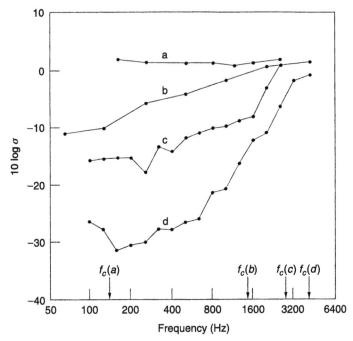

Fig. 3.32. Examples of radiation index in air: (a) 140 mm thick concrete; (b) diesel engine block; (c) 13 mm thick plaster(gypsum)board with lath gridwork; (d) steel pipe of 720 mm diameter and 1.3 mm wall thickness: f_c = critical frequency (Cremer *et al.*, 1988).

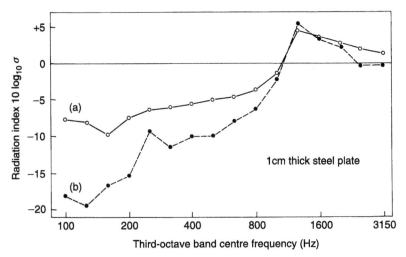

Fig. 3.33. Comparison of the radiation efficiencies of a plate under: (a) airborne sound excitation; (b) mechanical point force excitation (Macadam, 1976).

highly non-uniform structures and highly damped structures subject to localised excitations in which the spatial distribution of mean square vibration is highly non-uniform.

3.10 Sound Radiation due to Concentrated Forces and Displacements

Sound radiation from plates has so far been considered without reference to the location or spatial distribution of excitation. The state of equipartition of modal energy, a concept introduced in the previous section, is likely to obtain if all the modes are lightly damped and if the excitation has no particularly marked spatial structure that allows it preferentially to excite certain classes of modes. One example of a highly structured excitation is an incident plane wave which we shall discuss in Chapter 5.

Another form of excitation that, in theory, excites some modes in preference to others is the point force. The generalised modal force is given by the product of the physical force and the non-dimensional modal displacement at the point of excitation. In practice, the higher-order mode shapes of structures are generally not so well-known that the better excited modes can be distinguished from those of less excited. However, it is known that the mean-square vibration level at the location of the excitation point generally exceeds the spatial average mean-square vibration level on which the radiation efficiency is based; the reason is as follows. As discussed in Section 2.4.2, the velocity response at point $\mathbf{r}$ of a plate structure to single-frequency force excitation at a point ($\mathbf{r}_0$) may be written

$$v(\mathbf{r}, t) = \exp(j\omega t) \sum_n \frac{\widetilde{F}\phi_n(\mathbf{r}_0)\phi_n(\mathbf{r})}{\widetilde{Z}_n} \tag{3.114}$$

where ϕ_n is the non-dimensional modal velocity distribution (mode shape) of mode n, and $\widetilde{F}$ is the force amplitude. $\widetilde{Z}_n$ is the modal impedance, given by

$$\widetilde{Z}_n = j M_n(\omega - \omega_n^2/\omega) + M_n \eta_n \omega_n^2/\omega \tag{3.115}$$

where M_n is the generalised modal mass given by

$$M_n = \int_S m(\mathbf{r})\phi_n^2(\mathbf{r}) \, dS \tag{3.116}$$

where $m(\mathbf{r})$ is the mass per unit area and η_n is the modal loss factor. The mean-square response at the driving point is

$$
\overline{v^2(\mathbf{r}_0)} = \frac{1}{2}\mathrm{Re}\left\{\sum_n \frac{\widetilde{F}\phi_n^2(\mathbf{r}_0)}{\widetilde{Z}_n}\sum_m \frac{\widetilde{F}^*\phi_m^2(\mathbf{r}_0)}{\widetilde{Z}_m^*}\right\}
$$

$$
= \frac{|\widetilde{F}|^2}{2}\mathrm{Re}\left\{\sum_m\sum_n \frac{\phi_n^2(\mathbf{r}_0)\phi_m^2(\mathbf{r}_0)}{\widetilde{Z}_n\widetilde{Z}_m^*}\right\}
$$

(3.117)

and the mean-square response at any other point is

$$
\overline{v^2(\mathbf{r})} = \frac{1}{2}\mathrm{Re}\left\{\sum_n \frac{\widetilde{F}\phi_n(\mathbf{r}_0)\phi_n(\mathbf{r})}{\widetilde{Z}_n}\sum_m \frac{\widetilde{F}^*\phi_m(\mathbf{r}_0)\phi_m(\mathbf{r})}{\widetilde{Z}_m^*}\right\}
$$

(3.118)

The spatial-average mean-square response is

$$
\langle\overline{v^2}\rangle = \frac{1}{S}\int_S \overline{v^2(\mathbf{r})}\,dS = \frac{|\widetilde{F}|^2}{2S}\sum_n \frac{\phi_n^2(\mathbf{r}_0)A_n}{|\widetilde{Z}_n|^2}
$$

(3.119)

where $A_n = \int \phi_n^2(\mathbf{r})\,dS$, and the cross-terms disappear because the modes are assumed to be the normal (orthogonal) modes of the structure. The driving-point response exceeds the spatial-average response by virtue of the cross-terms ($m \neq n$) in Eq. (3.117).

   An alternative way of understanding this result is to consider Eq. (3.114) at, and remote from, the driving point. At the driving point, the amplitudes of the modal responses contributing to the summation are determined by the square of the modal amplitudes at $\mathbf{r}_0$ and the proximity of the driving frequency ω to the natural frequency ω_n through Z_n; the phases of Z_n^{-1} lie in the range of $\pi/2$ to $-\pi/2$ [Fig. 3.34(a)]. At any other point the phase can, in addition, take values of π or $-\pi$, by virtue of the standing-wave nature of the modes, i.e., $\phi_n(\mathbf{r})$, may be positive or negative relative to $\phi_n(\mathbf{r}_0)$ [Fig. 3.34(b)]. Hence the sum expressed in Eq. (3.114) evaluated at the driving point is likely to exceed that at any other point.

   In evaluating a modal average radiation efficiency it is tacitly assumed that the modal vibrations are uncorrelated so that the modal radiation efficiencies can be mathematically averaged[4], the resulting average being multiplied by the spatial-average mean-square velocity in the application of Eq. (3.27). This is not true in

----

[4] Or weighted appropriately, if $\langle\overline{v^2}\rangle$ does not contain equal contributions from all the resonant modes.

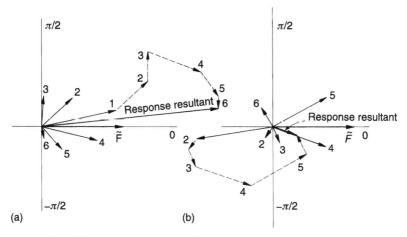

Fig. 3.34. Response phasors at: (a) the driving point; (b) a remote point.

the case of point excitation, even if the exciting force varies randomly with time, because $F(t)$ is common to all modal responses. We can view the necessity of including an 'extra' point force radiation contribution as being due to the response 'concentration' associated with spatially correlated modal motions in the vicinity of the driving point.

The contribution of the point force radiation, relative to that associated with the spatial-average mean-square velocity, increases with the average modal damping. This can be qualitatively understood in terms of Fig. 3.34(a), because the rate of change of phase and amplitude of modal response with the ratio of forcing to natural frequency decreases with an increase of damping. Hence, relatively more modes will contribute significantly to the driving-point response, whereas the standing-wave cancellation effect seen in Fig. 3.34(b) will still operate, and the ratio of driving-point response to spatial-average response will increase. In the limit of high damping, the point force radiation will correspond closely to that produced by a point force acting on an infinite plate. This can be evaluated by using a wavenumber transform technique on the inhomogeneous form of the bending-wave equation corresponding to Eq. (1.39);

$$D\left[\frac{\partial^4 \eta}{\partial x^4} + \frac{2\partial^4 \eta}{\partial x^2 \partial z^2} + \frac{\partial^4 \eta}{\partial z^4}\right] + m\frac{\partial^2 \eta}{\partial t^2} = \tilde{F}\delta(x-x_0)\delta(z-z_0)\exp(j\omega t) \quad (3.120)$$

where $\delta(x-x_0)$ and $\delta(z-z_0)$ are Dirac delta functions that concentrate the force per unit area onto point (x_0, z_0). The Dirac delta function possesses the properties that $\int_{-\infty}^{\infty} \delta(x-x_0)\,dx = 1$ and $\int_{-\infty}^{\infty} f(x)\delta(x-x_0)\,dx = f(x_0)$: it has the dimensions of the inverse of its argument. Hence, $\tilde{F}$ is the complex amplitude of the total applied force and the units of the forcing term on the right-hand side of Eq. (3.120) are those of force per unit area.

By analogy with the one-dimensional transform expressed in Eq. (3.86), a two-dimensional transform of a two-dimensional distribution of force per unit area may be defined:

$$\tilde{F}(k_x, k_z) = \int_{-\infty}^{\infty} \tilde{F}\delta(x - x_0)\delta(z - z_0) \exp(-jk_x x) \exp(-jk_z z) \, dx \, dz$$

$$= \tilde{F} \exp(-jk_x x_0) \exp(-jk_z z_0) \tag{3.121}$$

$$= \tilde{F} \text{ for } x_0 = 0, \quad z_0 = 0$$

Hence application of the transform to Eq. (3.120) yields

$$D[(k_x^2 + k_z^2)^2 - \omega^2 m]\tilde{\eta}(k_x, k_z) = \tilde{F} \tag{3.122}$$

Now we can replace the inertia term in Eq. (3.122) by substituting from Eq. (1.41) thus:

$$D[(k_x^2 + k_z^2)^2 - k_b^4]\tilde{\eta}(k_x, k_z) = \tilde{F} \tag{3.123}$$

This equation expresses the response of the plate to a plane travelling-wave excitation having wavenumber components k_x and k_z. If required, damping can be included by making D complex in the form $D' = D(1 + j\eta)$, where η is the plate loss factor.

The fluid wave impedance associated with a plane travelling wave having wavenumber components k_x and k_z, is, of course, the same as that given by Eq. (3.94), with k_x replaced by $(k_x^2 + k_z^2)^{1/2}$:

$$[\tilde{P}(k_x, k_z)]_{y=0} = \frac{\pm \omega \rho_0}{(k^2 - k_x^2 - k_z^2)^{1/2}} \tilde{V}(k_x, k_z) \tag{3.124}$$

Now the surface velocity transform is related to the surface displacement transform by

$$\tilde{V}(k_x, k_z) = j\omega\tilde{\eta}(k_x, k_z) \tag{3.125}$$

Hence the surface pressure transform can be written as

$$[\tilde{P}(k_x, k_z)]_{y=0} = \frac{\pm j\omega^2 \rho_0}{(k^2 - k_x^2 - k_z^2)^{1/2}} \frac{\tilde{F}(k_x, k_z)}{D[(k_x^2 + k_z^2)^2 - k_b^4]} \tag{3.126}$$

The power radiated by the whole plate can now be expressed in terms of the product of the surface pressure transform and complex conjugate of the surface velocity transform as in Eqs. (3.97) and (3.98). However, in this case integration over infinite ranges of x, z and k_x, k_z are necessary. Following the arguments applied to the one-dimensional integral we can reduce the integral to

$$\overline{P} = \frac{\rho_0 c \omega^2 |\widetilde{F}|^2}{8\pi^2 D^2} \int_{-k}^{k} \int_{-k}^{k} \frac{k \, dk_x dk_z}{[(k_x^2 + k_z^2)^2 - k_b^4]^2 (k^2 - k_x^2 - k_z^2)^{1/2}} \tag{3.127}$$

If we restrict our attention to frequencies well below the critical frequency of the plate, $k \ll k_b$, and since $(k_x^2 + k_z^2)^{1/2} \leq k$ in the range of integration, we can neglect $(k_x^2 + k_z^2)^2$ compared with k_b^4 in the denominator of the integral, to give

$$\overline{P} \approx \frac{\rho_0 c \omega^2 |\widetilde{F}|^2}{8\pi^2 D^2 k_b^8} \int_{-k}^{k} \int_{-k}^{k} \frac{dk_x dk_z}{(k^2 - k_x^2 - k_z^2)^{1/2}} \tag{3.128}$$

This integral is most readily solved by transforming to cylindrical coordinates so that $k_r^2 = k_x^2 + k_z^2$, $k_x = k_r \cos\phi$ and $k_z = k_r \sin\phi$. The element $dk_x dk_z$ transforms to the element $(k_r d\phi) dk_r$, and ϕ ranges from 0 to 2π. The integration over ϕ is straightforward and the integral reduces to

$$\overline{P} = \frac{\rho_0 c \omega^2 k |\widetilde{F}|^2}{4\pi D^2 k_b^8} \int_0^k \frac{k_r dk_r}{(k^2 - k_r^2)^{1/2}} = \frac{\rho_0 c \omega^2 k^2 |\widetilde{F}|^2}{4\pi D^2 k_b^8} \tag{3.129}$$

We replace k_b^8 by $(m\omega^2/D)^2$ to give

$$\overline{P} = \frac{\rho_0 c k^2 |\widetilde{F}|^2}{4\pi m^2 \omega^2} = \frac{\rho_0 |\widetilde{F}|^2}{4\pi c m^2} \tag{3.130}$$

This remarkable simple result indicates that the power generated by the excitation well below the critical frequency of an infinite or very highly damped, bounded plate is independent of frequency and plate stiffness, and inversely proportional to the square of the mass per unit area. Above the critical frequency, the whole plate surface radiates and all the input power is theoretically converted into sound in the absence of internal damping.

If we assume that this expression holds reasonably well for finite, damped plates, we can obtain a rough estimate of the proportion of sound power generated by this driving-point region and by the distributed vibration field at frequencies below the critical frequency. The time-average power injected into a finite plate by a point force is, on a frequency- or modal-average basis, given by Eqs. (2.4) and (2.68a) as

$$\overline{P}_{in} = \tfrac{1}{2} |\tilde{F}|^2 \operatorname{Re}(\tilde{Y}) = \tfrac{1}{2} |\tilde{F}|^2 / 8(mD)^{1/2} \qquad (3.131)$$

The total energy of vibration is equal to the sum of the modal energies and, for a uniform plate, is

$$\overline{E} = mS\langle \overline{v^2} \rangle \qquad (3.132)$$

According to the definition of loss factor, the equilibrium energy corresponds to a balance between power in and power dissipated:

$$\tfrac{1}{2} |\tilde{F}|^2 / 8(mD)^{1/2} = \eta \omega m S \langle \overline{v^2} \rangle \qquad (3.133a)$$

or

$$\langle \overline{v^2} \rangle = \tfrac{1}{2} |\tilde{F}|^2 / 8(mD)^{1/2} m \eta \omega S \qquad (3.133b)$$

The sound power radiated by multi-modal vibration is, by definition,

$$\overline{P}_s = \rho_0 c S \sigma \langle \overline{v^2} \rangle = \tfrac{1}{2} \rho_0 c |\tilde{F}|^2 \sigma / 8(mD)^{1/2} m \eta \omega \qquad (3.134)$$

Hence the ratio of powers radiated from the driving point and the reverberant vibration field is

$$\frac{\overline{P}}{\overline{P}_s} = \frac{4}{\pi} \left(\frac{\omega}{\omega_c} \right) \left(\frac{\eta}{\sigma} \right) \qquad (3.135)$$

The ratio is seen to increase with the ratio of plate loss factor to radiation efficiency. The apparent proportionality to the frequency ratio ω/ω_c is misleading, since σ varies approximately as $(\omega/\omega_c)^n$, where $1/2 < n < 1$ over much of the subcritical frequency range. For many practical structures the ratio η/σ will substantially exceed unity, especially at the lower frequencies of practical interest, and the ratio $\overline{P}/\overline{P}_s$ may well take a value of about unity. Of course, the point force contribution is proportionally more important for artificially damped structures, and Eq. (3.130) represents a lower limit for radiation by a point-excited plate, however, well damped it is.

Another case of concentrated excitation that is perhaps of even greater importance than that of the point force is that of the line force. Excitation of vibrations in plate structures by motion of the plate boundaries, perhaps by much stiffer frame structures, is a common practical situation. As opposed to a point force, a line force generates a one-dimensional plane wave in a plate, and hence only a one-dimensional wavenumber transform is necessary. We can proceed directly to the plate velocity transform by applying Eq. (1.37) to a plate, thus:

$$
\eta(x, t) =
\begin{cases}
\dfrac{-j\widetilde{F}'\exp(j\omega t)}{4Dk_b^3}[\exp(-jk_bx) - j\exp(-k_bx)], & x \geq 0 \\[4mm]
\dfrac{-j\widetilde{F}'\exp(j\omega t)}{4Dk_b^3}[\exp(jk_bx) - j\exp(k_bx)], & x \leq 0
\end{cases}
\tag{3.136}
$$

where $\widetilde{F}'$ is the force per unit length. Now

$$
\widetilde{V}(k_x) = j\omega\widetilde{\eta}(k_x) = \frac{\omega\widetilde{F}'}{4Dk_b^3}\left\{ \int_{-\infty}^{0} [\exp(jk_bx) - j\exp(k_bx)]\exp(-jk_xx)dx \right.
$$
$$
+ \int_{0}^{\infty} [\exp(-jk_bx) - j\exp(-k_bx)]
$$
$$
\left. \times \exp(-jk_xx)\, dx \right\}
\tag{3.137}
$$

An alternative and more straightforward approach to the evaluation of this velocity transform is to transform the equation of motion of the plate directly, in which the applied force is represented as a line delta function $F'\delta(0)$: this procedure is the one-dimensional equivalent of the analysis leading to Eq. (3.123). The result is

$$
|\widetilde{V}(k_x)|^2 = \frac{\omega^2|\widetilde{F}'|^2}{D^2(k_x^4 - k_b^4)^2}
\tag{3.138}
$$

Applying Eq. (3.100) the power radiated per unit length is

$$
\overline{P} = \frac{\rho_0 c k \omega^2 |\widetilde{F}'|^2}{4\pi D^2}\int_{-k}^{k} \frac{dk_x}{(k_x^4 - k_b^4)^2(k^2 - k_x^2)^{1/2}}
\tag{3.139}
$$

[Evaluate the contribution to radiated power of the bending near-field alone and explain the shortfall.] At frequencies below ω_c, we can neglect k_x^4 in comparison with k_b^4 in the denominator of the integrand to give

$$
\overline{P} = \frac{\rho_0 c k \omega^2 |\widetilde{F}'|^2}{4D^2 k_b^8} = \frac{\rho_0 |\widetilde{F}'|^2}{4\omega m^2}
\tag{3.140}
$$

This expression is very similar to that in Eq. (3.130) for a point force, the main difference being the presence of frequency in the denominator, which shows that the line force is more effective at lower frequencies. As with the point force, we can investigate the relative magnitudes of radiation from the excitation region and from the multi-mode vibration of the whole surface.

The time-average power input from a uniform line force to an infinite plate is

$$\overline{P}_{in} = \tfrac{1}{2}|\widetilde{F}'|^2 l/4D^{1/4}\omega^{1/2}m^{3/4} \tag{3.141}$$

where l is the length of the panel over which it acts. We assume this is distributed among the modes of a finite plate to give a mean-square velocity

$$\langle \overline{v^2} \rangle = \tfrac{1}{2}|\widetilde{F}'|^2 l/4D^{1/4}\omega^{3/2}m^{7/4}\eta S \tag{3.142}$$

The sound power radiated by multi-mode vibration is therefore

$$\overline{P}_s = \tfrac{1}{2}\rho_0 c\sigma|\widetilde{F}'|^2 l/4D^{1/4}\omega^{3/2}m^{7/4}\eta \tag{3.143}$$

and the ratio of the powers radiated by the excitation region and the whole panel is

$$\frac{\overline{P}}{\overline{P}_s} = 2\left(\frac{\omega}{\omega_c}\right)^{1/2}\left(\frac{\eta}{\sigma}\right) \tag{3.144}$$

The parameters are the same as those appropriate to point force excitation, but the frequency dependence is different. Again the variation of subcritical radiation efficiency as $(\omega/\omega_c)^n$ must be remembered.

The foregoing expressions for force-generated radiation may also be expressed in terms of the plate velocity v_0 at the excitation point, by replacing the force by the product of velocity and appropriate impedance. In the case of point force excitation, Eq. (3.130) becomes

$$\overline{P} = \frac{16\rho_0 |\tilde{v}_0|^2 D}{\pi cm} = \frac{16\rho_0 |\tilde{v}_0|^2 c^3}{\pi \omega_c^2} \tag{3.145}$$

Hence the power increases linearly with bending stiffness and inversely as panel mass; again a stiff, light structure, like a honeycomb sandwich panel, is seen to be an effective converter of vibration into sound. The equivalent expression for line velocity excitation is

$$\overline{P} = \frac{2\rho_0 D^{1/2}|\tilde{v}_0|^2}{m^{1/2}} = \frac{2\rho_0 c^2|\tilde{v}_0^2|}{\omega_c} \tag{3.146}$$

A useful application of the various expressions for force-generated radiation is to the estimation of the effect of locally constraining a plate structure on its

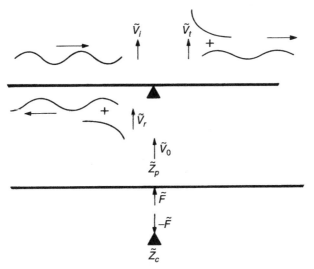

Fig. 3.35. Scattering of bending waves by a plate constraint.

sound radiation characteristics. Imagine a bending wave of velocity $\tilde{v}_i$ in a plate to be normally incident upon a line constraint of impedance $\tilde{Z}_c$ as shown in Fig. 3.35. The reaction force of the constraint is given by

$$\tilde{F} = -\tilde{Z}_c \tilde{v}_0 \tag{3.147}$$

and the response of the plate to this force is

$$\tilde{v} = \tilde{F}/\tilde{Z}_p \tag{3.148}$$

where $\tilde{Z}_p$ is the line force impedance of the plate. Scattered waves having normal velocities $\tilde{v}_r$ and $\tilde{v}_t$ are reflected and transmitted:

$$\tilde{v}_0 = \tilde{v}_i + \tilde{v}_r = \tilde{v}_t \tag{3.149}$$

where $\tilde{v}_r = \tilde{v}$. Combining Eqs. (3.147–3.149) gives

$$\tilde{F} = -\tilde{v}_i [\tilde{Z}_p \tilde{Z}_c / (\tilde{Z}_p + \tilde{Z}_c)] \tag{3.150}$$

and

$$\tilde{v}_0 = \tilde{v}_i [\tilde{Z}_p / (\tilde{Z}_p + \tilde{Z}_c)] \tag{3.151}$$

If $\left|\widetilde{Z}_p/\widetilde{Z}_c\right| \gg 1$, the incident wave is very weakly scattered by the constraint $\tilde{v}_0 \approx \tilde{v}_i$ and $F \approx -\tilde{v}_i \widetilde{Z}_c$. On the other hand, if $\left|\widetilde{Z}_p/\widetilde{Z}_c\right| \ll 1$, the incident wave is strongly scattered; $|\tilde{v}_0| \ll |\tilde{v}_i|$ and $F \approx -\tilde{v}_i \widetilde{Z}_p$. The reaction force can be considered in exactly the same way as an applied force and the relevant equations applied to estimate the additional sound radiation due to the presence of the constraint. Nikiforov (1981) used this type of analysis to evaluate the radiation efficiency of plates.

The practical conclusion to be drawn from this section is that the generation of inhomogeneous, bending-wave near fields by applied concentrated forces, or by the action of constraints, on otherwise freely vibrating plates, gives rise to the radiation of sound power in addition to that produced by the reverberant modal vibration field of the plate surface. In the case of mechanically excited structures, the sound power radiated by the excitation-induced bending near fields at subcritical frequencies forms a significant proportion of the whole, and sets a limit to the reduction in radiation that can be effected by damping a plate. Above the critical frequency the influence of force distribution and constraints is far less important because the whole surface radiates effectively, and the reduction of resonant modal vibration levels by damping is far more effective.

3.11 Sound Radiation by Non-Uniform Plate Structures

3.11.1 Beam-Stiffened Plates

The plate-like components of many practical structures are not uniform, homogeneous and isotropic like those so far considered. In order to increase their static stiffness, plates are often stiffened by discrete linear elements (we shall call these 'beams'), or corrugated, or of sandwich construction. Stiffening affects their flexural wave dispersion characteristics, which alters the ratio of structural to acoustic wavenumbers that we have seen is a major parameter in determining radiation efficiency. It is common engineering design practice to stiffen thin plates and shells by means of attaching beam-type elements, usually in parallel line arrays or rectangular grid-like arrays. The effect on the vibrational behaviour of a structure depends on the material properties, the geometric form and dimensions, and the disposition and form of connection of the stiffeners. Stiffeners that are not symmetrically disposed about the median plane of a flat plate introduce coupling between in-plane and bending waves. Effects on curved shell dynamics are even more complicated (see, for example, Egle and Sewall, 1968; Egle and Soder, 1969; ESDU, 1983). Thus, many extra parameters are introduced by stiffening, and the dynamic and acoustic effects are impossible to categorise in a simple and generic manner. Consequently, the following qualitative account of the effects of stiffening on sound radiation is generalised and offered only as broad guidance that should be supplemented by consultation of the cited references. It will be assumed that 'line' stiffeners in the form of beams are attached in a spatially periodic array. Non-periodic arrays may offer certain advantages in controlling structural wave energy propagation within a stiffened structure due to a phenomenon known as 'Anderson localisation' (Hodges and Woodhouse, 1989). This topic falls outside the scope of this book.

The effects of stiffening may be analysed either in terms of modes and their natural frequencies or in terms of wave dispersion characteristics. The constraints offered by stiffeners on plate normal displacement, rotation and twist partially reflect incident bending waves, introduce bending near fields in the vicinity of the constraints and, in the case of multiple stiffeners, produce multiple scattering effects that have profound effects on bending wave propagation behaviour, as discussed below. Consequently, mode shapes take complicated forms and the distribution of natural frequencies can be very irregular. At 'low' frequencies where the plate bending wavelength greatly exceeds the spacing between adjacent stiffeners, it is acceptable to consider the bending stiffness of discrete stiffeners to be 'smeared' (considered as a spatially uniform average) over the surface of the structural continuum that carries them. Consequently, low order mode shapes are similar to those of the unstiffened structure, but the natural frequencies are raised. At 'high' frequencies, where the bending wavelength is considerable less than the spacing, it is tempting to think of the individual, inter-stiffener panels as smaller bounded plates. However, unless the stiffeners have very high bending and torsional stiffness relative to plate bending stiffness, the boundary conditions of any one panel are frequency dependent because they are determined by the dynamic properties of the embracing stiffeners and of the whole of the rest of the structure. Frequency-dependent boundary conditions do not allow the existence of an orthonormal set of natural modes. It is thus dangerous to consider a stiffened plate simply as an array of smaller plates having simple boundary conditions.

We have seen that the relative phases of vibration of different regions of a plate greatly affects sound radiation through the cancellation phenomenon. The relative phases of free bending vibration in an array of panels separated by beam stiffeners depends crucially upon the dynamic properties of the separating stiffeners. Under forced vibration, it also depends upon the spatial phase distribution of the excitation. Consequently, the effect of stiffeners on sound radiation is quite different in the cases of panels excited by localised forces, acoustic fields and boundary layers. Maidanik (1962) suggests that the effect of the attachment of stiffeners to a panel is effective to produce many smaller panels, and therefore to increase its radiation efficiency below the critical frequency by a factor equal to $1 + 2P/B$, where P is the total length of the stiffeners and B the length of the panel perimeter. It is only an approximate correction factor, which should not be allowed to raise the radiation efficiency above unity, but it does indicate that the radiation index of a densely stiffened panel is likely to be close to zero dB even below the critical frequency. This simplistic model is, at best, a rough guide to the *maximum* frequency-average effect of stiffeners on radiation efficiency. Figures 3.36(a,b) show the measured radiation efficiencies of a model of a stiffened ship deck structure, in which the effect of stiffening can clearly be seen (von Venzke *et al.*, 1973). The effects of plate stiffening on airborne sound transmission are addressed in Section 5.12.

The effects on structural wave propagation, and on associated sound radiation, of *spatially periodic* constraint offered by stiffeners (and also of other forms of periodic constraint, such as the supports of railway lines), is best understood in terms of their effects on wave dispersion curves and associated phase velocities. Periodic constraint produces the phenomenon of 'stop' and 'pass' frequency bands. In lightly damped, *in-vacuo* structures, wave propagation is almost completely blocked in stop bands and waves are free to propagate in pass bands. Fluid loading by liquids provides a mechanism whereby the constraints imposed by stiffeners can be 'by-passed' by energy propagating in the fluid.

The effect of periodically spaced stiffeners is graphically illustrated by the example of the bending-wave dispersion characteristics of a periodically constrained (infinite) plate, shown in Fig. 3.37 (Mead, 1990). A 6 mm thick steel plate is reinforced at 300 mm intervals by a set of parallel beams that are assumed to act as uniform line simple supports which separate the plate into an infinite array of identical bays: the system is essentially two-dimensional. The periodicity of an *undamped* system requires that the *amplitude* of any *propagating* time- and space-harmonic wave variable on any one line parallel to the stiffeners in any one bay is the same at all equivalent positions in all other bays; but, the *phase* of the variable may vary from bay to bay by an equal increment. In the frequency ranges where the plotted variable μ_{xf} is purely real (in pass bands) it represents this phase increment: it is the non-dimensional wavenumber that is the solution to the dispersion equation of the free system at any frequency. Phase difference is ambiguous; therefore $\mu_{xfm} = \mu_{xf} + /- 2m\pi$ (*m* any integer) are also solutions. Each value of *m* corresponds to a separate pass band. In the frequency ranges where μ_{xfm} is purely imaginary (stop bands), any local disturbance does not propagate freely but decays exponentially with distance from an excitation point, the decay rate increasing with μ_{xf}.

The dispersion line of the non-dimensional acoustic wavenumber (in air) is shown in Fig. 3.37 together with that of the bending wave in the unstiffened plate. In the frequency range shown, the latter greatly exceeds the former, indicating that the critical frequency lies far higher and that the freely propagating, subsonic, bending waves in the infinite, *unstiffened* plate do not generate radiated

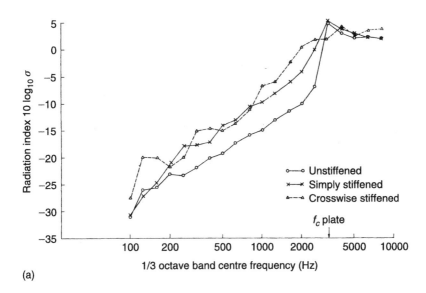

(a)

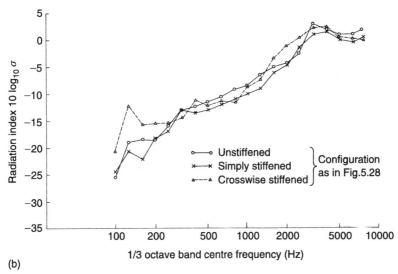

(b)

Fig. 3.36. (a) Effect of stiffening on the radiation efficiencies of point-excited plates; (b) effect of stiffening on the radiation efficiencies of plates excited by a diffuse sound field (von Venzke *et al.*, 1973).

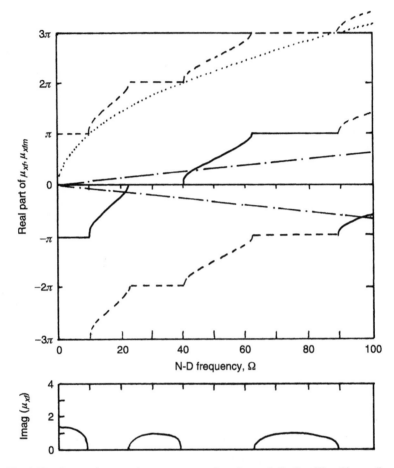

Fig. 3.37. Curves of propagation constants μ_{xf} for a plate periodically stiffened in one direction (no fluid loading): ____ μ_{xf}; - - - - - μ_{xfm} for $m = +1$ and -1; ...... unstiffened plate; _._. acoustic plane wave (Mead, 1990).

sound power. However, parts of the $m = 0$ and $m = -1$ stiffened-plate dispersion curves indicate supersonic phase speeds, and hence sound power radiation. The multiple scattering of bending waves caused by the stiffeners is seen to produce subcritical sound radiation, just as that due to the scattering by the boundaries of unstiffened finite plates considered earlier. The radiation characteristics of orthogonally stiffened structures are extremely complex and beyond the scope of this book.

The natural frequencies and modes of a *finite length* of plate containing N bays (panels) may be determined by superimposing horizontal lines on Fig. 3.37

corresponding to phase closure, i.e., at values of $\mathrm{Re}\{\mu_{xfm}\}$ equal to $m\pi/N$. The intersections with the dispersion curves of $\mathrm{Re}\{\mu_{xfm}\}$ curves indicate the natural frequencies of the modes of which there are $N-1$ in each pass band. The mode shapes are not generally spatially sinusoidal except for the lowest frequency modes in each pass band. The lowest and highest frequency modes having frequencies in the lowest band are shown in Fig. 3.25. Modes that have substantial contributions from the supersonic wavenumber components identified from the infinite plate model radiate the most effectively. Localised mechanical excitation of a periodically stiffened plate in a stop band generates a non-propagating near field. This radiates sound in a manner similar to the near bending-field region immediately surrounding the driving point of an infinite plate.

3.11.2   Corrugated Plates

Corrugation, which causes plates to be orthotropic, greatly increases the bending stiffness in one direction. The resulting wavenumber vector diagram is quite different from that of an isotropic plate shown in Fig. 3.26, the constant frequency contours not being quarter-circles. The equation of flexural wave motion of an orthotropic plate is

$$D_x\frac{\partial^4\eta}{\partial x^4} + 2(D_xD_z)^{1/2}\frac{\partial^4\eta}{\partial x^2\partial z^2} + D_z\frac{\partial^4\eta}{\partial z^4} + m\frac{\partial^2\eta}{\partial t^2} = 0 \qquad (3.152)$$

where D_x and D_z are the bending stiffnesses in the two orthogonal directions: compare Eq. (1.39). Substitution of

$$\eta(x, z, t) = \tilde{A}\exp(-jk_xx)\exp(-jk_zz)\exp(j\omega t) \qquad (3.153)$$

yields

$$[(D_x)^{1/2}k_x^2 + (D_z)^{1/2}k_z^2]^2 = \omega^2m \qquad (3.154)$$

The resulting wavenumber diagram for a rectangular, orthotropic plate takes the form shown in Fig. 3.38. It is seen that some plate modes have supersonic wavenumbers below the critical frequency ω_c of the basic plate material and hence radiate well. However, in any one frequency band below ω_c, these modes have a lower modal density than the subsonic modes; hence, if all modes having natural frequencies within a band are assumed to vibrate with equal energy, the modal average radiation efficiency will remain below unity, although it will be higher for a corrugated plate than for an uncorrugated plate of the same material and size (Heckl, 1960).

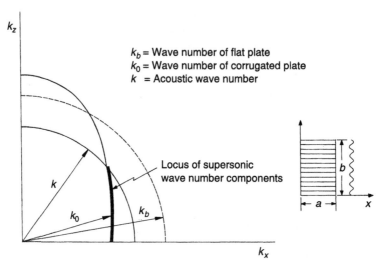

Fig. 3.38. Constant-frequency loci of flexural wavenumbers in a corrugated and an equivalent flat plate below the flat plate critical frequency.

Figure 3.39(a) compares measured radiation efficiencies of a 1 mm thick, aluminium plate when unstiffened, beam stiffened and in corrugated form (Drotleff, 1996): the geometric forms are shown in Fig. 3.39(b). The structures were excited at three points by uncorrelated broadband forces. The structural loss factors ranged between 0.05 and 0.1. Above 2 kHz, the radiation damping of the corrugated plate dominates its total damping. The very substantial influence of stiffeners and corrugation on radiation efficiency is clearly seen. Drotleff also shows the change in radiation efficiency effected by covering these base structures with a form of multiplayer insulation sheet typically used in road vehicles (see Section 5.12). One practical implication of his results is that it is not appropriate to evaluate the effectiveness of such layers by tests on uniform plates, if they are to be applied to non-uniform plates.

3.11.3   Sandwich Plates

The dispersion characteristic of an isotropic sandwich plate that consists of two plates separated by a central layer of rather lower material stiffness is very much affected by the elastic properties and thickness of this layer. The shear stiffness of the layer is normally rather low, in which case transverse wave propagation is controlled by the whole section-bending stiffness at low frequencies, by the central-layer shear stiffness at intermediate frequencies, and by the individual

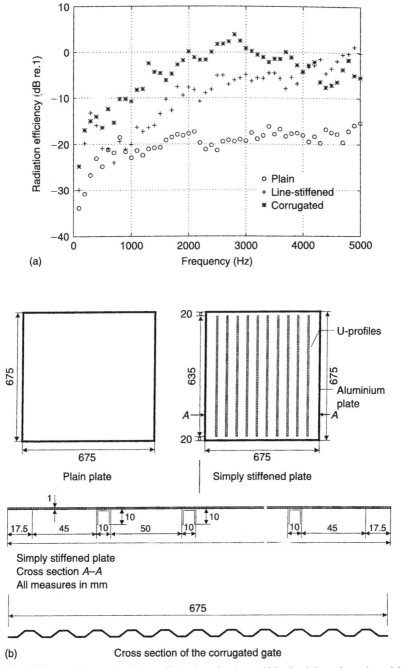

Fig. 3.39. (a) Measured radiation efficiencies of a 1 mm thick aluminium plate when plain, line-stiffened and in corrugated form (Drottleff, 1996); (b) geometric parameters of the plates.

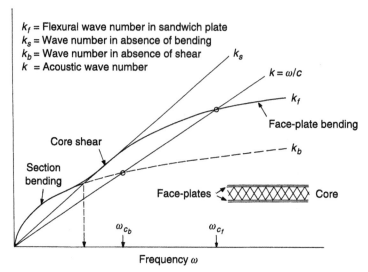

Fig. 3.40.   Typical dispersion curves for a sandwich plate.

faceplate-bending stiffness at very high frequencies [see Kurtze and Watters (1959) and Section 5.12]. A typical dispersion curve is sketched in Fig. 3.40, which shows that the critical frequency is higher than for the equivalent uniform plate, and hence the radiation efficiency may be reduced over part of the frequency range (Cremer *et al.*, 1988). Unfortunately, elastic layers can produce plate-layer-plate dilatational resonances, which, if not controlled by suitable damping, can greatly increase radiation and reduce the transmission loss (Smolenski and Krokosky, 1973).

The dilatation problem can be avoided by the use of honeycomb panels, which consist of very thin faceplates separated by cores that are stiff in transverse compression but weak in shear, as commonly used in aircraft and spacecraft to maximise stiffness-to-weight ratios (Heron, 1979, 1981). The critical frequency, and therefore the radiation efficiency of such structures is highly dependent upon the shear stiffness of the core, as shown by Fig. 3.41.

3.11.4 Composite Sound Insulation Panels

The sound insulation treatments applied to the structural envelopes of vehicles such as cars, trains and aircraft usually take the form of composite 'panels' in which the primary structure is covered by an insulation (or 'trim') layer comprising a dense impermeable cover sheet separated from the primary structure

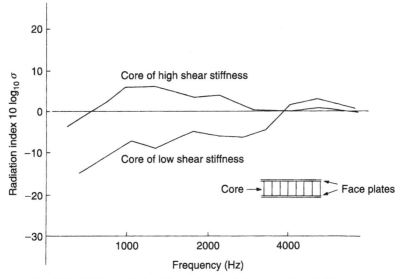

Fig. 3.41.   Typical radiation efficiency curves for honeycomb sandwich panels.

by a layer of fibrous material such as mineral wool or poroelastic material such as plastic foam (also known as a 'decoupler'). These treatments are designed to reduce airborne sound transmission and radiated structure-borne sound. Their radiation efficiencies and insulation effectiveness depend on the form of excitation (e.g., localised mechanical force, structure-borne or airborne sound) and also on the form of primary structure (Drotleff, 1996). They radiate most effectively when excited by airborne sound transmitted through the underlying primary structure and least effectively when excited by localised inputs. Generally speaking, the insulation performance falls as the radiation efficiency of the primary structure increases; it is thus less effective when applied to stiffened or curved plates than on uniform flat plates. This is unfortunate in terms of noise control. Further materials relating to sound insulation trim is presented in Section 5.12.

3.12   Sound Radiation by Curved Shells

The influence of shell curvature on sound radiation derives primarily from its effect on the flexural wave dispersion characteristics. Below the ring frequency, curvature increases flexural-wave phase velocities through the mechanism of mid-plane (membrane) strain, with a consequent increase of radiation efficiency.

Associated with this increase in wave speed is a reduction in the modal density. The complexity of behaviour of curved shells precludes any general analysis from presentation in this book. However, one class of curved-shell structures that is of considerable practical importance is that of circular cylindrical shells, which may be used as idealised models of such structures as industrial pipes, air conditioning ducts, aircraft and rocket bodies and submarine hulls. The general characteristics of cylindrical shell vibration have been presented in Chapter 1 and an analysis of fluid loading of cylinders is presented in Chapter 4. Here we confine our attention to a comparison of flat-plate and circular cylindrical-shell flexural wavenumber behaviour, and a qualitative discussion of the consequent difference between radiation behaviours.

The surface radial velocity of a wave propagating axially in a cylindrical shell of infinite length may be represented as

$$v_n(z, \phi, t) = \tilde{v}_n \cos n\phi \exp[j(\omega t - k_z z)], \qquad n = 0, 1, 2, \dots \qquad (3.155)$$

for the cylindrical coordinate system is shown in Fig. 3.42. Analysis presented in Chapter 4 shows that sound energy can be radiated only if the surface *axial* wavenumber k_z is less than the acoustic wavenumber k. This is contrast with flexural waves in an infinite, uniform, flat plate that radiate sound energy only if the bending wavenumber $k_b = (k_x^2 + k_z^2)^{1/2}$ is less than k. In a cylinder the wavenumber analogous to k_x is the circumferential wavenumber $k_s = n/a$, where a is the cylinder radius. The cosinusoidal variation with ϕ in Eq. (3.155) results from the interference between circumferential wavenumber components travelling in opposite directions around the cylinder: that is to say, Eq. (3.155) represents the interference field of two helical waves of equal and opposite circumferential

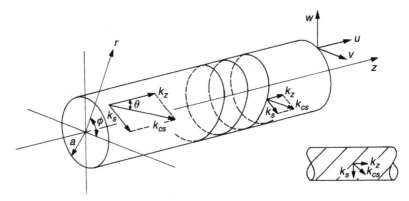

Fig. 3.42.   Circular cylindrical shell coordinates, displacements, wavenumber vectors and helical wavefronts.

wavenumber and equal axial wavenumber. It might be thought, therefore, that radiation from a wave represented by Eq. (3.155) would be analogous to that from a baffled flat strip of width $2\pi a$, carrying a sinusoidal standing wave across its width, in which case there exist supersonic wavenumber components in the width direction, as explained in Section 3.8. However, the edge effect between the strip and the baffle has no counterpart on the cylinder because the surface is continuous, and the equivalent 'edges' are coincident.

Under the conditions $k_z < k$, $k_s - n/a > k$, and $k_z^2 + k_s^2 > k^2$, which are not relevant to the $n = 0$ mode, adjacent zones of positive and negative volume velocity distributed around the circumference, seen in Fig. 3.43, do not completely cancel as they would in a plane travelling wave on an *infinite* plate below the critical frequency with $k_z < k$, but $k_x > k$ and $k_z^2 + k_s^2 > k^2$. However, their close proximity, in terms of an acoustic wavelength, makes their radiation very inefficient. The $n = 0$ 'breathing' mode radiates as a line monopole; the $n = 1$ 'beam bending' mode radiates as a line dipole; the $n = 2$ 'ovalling' mode radiates as a line quadrupole, and so on, the efficiency of radiation at any frequency decreasing with increase in the order of the equivalent source.

The corresponding expressions for power radiated per unit length, for $ka \ll 1$ and $k_z \ll k$, are (Skudrzyk, 1971)

$$\overline{P}_{n=0} = \tfrac{1}{2}\pi^2 \rho_0 ca(ka)\,|\tilde{v}_0|^2 \tag{3.156a}$$

$$\overline{P}_{n=1} = \tfrac{1}{4}\pi^2 \rho_0 ca(ka)^3\,|\tilde{v}_1|^2 \tag{3.156b}$$

$$\overline{P}_{n=2} = \tfrac{1}{64}\pi^2 \rho_0 ca(ka)^5\,|\tilde{v}_2|^2 \tag{3.156c}$$

The value of circumferential wavenumber k_s for a given n decreases with increase of cylinder radius. Hence, large-radius cylinder modes of order n can satisfy the condition

$$k_z^2 + k_s^2 < k^2 \tag{3.157}$$

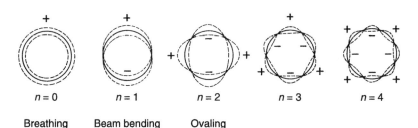

Fig. 3.43. Cross-sectional radial displacement mode shapes of a circular cylindrical shell.

at frequencies for which the equivalent modes of smaller cylinders, having the same axial wavenumber, give $k_z^2 + k_s^2 > k^2$. In the former case, inter-zone cancellation does not occur and the cylinder radiates with a radiation resistance close to $\rho_0 c$ per unit area. This shows that it is not sufficient that the frequency at which a cylinder vibrates in the bending mode ($n = 1$) should exceed the critical frequency based upon the bending-wave phase velocity in order to radiate efficiently; it is also necessary that

$$(k^2 - k_z^2)^{1/2}a > 1 \tag{3.158}$$

For thin-walled cylinders in which the wall thickness is much smaller than the radius, the $n = 1$ bending-mode axial wavenumber is

$$k_z = (2\rho\omega^2/a^2 E)^{1/4} \tag{3.159}$$

The transition from inefficient to efficient radiation occurs rather rapidly, as Fig. 3.44 indicates, so that for most practical purposes sound radiation from transversely vibrating slender bodies can be considered to be negligible if Eq. (3.158)

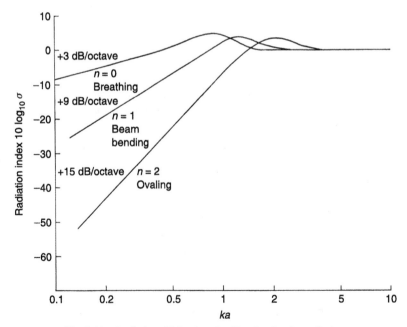

Fig. 3.44.   Radiation efficiencies of uniformly vibrating cylinders.

is not satisfied. Similarly, only when Eq. (3.157) is satisfied will radiation from cross-sectional distortion be significant.

As we have seen in the case of flat-plate radiation, wavenumber diagrams can be very useful in helping to identify those forms of vibration most effective in radiating sound. The bending-wave equation for flat plates produces a circular constant-frequency locus, as shown in Fig. 3.26. The curvature of a cylindrical shell produces coupling between radial, axial and circumferential motions, and there are consequently three coupled equations of motion and three classes of propagating waves (see Section 1.9). Although only the radial motion of a shell determines the sound radiation, the form of radial motion, and the associated dispersion characteristics, are significantly affected by mid-plane strains, especially at frequencies well below the ring frequency $f_r = c_l'/2\pi a$. These so-called membrane effects considerably raise the phase velocities of the class of waves whose displacement is predominantly radial (for convenience, we refer to these as 'flexural waves'), so much that in some cases, these waves have supersonic phase velocities at frequencies well below the critical frequency based upon the shell wall considered as a flat plate. Above the ring frequency, curvature effects largely disappear and the shell vibrates like an equivalent flat plate.

An approximate form of dispersion equation due to Heckl (1962b), presented previously as Eq. (1.53), is

$$\Omega^2 \approx \frac{(k_z a \beta^{1/2})^4}{[(k_z a \beta^{1/2})^2 + (n\beta^{1/2})^2]^2} + [(k_z a \beta^{1/2})^2 + (n\beta^{1/2})^2]^2 \qquad (3.160)$$

in which $\beta^2 = h^2/12a^2$, h is the wall thickness and $\Omega = f/(c_l'/2\pi a)$. This produces a wavenumber diagram of the form shown in Fig. 3.45. The membrane effect on wave speed is seen in the bending of frequency loci toward the origin in the range $0 < \Omega < 0.5$. A strange consequence of this behaviour is that at one frequency two helical waves of the same circumferential wavenumber, but different axial wavenumber, can propagate. Waves of low circumferential wavenumber involve greater membrane strain energy in proportion to flexural strain energy than waves of higher circumferential wavenumber. This leads to a rather unexpected variation of natural frequency with axial and circumferential wavelength.

Superimposition of the circular-arc acoustic wavenumber locus onto the structural wavenumber plots of Fig. 3.45 indicates the possibility of existence of supersonic structural waves at subcritical frequencies ($k_z^2 + k_s^2 < k^2$). Such waves clearly dominate low-frequency sound radiation, if they are generated by exciting forces. In a shell of finite length with simply supported edge conditions at the ends, natural modes are indicated by the intersection of the grid lines given by $k_s = n/a$ and the orthogonal set given by $k_z = m\pi/L$, where m is any positive integer and L the cylinder length. Modes for which $k_z^2 + k_s^2 > k^2$, $k_z^2 > k^2$ and

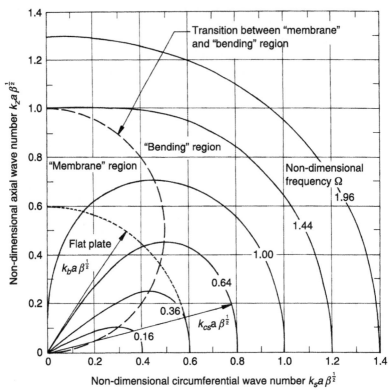

Fig. 3.45. Universal constant-frequency loci for flexural waves in thin-walled circular cylindrical shells ($n > 1$).

$k_s^2 < k^2$ can radiate like plate edge modes: inter-cell cancellation does not occur at the ends (see Fig. 3.46). This edge mode radiation model has been confirmed by Stepanishen (1978, 1982) who presents general integral expressions for the normalised modal radiation impedance and approximate algebraic expressions for 'high' and 'low' frequency. Some specific results for a ratio of cylinder length to radius of 3.0 are presented in Fig. 3.47. Fluid-loading aspects of cylindrical shell vibration are discussed in Chapter 4.

A statistical analysis on the basis of equipartition of modal energy in bands, analogous to that performed by Maidanik for flat plates, yields a general expression for modal average radiation efficiency (Manning and Maidanik, 1964). A similar form of analysis by Szechenyi (1971a), which is applicable to radiation by large diameter cylinders at *values of ka much greater than unity*, shows that two major frequency parameters influence the radiation efficiency. We have

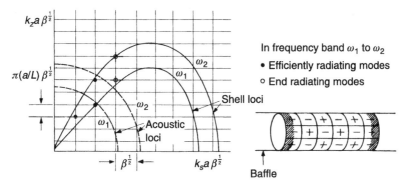

Fig. 3.46. End-radiating modes of a baffled cylindrical shell and regions of uncancelled volume velocity.

already met one of them, f/f_c, in connection with flat plates; the other is the ratio of critical frequency to ring frequency f_c/f_r. This parameter can also be expressed as

$$f_c/f_r = (c^2/1.8hc_l')/(c_l'/2\pi a) = (1/\beta)(c/c_l')^2 \tag{3.161}$$

Large-diameter, thin-wall shells have values of f_c/f_r much greater than unity, whereas industrial pipes typically have values of the order of, or considerably less than, unity. The significance of the value *of* f_c/f_r is that if it is greater than unity, there is a range of frequency between f_r and f_c in which the shell curvature effects on flexural wave speed disappear, and the cylinder radiates as a flat plate.

Hence the radiation efficiency falls above f_r and then rises again as f approaches f_c in the manner associated with flat-plate radiation. Figure 3.48 presents Szechenyi's radiation efficiency curves. A radiation index curve derived from measurements of shaker-induced vibration of a 0.9 m diameter, 1.80 m long, closed, steel cylindrical shell having a wall thickness of 4.8 mm thick wall (Fahy, 1969b) is compared in Fig. 3.49 with the theoretical results of Szechenyi (1971) and Manning and Maidanik (1964). The agreement is fairly satisfactory, although the slope of Szechenyi's line is 4 dB/octave and the average of the measured curve is nearer 5 dB/octave. Above f_r, where curvature effects diminish, the measured peak is characteristic of that of a flat plate.

Large diameter, thin-walled, circular cylindrical shells, such as those of aircraft fuselages, are usually stiffened by substantial ring frames disposed at regular axial intervals and also by much weaker, longitudinally oriented stiffeners (stringers) disposed at regular intervals around the shell circumference. The rather thicker pressure hulls of submarines are also stiffened by densely spaced ring frames.

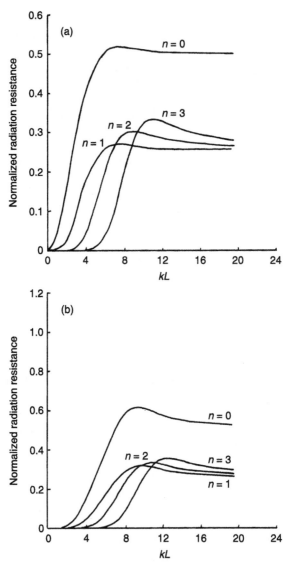

Fig. 3.47. Normalised self-radiation resistance of a modes of a cylinder of length-to-radius ratio of 3.0: L = cylinder length; n = circumferential order: (a) axial order = 1; (b) axial order = 2 (Stepanishen, 1982).

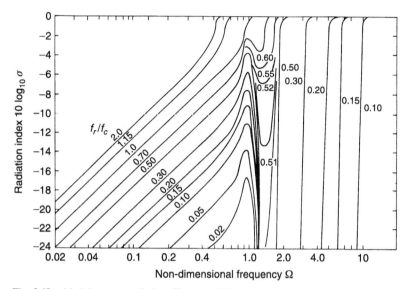

Fig. 3.48.   Modal-average radiation efficiency of thin-walled, large diameter, circular cylindrical shells (Szechenyi, 1971).

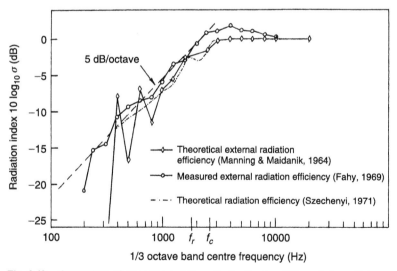

Fig. 3.49.   Comparison of measured and theoretical estimates of the radiation efficiency of a circular cylindrical shell (Manning and Maidanik, 1964; Fahy, 1969; Szechenyi, 1971).

The ring frames have the principal effect of considerably increasing the cut-off frequencies (and natural frequencies) of a range of the higher order ($n \geq 2$) flexural modes of thin-walled shells because they are much stiffer in in-plane bending than the plain shell. Above the lowest natural frequency of a section of plain shell bounded by two adjacent rings, the effect of the rings is largely to act as effectively rigid boundaries for each inter-ring section of the shell. This behaviour is illustrated by Fig. 3.50 (ESDU, 1983). A useful general analysis of the free vibration of orthogonally stiffened circular cylindrical shells is presented by Egle and Sewall (1968). Little quantitative data on the radiation efficiency of stiffened shells is available. However, given a knowledge of the free vibration mode shapes

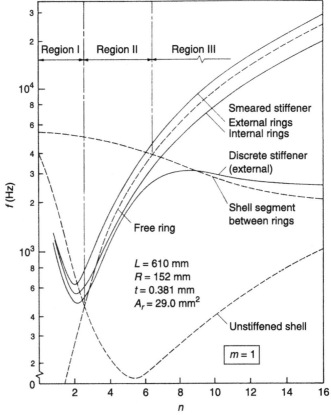

Fig. 3.50. Theoretical dispersion curves of a thin-walled, ring-stiffened, circular cylindrical shell: length = 610 mm, radius = 152 mm, wall thickness = 0.381 mm, cross-sectional area of ring stiffeners = 29 mm², axial mode order $m = 1$, circumferential mode order $= n$ (ESDU, 1983).

and natural frequencies, the method of analysis presented by Stepanishen (1982) may be extended to predict modal radiation impedances.

One of the more important acoustic effects of aircraft fuselage ring stiffeners is their influence on the coupling between the shell and the acoustic modes of the passenger compartment via their influence on modal frequencies and associated shapes. Ring frame design is of particular importance for propeller-driven aircraft because two or three lowest harmonics of propeller pressure fields often correspond quite closely with low order ring bending frequencies: clearly, rings must be designed to avoid close proximity.

The value of the parameter f_r/f_c of most industrial pipes (not including thin-walled heating and air conditioning ducts) is far greater than that of aircraft fuselages, and, in many cases, of the order of unity; this makes their radiation characteristics quite different. Pipes such as those found in petrochemical plants and gas distribution systems that are designed to transport fluid at pressures much greater than atmospheric generally have much smaller diameters, and much greater ratios of wall thickness to diameter (and hence wall thickness parameter β), than aircraft fuselages. However, many still qualify as 'thin-walled' shells. As mentioned in Section 1.9, their ring frequencies generally lie between 1 and 5 kHz which embraces the range of greatest human subjective sensitivity.

Figures 1.22 and 1.23 indicate that, at values of frequencies below the ring frequency, the influence of membrane strain (curvature) on the dispersion curves of the lower order shell waveguide modes is to reduce their axial wavenumber to values well below the 'equivalent' flat plate values; this influence increases with decrease of β [see also Eq. (1.49)]. Hence, the influence on a mode of given circumferential order n of industrial pipes is proportionately smaller than that on the corresponding mode of aircraft fuselages. However, the much smaller value of f_c/f_r of industrial pipes ensures that the axial wavenumbers of the low order modes of pipes are less than the acoustic wavenumber in the external air. This can be appreciated by superimposing the non-dimensional acoustic wavenumber line given by $ka\beta^{1/2} = \Omega(c_l''/c)\beta^{1/2}$, on Fig. 1.23. The intersection of this line with the dispersion curve of flexural waves in a flat plate marks the non-dimensional critical frequency Ω_c. As the wall thickness parameter β increases, Ω_c decreases in inverse proportion. The effect of the combination of smaller diameter and larger value of β is to increase the modal-average radiation efficiency in the range Ω less than unity.

At this point the reader should be warned that the *modal-average* radiation efficiency computed for industrial-type pipes can be very misleading. Pipe vibration and associated sound radiation are strongly influenced by the form of excitation. Vibration induced by connected mechanical equipment excites only the $n = 1$ pipe bending mode in the frequency range below the cut-off frequency of the $n = 2$ 'ovaling' mode ($\Omega \approx 2.7\beta$). Pipes are so stiff in bending that the bending mode has a very low critical frequency. Therefore the $n = 1$ curve in Fig. 3.44

is appropriate, and in the frequency range $ka < 1$, the radiation efficiency is given by $\sigma \approx \pi(ka)^3/4$ (+9 dB per octave). The higher order shell modes are very inefficient radiators in the frequency range $ka < n$ and they may also be mechanically excited at resonance near their cut-off frequencies. Consequently, they contribute substantially to surface vibration but little to radiation. In such cases, the radiation efficiency can fall well below the bending mode value. In all cases, however, the radiation efficiency asymptotes to unity for $ka \gg 1$.

Many industrial pipes are excited into vibration by internal sound sources such as flow control valves. Sources acting within the pipes generate acoustic waveguide modes. Each mode, except the zero order, plane wave, mode has a cut-off frequency below which it does not propagate. The non-dimensional cut-off of frequency of the lowest order non-plane mode is given by $ka = 1.84$ which corresponds to $\Omega_{10} = 1.84c/c_l'$: the value is 0.125 for steel pipes in atmospheric air. Below this frequency, the only propagating acoustic mode that excites a pipe wall is the plane wave which, in principle, can excite only the $n = 0$ breathing mode of a straight, uniform, pipe. In practice, circumferential non-uniformity of the pipe wall, together with features such as bends, supports and attached ancillary components, allow a plane wave to excite higher order modes of a pipe wall. In particular, the presence of radiused bends strongly promote the shell modes of circumferential orders 2 and 3 above their respective cut-off frequencies. Theoretical assessment of the response of pipework systems to plane wave excitation is therefore very problematic. The breathing stiffness impedance of a straight pipe well below the ring frequency is proportional to wall thickness and inversely proportional to the product of frequency and the square of the diameter: it is generally so high that the associated sound radiation is weak. This does not, of course, mean that the radiation efficiency is also negligible.

Pipe wall response to propagating acoustic waveguide modes is dominated by *coincidence* between their axial wavenumbers and those of shell modes having the same circumferential order (Norton and Karczub, 2003). [This phenomenon is explained more fully in Chapter 5.] Coincidence with the low order shell modes of industrial-type pipes that tend to dominate the response below the ring frequency occurs just above the corresponding acoustic mode cut-off frequencies. Hence the axial wavenumbers are small and much less than the acoustic wavenumber. The radiation efficiencies are therefore close to that of the pipe wall mode when executing *axially uniform* motion. The lowest coincidence frequency involving the pipe bending mode occurs just above Ω_{10} where pipe wall response spectra are generally observed to exhibit a rapid rise. The acoustic mode cut-off frequencies are such that the radiation efficiencies of the coincident structural modes are close to unity. Below Ω_{10}, the radiation efficiency should, in principle, correspond to that of the pipe wall breathing mode. However, sound sources such as control valves generate strong turbulence in the fluid immediately downstream of the valve orifice and this, together with the effects of bends and other wave-scattering

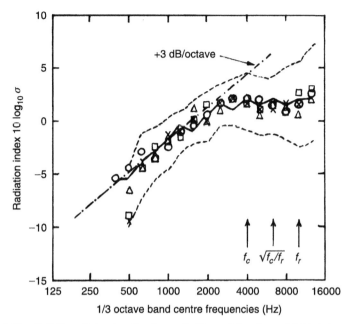

Fig. 3.51. Radiation efficiencies of a pipe excited by an internal sound field at flow speeds of $0, 30, 60, 90$ and $120\,\mathrm{ms}^{-1}$. (Holmer and Heymann, 1980).

features, appear to be effective in exciting the pipe bending mode. An example of pipe radiation efficiency measured on a straight pipe excited by a horn driver is presented in Fig. 3.51. The average rate of rise of the curve below Ω_{10} is about 4 dB per octave, whereas the corresponding rates for the pipe breathing and bending modes are 3 and 9 dB per octave, respectively. It is likely that the axisymmetric driver did not strongly excite the bending mode, although, in common with all other pipes, the radiation efficiency tends to asymptote to unity above Ω_{10}. Coincident response of industrial pipes is usually resonance controlled because such pipes are generally long in proportion to their radius and the resonance frequencies of wall vibration modes of a given n tend to cluster densely in the frequency range immediately above the corresponding structural cut-off frequency. However, it is not easy to increase the damping of many forms of industrial pipe because of the high stiffness associated with their relatively thick walls.

Figure 3.52 shows radiation efficiencies measured on three straight pipe lengths in a laboratory (Rennison, 1977). The radiation efficiency below Ω_{10} is seen to fall well below that of the breathing mode. The reason appears to be that, well below the ring frequency, a pipe is extremely stiff in pure radial

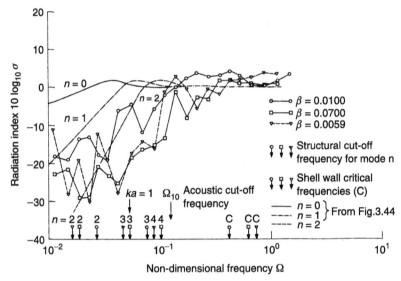

Fig. 3.52. Radiation indices of steel pipes excited by sound in the contained air (Rennison, 1977).

displacement, and $n = 1$ beam bending vibration and higher-order circumferential modes of the pipe were somehow more easily excited by the internal sound field. The cut-off frequencies of the higher order wall modes, immediately above which multiple resonances in these modes occur, can be calculated from Table 1.3 in Chapter 1, and are indicated on Fig. 3.52. As explained above, these modes are very inefficient radiators near their cut-off frequencies and, since they dominate the response energy in this frequency region, they produce the associated dips in the radiation efficiency curves seen in the figure. These show that it is virtually impossible to theoretically estimate the radiation efficiency or radiated sound power of operational industrial pipes below Ω_{10}. However, measurements made under field conditions on piping systems suggest that the assumption of breathing mode dominance provides a very conservative upper limit.

One of the results of the influence of wall curvature on cylinder wall wave speed is that the addition of stiffeners and other constraints to thin-walled cylinders does not increase their radiation efficiencies as much as in the case of a flat plate. Because some resonant shell modes are created from supersonic travelling-wave components, the application of damping to mechanically excited thin-wall cylindrical shells is likely to reduce radiated power more than it would on a flat plate of the same thickness. However, it should not be assumed that the addition of damping will effectively reduce sound radiated from industrial pipes excited by disturbances in the contained fluids, because the beam bending and ovaling ($n = 2$) modes tend to dominate radiation and these are not easily damped. The application of stiff constraints in the form of multiple supports and rings, together with insulation cladding is generally more efficacious. Sound transmission through cylindrical shells is discussed in Chapter 5.

3.13 Sound Radiation by Irregularly Shaped Vibrating Bodies

Although the analyses of radiation from uniform flat plates and cylinders are useful in revealing the major controlling parameters, many vibrating structures do not even approximate to such simple forms, making analytic solutions impossible. Therefore the general formulation for the sound radiation into an infinitely extended fluid domain V_a by a body of arbitrary shape and surface S_a, as shown in Fig. 3.53, is now derived.

The sound field generated by a harmonically vibrating object, or part thereof, immersed in an otherwise unbounded volume of fluid V_a can be derived by solving the homogeneous Helmholtz equation

$$\nabla^2 \tilde{p}(\mathbf{r}) + k^2 \tilde{p}(\mathbf{r}) = 0 \tag{3.162}$$

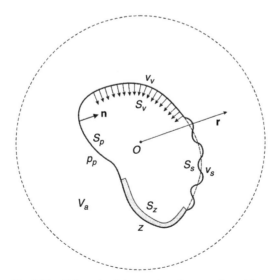

Fig. 3.53. Schematic of the exterior vibroacoustic problem.

subject to the boundary conditions imposed by the surface S_a. In general, the surface can be classified in terms of three different types of boundary condition. First, surface S_v on which the component of the velocity $\tilde{v}_v(\mathbf{r}_v)$ normal to the surface is prescribed so that the component of fluid particle velocity in the direction normal to, and *inwards* from, the radiating surface is given by $\tilde{v}_n(\mathbf{r}_v) = \tilde{v}_v(\mathbf{r}_v)$ (Neumann boundary condition). This type of boundary condition (but unprescribed) applies also on surface S_s where the radiating object is modelled as a flexible structure in which case the component of fluid particle velocity in direction normal to the radiating surface is equal to the transverse velocity of the vibrating structure so that $\tilde{v}_n(\mathbf{r}_s) = \tilde{v}_s(\mathbf{r}_s)$. Second, surface S_p on which the pressure $\tilde{p}_p(\mathbf{r}_p)$ is prescribed so that $\tilde{p}(\mathbf{r}_p) = \tilde{p}_p(\mathbf{r}_p)$ (Dirichlet boundary condition). Third, surface S_z which is characterised by normal specific acoustic impedance $\tilde{z}(\mathbf{r}_z)$, so that $\tilde{p}(\mathbf{r}_z)/\tilde{v}_n(\mathbf{r}_z) = \tilde{z}(\mathbf{r}_z)$ (mixed boundary condition). The link between the normal component of the complex fluid particle velocity and the gradient of the complex sound pressure normal to the surface S_a is given by Eq. (1.5) as

$$\tilde{v}_n(\mathbf{r}_{S_a}) = \frac{j}{\omega\rho_0}\left(\frac{\partial\tilde{p}}{\partial n}\right)_{\mathbf{r}=\mathbf{r}_{S_a}} \tag{3.163}$$

where $\mathbf{r}_{S_a}$ defines the radiating surface S_a and n is the local coordinate normal to S_a.

When a solid vibrates in an inviscid fluid medium, it displaces fluid volume at its boundary and applies normal forces associated with the corresponding rates of change of momentum of the fluid local to the surface. Assuming arbitrary distributions within the fluid of rates of volume displacement q per unit volume (monopole) and distributions of external forces per unit volume f (dipole), incorporation of these quantities into the equations of conservation of mass [Eq. (1.4)] and momentum [Eq. (1.5)] leads to the inhomogeneous wave equation

$$\nabla^2 p(\mathbf{r}, t) - \frac{1}{c^2}\frac{\partial p(\mathbf{r}, t)}{\partial t} = \nabla \cdot \mathbf{f}(\mathbf{r}, t) - \rho_0\frac{\partial q(\mathbf{r}, t)}{\partial t} \tag{3.164}$$

where ∇ is the divergence operator and ∇^2 is the Laplacian operator. The terms on the right-hand side represent the distributed kinetic (force) and kinematic (volumetric) sources of sound (Dowling and Ffowcs-Williams, 1983). The assumption of harmonic time dependence gives

$$\nabla^2\tilde{p}(\mathbf{r}) + k^2\tilde{p}(\mathbf{r}) = -\tilde{Q}(\mathbf{r}) \tag{3.165}$$

where $\tilde{Q}(\mathbf{r}) = j\rho_0\omega\tilde{q}(\mathbf{r}) - \nabla \cdot \tilde{\mathbf{f}}(\mathbf{r})$. Irrespective of the physical nature of $\tilde{Q}(\mathbf{r})$, the solution to Eq. (3.165) can be constructed from the pressure field produced by a point monopole source which satisfies Eq. (3.165) with $Q(\mathbf{r}) = \delta(\mathbf{r} - \mathbf{r}_0)$, where $\delta(\mathbf{r} - \mathbf{r}_0)$ is the three-dimensional Dirac delta function and $\mathbf{r}_0$ specifies the location of the source. This is known as the acoustic Green's function $\widetilde{G}(\mathbf{r}, \mathbf{r}_0, \omega)$ (Morse and Feshbach, 1953), which is a solution to the inhomogeneous scalar Helmholtz equation written in the form

$$\nabla^2 \widetilde{G}(\mathbf{r}, \mathbf{r}_0, \omega) + k^2 \widetilde{G}(\mathbf{r}, \mathbf{r}_0, \omega) = -\delta(\mathbf{r} - \mathbf{r}_0) \qquad (3.166)$$

where $\mathbf{r}$ is the observation point. Green's functions may satisfy various forms of boundary condition. In the following analysis, we employ the Green's function that applies to unbounded fluid. In the absence of scattering boundaries, the Green's function must also satisfy the condition that only waves travelling outward from the point source are allowed, and that the pressure tends to zero at an infinite distance from the source:

$$\lim_{|\mathbf{r}| \to \infty} |\mathbf{r}| \left(\frac{\partial \tilde{p}(\mathbf{r})}{\partial |\mathbf{r}|} + jk\tilde{p}(\mathbf{r}) \right) = 0 \qquad (3.167)$$

This condition is known as the Sommerfeld radiation condition at infinity (Sommerfeld, 1949). Thus,

$$\lim_{|\mathbf{r} - \mathbf{r}_0| \to \infty} |\mathbf{r} - \mathbf{r}_0| \left(\frac{\partial \widetilde{G}(\mathbf{r}, \mathbf{r}_0)}{\partial |\mathbf{r} - \mathbf{r}_0|} + jk\widetilde{G}(\mathbf{r}, \mathbf{r}_0) \right) = 0 \qquad (3.168)$$

Advanced acoustics texts (e.g., Morse and Ingard, 1968; Pierce, 1989) show that in free space, the solution of Eq. (3.166) subject to the boundary condition in Eq. (3.168) is

$$\widetilde{G}(\mathbf{r}, \mathbf{r}_0, \omega) = \frac{e^{-jk|\mathbf{r} - \mathbf{r}_0|}}{4\pi |\mathbf{r} - \mathbf{r}_0|} \qquad (3.169)$$

a solution that corresponds to Eq. (3.2) for the pressure field produced in free space by a small pulsating sphere ($ka \ll 1$). It should be noted that this function is singular for $\mathbf{r} = \mathbf{r}_0$. This function is known as the 'free-space' Green's function.

The Kirchhoff–Helmholtz integral equation for the derivation of the radiated acoustic field of objects with irregular shapes is derived using Green's second identity which states that, for any two scalar functions φ and ψ which are sufficiently smooth and non-singular in a domain V enclosed by a surface S,

$$\int_S \left(\varphi \frac{\partial \psi}{\partial n} - \psi \frac{\partial \varphi}{\partial n} \right) dS = \int_V \left(\varphi \nabla^2 \psi - \psi \nabla^2 \varphi \right) dV \qquad (3.170)$$

where the normal to the surface **n** has positive orientation pointing outwards from the volume V (see, for example, Kreyszig, 1999). Although the radiation problem at hand assumes an unbounded fluid domain, this integral expression is applied assuming ψ as the free field Green's function $\widetilde{G}(\mathbf{r}, \mathbf{r}_a)$ defined in Eq. (3.169) and φ is the sound pressure $\tilde{p}(\mathbf{r}_a)$ in the volume V_a to be bounded by the radiating surface $S_a = S_v \cup S_s \cup S_z \cup S_p$ and by a much bigger imaginary surface S_{R2} as shown in Fig. 3.54. In order to simplify the mathematical formulation, is taken to be a sphere of radius $R_2 \to \infty$ so that the Sommerfeld's radiation condition applies. The centre position of the sphere S_{R2} is chosen at the position **r** and, in order to avoid the singularity problem of the free field Green's function $\widetilde{G}(\mathbf{r}, \mathbf{r}_a, \omega)$ when $\mathbf{r}_a = \mathbf{r}$, another sphere S_{R1} concentric with S_{R2} and with radius $R_1 \to 0$ is defined. Since the point **r** is excluded from domain V_a, both $\tilde{p}$ and $\widetilde{G}$ satisfy Eq. (3.162) in V_a. Therefore, as shown in Fig. 3.54(a), the acoustic domain V_a is bounded by three surfaces: the radiating surface S_a, the infinitesimally small spherical surface S_{R1} and the infinitely large spherical surface S_{R2}, both the latter centred on **r** (see, for example, Desmet, 1998; Wu, 2000). Green's second identity is therefore applied by replacing the function φ by the sound pressure function $p(\mathbf{r}_a)$ that satisfies the homogeneous Helmholtz Eq. (3.162) in the volume V_a and the three types of boundary condition on the surface S_a listed above, and substituting the free space Green's function $\widetilde{G}(\mathbf{r}, \mathbf{r}_a, \omega)$ for the function ψ so that the following integral equation is obtained:

$$\int_{S_a + S_{R_1} + S_{R_2}} \left(\tilde{p}(\mathbf{r}_a) \frac{\partial \widetilde{G}(\mathbf{r}, \mathbf{r}_a)}{\partial n} - \widetilde{G}(\mathbf{r}, \mathbf{r}_a) \frac{\partial \tilde{p}(\mathbf{r}_a)}{\partial n} \right) dS = 0 \qquad (3.171)$$

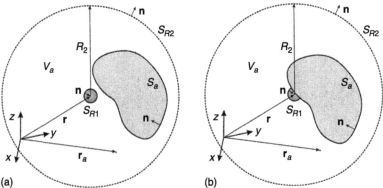

(a)   (b)

Fig. 3.54. Schematic of the exterior problem where the fluid domain is bounded by the radiating surface S_a, the spherical surface S_{R2} and the spherical surface S_{R1} which is either: (a) outside or (b) intersected by the bounding surface S_a.

The volume integral and the right-hand side of Eq. (3.170) has value zero because both $\varphi = \tilde{p}(\mathbf{r}_a)$ and $\psi = \widetilde{G}(\mathbf{r}, \mathbf{r}_a)$ satisfy Eq. (3.162) in V_a and therefore both $\varphi \nabla^2 \psi$ and $\psi \nabla^2 \varphi$ are equal to $-k^2 \widetilde{G}(\mathbf{r}, \mathbf{r}_a) \tilde{p}(\mathbf{r}_a)$.

We consider first the case shown in Fig. 3.54(a) where the point $\mathbf{r}$ is located within the volume V_a. When the position defined by $\mathbf{r}_a$ is on the spherical surface S_{R1} such that $|\mathbf{r} - \mathbf{r}_a| = R_1$ then considering a spherical coordinate system (R_1, ϕ, θ), the integral over S_{R1} gives

$$
\int_{S_{R_1}} \left(\tilde{p}(\mathbf{r}_a) \frac{\partial \widetilde{G}(\mathbf{r}, \mathbf{r}_a)}{\partial n} - \widetilde{G}(\mathbf{r}, \mathbf{r}_a) \frac{\partial \tilde{p}(\mathbf{r}_a)}{\partial n} \right) dS_{R_1}
$$

$$
= \int_0^{2\pi} \left(\int_0^{\pi} \left\{ -\tilde{p}(R_1, \phi, \theta) \frac{\partial}{\partial R_1} \left(\frac{e^{-jkR_1}}{4\pi R_1} \right) \right. \right.
$$

$$
\left. \left. + \left(\frac{e^{-jkR_1}}{4\pi R_1} \right) \frac{\partial \tilde{p}(R_1, \phi, \theta)}{\partial R_1} \right\} R_1^2 \sin\theta d\theta \right) d\phi \qquad (3.172)
$$

$$
= \int_0^{2\pi} \left(\int_0^{\pi} \left\{ \tilde{p}(R_1, \phi, \theta)(1 + jkR_1) \frac{e^{-jkR_1}}{4\pi} \right. \right.
$$

$$
\left. \left. + \frac{R_1 e^{-jkR_1}}{4\pi} \frac{\partial \tilde{p}(R_1, \phi, \theta)}{\partial R_1} \right\} \sin\theta d\theta \right) d\phi
$$

which in the limiting case $R_1 \rightarrow 0$, such that $p(\mathbf{r}_a) \rightarrow p(\mathbf{r})$, gives

$$
\lim_{R_1 \to 0} \int_{S_{R_1}} \left(\tilde{p}(\mathbf{r}_a) \frac{\partial \widetilde{G}(\mathbf{r}, \mathbf{r}_a)}{\partial n} - \widetilde{G}(\mathbf{r}, \mathbf{r}_a) \frac{\partial \tilde{p}(\mathbf{r}_a)}{\partial n} \right) dS_{R_1}
$$

$$
= \tilde{p}(\mathbf{r}) \int_0^{2\pi} \int_0^{\pi} \frac{\sin\theta}{4\pi} d\theta d\phi = \tilde{p}(\mathbf{r}) \qquad (3.173)
$$

If the position defined by $\mathbf{r}_a$ is on the spherical surface S_{R2}, so that $|\mathbf{r} - \mathbf{r}_a| = R_2$, the integral over S_{R2} can be derived using Eq. (3.172) and substituting R_2 for

R_1 which gives

$$
\int_{S_{R_2}} \left(\tilde{p}(\mathbf{r}_a) \frac{\partial \tilde{G}(\mathbf{r}, \mathbf{r}_a)}{\partial n} - \tilde{G}(\mathbf{r}, \mathbf{r}_a) \frac{\partial \tilde{p}(\mathbf{r}_a)}{\partial n} \right) dS
$$

$$
= \int_0^{2\pi} \left(\int_0^{\pi} \left\{ \tilde{p}(R_2, \phi, \theta) \frac{\partial}{\partial R_2} \left(\frac{e^{-jkR_2}}{4\pi R_2} \right) \right. \right.
$$

$$
\left. \left. - \left(\frac{e^{-jkR_2}}{4\pi R_2} \right) \frac{\partial \tilde{p}(R_2, \phi, \theta)}{\partial R_2} \right\} R_2^2 \sin\theta d\theta \right) d\phi \qquad (3.174)
$$

$$
= -\int_0^{2\pi} \left(\int_0^{\pi} \left\{ \tilde{p}(R_2, \phi, \theta) + R_2 \left(\frac{\partial \tilde{p}(R_2, \phi, \theta)}{\partial R_2} \right. \right. \right.
$$

$$
\left. \left. \left. + jk\tilde{p}(R_2, \phi, \theta) \right) \right\} \frac{e^{-jkR_2}}{4\pi} \sin\theta d\theta \right) d\phi
$$

According to the Sommerfeld radiation condition [Eq. (3.168)], and assuming that the sound pressure vanishes in the limit $R_2 \to \infty$, the equation reduces to

$$
\lim_{R_2 \to \infty} \int_{S_{R_2}} \left(\tilde{p}(\mathbf{r}_a) \frac{\partial \tilde{G}(\mathbf{r}, \mathbf{r}_a)}{\partial n} - \tilde{G}(\mathbf{r}, \mathbf{r}_a) \frac{\partial \tilde{p}(\mathbf{r}_a)}{\partial n} \right) dS_{R_2} = 0 \qquad (3.175)
$$

Substituting Eqs. (3.173) and (3.175) into Eq. (3.171) yields:

$$
\int_{S_a} \left(\tilde{p}(\mathbf{r}_a) \frac{\partial \tilde{G}(r, \mathbf{r}_a)}{\partial n} - \tilde{G}(\mathbf{r}, \mathbf{r}_a) \frac{\partial \tilde{p}(\mathbf{r}_a)}{\partial n} \right) dS_a = -\tilde{p}(\mathbf{r}) \qquad (3.176)
$$

As shown in Fig. 3.54(b), we consider now the point at position $\mathbf{r}$ is located on the closed surface S_a so that, for $R_1 \to 0$, the surface R_1 becomes a hemisphere. In this case, provided the surface S_a is smooth such that the normal direction to S_a is uniquely defined at any position on the surface, when the point defined by $\mathbf{r}_a$ is on the hemispherical surface S_{R1}, the integral in Eq. (3.173) gives

$$
\lim_{R_1 \to 0} \int_{S_{R_1}} \left(\tilde{p}(\mathbf{r}_a) \frac{\partial \tilde{G}(\mathbf{r}, \mathbf{r}_a)}{\partial n} - \tilde{G}(\mathbf{r}, \mathbf{r}_a) \frac{\partial \tilde{p}(\mathbf{r}_a)}{\partial n} \right) dS_{R_1}
$$

$$
= \tilde{p}(\mathbf{r}) \int_0^{2\pi} \int_0^{\pi/2} \frac{\sin\theta}{4\pi} d\theta d\phi = \frac{1}{2} \tilde{p}(r) \qquad (3.177)
$$

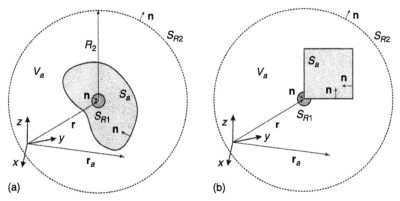

Fig. 3.55. Schematic of the exterior problem where the fluid domain is bounded by the radiating surface S_a, the spherical surface S_{R2} and the spherical surface S_{R1} which is either: (a) inside S_a or (b) intersected by a protrusion in the bounding surface S_a.

(Note that the integration in $d\theta$ is between 0 and $\pi/2$.) Therefore, substituting Eqs. (3.177) and (3.175) into Eq. (3.171), gives

$$\int_{S_a} \left(\tilde{p}(\mathbf{r}_a) \frac{\partial \tilde{G}(\mathbf{r}, \mathbf{r}_a)}{\partial n} - \tilde{G}(\mathbf{r}, \mathbf{r}_a) \frac{\partial \tilde{p}(\mathbf{r}_a)}{\partial n} \right) dS_a = -\frac{1}{2} \tilde{p}(\mathbf{r}) \qquad (3.178)$$

Finally, considering the case shown in Fig. 3.55(a) where the point at position $\mathbf{r}$ is located within the closed surface S_a such that for $R_1 \rightarrow 0$ it is not located in the domain V_a, then, when the point defined by $\mathbf{r}_a$ is on the spherical surface S_{R1}, the integral of Eq. (3.173) is bound to be 0 so that, recalling Eq. (3.175) the integral in Eq. (3.171) is given by

$$\int_{S_a} \left(\tilde{p}(\mathbf{r}_a) \frac{\partial \tilde{G}(\mathbf{r}, \mathbf{r}_a)}{\partial n} - \tilde{G}(\mathbf{r}, \mathbf{r}_a) \frac{\partial \tilde{p}(\mathbf{r}_a)}{\partial n} \right) dS_a = 0 \qquad (3.179)$$

The three cases discussed above can be summarised in one formula which, using Eq. (3.163), gives the so called Kirchhoff–Helmholtz integral equation:

$$c(\mathbf{r}) \tilde{p}(\mathbf{r}) = \int_{S_a} \left(\tilde{p}(\mathbf{r}_a) \frac{\partial \tilde{G}(\mathbf{r}, \mathbf{r}_a)}{\partial n} + j \rho_0 \omega \tilde{G}(\mathbf{r}, \mathbf{r}_a) \tilde{v}_n(\mathbf{r}_a) \right) dS_a \qquad (3.180)$$

with

$$
c(\mathbf{r}) = \begin{cases} -1 & \mathbf{r} \in V_a \\ 0 & \mathbf{r} \notin V_a \\ -1/2 & \mathbf{r} \in S_a \end{cases} \tag{3.181}
$$

This equation states that for any harmonic sound pressure field that satisfies the homogeneous Helmholtz equation and the Sommerfeld radiation condition, the sound pressure at any point $\mathbf{r}$ of the unbounded domain V_a is determined by the combination of the pressure distribution $p(\mathbf{r}_a)$ and the normal velocity distribution $v(\mathbf{r}_a)$ on the closed surface S_a.

If, as shown in Fig. 3.55(b), the point defined by $\mathbf{r}_a$ is located at a corner of the closed surface S_a so that the normal to the surface cannot be unambiguously defined, then

$$
c(\mathbf{r}) = 1 + \frac{1}{4\pi} \int_{S_a} \frac{\partial}{\partial n}\left(\frac{1}{|\mathbf{r} - \mathbf{r}_a|}\right) dS_a \tag{3.182}
$$

which indicates the exterior solid angle, expressed as a fraction of 4π, of the surface S_a at the corner position $\mathbf{r}$.

Therefore, the direct boundary integral formulation for exterior problems with harmonic time dependence relates the acoustic pressure in any point in the free field to the pressure and normal velocity on the boundary surface of any body immersed in the fluid. Although the velocity term in Eq. (3.180) involves surface velocity, it is actually a surface *acceleration* term, which more accurately indicates the physical nature of its influence on sound pressure in the fluid.

The Kirchhoff–Helmholtz integral equation is, on first acquaintance, rather puzzling. Surely, the sound field generated by a harmonically vibrating body in free space is uniquely determined by the geometry of its surface, the acoustic properties of the fluid and the distribution of normal velocity over that surface; so why is there a surface pressure term in Eq. (3.180)? The answer lies in the particular form of the free-space Green's function which is relevant at a point monopole operating in *free space*. In fact, the velocity term in Eq. (3.180) is identical to that in Eq. (3.2). However, although the distribution of surface motion may be considered to comprise an array of elementary sources of volume acceleration, each element is *operating in the presence of the remainder of the body* which scatters the radiated sound and causes it to differ from that of an isolated monopole. The surface pressure term represents the effect of this scattering process. In the special case of vibration of an infinitely extended plane surface, the scattering is such that the pressure term equals the acceleration term and the sound pressure per unit elemental source strength is doubled, as represented by Eq. (3.3). Integration over the surface yields the Rayleigh integral (3.4).

In order to properly take into account this scattering effect, the numerical evaluation of the sound field radiated by an object has to be carried out into two steps. First, the Kirchhoff–Helmholtz integral equation is solved for positions on the radiating surface itself assuming $c = -1/2$ in order to evaluate either the boundary sound pressure $p(\mathbf{r}_a)$ or normal particle velocity $v_n(\mathbf{r}_a)$, depending on the type of imposed boundary condition. This operation must be handled with care since the integral involves collocation of the observation and source points used to model the point sources on the radiating surface in which case the free-space Green's function $\widetilde{G}(\mathbf{r}_a, \mathbf{r}_a)$ is singular. In general, for smooth surfaces the singularity is weak and thus the Kirchhoff–Helmholtz integral equation converges (Wu, 2000). Having solved for the surface boundary conditions, the acoustic field within the volume is then evaluated from the Kirchhoff–Helmholtz integral equation assuming $c = 1$ and using in the right-hand side integral the boundary sound pressure $p(\mathbf{r}_a)$ and normal particle velocity $v_n(\mathbf{r}_a)$ derived in the first step. This solution procedure is usually implemented numerically by means of the boundary element method (BEM) which is presented in Chapter 8.

It is also important to note that, as shown in Chapter 7, the same integral formulation applies to interior problems such as the evaluation of the acoustic field generated within an enclosure. This raises a problem since the interior sound field is characterised by a resonance frequencies which do not exist in the correspondent exterior problem. This is purely a mathematical problem which is normally overcome in the first step of the solution procedure by adding to the Kirchhoff–Helmholtz integral additional constraints related to the acoustic field inside the externally radiating body where the sound pressure and particle velocity is known to be zero. This procedure is known as the CHIEF method which is explained in Chapter 8.

When, as shown in Fig. 3.56, acoustic sources operate within the volume of fluid, the total sound pressure field is given by the superposition of the *free-field* sound pressure generated by the sources in absence of the scattering body(ies) and the scattered/radiated sound pressure field determined by the pressure and normal velocity on the boundary surface(s) of the body(ies) immersed in the fluid as expressed by the integral equation (Junger and Feit, 1986):

$$\tilde{p}(\mathbf{r}) = \frac{1}{c(\mathbf{r})} \int_{S_a} \left(\tilde{p}(\mathbf{r}_a) \frac{\partial \widetilde{G}(\mathbf{r}, \mathbf{r}_a)}{\partial n} + j\rho_0\omega \, \widetilde{G}(\mathbf{r}, \mathbf{r}_a)\tilde{v}_n(\mathbf{r}_a) \right) dS_a$$

$$+ \int_{V_s} \widetilde{Q}(\mathbf{r}_s) \, \widetilde{G}(\mathbf{r}, \mathbf{r}_s) \, dV_s \tag{3.183}$$

This equation is often referred as the 'direct boundary integral formulation' since the boundary variables have direct physical meaning. As for the radiation problem, the evaluation of the sound field is carried out in two steps: first,

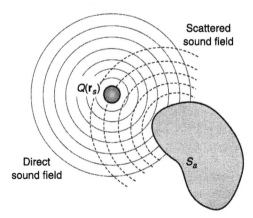

Fig. 3.56. Schematic of the exterior vibroacoustic problem involving volume-distributed sources.

Eq. (3.183) is solved for positions on the scattering/radiating surface assuming $c = -1/2$ and second, Eq. (3.183) is solved using in the right-hand side integral the boundary sound pressure and normal particle velocities derived in step 1. In the case of scattering by a rigid body, v_n is set to zero in Eq. (3.183). In case where active acoustic sources insonify a flexible structure, Eq. (3.183) must be solved in conjunction with the equation(s) governing the dynamics of the structure. This is termed as the 'fully coupled' vibroacoustic problem which is addressed in Chapter 8.

Acoustic problems which involve an open boundary surface S_a (Fig. 3.57) are normally treated with the 'indirect boundary integral formulation' (Filippi, 1977). A brief description of this alternative approach is given below without entering into the details of the formulation. In the indirect boundary integral approach, the sound pressure at any point of the acoustic field is related to the distribution of a single layer potential $\tilde{\sigma}(\mathbf{r}_a)$ and a double layer potential $\tilde{\mu}(\mathbf{r}_a)$ on the boundary surface S_a such that

$$\tilde{p}(\mathbf{r}) = \int_{S_a} \left(\tilde{\mu}(\mathbf{r}_a) \frac{\partial \tilde{G}(\mathbf{r}, \mathbf{r}_a)}{\partial n} - \tilde{\sigma}(\mathbf{r}_a) \tilde{G}(\mathbf{r}, \mathbf{r}_a) \right) dS_a \qquad (3.184)$$

The single layer potential $\tilde{\sigma}(\mathbf{r}_a)$ represents the difference of the sound pressure gradients in direction normal to the surface S_a on the two sides of the boundary surface S_a; that is

$$\tilde{\sigma}(\mathbf{r}_a) = \frac{\partial \tilde{p}(\mathbf{r}_a^+)}{\partial n} - \frac{\partial \tilde{p}(\mathbf{r}_a^-)}{\partial n} \qquad (3.185)$$

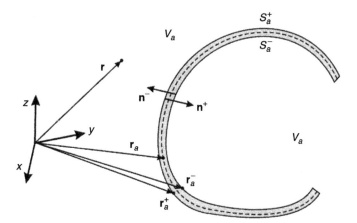

Fig. 3.57. Schematic of the exterior vibroacoustic problem involving an open boundary.

where $(\mathbf{r}_a^+)$ and $(\mathbf{r}_a^-)$ indicate the positions on the two sides of the boundary surface. This potential function can therefore be interpreted as a distribution of monopole sources on the median surface. This term is zero if the surface takes the form of a solid plate or shell. The double layer potential $\tilde{\mu}(\mathbf{r}_a)$ represents the sound pressure difference between the two sides of the boundary surface S_a; that is

$$\tilde{\mu}(\mathbf{r}_a) = \tilde{p}(\mathbf{r}_a^+) - \tilde{p}(\mathbf{r}_a^-) \tag{3.186}$$

Thus it can be interpreted as a distribution of dipole sources on the median surface.

Application of the numerical boundary element method (see Chapter 8) to the analysis of sound radiation from complex structures and machinery that vibrate in many modes and radiate over a wide frequency range is very demanding of CPU time (and RAM for in-core solution). The computation of the radiation from an internal combustion engine up to 4 kHz demands thousands of surface elements and takes minutes for one frequency point on a typical workstation. On the other hand, the implementation of the Rayleigh integral is straightforward and orders of magnitude faster. In an assessment of the practicability of applying computational procedures based on these two formulations to the evaluation of sound power radiated by automotive engine components, Seybert *et al.* (1998) compare the results of application to two forms of automotive sump (oil-pan), mounted on the same engine, on each of which the complex cross spectra of operational acceleration measurements had been measured at over 500 points. Provided that the BE model incorporated a rigid baffle to simulate the scattering of radiated sound by the presence of the rest of the engine, the predicted sound power spectra agreed very closely with the Rayleigh solution and experimental estimates based upon sound intensity measurement. It was concluded that the simpler Rayleigh model may well give acceptably accurate results in many practical cases involving individual radiating components of multi-component structure, especially since it makes the computational trials of many trial variants in the search for the quietest design more economic. However, use of the Rayleigh integral is not likely to be so reliable in predicting sound pressures at *specific points* in the radiation field.

Even though the average acceleration levels on the 2.1 mm thick steel sump were higher than on the 10.7 mm thick aluminium sump, the sound power level of the former was about 5 dB lower than that of the latter. This was because the critical frequency of the aluminium sump walls was 1190 Hz and that of the steel was 6050 Hz. A comparison of the theoretical radiation efficiencies is presented in Fig. 3.58.

It is possible, by means of the application of the BEM, to evaluate an alternative form of Green's function that has *zero normal derivative* on the surface of the body. The body of interest is assumed to be rigid and it is insonified by a unit strength point monopole source located at a field point of interest: the computation would have to be repeated for each field point of interest with $\partial p/\partial n = 0$ at all points on the body, the transfer function between surface pressure and monopole volume velocity is equal to $j\omega\rho_0 g(\mathbf{r}, \mathbf{r}_a)$ where $\partial g(\mathbf{r}, \mathbf{r}_a)/\partial n = 0$ everywhere on the body. Replacing $G(\mathbf{r}, \mathbf{r}_a)$ by $g(\mathbf{r}, \mathbf{r}_a)$ in Eq. (3.180), the surface pressure term disappears and the radiated field pressure can be expressed purely in terms of the distribution of normal surface acceleration.

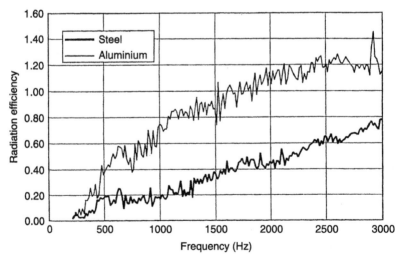

Fig. 3.58. Theoretical radiation efficiencies of a steel and an aluminium sump (oil pan) calculated by BEM (Seybert *et al.*, 1998).

This reduced form of Kirchhoff–Helmholtz equation provides the basis for an experimental evaluation of this form of Green's function that exploits the previously mentioned principle of acoustic reciprocity whereby the relation between the sound pressure received at any point generated by a harmonic point monopole is invariant with respect to exchange of the source and receiver positions, *irrespective of the geometrical and acoustic properties of the environment*, provided that all partaking systems behave *linearly* and that the fluid is quiescent in the absence of sound (Lord Rayleigh, 1896).

In the practical implementation of this principle, a concentrated, omnidirectional, source of known fluid volume acceleration (approximating a point monopole) is placed at a field point of interest and the complex transfer function expressing the ratio of the complex amplitude of sound pressure to the complex amplitude of source volume acceleration is evaluated at a set of points located on small discrete surface elements distributed over the non-vibrating body, as illustrated by Fig. 3.59. [In principle, the body should be rigid, but the acoustically induced response of most engineering structures of practical interest is sufficiently weak not to alter the surface pressure to an unacceptable degree.] This transfer function is identical to that between the volume acceleration of a small surface element on the vibrating body and the sound pressure at the field point of interest.

If the amplitudes and phases of acceleration of the forementioned set of points on a harmonically vibrating body relative to any one point are known, the contribution to the complex pressure amplitude at the field point **r** of each region of surface associated with each surface point $\mathbf{r}_s$ may be evaluated by

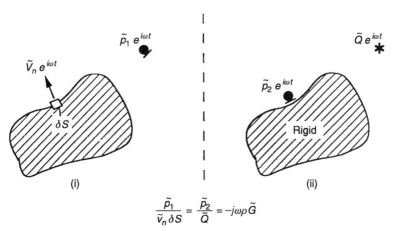

$$\frac{\tilde{p}_1}{\tilde{v}_n \delta S} = \frac{\tilde{p}_2}{\tilde{Q}} = -j\omega\rho \tilde{G}$$

Fig. 3.59. Ilustration of the experimental determination of a Green's function having zero normal derivative at the surface of a body (Fahy, 2004).

implementing the following equation:

$$\tilde{p}(\mathbf{r}) = \sum_{i=1}^{N} \widetilde{H}_i(\mathbf{r}|\mathbf{r}_s)\widetilde{A}_i(\mathbf{r}_s) \tag{3.187}$$

in which $\widetilde{A}_i$ is the volume acceleration of element i and $\widetilde{H}_i$ is the associated transfer function determined by experiment (or BEM calculation). A practical example is presented in Fig. 6.5 (Pankoke, 1993).

There are three substantial practical problems associated with this experimental technique. First, it is not simple to decide upon an appropriate discretisation of the body surface because the spatial distribution of amplitude and phase of surface acceleration is often very irregular as on the non-uniform structures such as an aircraft fuselage or a car door. Second, many structures undergo non-harmonic, multi-mode, vibration in which regions separated by distances of more than about 300 mm exhibit uncorrelated motion. It is not appropriate in this textbook to discuss the means of overcoming these problems: reference should be made to reviews of applications of vibroacoustic reciprocity (Fahy, 1995, 2004). Further descriptions of the experimental application of reciprocity to vibroacoustic problems will be found in Section 6.4. It should be mentioned that reciprocity-based techniques find many applications in modern noise control technology, especially in relation to vehicle acoustics.

Problems

3.1 By modelling the human mouth as a monopole source of volume velocity, evaluate the magnitude of the volume velocity produced by a singer of the note middle C_4 if the sound pressure level produced in the open air at a distance of 20 m is 60 dB.

3.2 Evaluate the error in on-axis pressure incurred by replacing the actual radial distance by the average radial distance in the integral of Eq. (3.4) in the case of a circular piston of diameter 150 mm at average distances of 1, 10 and 50 m and frequencies of 1 and 10 kHz.

3.3 By imagining a spherical wavefield to emanate from any observation point in the far field of a vibrating-plane radiator, and considering the intersection with the plane of wavefronts separated by $\lambda/2$, show how the regions providing contributions of opposite phase to the integral of Eq. (3.4) can be identified.

3.4 Prove the relationship given in Eq. (3.23).

3.5 Evaluate η_{rad}/σ from Eq. (3.28) for a steel plate of 3-mm thickness in air and water at 500 Hz.

3.6 By assuming that the (1,1) mode of a square panel radiates as a point volume velocity source, evaluate the radiation loss factor of a 300 mm × 300 mm steel panel of 1 mm thickness vibrating at its fundamental natural frequencies in air and in water. Assume simply supported boundary conditions in a baffle. What is the effect of fluid loading on natural frequency? [See Chapter 4.]

3.7 Check that the 'bandwidth' of the major peak in the wavenumber modulus-squared spectrum of a one-dimensional sinusoidal mode of a panel of width a is given approximately by $\Delta = 2\pi/a$ and is independent of the modal 'wavelength'.

3.8 Give a qualitative explanation of the fact that the radiation efficiencies of odd–odd modes of a rectangular panel of side lengths a and b exceed those of other modes when $ka, kb \ll 1$.

3.9 Demonstrate that the modes having equal modal-wavenumber components $k_x = p\pi/\alpha$, $k_y = q\pi/b$ in simply supported rectangular panels of the same thickness and material but of different sizes have the same natural frequencies. What are the necessary conditions on the corresponding sidelength ratios of the two panels for this to be possible?

3.10 Check the conclusions from Eqs. (3.73–3.78) concerning response concentration at the driving point by explicitly evaluating the significant modal response phasors at various positions on a simply supported, 2-mm thick rectangular steel plate of side lengths 500 and 700 mm when excited at an arbitrary point at a frequency of 500 Hz. Assume a uniform modal loss factor of 10^{-2}.

3.11 Estimate the sound power radiated into air by a 3-mm thick steel plate at 500 Hz due to the presence of a concentrated mass of 0.1 kg. The incident wavefield can be assumed to produce an acceleration level of 40 dB *re*.1g rms at 500 Hz. Equations (2.68a), (3.104) and (3.110) apply.

3.12 Calculate the approximate critical frequency for bending waves travelling parallel to the corrugations of a 1-mm thick aluminium alloy plate of which the corrugations have an amplitude of 10 mm and a wavelength of 25 mm. Equation (3.11) applies. Assuming that an orthotropic model applies, plot the bending wavenumber locus for a frequency of 2 kHz and superimpose the corresponding acoustic wavenumber locus in air at 20°C. What is the limiting angle of bending-wavenumber propagation for subsonic waves?

3.13 Derive Eq. (3.118).

3.14 Using Eq. (3.120), evaluate the ratio of critical frequency to ring frequency for typical aircraft fuselage structures and industrial pipes.

3.15 By reference to the Kirchhoff–Helmholtz integral equation, or otherwise, explain why the pressure outside, but in the plane of, an unbaffled flat plate radiator is zero.

4 | Fluid Loading of Vibrating Structures

4.1 Practical Aspects of Fluid Loading

In this book the term 'fluid loading' refers to the force field that a fluid exerts on a vibrating structure with which it is in contact, together with its effects on the vibrational and acoustic behaviour of the structure. There are many cases of practical interest to the engineer in which the interaction between vibrating structures and contiguous fluids has a profound influence upon the magnitude and frequencies (and even the stability) of the structural vibration. Examples include dams, chimney stacks, heat exchanger tubes, ships and their propellers, off-shore platforms, aircraft and electrical transmission cables. In many of these cases the fluid itself is responsible for the vibration, and simultaneously reacts to the vibration. For instance, sea waves may excite the leg of an off-shore platform into vibration, and the reaction of the surrounding water is such as to alter the natural frequencies of the leg from their *in-vacuo* values. In this example, the fluid–structure interaction is passive in the sense that the small-amplitude motion of the structure does not significantly alter the exciting forces. In other cases, such as flutter of aircraft wings and galloping of transmission lines, the motion profoundly influences the exciting forces, so that a feedback loop exists and dynamic instability may occur. These examples belong to the fields of aero- or hydro-elasticity, which are not within the scope of this book.

In many of the examples cited above, the fluid reaction forces can be estimated with sufficient accuracy by assuming that the fluid is incompressible, meaning that its density is unchangeable. Mathematically, this assumption leads to the Laplace equation for pressure

$$\frac{\partial^2 p}{\partial x^2} + \frac{\partial^2 p}{\partial y^2} + \frac{\partial^2 p}{\partial z^2} = 0 \tag{4.1}$$

Comparison with the acoustic wave Eq. (1.3) shows that this is equivalent to assuming that the speed of sound c is infinite, which is compatible with the concept of an incompressible fluid, since $c^2 = (\partial p/\partial \rho)_0$, the subscript 0 meaning 'evaluated at the equilibrium condition'. The reason why incompressible fluid solutions are adequate for the estimation of fluid reactions to structural vibration in many of the cases cited above is that only the pressures on the vibrating surface are of interest and the frequency of oscillation is rather small; more precisely, that the non-dimensional parameter $kl = 2\pi l/\lambda \ll 1$, where l is a geometric dimension typical of the physical extent of the structure. This condition is more likely to be satisfied in liquids than in gases. A small value of kl generally implies that the vibrating structure is not able to efficiently compress the fluid on its surface and therefore cannot produce a strong acoustic field. During vibration, the fluid local to the surface must be displaced in an oscillatory fashion, and the surface pressures are primarily associated with the oscillatory momentum changes of the effectively incompressible fluid.

In the case of a loudspeaker, or other structures that vibrate at frequencies for which the acoustic wavelength is not very much greater than a typical struc- tural dimension, an incompressible-fluid model is generally inadequate. The fluid loading must be estimated by solving the fluid wave equation subject to the boundary conditions imposed by the vibrating surface (and any other sur- faces bounding the fluid). In cases where a vibrating structure comprises part of an enclosure, or is situated within an enclosure, fluid loading is strongly influenced by reflection of radiated sound from the enclosure boundaries, unless there is sufficient sound absorption material in the enclosure to effectively sup- press reflections and acoustic mode resonances. The practical significance of fluid loading on a vibrating structure is that it changes both its natural frequen- cies (by inertial or elastic forces) and damping (by radiation of sound energy). It can thereby influence the vibrational response to excitation forces and the consequent radiation of sound energy. Naturally, fluid-loading effects are most strongly exhibited by structures in contact with dense fluids, because the fluid forces are generally proportional to mean fluid density. Vibrational wave speeds in the submerged regions of ships' hulls, and in liquid-filled pipes, are greatly influenced by fluid loading. However, even air can produce some striking fluid-loading effects, as can be seen by mounting a small loudspeaker at one end of a tube that is terminated by a closed or open end, and then monitoring the loud-speaker displacement as a function of driving frequency. Its compressive stiffness also profoundly affects the vibrational response of structures that enclose small volumes of air, such as double leaf windows, partitions and 'floating' floors. This chapter presents methods of mathematical analysis of fluid loading, and discusses its effects on vibrating plate and cylindrical shell structures and sound radiation therefrom.

4.2 Pressure Fields on Vibrating Surfaces

A large proportion of sound sources of practical importance radiate through the actions of vibrating surfaces on the surrounding fluid. In most cases, the geometric forms of the surfaces, and the spatial and temporal distributions of vibration, are very complex. If one wishes to estimate the sound power radiated under free-field conditions, in which the surrounding fluid is essentially unbounded and free of strongly reflective objects, it is possible to employ in the mathematical analysis of the radiated field certain approximations that greatly ease the analytical problems. In the absence of significant sound absorption by the fluid, sound power transmitted through any surface completely surrounding a source is conserved; the far radiation field can be determined at a distance, very large compared with the typical dimension of a vibrating object, from which the sound power of the source can be evaluated.

However, if it is desired to evaluate the fluid loading *on* a vibrating surface, the luxury of the far-field approximation is not available. Consequently, there exist far fewer analyses of surface fields than of far fields. It is, however, necessary in many cases of practical importance to be able to evaluate acoustic fluid-loading effects on vibrating structures. Among these we may cite underwater transducer design, loudspeaker cone design, analysis of sound propagation in tubes and pipes, analysis of blast resistance of windows, evaluation of heat exchanger tube vibration, fatigue estimation for metal foil heat insulation in reactors, analysis of tympanic musical instruments, noise control studies of paper-handling machinery, and the analysis of microphone membrane dynamics. Clearly, it is not possible within the scope of this book to discuss such diverse problems in detail. Therefore, the basic principles and certain elementary examples are analysed in order to illustrate the general nature of fluid-loading effects.

In principle, the sound field generated by a vibrating surface can be evaluated by solving the Kirchhoff–Helmholtz integral equation introduced in Chapter 3.

Computer-based numerical methods of solution are introduced in Chapter 8. However, although such methods are widely used to evaluate fluid-loading effects for the purposes of engineering design and development, they are not well suited to provide insight into the physical characteristics and parametric dependence of fluid loading on vibrating structures. Therefore, this chapter presents a range of simple, ideal models which may be treated by analytical mathematical proce- dures to yield qualitative understanding of the generic features of fluid loading and its effects.

Where the vibrating surface is planar and infinitely extended, the Rayleigh integral already encountered in Chapter 3 applies:

$$\tilde{p}(\mathbf{r})e^{j\omega t} = \frac{j\omega\rho_0}{4\pi}e^{j\omega t}\int_S\left[2\tilde{v}_n(\mathbf{r}_s)\frac{e^{-jkR}}{R}\right]dS \qquad (4.2)$$

In this case, a small surface element of area δS vibrating with normal velocity $\tilde{v}_n$ radiates like a point source of strength $\tilde{Q} = 2\tilde{v}_n\delta S$ in free space. Considerations of symmetry show that the presence of the rest of the plane surface, although sub- ject to acoustic pressures, cannot influence a spherically spreading field because there would be no particle velocity normal to that plane. The Rayleigh integral can strictly be applied only to *infinitely extended planar* surfaces, but it provides a good estimate for finite, plane, vibrating surfaces of dimensions larger than an acoustic wavelength, except for the source regions close to a free edge, such as on an unbaffled loudspeaker cone. In order to determine the fluid loading on a vibrating surface it is necessary to evaluate $p(\mathbf{r})$, with $\mathbf{r}$ indicating points on the surface. If we consider the pressure produced at a point on the surface caused by vibration of the immediately surrounding area as $kR = k|\mathbf{r} - \mathbf{r}_s|$ goes to zero, the integrand in Eq. (4.2) is seen to be singular; that is it goes to infinity, because $e^{-jkR}/R \to 1/R - jk$ as $R \to 0$. However, the integral itself remains finite. This may be seen by assuming $\tilde{v}_n$ to remain constant over the small local region and considering the contribution to pressure of annular strips surrounding the points of interest.

Substitution of the small kR approximation for e^{-jkR}/R into Eq. (4.2) demon- strates that the influence of elemental volume velocity $\tilde{v}_n\delta S$ on the pressure at distances very small compared with an acoustic wavelength ($2\pi R/\lambda \ll 1$) can be separated into two components. The $\sin(kR)$ component, which is in phase with $\tilde{v}_n$, is independent of distance; the $\cos(kR)$ component, which is in quadra- ture with $\tilde{v}_n$, increases as the separation distance decreases [note the j before the integral in Eq. (4.2)]. We may broadly interpret these dependencies as indica- tions that on vibrating planar surfaces of small extent compared with an acoustic wavelength, the pressure at any point that is in phase with the surface velocity at that point is determined essentially by the integral of the real components of ele- mental volume velocities over the surrounding region within the small kR limit;

that is, the *net* volume velocity of the surface. The quadrature component, on the other hand, is essentially determined by the motion at the point and is much less sensitive to the distribution of volume velocity over the surrounding area. The $1/R$ dependency of this latter component of the field, which is characteristic of incompressible-fluid pressure fields that satisfy the Laplace equation, justifies its description as the 'near-field' component of the surface pressure field.

As an example of this characteristic, consider a rigid circular disc of radius a vibrating in a coplanar rigid baffle. The pressure at the centre due to the motion of an annulus of radius R is given by Eq. (4.2) as

$$\delta \tilde{p}(0) = (j\omega\rho_0/4\pi R)2\tilde{v}_n e^{-jkR}2\pi R\delta R \tag{4.3}$$

The total pressure at the centre is given by the integral over the limits 0 to a:

$$\tilde{p}(0) = \rho_0 c\tilde{v}_n(1 - e^{-jka}) \tag{4.4}$$

(Plot the real and imaginary components of $\tilde{p}(0)/\rho_0 c\tilde{v}_n$ as a function of ka.) For $ka \ll 1$, Eq. (4.4) becomes

$$\tilde{p}(0) = \rho_0 c\tilde{v}_n[(ka)^2/2 + jka] \tag{4.5}$$

Even at low ka, the surface pressure component in quadrature with the velocity is not actually quite uniform over a piston in a baffle, and an integration of the pressure yields an expression for the total fluid reaction force (Pierce, 1989):

$$\tilde{F} = \rho_0 c\tilde{v}_n\pi a^2[(ka)^2/2 + j(8/3\pi)ka] \tag{4.6}$$

Note that the average resistive component of pressure is equal to that at the centre of the piston. The ratio $\tilde{F}/\tilde{v}_n$ is equivalent to the mechanical impedance discussed in Chapter 2. The ratio of average acoustic pressure $\tilde{F}/\pi a^2$ to the rate of displacement of fluid volume (volume velocity) is known as the 'acoustic radiation impedance', given by

$$\tilde{Z}_{rad} = \tilde{F}/(\pi a^2)^2\tilde{v}_n = (\rho_0 c/\pi a^2)[(ka)^2/2 + j(8/3\pi)ka] \tag{4.7}$$

The real part of $\tilde{Z}_{rad}$ is termed the 'acoustic radiation resistance', and the imaginary part is termed the 'acoustic radiation reactance'. Examination of Eq. (4.7) reveals that the reactive reaction of the fluid is inertial in nature, a mass equal to $(8/3)\rho_0 a^3$ is apparently being added to that of the piston. This is associated with the kinetic energy of fluid motion in the proximity of the piston. The resistive component is associated with energy radiation, which produces radiation damping.

If such a piston is mounted on a damped spring suspension and excited by a force, the *in-vacuo* mechanical impedance is

$$\tilde{Z} = j(\omega M - S/\omega) + B \qquad (4.8)$$

where M is the piston mass, and S and B are, the suspension spring stiffness and viscous damping coefficient, respectively. Under the action of fluid loading, the equation of motion of the piston in response to an external force $\tilde{F}_0 \exp(j\omega t)$ may be written

$$\tilde{Z}\tilde{v}_n = \tilde{F}_0 - (\pi a^2)^2 \tilde{Z}_{rad} \tilde{v}_n \qquad (4.9)$$

Equation (4.9) may be physically interpreted as representing the response of a fluid-loaded structure to an applied force, thus;

$$\tilde{v}_n = \tilde{F}_0/[\tilde{Z} + (\pi a^2)^2 \tilde{Z}_{rad}] \qquad (4.10)$$

Note the different dimensions of mechanical and acoustic radiation impedances. The effective mass of the fluid-loaded structure is $M + (8/3)\rho_0 a^3$ and the effective viscous damping coefficient is $B + 1/2\rho_0 c\pi a^2(ka)^2$. Such a combination of mechanical and radiation impedances can be made for any mode of a structure vibrating in a fluid.

The radiation resistance of a baffled piston at low ka is very small compared with the value of $\rho_0 c/\pi a^2$ that is experienced by the piston at values of $ka \gg 1$. However, a piston radiating into an anechoically terminated tube of the same diameter at low ka also experiences a radiation resistance $\rho_0 c/\pi a^2$. The reason for the difference is that, in the latter case, the compressed fluid cannot 'escape' sideways as it can from a small baffled piston.

Suppose now that the small baffled piston in our example is surrounded by an annular ring that vibrates in opposite phase to the piston (Fig. 4.1). Let the inner and outer radii of the ring be a_1 and a_2. The pressure at the centre of the piston is

$$\tilde{p}(0) = j\omega\rho_0\tilde{v}_n\left[\int_0^{a_1} e^{-jkR}dR - \int_{a_1}^{a_2} e^{-jkR}dR\right]$$

$$= \rho_0 c\tilde{v}_n[1 - 2\exp(-jka_1) + \exp(jka_2)] \qquad (4.11)$$

For $ka_2 \ll 1$, this expression, correct to second order in ka_2, is

$$\tilde{p}(0) = \rho_0 c\tilde{v}_n[(ka_1)^2 - (ka_2)^2/2 + jk(2a_1 - a_2)] \qquad (4.12)$$

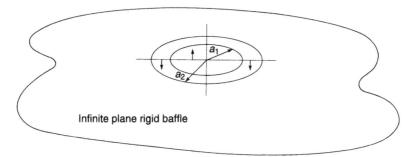

Fig. 4.1. Pistons vibrating in antiphase.

If the areas of the piston and ring are equal, so that the *net* volume velocity $\int \tilde{v}_n \, dS$ is zero, Eq. (4.12) reduces to

$$\tilde{p}(0) = j\rho_0 c\tilde{v}_n ka_2(\sqrt{2} - 1) \tag{4.13}$$

This shows that the resistive component of $\tilde{p}(0)$ is negligible. In fact, the radiation resistance of the whole piston-ring system is also negligible for small ka_2, although the radiation reactance remains of the same order as that of a uniformly moving piston of radius a_2. These examples serve to introduce two extremely important characteristics of acoustic fields generated by harmonically vibrating surfaces. First, the resistive component of the surface pressure field, and indeed the radiated power, are functions of the distribution of vibration amplitude and phase over the *whole* surface, whereas the reactive component tends to be controlled mainly by the *local* motion of the surface, and is related to the kinetic energy imparted to the local fluid. Second, at frequencies for which the acoustic wavelength greatly exceeds the surface dimensions of a radiator, the radiated power is determined essentially by the net volume velocity and is very insensitive to the detail of the spatial distribution of normal surface velocity.[1]

The radiation impedance of a piston of arbitrary ka is of considerable practical significance in various areas of applied acoustics. For example, the low frequency radiation impedance at the exit of a circular-section duct approximates closely to that of a piston, as does that of a baffled loudspeaker cone below the frequency at which it starts to 'break up' into complex vibration modes. The details of the analysis, which is based upon Eq. (4.2), may be found in many

---

[1] In cases of structures excited by broad forces, the correlation of vibrational velocities at pairs of points decreases with their separation distance and with the excitation bandwidth. Although pairs of surface elements undergoing mutually uncorrelated vibration do induce pressures on each other, their sound *powers* are independent of each other's motion.

other textbooks, e.g., Pierce (1989). The result is as follows:

$$\tilde{Z}_{rad} = R_{rad} + jX_{rad} \tag{4.14}$$

where

$$R_{rad} = \frac{\rho_0 c}{\pi a^2}\left[1 - \frac{2J_1(2ka)}{2ka}\right], \quad X_{rad} = \frac{\rho_0 c}{\pi a^2}\left[\frac{2H_1(2ka)}{2ka}\right]$$

where J_1 is the first order Bessel function of order one and H_1 is the Struve function of order one (Watson, 1966). The non-dimensional form of these components, known as impedance ratios are given by $R'_{rad} = R_{rad}(\pi a^2/\rho_0 c)$ and $X'_{rad} = X_{rad}(\pi a^2/\rho_0 c)$. They are plotted as functions of ka in Fig. 4.2. the acoustic wavelength equals the piston diameter when $ka = \pi$. After an initial variation as $(ka)^2$ at low ka, as indicated by Eq. (4.7), the resistance increases almost linearly with ka and then turns, at about the frequency where the wavelength equals the diameter, to asymptote in an oscillatory fashion to the plane-wave resistance. The reactive component, which is inertial in nature, peaks at about half this frequency and then decreases asymptotically to zero at large ka. Figure 4.2 also shows the radiation impedance of an unbaffled piston (Rschevkin, 1963) which is a good approximation to an unbaffled cone loud-speaker. Note that, in correspondence with our earlier qualitative discussion of

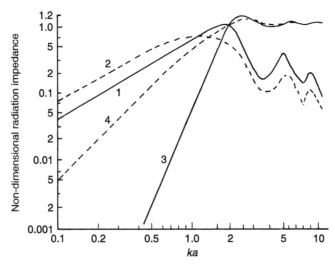

Fig. 4.2. Non-dimensional radiation impedance ratios of baffled and unbaffled circular pistons: (1) X'_{rad} for unbaffled piston; (2) X'_{rad} for baffled piston; (3) R'_{rad} for unbaffled piston; (4) R'_{rad} for baffled piston (Rschevkin, 1963).

the extent of a vibrating surface that significantly affects the reactive and resistive components of local surface pressure, the lack of a baffle affects the low frequency radiation resistance (and hence, radiated power per unit mean square velocity) much more than the reactance. This is one of the principal reasons for mounting loudspeakers in cabinets.

In most practical cases, the accurate evaluation of fluid loading using the integral equation is not possible by analytical techniques because the surface geometries and forms of motion are too complex. Therefore, numerical techniques of evaluating the integrals have been developed, which are discussed in detail in Chapter 8. However, there is a class of surface geometries for which analytic solutions are available; these are surfaces that conform to constant coordinate surfaces in coordinate systems in which the acoustic-wave equation is separable, the most common being rectangular, cylindrical and spherical. The reason why analytic solutions are possible relates to the general form of the Kirchhoff–Helmholtz integral equation, which shows that if, instead of the point source, free-space, Green's function, a Green's function suitable to the coordinate system having *zero normal derivative dG/dn* on the surface is used, only the surface normal velocity term appears; the surface pressure need not then be known, as in the Rayleigh integral for plane surfaces.

An equivalent approach, more readily understood from a physical point of view, is the boundary-matching method, whereby the acoustic particle velocity normal to the surface, expressed in terms of a conformal coordinate system, is matched to the normal velocity of the surface. As an example of considerable practical importance, we consider the fluid loading of an infinitely long circular cylindrical body for which the distribution of normal surface velocity is cosinusoidal in both the axial and circumferential directions. The Helmholtz equation in appropriate cylindrical coordinates is

$$\frac{\partial^2 p}{\partial r^2} + \frac{1}{r}\frac{\partial p}{\partial r} + \frac{1}{r^2}\frac{\partial^2 p}{\partial \phi^2} + \frac{\partial^2 p}{\partial z^2} + k^2 p = 0 \qquad (4.15)$$

The equation has separable solutions $p(r, \phi, z) = p_1(r)p_2(\phi)p_3(z)$. The surface normal velocity distribution is assumed to be

$$v_n(z, \phi, t) = \tilde{v}_n \cos n\phi \cos k_z z \exp(j\omega t) \qquad (4.16)$$

in which n/a is the circumferential wavenumber and k_z is the axial wavenumber. The radial component of acoustic particle velocity u_r is given by the fluid momentum equation as

$$\frac{\partial p}{\partial r} = -\rho_0 \frac{\partial u_r}{\partial t} = -j\omega\rho_0 u_r \qquad (4.17)$$

The acoustic field must be periodic in the axial and azimuthal coordinates with the same periods as the surface vibration. (Why?) Hence $p_3(z) = \cos k_z z$ and $p_2(\phi) = \cos n\phi$. Equation (4.15) becomes

$$\frac{\partial^2 p_1(r)}{\partial r^2} + \frac{1}{r}\frac{\partial p_1(r)}{\partial r} + \left[k^2 - k_z^2 - \left(\frac{n}{r}\right)^2\right] p_1(r) = 0 \qquad (4.18)$$

This is Bessel's equation of which the solutions are linear combinations of Bessel functions of the first and second kind (Watson, 1966):

$$\tilde{p}_1(r) = \tilde{A} J_n[(k^2 - k_z^2)^{1/2} r] + \tilde{B} Y_n[(k^2 - k_z^2)^{1/2} r] \qquad (4.19)$$

The ratio of the constants $\tilde{B}/\tilde{A}$ can be determined from the radiation condition as r tends to ∞. Writing the argument $(k^2 - k_z^2)^{1/2}$ as x,

$$\lim_{x\to\infty} J_n(x) \to (2/\pi x)^{1/2} \cos[x - (2n+1)\pi/4] \qquad (4.20a)$$

$$\lim_{x\to\infty} Y_n(x) \to (2/\pi x)^{1/2} \sin[x - (2n+1)\pi/4] \qquad (4.20b)$$

At a sufficiently large radial distance from the axis of the cylinder, and with $k_z \neq k$, the radiated field must tend to form of a plane wave: $\tilde{p}(r) \to \tilde{A}\exp(-jkr) = \tilde{A}(\cos kr - j \sin kr)$ as $r \to \infty$. By analogy with Eq. (4.19) for $kr \to \infty$ we see that $\tilde{B} = -j\tilde{A}$, and hence

$$\tilde{p}(r, \phi, z) = \tilde{A}[J_n((k^2 - k_z^2)^{1/2} r) - j Y_n((k^2 - k_z^2)^{1/2} r)] \cos n\phi \cos k_z z \quad (4.21)$$

The combination $J_n - j Y_n$ is the Hankel function H_n of the first kind.

At the surface of a cylinder ($r = a$), Eqs. (4.16), (4.17) and (4.21) link the surface pressure and normal velocity through

$$-j\omega\rho_0 \tilde{v}_n = \tilde{A}(k^2 - k_z^2)^{1/2} H_n'[(k^2 - k_z^2)^{1/2} a] \qquad (4.22)$$

where the prime denotes differentiation with respect to the argument of the function. Hence,

$$\tilde{A} = \frac{-j\omega\rho_0 \tilde{v}_n}{(k^2 - k_z^2)^{1/2} H_n'[(k^2 - k_z^2)^{1/2} a]} \qquad (4.23)$$

The axial variation of surface velocity is manifested in the term $(k^2 - k_z^2)^{1/2}$, which may be real or imaginary, and the circumferential variation determines the order of the Hankel function. If, at the frequency of oscillation, the axial wavenumber k_z exceeds the acoustic wavenumber $k = \omega/c$, the argument of the Hankel function is imaginary and the function can be replaced by a modified Hankel function of real argument

$$K_n(x) = (\pi/2) j^{n+1} H_n(jx) \tag{4.24}$$

The radial variation of pressure in the acoustic field is given by

$$\tilde{p}_1(r) = \frac{-j\omega\rho_0 \tilde{v}_n H_n[(k^2 - k_z^2)^{1/2} r]}{(k^2 - k_z^2)^{1/2} H'_n[(k^2 - k_z^2)^{1/2} a]} \tag{4.25}$$

and at the surface of the cylinder

$$\frac{\tilde{p}_1(a)}{\tilde{v}_n} = \frac{-j\omega\rho_0 H_n[(k^2 - k_z^2)^{1/2} a]}{(k^2 - k_z^2)^{1/2} H'_n[(k^2 - k_z^2)^{1/2} a]} \tag{4.26}$$

where $H_n(x)$ is replaced by $K_n(x)$ according to Eq. (4.24) when $k_z > k$. Equation (4.26) is an expression of cylinder surface *specific* acoustic impedance associated with a particular distribution of surface velocity and describes the fluid loading: it is a form of specific radiation impedance that we shall denote by $\tilde{z}_{rad}$.

Further discussion of the variation of the radiation impedance with the parameters involved is presented in Sections 4.7 and 4.8. A more complete treatment may be found in a specialised treatise on the subject of structure–fluid interaction (Junger and Feit, 1986). However, certain approximate expressions presented therein provide an indication of the general physical characteristics of the fluid loading. Where the axial wavenumber exceeds the acoustic wavenumber $k_z > k$ the radiation impedance is purely imaginary, representing a purely inertial fluid loading:

$$\left.\begin{array}{ll} \tilde{z}_{rad} \approx -j\omega\rho_0 a \ln\left[|k_z^2 - k^2|^{1/2} a\right] & n = 0 \\ \tilde{z}_{rad} \approx j\omega\rho_0 a/n & n \geq 1 \end{array}\right\} (k_z^2 - k^2)a^2 \ll (2n+1) \tag{4.27a}$$

$$\tilde{z}_{rad} \approx j\omega\rho_0/(k_z^2 - k^2)^{1/2} \qquad (k_z^2 - k^2)^{1/2} a \gg n^2 + 1 \tag{4.27b}$$

The inertial fluid loading is clearly the greatest for the axially symmetric breathing mode ($n = 0$) and decreases with increasing circumferential order n.

Hence we would expect the natural frequencies of cylindrical shells of low circumferential order to be most affected by fluid loading, provided their axial distribution of normal surface velocity satisfies the condition $k_z > k$. That this is indeed the case is shown by Fig. 4.3 (Pallett, 1972).

When the axial wavenumber is less than the acoustic wavenumber (that is, the axial wavelength exceeds the acoustic wavelength), the specific radiation impedance of a cylindrical surface is complex, possessing both real and positive imaginary (inertial) components. When $(k^2 - k_z^2)^{1/2}a \ll 1$, the resistive component of the breathing mode is dominant and the specific radiation resistance substantially exceeds $\rho_0 c$. That of the higher circumferential orders peaks at a value of about $2\rho_0 c$ at a frequency given by $(k^2 - k_z^2)^{1/2}a \approx n$ and then tends to $\rho_0 c$ (see Fig. 3.44). When $(k^2 - k_z^2)^{1/2}a \gg n^2 + 1$, the specific reactive impedance is given approximately by $j\omega\rho_0 a/2(k^2 - k_z^2)a^2$.

The evaluation of fluid loading on transversely vibrating slender structures such as wires, pipes and beams can be important in the analysis of the behaviour of stringed instruments, cross flow heat exchangers, acoustically induced fatigue failure of wire screens and other systems of practical importance. We have already seen in the case of the piston that where the typical dimension of the region of vibration is substantially smaller than an acoustic wavelength, it is the net volume displacement of fluid and not the details of the distribution of displacement that primarily determines the radiated acoustic field. In the case of transverse vibration of a slender body, in the absence of distortion of the cross section normal to the long axis, the *net* volume displacement of fluid at frequencies where the acoustic wavelength greatly exceeds a typical cross-section dimension is virtually zero, as in the case of the piston and concentric annulus discussed earlier. The fluid has time to 'slip around' the circumference toward the opposite side, to 'avoid' being compressed: however it does possess oscillatory kinetic energy. Although slender bodies are not all circular in cross section, it is reasonable to analyse the case of a transversely vibrating circular cylindrical rigid body in order to achieve an appreciation of the general nature of fluid loading in such cases: note that $k_z = 0$ and $n = 1$ in this case. In the limit $(k^2 - k_z^2)^{1/2}r \rightarrow 0$, we can rewrite Eq. (4.21) in the asymptotic form

$$\tilde{p}(r, \phi, z) = \frac{j2\tilde{A}}{(k^2 - k_z^2)^{1/2}r} \cos\phi \cos k_z z \qquad (4.28)$$

and from (4.17),

$$j\omega\rho_0 \tilde{v}_1 = \frac{j2\tilde{A}}{(k^2 - k_z^2)^{1/2}a^2} \qquad (4.29)$$

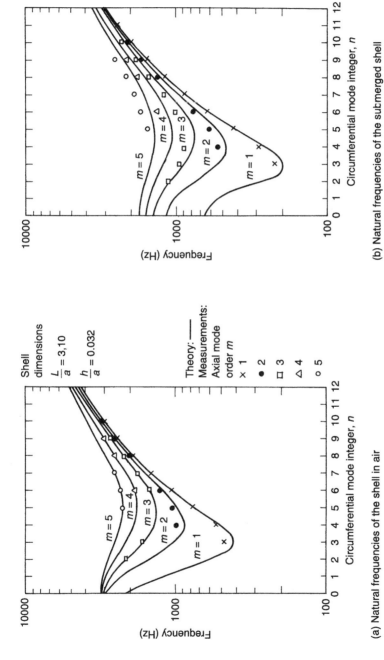

(a) Natural frequencies of the shell in air

(b) Natural frequencies of the submerged shell

Fig. 4.3.   Effect of water loading on the natural frequencies of an aluminium circular cylindrical shell with closed ends (Pallett, 1972).

Hence

$$\tilde{p}(a, \phi, z) = j\omega\rho_0 a \tilde{v}_1 \cos\phi \cos k_z z \qquad (4.30)$$

This result is equivalent to Eq. (4.27a) with $n = 1$. The transverse fluid reaction force per unit length on the body is given by

$$\tilde{F}(z) = a \int_0^{2\pi} \tilde{p}(a, \phi, z) \cos\phi \, d\phi = j\omega\pi a^2 \rho_0 \tilde{v}_1 \cos k_z z \qquad (4.31)$$

The mechanical impedance per unit length due to fluid loading is

$$\tilde{Z} = \tilde{F}(z)/\tilde{v}_1 \cos k_z z = j\omega\pi a^2 \rho_0 = j\rho_0 c\pi a(ka) \qquad (4.32)$$

which is inertial, giving a mass per unit length equal to that of a cylinder of fluid of radius a. A water-filled, thin-walled tube vibrating in water at low frequencies experiences an external inertial loading equal to that produced by the mass of the contained water.

This analysis suggests that no acoustic power would be radiated from such a vibrating body. However, a more exact analysis shows that, to third order in ka, the impedance is

$$\tilde{Z} = \rho_0 c \left[j\pi a(ka) + \frac{1}{2}\pi^2 a(ka)^3 \right] \qquad (4.33)$$

The second resistive term is much smaller than the inertial term for $ka \ll 1$, and sound radiation is very inefficient [see Eq. (3.156b)].

4.3   Wave Impedances of Structures and Fluids

In Chapter 2 we considered impedances associated with localised forces and moments applied at fixed points on structures. In this book we are particularly concerned with coupling between waves in structures and in fluids with which they are in contact; such coupling clearly involves distributions of interaction forces over extended surfaces. Hence it is useful to define another form of impedance in which a distribution of force is implicitly assumed: this is the 'wave impedance'. A discussion of fluid-loading effects on point impedance is delayed until the end of this chapter. The mathematical justification for the use of wave impedance is that any spatial distribution of a variable, such as force or displacement, can be analysed into, and synthesised from, a superposition of an infinite number of elementary distributions, each of which is sinusoidal in space and has a particular wavenumber, as described in Section 1.1. When structures and fluids undergo coupled vibration, the spatial distributions of normal displacements at the interfaces are common; hence the wavenumber spectra of the displacements are also common.

A wave impedance of a structure, or a volume of fluid, can be associated with a specific wavenumber, or frequency and phase speed combination, $k = \omega/c_{ph}$. In the case of a *uniform* structure, it is evaluated mathematically by applying a force to the structure in the form of a sinusoidal travelling wave and deriving the structural response from the equation(s) of motion. It should be noted carefully that the response to such a force takes the form of a simple travelling wave of wavenumber *equal to that of the force only if the structure is both uniform and infinitely extended*; if not, scattering of the forced wave occurs at the boundaries or discontinuities, and the wavenumber spectrum of the response is spread. It should also be remembered that the wavenumber spectrum of a point force is uniformly distributed across all wavenumbers from minus infinity to plus infinity, just as the frequency spectrum of an infinitely narrow temporal pulse is distributed uniformly from minus infinity to plus infinity.

As an example of the derivation of wave impedance we shall consider an infinitely extended, undamped, uniform plate subject to a harmonic transverse force in the form of a plane travelling wave. The one-dimensional bending equation for an *in-vacuo* plate of thickness h subject to a force $\tilde{f}$ per unit area, is

$$D\frac{\partial^4 \eta}{\partial x^4} + m\frac{\partial^2 \eta}{\partial t^2} = \tilde{f} \exp[j(\omega t - \kappa x)] \tag{4.34}$$

where $D = Eh^3/12(1 - v^2)$. Because the plate is uniform and infinite, the solution must take the form

$$\eta(x, t) = \tilde{\eta} \exp[j(\omega t - \kappa x)]$$

Substitution of this form into Eq. (4.34) yields

$$(D\kappa^4 - m\omega^2)\tilde{\eta} = \tilde{f} \tag{4.35}$$

The wave impedance of the plate is defined as the ratio of the complex amplitude of force per unit area to the complex velocity amplitude:

$$\tilde{z}_{wp} = \tilde{f}/j\omega\tilde{\eta} = -j(D\kappa^4 - m\omega^2)/\omega \tag{4.36}$$

In Section 1.7 it was shown that the free-plate bending wavenumber $k_b = (\omega^2 m/D)^{1/4}$; hence $\tilde{z}_{wp} = 0$ when $\kappa = k_b$. Excitation by a travelling force wave having $\kappa = k_b$ may be likened to excitation of a simple undamped oscillator at its natural frequency; an infinitely small force produces an infinitely large response. Equation (4.36) also reveals that the wave impedance of the plate is springlike for $\kappa \gg k_b$, and masslike for $\kappa \ll k_b$.

If the plate is damped according to the complex modulus introduced in Chapter 2,

$$\tilde{z}_{wp} = -j(D\kappa^4 - m\omega^2)/\omega + D\kappa^4\eta/\omega \qquad (4.37)$$

Hence $\tilde{z}_{wp}$ never equals zero, but when $\kappa = k_b$ it is purely real and equal to $Dk_b^4\eta/\omega$. Since η is normally of the order 10^{-2}, the real part of the impedance is dominant only when the excitation force wavenumber is very close to k_b. We now assume the plate to be in contact on one side with a semi-infinitely extended, two-dimensional layer of fluid ($y > 0$) of which it forms the boundary. We assume further that a bending wave of arbitrary wavenumber k_x is generated in the plate by a force wave applied to the other side, as illustrated in Fig. 4.4. The acoustic pressure field in the fluid is found by using the solution of the two-dimensional wave Eq. (1.6), together with the fluid momentum Eqs. (1.5a,b) at the plate–fluid interface. The x-wise variation of all acoustic variables must follow that of the force and plate displacement. From Eq. (1.8) we find that $k_y = \pm(k^2 - k_x^2)^{1/2}$. The appropriate sign of the square root is determined by the physics of the model. When $k_x < k$, the positive sign corresponds to a plane sound wave travelling away from the plate surface, as seen from Eq. (1.7); no wave can propagate toward the plate surface and therefore the negative sign is disallowed. On the other hand, when $k_x > k$, k_y becomes imaginary, and the disturbance decays exponentially with distance from the plate surface; in this case the negative sign of the square root must be selected so that $k_y = -j(k_x^2 - k^2)^{1/2}$. Application of the fluid momentum equation in the direction normal to the surface yields the

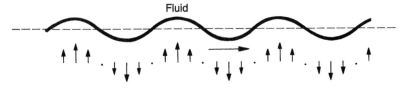

Applied force $f(x, t) = \tilde{f} \exp[j(\omega t - k_x x)]$

Fig. 4.4.   Excitation of a plate by a travelling force wave.

acoustic pressure at the plate surface ($y = 0$) as

$$p(x, 0, t) = \tilde{p} \exp[j(\omega t - k_x x)] = \frac{\omega \rho_0 \tilde{v}}{\pm (k^2 - k_x^2)^{1/2}} \exp[j(\omega t - k_x x)] \quad (4.38)$$

where $\tilde{v} = j\omega\tilde{\eta}$ is the complex amplitude of the plate velocity. Hence the wave impedance of the fluid space, which is effectively a specific acoustic radiation impedance, is

$$\tilde{z}_{wf} = (\tilde{p})_{y=0}/\tilde{v} = \pm\omega\rho_0/(k^2 - k_x^2)^{1/2} \quad (4.39)$$

When $k_x < k$,

$$\tilde{z}_{wf} = \rho_0 c/[1 - (k_x/k^2)]^{1/2} \quad (4.40)$$

This impedance is purely real (resistive) and positive, indicating that the vibrating plate does work on the fluid, the resulting energy being radiated away in the form of acoustic plane waves of which the wavenumber vector makes an angle of $\cos^{-1}(k_x/k)$ to the plate plane; the fluid loading thus acts on the plate in the form of damping. The impedance approaches infinity as k_x approaches k, and it asymptotes to $\rho_0 c$ as k_x approaches zero. The physical interpretation of this result is that the fluid impedance approaches that of a progressive plane wave travelling purely in the y-direction as the ratio of phase speed of the forcing field to the speed of sound tends to infinity; or as the ratio of the wavelength of the plate motion to the wavelength of sound tends to infinity. When $k_x > k$,

$$\tilde{z}_{wf} = j\omega\rho_0/(k_x^2 - k^2)^{1/2} = j\rho_0 c/[(k_x/k)^2 - 1]^{1/2} \quad (4.41)$$

Equation (4.41) indicates that when the phase speed of the forcing field is less than the speed of sound in the fluid, the fluid impedance is purely reactive, and the fluid loading is inertial in nature, corresponding in a numerical sense to a fluid layer of thickness $(k_x^2 - k^2)^{-1/2}$ moving with a uniform transverse velocity equal to that of the plate. Both the inertial and damping loading effects are greatest when $k_x \approx k$ and are, in principle, infinite when $k_x = k$. The latter conclusion is a product of the idealised mathematical model, which need not worry us when dealing with real, bounded structures that possess finite impedance.

When considering excitation of a coupled structure-fluid system it is useful to combine the impedances of the two to form the impedance of the fluid-loaded structure, as in the earlier example of the piston. This can be done by reference to the equation of motion of the fluid-loaded structure, which for the plate is

$$D\frac{\partial^4 \eta}{\partial x^4} + m\frac{\partial^2 \eta}{\partial t^2} = \tilde{f} \exp[j(\omega t - k_x x)] - p(x, 0, t) \quad (4.42)$$

The last term represents the acoustic pressure at the interface which is produced by plate motion. In terms of impedances, Eq. (4.42) can be written $\tilde{v}\tilde{z}_{wp} = \tilde{f} - \tilde{v}\tilde{z}_{wf}$ or

$$\tilde{f}/\tilde{v} = \tilde{z}_{wp} + \tilde{z}_{wf} \tag{4.43}$$

Hence the wave impedance of the coupled system equals the sum of the two wave impedances.

We have briefly considered the fluid loading associated with waves generated in a plate by an applied force wave. The reaction of a fluid to vibration of a structure with which it is in contact must clearly also influence the process *of free-wave* propagation in the structure. In an undamped wave-bearing system, the equation that has to be solved for the free propagation wavenumber corresponds to setting the wave impedance equal to zero; that is, zero input force is equivalent to free vibration. The resulting equation is the dispersion equation. In the case of the fluid-loaded, undamped plate of impedance given by Eq. (4.43), we cannot state *a priori* that the fluid impedance will be wholly reactive, because its nature depends upon k, which is the quantity sought. The general solution of the problem is beyond the scope of this book. However, reference to Fig. 1.17 in Section 1.8 shows that below the critical frequency the *in-vacuo* bending wavenumber is greater than the acoustic wavenumber in the fluid. We have seen that if the plate wavenumber exceeds that of the sound in the fluid, the reaction is masslike. Recalling that the *in-vacuo* free bending wavenumber is given by $k_b = (\omega^2 m/D)^{1/4}$, it is clear that additional mass will increase k_b and reduce the phase speed of free waves in the plate. Hence we may obtain an approximate solution for the *low-frequency,* fluid-loaded free wavenumber k_b' by assuming that $k_b' \gg k$. From Eq. (4.41),

$$\tilde{z}_{wf} \approx j\omega\rho_0/k_b' \tag{4.44}$$

and substitution from Eqs. (4.36) and (4.44) into Eq. (4.43), together with setting the total wave impedance to zero, yields

$$Dk_b'^4 - \omega^2(m + \rho_0/k_b') = 0 \tag{4.45}$$

Although it is not simple to solve Eq. (4.45) for k_b' explicitly, since it is a fifth-order equation in k_b', it is clear that the term ρ_0/k_b' represents an addition to the plate mass per unit area m, just as the term $8\rho_0 a^3/3$ does for a small piston, and that this contribution increases as k_b' falls. Since, *in-vacuo*, $Dk_b^4 = \omega^2 m$, and fluid mass loading reduces the phase speed and increases the bending wavenumber, we may assume that $Dk_b'^4 \gg \omega^2 m$ to give

$$k_b' \approx (\omega^2 \rho_0 / D)^{1/5} \qquad (4.46)$$

This rather surprising result indicates that, in dense fluids, at frequencies very much below the critical frequency, the plate mass has no influence on the free bending wavenumber because the inertial fluid loading is dominant. The corresponding phase speed is given by

$$c_{ph} \approx (\omega^3 D / \rho_0)^{1/5} \qquad (4.47)$$

Further qualitative observations may be made about fluid-loaded, free flexural waves in infinite, thin plates. It is clearly not possible for a wave to propagate freely with a wavenumber equal to the acoustic wavenumber because the fluid wave impedance is infinite. In fact, detailed mathematical analysis shows that the only physically significant purely *real* solution for the free flexural wavenumber represents a surface wave that travels sub-sonically even above the critical frequency (Strawderman *et al.*, 1979), because as k_x approaches k, the inertial loading is seen from Eq. (4.41) to increase indefinitely. Above the critical frequency, a supersonic, so-called 'leaky wave', can propagate. A comprehensive treatment of this fluid-loading problem is presented by Crighton (1988).

4.4 Fluid Loading of Vibrating Plates

We have seen that, for infinitely extended flat plates carrying sinusoidal plane waves of transverse displacement, there is a strict division of the nature of the fluid reaction between purely inertial loading for subsonic phase speeds, and purely resistive or damping-like loading for supersonic phase speeds. Real structures are bounded, and we must now turn to the question of fluid loading of bounded plates.

So far in this chapter we have considered the analysis of fluid loading by means of a Green's function approach for a source region of finite extent, and a boundary-matching approach for vibrations of surfaces of infinite extent. The latter approach can be applied to vibrating surfaces of limited extent, provided that they form part of an infinitely extended surface for which the geometry conforms to a coordinate system in which the wave equation is separable. Examples include a flat vibrating panel set in an otherwise infinite rigid planar baffle and a vibrating tube set in coaxial rigid tubular extensions of semi-infinite extent. The cynic might observe that we seem to be nowhere closer to reality,

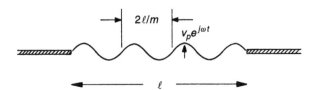

Fig. 4.5.   Modal vibration of a baffled plate.

because infinitely extended surfaces of regular geometry do not exist, whether vibrating or not. It transpires, however, that in many cases the inertial component of fluid loading on a region of structure tends to be related to the kinetic energy of the *local* fluid and is rather insensitive to the state of vibration at remote points.

We have already encountered the concept of spatial frequency (wave number) analysis and synthesis of spatially distributed variables. This mathematical technique is now applied to the analysis of fluid loading on the two-dimensional plate–baffle system shown in Fig. 4.5. It is assumed that the simply supported plate vibrates in one of its *in-vacuo* modes in which the normal velocity distribution is given by

$$v(x, t) = \begin{cases} \tilde{v}_p \sin(m\pi x/l)e^{j\omega t}, & 0 < x < l \\ 0, & 0 > x > l \end{cases} \tag{4.48}$$

The wavenumber transform of v is

$$\tilde{V}(k_x) = \tilde{v}_p \int_0^l \sin(m\pi x/l) \exp(-jk_x x) \, dx \tag{4.49}$$

In this case we can simplify the integration by expressing the standing wave as the sum of two travelling waves. Letting $k_m = m\pi/l$,

$$\begin{aligned} \tilde{V}(k_x) &= -\frac{j}{2}\tilde{v}_p \int_0^l [\exp(jk_m x) - \exp(-jk_m x)] \exp(-jk_x x) dx \\ &= -\frac{1}{2}\tilde{v}_p \left[\frac{\exp[j(k_m - k_x)x]}{k_m - k_x} + \frac{\exp[-j(k_m + k_x)x]}{k_m + k_x} \right]_0^l \\ &= -\frac{1}{2}\tilde{v}_p \left[\frac{\exp[j(k_m - k_x)l]}{k_m - k_x} + \frac{\exp[-j(k_m + k_x)l]}{k_m + k_x} - \frac{2k_m}{k_m^2 - k_x^2} \right] \end{aligned} \tag{4.50}$$

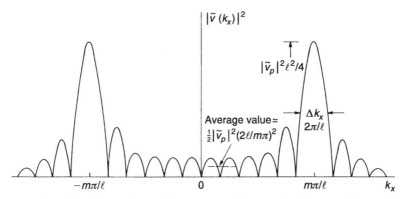

Fig. 4.6. Wavenumber modulus spectrum of plate velocity (diagrammatic).

The modulus of the transform is

$$|\tilde{V}(k_x)| = |\tilde{v}_p| \left(\frac{2\pi ml}{(k_xl)^2 - (m\pi)^2} \right) \sin \left(\frac{k_xl - m\pi}{2} \right) \qquad (4.51)$$

Its square is plotted in Fig. 4.6. The wavenumber spectrum of the surface acoustic field generated by the wavenumber component $\tilde{V}(k_x)$ is given by Eqs. (3.94) and (4.39) as

$$[\tilde{P}(k_x)]_{y=0} = \tilde{V}(k_x)\tilde{z}_{wf}(k_x) \qquad (4.52)$$

At this stage we have to specify the frequency of vibration. The analysis of the wavenumber spectrum was solely concerned with the variation of v with x: the result, Eq. (4.50), contains no frequency term. The frequency becomes vital once the question of the type of acoustic field generated by the vibration arises. This is because the form, and indeed associated physical nature, of fluid loading represented by $\tilde{z}_{wf}$ depends upon how the wavenumber spectrum $\tilde{V}(k_x)$ is distributed with respect to the acoustic wavenumber k *at the particular frequency of vibration*. For $|k_x| > k$, the corresponding component of $\tilde{z}_{wf}(k_x)$ is purely imaginary and is given by Eq. (4.41); for $|k_x| < k$ the component is purely real and is given by Eq. (4.40). The modulus of k_x is used because the bounded plate carries waves travelling in both directions. These conditions are illustrated in Fig. 4.6, which indicates that, because the truncation or windowing of a sinusoid of 'wavelength' λ spreads its wavenumber spectrum around the wavenumber $2\pi/\lambda$, the fluid loading generally comprises both reactive and resistive components at all frequencies.

Equation (4.52) represents the wavenumber spectral decomposition of the surface pressure. The actual surface pressure amplitude distribution $\tilde{p}(x)$ is given by the inverse Fourier transformation of $\tilde{P}(k_x)$

$$\tilde{p}(x,0) = \frac{1}{2\pi} \int_{-\infty}^{\infty} [\tilde{P}(k_x)]_{y=0} \exp(jk_xx)\, dk_x \tag{4.53}$$

The integral must be split into three parts, thus:

$$\tilde{p}(x,0) = \frac{\rho_0 c}{2\pi} \int_{-k}^{k} \tilde{V}(k_x)\left(1 - k_x^2/k^2\right)^{-1/2} \exp(jk_xx)\, dk_x$$

$$+ \frac{j\rho_0 c}{2\pi} \int_{k}^{\infty} \tilde{V}(k_x)\left(k_x^2/k^2 - 1\right)^{-1/2} \exp(jk_xx)\, dk_x \tag{4.54}$$

$$+ \frac{j\rho_0 c}{2\pi} \int_{-\infty}^{-k} \tilde{V}(k_x)\left(k_x^2/k^2 - 1\right)^{-1/2} \exp(jk_xx)\, dk_x$$

The first integral represents the resistive component of the fluid loading and the other two represent the reactive part. These integrals do not have general closed-form solutions; and, anyway, a knowledge of the detailed distribution of surface pressure is not always necessary. Of more common interest are the radiated sound power and the effective mass added to the plate by the reactive component of the fluid loading.

The sound power radiated from bounded plates has been treated in Chapter 3 by means of a Rayleigh integral approach (Wallace, 1972). Extension into three dimensions of the two-dimensional wavenumber spectrum approach of Heckl, outlined in Eqs. (3.90–3.103), produces very similar results. This form of analysis can be adopted to provide an estimate of the inertial loading on plates vibrating well below the critical frequency. The reactive power generated by a vibrating plate in a fluid is put into near-field kinetic energy. Over alternate quarter cycles of oscillation, fluid kinetic energy is created, and over the other two quarter-cycles the same energy returns to the plate, so that over a half-period, or multiple thereof, zero net work is done by the plate (Fig. 4.7). The effective mass per unit area associated with the near field can be obtained from the imaginary part of the product of the complex pressure and the associated normal fluid velocity by the relation

$$\frac{1}{4} m_e l v_p^2 = \frac{1}{2\omega} \text{Im} \int_0^l \left\{ \tilde{p}(x,0)\tilde{v}(x) \right\} dx \tag{4.55}$$

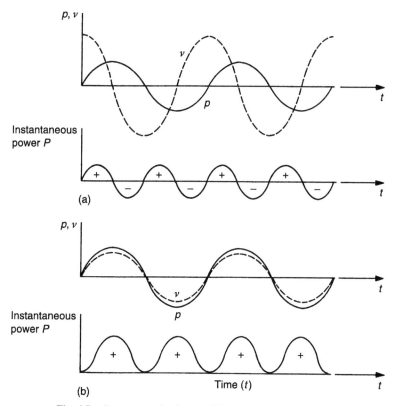

Fig. 4.7. Power transmitted to the fluid: (a) reactive; (b) resistive.

for a one-dimensional vibration field in which $\tilde{v}_p$ is assumed to be purely real. The extra factor of one-half appears on the left-hand side because of the sinusoidal spatial variation of v with x.

Substitution of the expressions for $\tilde{p}(x, 0)$ and $\tilde{v}(x)$ in terms of wavenumber transforms, and integration over the length of the plate, in a manner similar to that presented in Eqs. (3.90–3.103), yields the following expression for the effective mass per unit area of the plate:

$$m_e = \frac{2\rho_0}{\pi l v_p^2} \int_k^\infty \frac{|\tilde{V}(k_x)|^2 dk_x}{(k_x^2 - k^2)^{1/2}} \tag{4.56}$$

The plot of $|\tilde{V}(k_x)|^2$ in Fig. 4.6 suggests that the main contribution to the integral is associated with wavenumbers around $k_x = m\pi/l$. Comparison with the contribution in the region $k_x \approx k$ shows that the former is generally the larger, except when m is 1 or 2, and for all m when $m\pi/l$ approaches k. Hence we may evaluate m_e for values of $m\pi/l \gg k$ by substituting $m\pi/l$ for k_x in the denominator of the integrand, removing it from the integrand, and extending the range of integration from zero to infinity. Substitution into Eq. (4.56) of the expression for $|\widetilde{V}(k_x)|$ in Eq. (4.51), and performance of the integration, yields

$$m_e \approx \rho_0/(m\pi/l) = \rho_0/k_m \qquad (4.57)$$

Note that this expression is equivalent to that given by Eq. (4.44) for a plane flexural wave travelling in an unbounded uniform plate. This equivalence indicates that the near-field inertial loading at subcritical frequencies is not very sensitive to the boundary conditions of the plate, unlike the resistive loading. This result reinforces our previous conclusion that reactive fluid loading is associated essentially with local vibration. For modes that have natural frequencies just below the critical frequency, the inertial component of fluid loading of this one-dimensional plate will be somewhat larger than that given by Eq. (4.57), because the peak in the modal wavenumber spectrum coincides with an asymptotic trend of the denominator of Eq. (4.56) toward zero. The physical interpretation of the trend is that a greater depth of fluid is being displaced in a reactive oscillatory fashion. The inertial loading for plate modes that have natural frequencies above the critical frequency is smaller than that given by Eq. (4.57) because the range of integration (k to ∞) excludes the peak in the modal wavenumber spectrum.

The inertial loading of two dimensional plate modes of order (p, q) at frequencies well below the critical frequency will also be given by Eq. (4.57), with k_m equal to $[(p\pi/a)^2 + (q\pi/b)^2]^{1/2}$. The localised nature of inertial fluid loading leads us to expect that a similar expression would apply to the modes of cylindrical shells, provided the axial wavenumber exceeds the acoustic wavenumber. Indeed this is so, and Eq. (4.27a) with $k < k_z \ll n/a$ confirms our expectations. The inertial loading on very low order modes of a rectangular plate are not accurately given by Eq. (4.57) because the width of the main peak of the modal wavenumber spectrum extends over a range $\pm 2\pi/l$ about $m\pi/l$, and when $m = 2$ it extends to $k_x = 0$, hence including the range for which $k \approx k_x$, and the loading is underestimated by Eq. (4.57). The wavenumber spectrum of the fundamental mode ($m = 1$) has its maximum at $k_x \approx 0$. In addition to making this mode far more efficient as a radiator at subcritical frequencies than the other low-order modes, it also produces a rather high inertial loading, which can be estimated fairly accurately by treating the plate as a piston oscillating with an equivalent volume velocity, and using piston radiation impedance expressions.

The results of detailed analysis of fluid loading on rectangular plates by Lomas and Hayek (1977) confirm that Eq. (4.57) provides a reasonable estimate.

4.5 Natural Frequencies of Fluid-Loaded Plates

In cases where it is necessary to include fluid-loading effects in the estimation of natural frequencies of structures such as ship hulls, sonar transducers, liquid-filled pipes and liquid-cooled nuclear reactor internals, it is important to distinguish between those in which the fluid is contained within a region bounded wholly, or largely, by the structure under consideration, and those in which the structure is enveloped by the fluid, and the fluid is considered virtually unbounded. The essential physical difference between these two cases may be described qualitatively as follows. A wholly, or largely, bounded fluid possesses natural frequencies and modes proper to its geometry and physical properties: energy is stored in the standing wave components of the modes. In cases where the bounding structure is not considered to be rigid, the total system of structure plus fluid should, in principle, be analysed as a coupled system. The sound waves generated in a volume by vibration of the bounding surface (or part thereof) are reflected from the bounding surfaces and exert reactive loadings on the structure which may be *either* inertial or elastic in nature. The resistive component of fluid loading (radiation damping) is also affected by fluid containment (see Chapters 7 and 8). This case is exemplified by the liquid-filled pipe, as explained in Chapter 7. In cases where the structure is in contact with an otherwise effectively unbounded fluid (for example, a submarine in the sea), only the near acoustic field stores energy and the reactive fluid-loading forces are associated only with the near-field pressures which manifest themselves as inertial loadings, as we have seen in the previous section.

The natural frequencies of a strongly coupled structure–fluid system are different from those of the two components when uncoupled. Even so, they may generally be divided into two sets—those in which the vibrational energy is stored primarily in the fluid, and those where it resides primarily in the structure. Fortunately for the analyst, the corresponding coupled-mode shapes observed in the dominant component often differ relatively little from mode shapes identified in the uncoupled state, except in cases where the fluid is confined within a volume that is narrow in one or two principal dimensions. This insensitivity of mode shape is even more apparent when a fluid is largely unbounded; the structural mode shapes remain almost unchanged, and the natural frequencies fall below their *in-vacuo* values in proportion to the square root of the ratio of the loaded to unloaded modal masses. The analysis of reactive loading on plate bending waves having wavenumbers much greater than an acoustic wavenumber has shown that the effective added mass per unit area is given approximately by ρ_0/k_m, where k_m is the primary effective wavenumber component of the vibration. Hence we may approximate the fluid-loaded structure natural frequencies by

$$\omega'_m \approx \omega_m (1 + \rho_0/mk_m)^{-1/2} \tag{4.58}$$

where ω_m is the *in-vacuo* natural frequency, k_m the primary modal wavenumber component, and m the average structural mass per unit area. The important conclusion from this result is that it is usually only the natural frequencies of the very low order modes of a plate structure that are significantly affected by fluid loading. It must be noted that the natural modes of plates and shells surrounded by largely unbounded fluids, in which energy can be radiated away to infinity, are not mathematically orthogonal (Davies, 1971; Mkhitarov, 1972; Stepanishen, 1982). The natural frequencies of fluid-loaded plates and shells are computed by Lomas and Hayek (1977) and Warburton (1978).

4.6 Effects of Fluid Loading on Sound Radiation from Point-Excited Plates

It is of considerable practical importance for those concerned with the vibration of plate and shell structures in contact with water to be able to estimate, and to have a qualitative feel for, the effects of the water loading on the generation of structural and acoustic waves by concentrated applied forces such as those generated by machinery mounts. The detailed mathematical analyses of fluid loading effects involve rather advanced analytical techniques, and in many cases quantitative results can be obtained only by numerical procedures; however, some general qualitative trends can be deduced. Maidanik (1966) shows that at frequencies well below the critical frequency, the sound power radiated by a thin plate, excited by a simple harmonic point force, into a semi-infinite volume of fluid present only on one side of the plate, is given by

$$\overline{P} = k^2 \beta^2 \left| \widetilde{F} \right|^2 / 4\pi \rho_0 c \tag{4.59}$$

when the fluid-loading parameter $\beta = \rho_0 c/m\omega$ is much less than unity: the associated directivity is uniform except very close to the plate where the pressure drops to zero. This condition is one of 'light fluid loading' and the result corresponds to Eq. (3.130), in which the effect of fluid loading on the plate vibrations was entirely neglected. When the fluid loading is heavy and $\beta \gg 1$, the radiated power becomes

$$\overline{P} = k^2 \left|\widetilde{F}\right|^2 / 12\pi\rho_0 c \tag{4.60}$$

and the directivity becomes that associated with a dipole source having its axis coincident with the applied force. Note that Eq. (4.60) does not contain any plate parameters, the plate mass not affecting the free-plate wavenumber, as seen in Eq. (4.46).

The expressions for power radiated per plate unit length, when excited by a line force per unit length $\widetilde{F}' \exp(j\omega t)$, are as follows (Rummerman, 2002):

$$\overline{P} = \begin{cases} k \left|\widetilde{F}'\right|^2 \beta^2 / 4\rho_0 c, & \text{for } \beta \ll 1 \tag{4.61} \\ k \left|\widetilde{F}'\right|^2 / 8\rho_0 c, & \text{for } \beta \gg 1 \tag{4.62} \end{cases}$$

As with the point force, increase of fluid loading causes the directivity to change from that of a line monopole to that of a line dipole. Fluid-loading effects on point and line force impedances are more difficult to calculate than those on radiation. Table 4.1 summarises the main results (Crighton, 1972, 1977).

TABLE 4.1

Low-Frequency Mobility Formulas for Fluid-Loaded Plates[a]

Line force	$\widetilde{Y}_F = \dfrac{\omega(1 - j\tan(\pi/10))}{5Dk_b^3\sigma^3}$	
	$\widetilde{Y}_{F_0} = \dfrac{\omega(1 - j)}{4Dk_b^3}$	$\sigma = (\rho_0/mk_b)^{1/5}$
Line moment	$\widetilde{Y}_M = \dfrac{k_b^3(1 + \cot(\pi/5))}{5m\omega\sigma}$	
	$\widetilde{Y}_{M_0} = \dfrac{k_b^3}{4m\omega}(1 + j)$	
Point force	$\widetilde{Y}_F = \left(\dfrac{1}{8(Dm)^{1/2}}\right) \dfrac{4}{5}\left(\dfrac{k}{vk_b}\right)^{2/5}\left(1 + j\tan\dfrac{\pi}{10}\right)$	$v = \rho_0 k/mk_b^2$
	$\widetilde{Y}_{F_0} = \left(\dfrac{1}{8(Dm)^{1/2}}\right)$	

[a]The subscript 0 indicates *in-vacuo* values. Frequency condition: $\sigma \gg 1$. *In-vacuo* bending wavenumber $k_b = (\omega^2 m/D)^{1/4}$.

4.7   Natural Frequencies of Fluid-Loaded, Thin-Walled, Circular Cylindrical Shells

An analysis of the fluid loading on an infinitely long, uniform, circular cylindrical body for which the normal surface velocity is time-harmonic and cosinusoidal in both axial and circumferential directions has been presented in Section 4.2. The associated specific wave impedance for a given circumferential order n is given by the expression in Eq. (4.26). As we have seen in the analysis of fluid loading on a bounded flat plate in Section 4.4, the expression for the wave impedance may be applied to the analysis of the fluid loading on the *in-vacuo* modes of a bounded, baffled structure by means of wavenumber decomposition of the mode shapes. This technique has been applied by Stepanishen (1978, 1982) to the calculation of fluid loading on the *in-vacuo* modes of finite length, simply supported, circular cylindrical shell, baffled by semi-infinite, rigid, co-axial extensions of the same diameter. The modes are coupled by the fluid loading because the axial wavenumber spectra of all modes are spread about the principal axial wavenumber and 'overlap', so that the surface pressure associated with a given axial wavenumber component of any one mode acts on the same wavenumber component of all other modes that share the same circumferential order. This is a complicating factor in the modal analysis of all fluid-loaded structures, but is generally not of great practical significance.

In addition to approximate asymptotic expressions for the 'high' and 'low' frequency radiation, Stepanishen applies numerical analysis to the calculation of the fluid-loading reactance of a shell having a length-to-radius ratio of 3.0. Examples of the results are presented in Figs. 4.8 and 4.9. The 'added' generalised (modal) mass M_{mnf}, which is approximately given by the specific reactance ratio at the *in-vacuo* shell mode natural frequency ω_{mn} multiplied by $\rho_0 c 2\pi La/\omega_{mn}$ is added to the *in-vacuo* modal mass M_{mn} and reduces the frequency by a factor $(1 + M_{mnf}/M_{mn})^{1/2}$. The cross-modal reactances are generally sufficiently small not to contribute significantly to M_{mnf}. An example of the effect of water loading on a shell is presented in Fig. 4.3 in which it is seen that the natural frequencies of modes of low circumferential order are the most strongly affected by the fluid loading.

4.8   Effects of Fluid Loading on Sound Radiation by Thin-Walled, Circular Cylindrical Shells

The effect of fluid loading on the vibration and sound radiation of a thin-walled circular, cylindrical shell of finite length subject to point force excitation

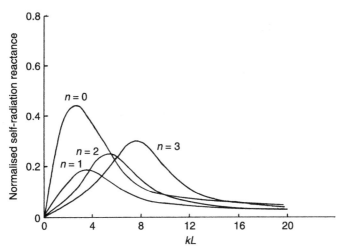

Fig. 4.8. Normalised self-radiation reactance of modes of a cylinder of length-to-radius ratio of 3.0: L = cylinder length; n = circumferential order; axial order = 1 (Stepanishen, 1982).

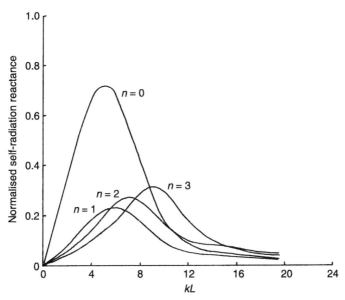

Fig. 4.9. Normalised self-radiation reactance of modes of a cylinder of length-to-radius ratio of 3.0: L = cylinder length; n = circumferential order; axial order = 2 (Stepanishen, 1982).

has been studied by Laulagnet and Guyader (1989) using a series expansion of *in-vacuo* shell modes to represent the structural vibration field. For the purpose of computing the radiated sound field, it is assumed that the shell of length L is simply supported at both ends and is extended to infinity at both ends by rigid co-axial baffles of the same diameter. The *in-vacuo* radial displacement mode shapes are therefore sinusoidal functions of the axial coordinate z given by $\sin(m\pi z/L)\sin(n\phi)$. The steel cylinder model employed in the numerical studies has a radius of 0.4 m, a length of 1.2 m and a wall thickness of 3 mm. The ring frequency is 2 kHz and the equivalent flat plate critical frequencies in air and water are 4.1 and 78.1 kHz, respectively. The assumed structural loss factor is 10^{-2}. Laulagnet and Guyader demonstrate that neglect of cross-modal coupling in the case of heavy fluid loading by water produces maximum errors of about 3 dB in spectral peaks of radiated power and negligible error in shell vibration energy. They consider that their qualitative conclusions regarding the relative contributions to sound power radiation by the various modes are not invalidated by such neglect.

The sound powers radiated by the various modes are individually computed and, by virtue of the neglect of cross-modal radiation coupling, the total radiated power is given by the sum of the powers radiated by the individual modes. The time-average sound power radiated by mode of order (m, n) excited by a generalised (modal) force of unit amplitude and frequency ω is given by

$$P_{mn} = (\omega^2/2)R_{mn}|\tilde{w}_{mn}|^2 \tag{4.63}$$

where $\tilde{w}_{mn}$ is the complex modal radial displacement amplitude per unit modal force and R_{mn} is the real part of the complex modal radiation impedance $\tilde{Z}_{mn}$. The latter is computed by applying the same technique of axial wavenumber decomposition of the mode shape employed by Stepanishen, together with the application of the expression in Eq. (4.26) which is effectively the specific radiation impedance associated with axial wavenumber component k_z and circumferential order n. The modal displacement amplitude per unit modal force is given by

$$\tilde{w}_{mn} = [M_{mn}(\omega_{mn}^2(1 + j\eta) - \omega^2) + j\omega\tilde{Z}_{mn}]^{-1} \tag{4.64}$$

where M_{mn} is the generalised (modal) mass, ω_{mn} is the *in-vacuo* modal natural frequency and η_{mns} is the modal structural loss factor. At a *fluid-loaded* modal resonance frequency ω_{mnfl}, the total imaginary part of the modal impedance (structural reactance plus inertial fluid reactance) is, by definition, zero. The time-average sound power radiated per unit modal force is then given by

$$\overline{W}_{mn}/|\tilde{F}_{mn}|^2 = \omega_{mnfl}^2 R_{mn}/2(\eta M_{mn}\omega_{mn}^2 + \omega_{mnfl}R_{mn})^2 \tag{4.65a}$$

which may be written in terms of the modal structural and radiation loss factors η_{mns} and η_{mnr} as

$$\overline{W}_{mn}\big/|\widetilde{F}_{mn}|^2 = [\eta_{mnr}\big/2(\eta_{mns}+\eta_{mnr})^2][\omega_{mnfl}\big/M_{mn}\omega_{mn}^2] \qquad (4.65b)$$

The sound power radiated at a fluid-loaded resonance frequency is seen to be proportional to the modal radiation loss factor and inversely proportional to the *square* of the sum of the structural and radiation loss factors. In common with other coupled dynamic systems, the coupling power (in this case, the sound power per unit modal force) is maximum when the two components of loss factors are equal. This maximum is given by

$$\overline{W}_{mn}\big/|\widetilde{F}_{mn}|^2_{max} = \omega_{mnfl}\big/8M_{mn}\eta_{mns}\omega_{mn}^2 \qquad (4.66)$$

In air, the fluid-loaded and *in-vacuo* natural frequencies are very little different, so Eq. (4.66) simplifies.

The authors place modes in three categories: (i) modes in which the radiation loss factor is much smaller than the structural loss factor and the radiated sound power at resonance is controlled by structural damping (SD modes); (ii) modes in which the structural loss factors are much smaller than the radiation loss factors and the radiated sound power at resonance is controlled by radiation damping (RD modes); (iii) modes that have approximately equal structural and radiation damping (ED modes). When they exist, ED modes tend to dominate the radiation at modal resonance frequencies.

As explained in Section 3.10, the radiation loss factor of a given mode depends upon the ratio of axial and circumferential structural wavenumbers to the acoustic wavenumber. In addition, the radiation loss factor is linearly proportional to the fluid density. In air, at frequencies well below the ring frequency, only a few isolated modes have radiation loss factors of the same order as the assumed structural loss factor; most are SD modes. The few (nearly) ED modes dominate the resonant radiation peaks. As the ring frequency is approached, more (nearly) ED modes operate and the radiated power per unit force increases, as shown in Fig. 4.10. The power *peaks* could be reduced by increasing structural damping, but by less than a factor of 6 dB per doubling because both forms of damping contribute to the denominator of Eq. (4.65).

Now we turn to radiation into water. Because the acoustic wavenumber in air exceeds that in water at the same frequency by a factor of about 4.3, the radiation efficiency of a given mode at a given frequency radiating into water is far less than that into air. But the fluid density ratio of 775 tends to outweigh the deficiency of radiation efficiency for the more strongly radiating modes, so that some have radiation loss factors substantially greater than their structural loss factors; they are the RD modes. However, unlike the case of air, the modes having the

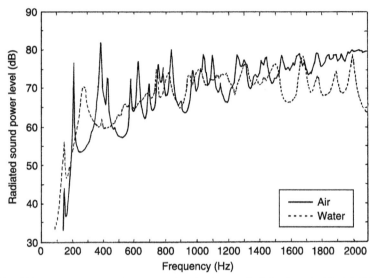

Fig. 4.10. Sound powers radiated by a unit point-force excited circular cylindrical shell into air and water (Laulagnet and Guyader, 1989).

highest radiation efficiencies do not dominate *resonant* modal radiation. Hence, the less efficient ED modes also contribute substantially to radiated power at resonance. Equation (4.65) indicates that the sound power of an RD mode *at resonance decreases* as the radiation loss factor *increases*: this conclusion is counter-intuitive. The effect of increasing damping on resonantly radiated power is not so clear cut as in air, since RD modes may be converted into ED modes, and ED modes may be converted into SD modes; the overall effect depends specifically on the modal mix.

So far we have considered only modal radiation at resonance. We now consider the total power radiated by a mode when excited by a broadband force of uniform spectral density $G_F(\omega)$. The expression corresponding to Eq. (4.65) is

$$\overline{W}_{mn}/G_F(\omega) = [\eta_{mnr}\pi/4(\eta_{mns} + \eta_{mnr})][\omega_{mnfl}/M_{mn}\omega_{mn}] \qquad (4.67)$$

Note carefully that the sum of the modal loss factors in the denominator is not squared as it was for radiation at modal resonance. This difference is highly significant; the sound power asymptotes to an upper limit when the radiation loss factor greatly exceeds the structural loss factor. In this case, RD mode radiation into water exceeds ED mode radiation; and increase of structural damping has less influence on broadband sound power than it does at resonance. In air, increased structural damping has greater effect than it does in water, but it is less than on mode radiation at resonance.

Remarkably, the normalised sound powers radiated into air and water by the particular point-excited structural model described are of similar order of magnitude, as shown in Fig. 4.10. However, the differences seen in the frequencies of the spectral peaks indicate that, at least in relation to resonant modal radiation, quite different modes dominate the radiation in the two cases. (It should be emphasised that the discussion presented above relates specifically to the particular forms of structural model and excitation: the conclusions are not generic. In particular, substantial increase of the wall thickness parameter is likely to produce significantly different conclusions. However, the analytical approach and the mode classification scheme is of general application and value in aiding understanding of radiation behaviour.)

4.9 Damping of Thin Plates by Porous Sheets

It has long been observed that the application of a thick sheet of porous material to a thin flat plate often greatly increases the damping of the plate vibration modes, in some cases to near critical values. The fact that this phenomenon is observed even in cases where a thin air layer separates the two media clearly rules out a purely mechanical mechanism. Theoretical models developed by Cummings (2001) demonstrate that the damping is caused by the generation within the porous layer of an acoustic near field that involves strong particle velocity fluctuations, thereby dissipating energy into heat through the action of viscosity. This enhances the radiation resistance of the structure. At any one frequency, the effect decreases with reduction of plate critical frequency, because of the concomitant reduction in the strength of the near field.

Problems

4.1 A baffled loudspeaker cone of 150 mm diameter and mass 5 milligrams has an *in-vacuo* natural frequency of 35 Hz. Estimate the change of natural frequency due to the inertial component of air loading. [Equation (4.7) applies.]

4.2 Estimate the approximate bending wave phase velocities at frequencies of 100 Hz and 2 kHz of a 5 mm thick steel plate when submerged in air and in water. Equation (4.44) applies. Estimate the difference between the natural frequencies of the $(2, 2)$ and $(10, 10)$ vibration modes of a simply supported rectangular 5 mm thick steel plate of dimensions $500\,\text{mm} \times 700\,\text{mm}$ when submerged in air and water.

4.3 The added mass per unit area due to fluid loading on a long, uniformly pulsating, cylinder of radius a can be shown to approach $m = -\rho_0 a \ln(ka)$ for $ka \ll 1$. Show that, in spite of the fact that this increases without limit as ka tends to zero, fluid loading, even by a liquid, is not likely in practice to introduce an extra ring frequency at a very low value of ka.

5 | Transmission of Sound through Partitions

5.1 Practical Aspects of Sound Transmission through Partitions

There are two main methods of inhibiting the transmission of sound energy from one region of fluid to another. In the first, sound energy is absorbed in transit by materials that are specially chosen to accept energy efficiently from waves in the contiguous fluid, and then efficiently to dissipate it into heat. Systems that utilise this principle include room wall sound absorbers, absorbent duct liners and splitter attenuators in ventilation systems. Alternatively, sound in transit may be reflected by means of introducing a large change of acoustic impedance into the transmission path. Examples include internal combustion engine exhaust expansion chambers, in which the changes of cross section are effective; hydraulic line silencers, in which the wave in the liquid encounters an acoustically 'soft' pipe section surrounded by pressurised gas; and partitions of solid sheets such as room walls and industrial noise control enclosures. Partition of adjacent fluid regions may, of course, not be total, in which case we use the terms 'barrier' and 'screen'. Many forms of sound transmission control system employ both methods in combination; examples include double-leaf partitions in buildings and sound insulation 'trim' in vehicles.

The design and construction of effective partitions is a central element in the practice of noise control by engineers and architects. An awareness of the basic physical principles and of good design practice is important to a wider group, including local authority planners, environmental health officers, buildings and works officers, and industrial management. In this chapter, the basic principles of the subject are illustrated by analyses of sound transmission through some simple idealised models of uniform single- and double-leaf partition constructions. A review of some of the more important extensions of these analyses to account more completely for features of practical systems is illustrated by the presentation of a range of typical experimental data. Close fitting covers are often installed around vibrating machinery in order to reduce radiated sound; and, it is common practice to cover the boundaries of vehicle passenger compartments with multi-layer sheets of sound insulation material in order to reduce sound transmission through them. These related topics are addressed in a largely qualitative manner. Finally, the transmission characteristics of some non-uniform and non-plane structures are discussed, including circular cylindrical shells and pipes.

5.2 Transmission of Normally Incident Plane Waves through an Unbounded Partition

The idealised system is shown in Fig. 5.1. A uniform, unbounded, non-flexible partition of mass per unit area m is mounted upon a viscously damped, elastic

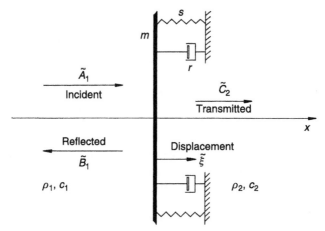

Fig. 5.1. Idealised model of normal incidence sound transmission through a single-leaf partition (Fahy, 1987).

suspension, having stiffness and damping coefficients per unit area of s and r, respectively. This represents an approximation to the fundamental mode of a large panel. The partition separates fluids of different characteristic specific acoustic impedances, $\rho_1 c_1$ and $\rho_2 c_2$. A plane sound wave of frequency ω is incident upon the partition from the region $x < 0$; the incident pressure field is written as

$$p_i(x, t) = \tilde{A}_1 \exp[j(\omega t - k_1 x)] \tag{5.1}$$

where $k_1 = \omega/c_1$. The pressure field of the wave reflected from the partition is written

$$p_r(x, t) = \tilde{B}_1 \exp[j(\omega t + k_1 x)] \tag{5.2}$$

The coefficients $\tilde{A}_1$ and $\tilde{B}_1$ are linked by the normal particle velocity at the left-hand surface of the partition, which moves with a normal velocity equal to $j\omega\tilde{\xi}$. Hence

$$\tilde{A}_1 - \tilde{B}_1 = j\omega\rho_1 c_1\tilde{\xi} \tag{5.3}$$

As we have seen in Sections 4.2 and 4.3, the mechanical impedance of an *in-vacuo* structure may be combined with the fluid-loading impedance associated with structural motion to form the total effective impedance of a 'fluid-loaded structure' as presented to mechanically applied forces [see Eqs. (4.10 and 4.43)]. We may enquire whether such a concept is relevant to acoustically applied forces.

The acoustic pressure field radiated in the negative x-direction by a displacement $\tilde{\xi}$, whatever the cause of partition motion, is

$$p_r^-(x, t) = \tilde{C}_1 \exp[j(\omega t + k_1 x)] \tag{5.4}$$

where $\tilde{C}_1 = -j\omega\rho_1 c_1\tilde{\xi}$. The corresponding wave radiated in the positive x-direction is

$$p_r^+(x, t) = \tilde{C}_2 \exp[j(\omega t - k_2 x)] \tag{5.5}$$

where $\tilde{C}_2 = j\omega\rho_2 c_2\tilde{\xi}$ and $k_2 = \omega/c_2$. These fields may be termed the 'radiated fields'.

According to Eqs. (5.1)–(5.3), the total pressure field on the left-hand side of the partition is

$$p^-(x, t) = \tilde{A}_1 \exp[j(\omega t - k_1 x)] + (\tilde{A}_1 - j\omega\rho_1 c_1\tilde{\xi}) \exp[j(\omega t + k_1 x)] \tag{5.6}$$

$$= 2\tilde{A}_1 \cos k_1 x \exp(j\omega t) - j\omega\rho_1 c_1\tilde{\xi} \exp[j(\omega t + k_1 x)]$$

Equation (5.6) may be rewritten, using Eq. (5.4), as

$$p^-(x,t) = 2\tilde{A}_1 \cos k_1 x \exp(j\omega t) + \tilde{C}_1 \exp[j(\omega t + k_1 x)] \qquad (5.7)$$

Now, the first term on the right-hand side of Eq. (5.7) represents the standing interference field created by the incidence upon and reflection from a completely immobile partition; we may term this as the 'blocked pressure field'. The second term represents the pressure field generated by partition motion. Hence the total field on the incident side equals the sum of the blocked field and the radiated field: the total field on the right-hand side is simply the radiated field represented by Eq. (5.5). The equation of motion of the partition is

$$m\ddot{\xi} + r\dot{\xi} + s\xi = p(x = 0^-, t) - p(x = 0^+, t) \qquad (5.8)$$

where $x = 0^-$ and $x = 0^+$ refer to the left- and right-hand faces of the partition. Substitution from Eqs. (5.5) and (5.6) gives

$$(-\omega^2 m + j\omega r + s)\tilde{\xi} = 2\tilde{A}_1 - j\omega \rho_1 c_1 \tilde{\xi} - j\omega \rho_2 c_2 \tilde{\xi} \qquad (5.9)$$

The fluid-loading (radiation) pressure terms on the right-hand side of this equation may be incorporated into the term on the left-hand side, which represents the *in-vacuo* partition properties, to give

$$[-\omega^2 m + j\omega(r + \rho_1 c_1 + \rho_2 c_2) + s]\tilde{\xi} = 2\tilde{A}_1 \qquad (5.10)$$

The fluid-loading terms represent radiation damping to be added to mechanical damping. If we express the left-hand side in terms of the partition velocity $\tilde{v} = j\omega\tilde{\xi}$, instead of the displacement $\tilde{\xi}$, we can rewrite this equation as

$$[j(\omega m - s/\omega) + (r + \rho_1 c_1 + \rho_2 c_2)]\tilde{v} = 2\tilde{A}_1 \qquad (5.11)$$

or

$$(\tilde{z}_p + \tilde{z}_f)\tilde{v} = 2\tilde{A}_1 \qquad (5.12)$$

where z_p and z_f are the partition (*in-vacuo*) and fluid-loading impedances, respectively. The forcing term $2\tilde{A}_1$ on the right-hand side is the blocked surface pressure field [compare with Eq. (4.43)]. Equation (5.12) proves that we may treat the problem as one of the response of a fluid-loaded structure to the surface pressure distribution of a blocked incident field. In fact, such a decomposition of the total field, which leads to this concept, is valid for any elastic structure immersed in a fluid; however, in most practical cases the analysis is far more complicated than in this simple one-dimensional idealisation.

Having obtained an expression for the velocity of the partition in terms of the amplitude of the incident pressure wave, we can now write an expression for $\tilde{C}_2$, the transmitted wave amplitude. Using Eqs. (5.5) and (5.12),

$$\tilde{C}_2 = \rho_2 c_2 \tilde{v} = 2\tilde{A}_1 \rho_2 c_2 / (\tilde{z}_p + \tilde{z}_f)$$

$$= \frac{2\tilde{A}_1}{j(\omega m - s/\omega)\rho_2 c_2 + (r/\rho_2 c_2 + \rho_1 c_1/\rho_2 c_2 + 1)} \quad (5.13)$$

The 'sound power transmission coefficient' τ is defined as the ratio of transmitted to incident sound powers. In this special case of normal incidence, it also equals the sound intensity transmission coefficient, given by

$$\tau = \frac{|\tilde{C}_2|^2 / 2\rho_2 c_2}{|\tilde{A}_1|^2 / 2\rho_1 c_1} = \frac{4n}{[(\omega m - s/\omega)/\rho_2 c_2]^2 + (\omega_0 m n/\rho_2 c_2 + n + 1)^2} \quad (5.14)$$

where $n = \rho_1 c_1/\rho_2 c_2$, and r has been replaced by $\omega_0 m \eta$, where η is the *in-vacuo* loss factor. The logarithmic form of the sound power transmission coefficient is the 'transmission loss' (*TL*) [also known as 'sound transmission loss' (*STL*)] defined by

$$TL = 10 \, \log_{10}(1/\tau) \quad \text{dB}$$

When this quantity relates to sound transmission between two rooms, it is termed as the 'sound reduction index' (R).

The transmission coefficient clearly has a maximum value at the undamped natural frequency of the partition given by $\omega_0 = (s/m)^{1/2}$. Three special cases may be identified:

(i) $\omega \ll \omega_0 = (s/m)^{1/2}$, well below the *in-vacuo* natural frequency:

$$\tau \approx \frac{4n}{(s/\omega \rho_2 c_2)^2 + (s\eta/\omega_0 \rho_2 c_2 + n + 1)^2} \approx \frac{4n}{(s/\omega \rho_2 c_2)^2 + (n + 1)^2} \quad (5.15)$$

because η is normally much less than unity. Now, $s/\omega \rho_2 c_2 = (\omega_0/\omega) \times (\omega_0 m/\rho_2 c_2)$ and $\omega_0 m/\rho_2 c_2$ is normally much greater than unity for typical structures at audio frequencies in gases, but not necessarily in liquids. If the fluid on both sides is air, Eq. (5.15) can, under this frequency condition, generally be reduced to

$$\tau \approx (2\rho_0 c\omega/s)^2 \tag{5.16}$$

The equivalent transmission loss is

$$TL = 20\log_{10} s - 20\log_{10} f - 20\log_{10}(4\pi\rho_0 c) \quad \text{dB} \tag{5.17}$$

where $f = \omega/2\pi$ Hz. The TL is seen to be determined primarily by the elastic stiffness of the mounting and is insensitive to mass and damping. It decreases with frequency by 6 dB per octave.

If the fluid impedance ratio n is very large, or if the mass per unit area of the partition is very low (e.g., thin plastic sheet), Eq. (5.17) is not valid. If $n \gg (\omega_0/\omega)(\omega_0 m/\rho_2 c_2)$, which means $\rho_1 c_1 \gg \omega_0^2 m/\omega$, then

$$\tau \rightarrow 4/n \tag{5.18}$$

and is independent of the mechanical properties of the partition. For example, if a partition separates air and water, $\tau \approx 1.1 \times 10^{-3}$, or $TL \approx$ 29.5 dB.

(ii) $\omega \gg \omega_0$, well above the natural frequency:

$$\tau \approx \frac{4n}{(\omega m/\rho_2 c_2)^2 + (n+1)^2} \tag{5.19}$$

because $\eta < 1$. If the fluid on both sides is air, then normally $\omega m/\rho_2 c_2 \gg 1$ and

$$\tau \approx (2\rho_0 c/\omega m)^2 \tag{5.20}$$

Correspondingly,

$$TL = 20\log_{10} m + 20\log_{10} f - 20\log_{10}(\rho_0 c/\pi) \quad \text{dB}$$

or

$$TL = 20\log_{10}(mf) - 42 \quad \text{dB} \tag{5.21}$$

The TL is seen to be determined primarily by mass per unit area, and is largely independent of damping and stiffness; it increases with frequency at 6 dB per octave and 6 dB per doubling of mass. Equation (5.21) is known as the 'normal incidence mass law'.

Very lightweight films at low frequencies may not behave according to Eq. (5.21) because $\omega m/\rho_2 c_2$ may not be much greater than unity. If $n \gg \omega m/\rho_2 c_2$, or $\omega m \ll \rho_1 c_1$, then τ is given by Eq. (5.18).

(iii) $\omega = \omega_0$, the natural frequency:

$$\tau = \frac{4n}{[\eta(\rho_2 c_2/\omega_0 m)^{-1} + (n+1)]^2} \tag{5.22}$$

If the fluid on both sides of the partition is the same and if $\eta \ll \rho_0 c/\omega m$, then

$$\tau \approx 1 \tag{5.23}$$

If $\eta \gg \rho_0 c/\omega m$, then

$$\tau \approx (2\rho_0 c/\eta\omega_0 m)^2 \tag{5.24}$$

The corresponding transmission losses are

$$TL = 0 \quad \text{dB} \tag{5.25}$$

and

$$TL = 20\log_{10} f_0 + 20\log_{10} m + 20\log_{10} \eta - 20\log_{10}(\rho_0 c/\pi) \quad \text{dB} \tag{5.26}$$

The latter value differs from the mass law value by $20\log_{10}(\eta)$ dB. Equations (5.23) and (5.25) indicate total transmission at resonance when radiation damping exceeds mechanical damping. Equation (5.26) shows that the mass, damping and stiffness (through f_0) all influence transmission at resonance, provided that the mechanical damping exceeds the radiation damping. If $n > 1$, then $\eta(\rho_2 c_2/\omega_0 m)^{-1}$ must be comparable with η for mechanical damping to have any effect.

It is tempting to use this model to evaluate the transmission characteristics of a bounded flexible panel vibrating in its fundamental mode of vibration, in which the phase of the displacement is uniform over the whole surface. Examples of practical concern include windows and the panels of enclosures. However, at the fundamental natural frequencies typical of glazing panels (10–30 Hz), the acoustic wavelength is so large (34–11 m) compared with the typical aperture dimension that the transmission is controlled as much by aperture diffraction as by the window dynamics: the partition acts like a piston of small ka in a baffle and therefore does not radiate (transmit) effectively. The same stricture applies to the transmission characteristics of 'acoustic' louvres installed in the walls of machinery plant rooms. At low audio frequencies a simple hole in the wall has a non-zero transmission loss. In these cases, it is the insertion loss (received sound pressure level with and without the insertion of the particular item) that is significant.

The results of the foregoing analysis suggest that in cases where the characteristic acoustic impedance of one medium is much greater than the other (e.g., air/water) the mechanical properties of a partition have little influence on the transmission, which is controlled simply by the ratio of the impedances. It should also be noted that all the expressions for τ are reciprocal in n, so that τ is independent of the direction of the normally incident plane wave.

5.3   Transmission of Obliquely Incident Plane Waves through an Unbounded Flexible Partition

Having established the principle of applying the blocked surface pressure as the forcing field on a fluid-loaded structure, we may now apply it to the case of an unbounded, thin, uniform, elastic plate upon which acoustic plane waves of frequency ω are incident at an arbitrary angle ϕ_1 as shown in Fig. 5.2.

The component of the incident wavenumber vector $\mathbf{k}$ that is directed parallel to the partition plane (sometimes called the trace wavenumber), is $k_z = k_1 \sin \phi_1$. Since the partition is uniform and unbounded, no point is different dynamically from any other; therefore, the flexural wave induced in the partition must have a wavenumber $k_z = k_1 \sin \phi_1$. The blocked pressure at the partition surface is

$$p_{bl}(x = 0^-, z, t) = 2\tilde{A}_1 \exp[j(\omega t - k_1 \sin \phi_1 z)] \qquad (5.27)$$

The coefficient $2\tilde{A}_1$ is equivalent to the applied force per unit area in Eq. (4.43); hence

$$2\tilde{A}_1 = (\tilde{z}_{wp} + \tilde{z}_{wf})\tilde{v} \qquad (5.28)$$

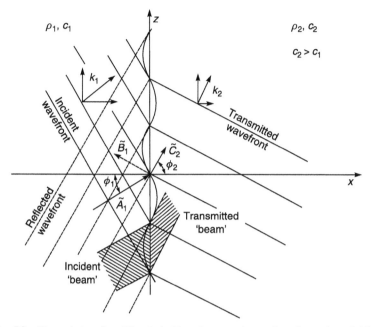

Fig. 5.2.   Transmission of an obliquely incident plane sound wave through an unbounded flexible partition (Fahy, 1987).

The fluid wave impedance is given by Eq. (4.38) as

$$\tilde{z}_{wf} = \rho_1 c_1 (1 - \sin^2\phi_1)^{-1/2} + \rho_2 c_2 \left[1 - (k_1\sin\phi_1/k_2)^2\right]^{-1/2}$$

$$= \tilde{z}_{wf_1} + \tilde{z}_{wf_2} \tag{5.29}$$

and the partition wave impedance is given by Eq. (4.37) as

$$\tilde{z}_{wp} = -(j/\omega)(Dk_1^4\sin^4\phi_1 - m\omega^2) + Dk_1^4\eta/\omega\sin^4\phi_1 \tag{5.30}$$

Note that the component of wave impedance $\tilde{z}_{wf_2}$ produced by the fluid to the right of the partition is only real if

$$\sin\phi_1 < k_2/k_1 = c_1/c_2 \tag{5.31}$$

Hence, energy transmission is limited to a range of ϕ_1 satisfying this condition. For instance, irrespective of the properties of a partition, plane-wave energy cannot be transmitted by a uniform partition from air into water at angles of incidence greater than $13.4°$. The transmitted pressure coefficient $\tilde{C}_2$ is related to the partition normal velocity $\tilde{v}$ by

$$\tilde{C}_2 = \tilde{z}_{wf_2}\tilde{v} \tag{5.32}$$

Equations (5.28) and (5.32) yield

$$\tilde{C}_2 = \frac{2\tilde{A}_1\tilde{z}_{wf_2}}{\tilde{z}_{wf_1} + \tilde{z}_{wf_2} + \tilde{z}_{wp}} \tag{5.33}$$

The sound intensity transmission coefficient is given by

$$\tau = \frac{|\tilde{C}_2|^2/2\rho_2 c_2}{|\tilde{A}_1|^2/2\rho_1 c_1} \tag{5.34}$$

However, this is not generally the ratio of sound power transmitted per unit area of partition to sound power incident per unit area of partition because of refraction when $c_1 \neq c_2$. Reference to Fig. 5.2 reveals that the widths of corresponding 'beams' on the two sides are in the ratio

$$\cos\phi_1/\cos\phi_2 = (1 - \sin^2\phi_1)^{1/2}/(1 - \sin^2\phi_2)^{1/2}$$
$$= (1 - \sin^2\phi_1)^{1/2}/[1 - (c_2\sin\phi_1/c_1)^2]^{1/2} \tag{5.35}$$

The sound *power* transmission coefficient is therefore given by

$$\tau_p = \frac{4|\tilde{z}_{wf_2}|^2}{|\tilde{z}_{wf_1} + \tilde{z}_{wf_2} + \tilde{z}_{wp}|^2}\left[\frac{\rho_1 c_1}{\rho_2 c_2}\right]\left[\frac{1 - (c_2\sin\phi_1/c_1)^2}{1 - \sin^2\phi_1}\right]^{1/2} \tag{5.36}$$

This rather complicated expression reduces to a much simpler form when the fluids on the two sides are the same. Then $\phi_1 = \phi_2 = \phi$, and

$$\tau_p = \tau = \left|\frac{\tilde{z}_{wf}}{\tilde{z}_{wf} + \tilde{z}_{wp}}\right|^2 \tag{5.37}$$

where $\tilde{z}_{wf} = \tilde{z}_{wf_1} + \tilde{z}_{wf_2} = 2\tilde{z}_{wf_1} = 2\tilde{z}_{wf_2}$. The explicit form of Eq. (5.37) is

$$\tau = \frac{(2\rho_0 c \sec\phi)^2}{[2\rho_0 c\sec\phi + (D/\omega)\eta k^4\sin^4\phi]^2 + [\omega m - (D/\omega)k^4\sin^4\phi]^2} \tag{5.38a}$$

In order to investigate the relative influences of partition mass, stiffness and damping, it is helpful to consider the conditions under which the incident wave is coincident with the flexural wave in the partition. The wavenumber of the wave induced in the partition by the incident field is, as we have seen, equal to the trace wavenumber $k_z = k \sin \phi$. The expression for the free flexural wavenumber in a plate is $k_b^4 = \omega^2 m / D$. Hence Eq. (5.38a) may be rewritten as

$$\tau = \frac{(2\rho_0 c / \omega m)^2 \sec^2 \phi}{[(2\rho_0 c / \omega m) \sec \phi + (k/k_b)^4 \eta \sin^4 \phi]^2 + [1 - (k/k_b) \sin^4 \phi]^2} \tag{5.38b}$$

The *coincidence* condition is

$$k \sin \phi = k_b = (\omega^2 m / D)^{1/4} \tag{5.39}$$

which corresponds to the disappearance of the reactive contribution to the denominator of Eqs. (5.38). Rewriting Eq. (5.39) as

$$\omega_{co} = (m/D)^{1/2} (c / \sin \phi)^2 \tag{5.40}$$

shows that for a given angle of incidence ϕ there is a unique coincidence frequency ω_{co}, and vice versa. However, since $\sin \phi$ cannot exceed unity, there is a lower limiting frequency for the coincidence phenomenon given by

$$\omega_c = c^2 (m/D)^{1/2} \tag{5.41}$$

where ω_c is known as the *critical frequency*, or lowest coincidence frequency. Equation (5.40) can therefore be rewritten as

$$\omega_{co} = \omega_c / \sin^2 \phi \tag{5.42a}$$

or

$$\sin \phi_{co} = (\omega_c / \omega)^{1/2} \tag{5.42b}$$

where ϕ_{co} is the coincidence angle for frequency ω. These relationships are illustrated graphically in Fig. 5.3. The nature of coincidence is further illustrated in Fig. 5.4. It is clear from Eq. (5.41) that lightweight, stiff partitions, such as honeycomb sandwiches having a high core shear stiffness, tend to exhibit lower critical frequencies than homogeneous partitions of similar weight but of lower stiffness. However, as shown in Section 5.12, it is possible to raise the critical frequencies of such panels by decreasing the core shear stiffness.

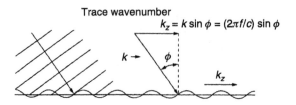

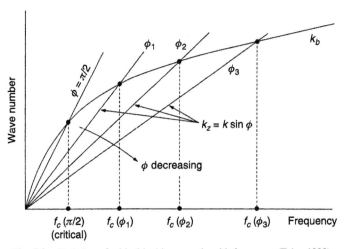

Fig. 5.3. Variation of critical incidence angle with frequency (Fahy, 1998).

In the case of uniform homogeneous flat plates of material density ρ_s, Eq. (5.41) can be written as

$$\omega_c = c^2(\rho_s h)^{1/2}[(Eh^3/12(1-v^2)]^{-1/2}$$

or

$$f_c = c^2/1.8hc_l' \quad \text{Hz} \tag{5.43}$$

where h is the plate thickness andM c_l' is the phase speed of longitudinal waves in the plate. Thus the product hf_c is a function only of the material properties of the fluid and solid media. This product is tabulated for a range of common materials in air at 20°C in Table 5.1. As an example, the critical frequency of 6-mm-thick steel plate in air is 2060 Hz. In water, the values would be greater by a factor of approximately nineteen than those for the same plate in air. Hence, in

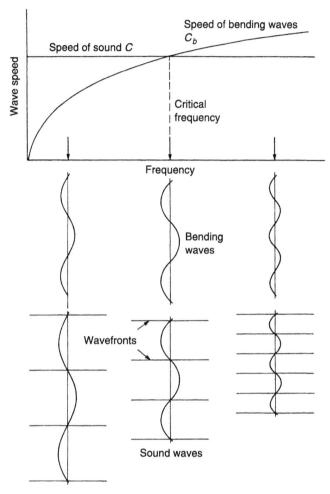

Fig. 5.4.   Coincidence diagram (Fahy, 1987). (Trace the acoustic wavefronts onto a transparent sheet and superimpose them on to the corresponding bending waves. Adjust the angle of the superimposed diagram to try to make the acoustic wavefronts coincide with the points of zero displacement of the bending waves to identify coincidence – if physically possible.)

marine engineering applications, frequencies greater than f_c are rarely of practical importance.

Returning to the transmission coefficient [Eq. (5.38)], it is now clear that the influence of the coincidence phenomenon, which corresponds to the disappearance of the reactive term in the denominator, will affect the value of τ at all

TABLE 5.1

Product of Plate Thickness and Critical
Frequency in Air $(20°C)^a$

Material	hf_c (ms^{-1})
Steel	12.4
Aluminium	12.0
Brass	17.8
Copper	16.3
Glass	12.7
Perspex	27.7
Chipboard	23^b
Plywood	20^b
Asbestos cement	17^b
Concrete	
Dense	19^b
Porous	33^b
Light	34^b
Plasterboard (gypsum)	≈ 34

[a] To obtain values in water, multiply by 18.9.
[b] Variations of up to $\pm 10\%$ possible.

frequencies in the range $\omega_c \leq \omega \leq \infty$. It is instructive to examine the variation of τ with the angle of incidence for a fixed frequency.

Consider first the range of frequency below the critical frequency of the partition. The ratio of the trace wavenumber of the exciting field to the free flexural wavenumber is given by

$$\frac{k_z}{k_b} = \frac{k \sin \phi}{(\omega^2 m/D)^{1/4}} \tag{5.44a}$$

which, from Eq. (5.40), may be written as

$$\frac{k_z}{k_b} = (\omega/\omega_c)^{1/2} \sin \phi \tag{5.44b}$$

The physical interpretation of the fact that for $\omega < \omega_c$, this ratio is necessarily less than unity is that the phase speed *of free* bending waves is *less* than the trace wave speed of the incident field at all angles of incidence. The influence of this condition on transmission is seen in the dominance of the inertia term ωm over the stiffness term $(D/\omega)k^4 \sin^4 \phi$ in the denominator of Eq. (5.38a).

Clearly the mechanical damping term, which is η ($\ll 1$) times the stiffness term, is also negligible compared with the inertia term. Hence the sound power transmission coefficient at frequencies well below the critical frequency is to a good approximation

$$\tau(\phi) = 1/[1 + (\omega m \cos \phi/2\rho_0 c)^2] \tag{5.45}$$

Provided that $\omega m \cos \phi \gg 2\rho_0 c$, which is normally true except for $\phi \to \pi/2$, the corresponding transmission loss is given by

$$TL(\phi) = 20 \log_{10}(\omega m \cos \phi/2\rho_0 c) \quad \text{dB} \tag{5.46}$$

Comparison with the normal incidence mass law [Eq. (5.21)] shows that

$$TL(0) - TL(\phi) \approx -20 \log_{10}(\cos \phi) \quad \text{dB} \tag{5.47}$$

and hence the difference increases as the angle of incidence approaches $\pi/2$ (grazing). The dependence of the TL on angle of incidence is particularly important in relation to the transmission of the noise of wheels through the side walls and windows of road and rail vehicles because the sound is incident at angles close to ninety degrees. An ISVR experimental study of transmission of external sound through a car window showed a difference of about 10 dB between TLs measured at angles of incidence of 75° and 85° in the range 3–8 kHz, which agrees with Eq. (5.47). Diffuse field testing is clearly inapplicable to such cases.

Now, the condition $k_z/k_b < 1$, although always true when $\omega < \omega_c$, is not restricted to this frequency range. Reference to Eq. (5.42a) shows that Eq. (5.44b) may be written as $k_z/k_b = (\omega/\omega_{co})^{1/2}$. Thus, the conclusions drawn above concerning the dominance of the inertia term apply, for a *given angle of incidence*, not just for $\omega \ll \omega_c$ but for $\omega \ll \omega_c/\sin^2 \phi$. As ω approaches, ω_{co} the magnitude of the stiffness term in the transmission expression approaches that of the inertia term, a maximum in the sound power transmission coefficient occurs at $\omega = \omega_{co}$, and

$$\tau = 1/(1 + \eta \omega_{co} m \cos \phi/2\rho_0 c)^2 \tag{5.48}$$

Comparison of this expression with that for purely mass-controlled transmission at the same frequency [Eq. (5.45)] shows that the difference between the corresponding sound reduction indices is at least $20 \log_{10} \eta$ dB. If $\eta > 2\rho_0 c/\omega_{co} m \cos \phi$, the transmission of sound energy in the vicinity of coincidence is controlled by mechanical damping.

At frequencies above ω_{co}, the stiffness term dominates in the transmission expression and

$$\tau \approx 1/[1 + (Dk^4 \sin^4 \phi \cos \phi / 2\rho_0 c\omega)^2] \qquad (5.49)$$

In most cases of sound transmission in air, the stiffness term greatly exceeds unity and hence the transmission loss for a given ϕ increases at approximately 18 dB per doubling of frequency: the damping exerts no influence in this range. The form of variation of the TL with frequency for constant ϕ is shown in Fig. 5.5.

An alternative view of this rather complicated behaviour is obtained by considering transmission over the whole range of angle of incidence *at fixed frequency*. Below the critical frequency, transmission at all angles is, of course, mass controlled. At any frequency above the critical frequency, Eq. (5.42b) determines the coincidence angle. If $\sin \phi$ is less than $\sin \phi_c$, i.e., $\phi < \phi_{co}$, Eq. (5.38) shows that the inertia term dominates: if $\phi > \phi_{co}$ then stiffness dominates. At $\phi = \phi_{co}$ damping is in control provided it is sufficiently large to exceed acoustic radiation damping. This behaviour is illustrated in Fig. 5.6.

In practice, sound waves are usually incident upon a partition from many angles simultaneously, e.g., the wall of a room or a window exposed to traffic noise. The appropriate transmission coefficient can, in principle, be derived from Eq. (5.38) by weighting according to the directional distribution of incident intensity and integrating over angle of incidence. In practice, the directional distribution of incident intensity is rarely known, and therefore, in room acoustics, an ideal diffuse field model is often assumed, in which plane waves are incident

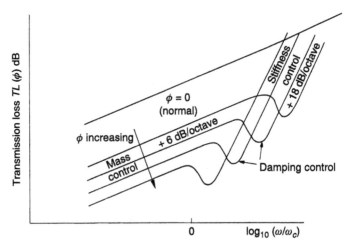

Fig. 5.5. Variation of TL with frequency for a single angle of incidence (Fahy, 1987).

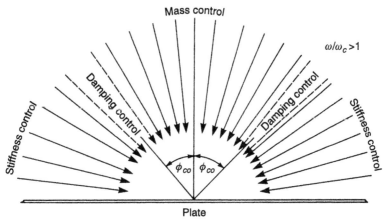

Fig. 5.6. Regions of incidence angle associated with mass, stiffness and damping control of plane wave sound transmission at a single supercritical frequency (Fahy, 1987).

from all directions with equal probability and with random phase. The appropriate weighting leads to the following expression for the diffuse field sound power transmission coefficient (Prove):

$$\tau_d = \frac{\int_0^{\pi/2} \tau(\phi) \sin\phi \cos\phi \, d\phi}{\int_0^{\pi/2} \sin\phi \cos\phi \, d\phi} = \int_0^{\pi/2} \tau(\phi) \sin 2\phi \, d\phi \qquad (5.50)$$

The $\cos\phi$ term arises from the variation with ϕ of the plane-wave intensity component normal to the partition, and the $\sin\phi$ term relates the total acoustic power carried by the incident waves to their angle of incidence. The general expression for τ is not amenable to analytic integration, but the restricted expression in Eq. (5.45) may be evaluated for frequencies well below the critical frequency. The resulting diffuse field transmission loss

$$TL_d \approx TL(0) - 10 \log_{10}[0.23 TL(0)] \quad \text{dB} \qquad (5.51)$$

The equivalent index of diffuse field sound transmission employed in the field of building acoustics is the 'Sound Reduction Index' R. This terms applies specifically to the TL of a partition that separates two rooms. It is generally found that experimental results of measurements of the sound reduction index of building partitions do not agree very well with Eq. (5.51), tending to higher values more in accord with an empirical expression

$$R_f = R(0) - 5 \quad \text{dB} \quad \text{or} \quad R_f \approx 20 \log_{10}(mf) - 47 \quad \text{dB} \qquad (5.52)$$

which is called the 'field incidence mass law': it corresponds to omission of sound waves incident at angles between 78 and 90°. Sharp (1978) states that there is no physical justification for this *ad hoc* restriction. As explained in the following section, Leppington *et al.* (1987) suggest, on the basis of a rigorous analysis of transmission through bounded partitions, that the effectiveness of this 'correction' is fortuitous. This view is supported by the work of Villot *et al.* (2001), who investigated the spatial windowing effect associated with the finite dimensions of panels.

Curves of $TL(0)$, TL_d and TL_f for subcritical frequencies are compared in Fig. 5.7. An expression for transmission loss at frequencies above the critical frequency was derived by Cremer (1942):

$$TL_d = TL(0) + 10\log_{10}(f/f_c - 1) + 10\log_{10}\eta - 2 \quad \text{dB} \qquad (5.53)$$

The dominant influence of coincidence transmission is seen in the presence of the loss factor term. The general form of the theoretical diffuse field transmission loss curve for infinite partitions is shown in Fig. 5.8. As mentioned above, deviations from this curve are observed in experimental results obtained on *bounded* panels. An example of a measured sound reduction index curve that exhibits such deviation and also exhibits a distinct coincidence dip is presented in Fig. 5.9;

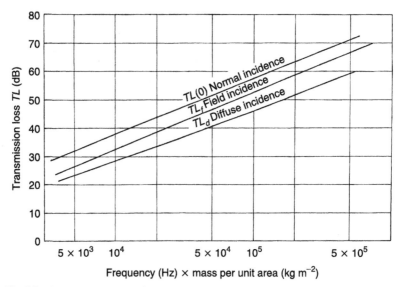

Fig. 5.7. Transmission losses of unbounded partitions at subcritical frequencies (Fahy, 1987).

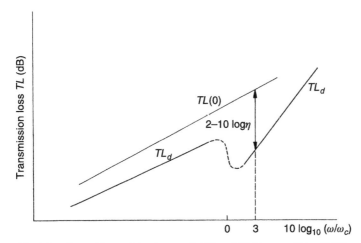

Fig. 5.8.   General form of the theoretical diffuse field *TL* curve (Fahy, 1987).

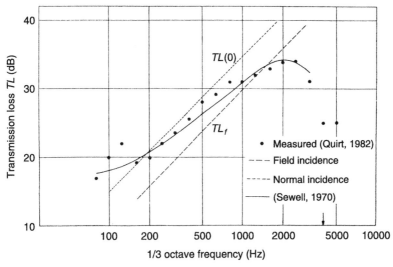

Fig. 5.9.   Transmission loss of 3 mm thick glazing (Quirt, 1982).

the dip is deep because the damping of glazing generally has fairly small values unless special edge or lamination treatment is employed. The reference to Sewell's analysis cited in the figure is discussed further at the end of the next section, together with other relevant publications.

5.4　Transmission of Diffuse Sound through a Bounded Partition in a Baffle

The two principal factors that can cause the diffuse-field transmission performance of a real, bounded panel in a rigid baffle to differ significantly from the theoretical performance of an unbounded partition are (i) the existence of standing-wave modes and associated resonance frequencies; (ii) diffraction by the aperture in the baffle that contains the panel. As we have seen in Chapter 3, the radiation efficiencies of modes of bounded plates vibrating at frequencies below the critical frequency are very much influenced by the presence of the boundaries, and are generally less than unity. It will be shown in Chapter 6 that the response of a structural mode to acoustic excitation is proportional to its radiation efficiency. We have also seen that the radiation efficiency of an *infinite* partition, excited by plane waves obliquely incident at angle ϕ, is $\sigma = \sec \phi$, which generally exceeds unity, and that the response is mass controlled below the critical frequency. Therefore, in comparing the transmission coefficients at *subcritical* frequencies of bounded and unbounded partitions of the same thickness and material, it would seem that the relatively low values of radiation efficiency of the low order bounded plate modes (due to radiation cancellation) would be offset by the enhanced response produced by modal resonance (which is absent in the infinite partition). However, it transpires from analyses that are mathematically too complex to be presented here that resonant 'amplification' is generally insufficient to make up for the low values of modal radiation efficiencies associated with modal *resonance* frequencies, except in small, very lightly damped panels.

Application of the multi-mode, diffuse-field response equation presented in Chapter 6 [Eq. (6.37)] shows that the ratio of the mass-controlled, field incidence, infinite-partition transmission coefficient τ_∞, to the modal-average bounded-plate, resonant-mode, transmission coefficient τ_r, for homogeneous plates of equal densities and thicknesses, is approximately

$$\tau_\infty / \tau_r \approx (hc'_l\omega/\sqrt{3}c^2)(\eta_{tot}/\sigma^2) \tag{5.54}$$

in which h is the plate thickness, σ is the modal-average radiation efficiency in a frequency band centred on ω and η_{tot} is the sum of the plate loss factors due to frictional mechanisms, transmission of vibrational energy into the support structure and sound radiation. Note that the sensitivity to the magnitude of the radiation efficiency; this arises because both response and radiation are proportional to σ. Well below the critical frequency, the modal-average radiation efficiency of a simply supported rectangular plate is given approximately by Cremer *et al.* (1988) as

$$\sigma \approx (P\lambda_c/\pi^2 A)(f/f_c)^{1/2} \qquad (5.55)$$

where P is the panel perimeter length, A the panel area and λ_c the wavelength at f_c, which is by definition equal in the plate and the fluid. Substitution of this expression into Eq. (5.54) yields

$$\tau_\infty/\tau_r \approx 200(A/P\lambda_c)^2 \eta_{tot} \qquad (5.56)$$

which is frequency independent (except for any such dependence of η_{tot}) and typically much greater than unity, except in the case of very lightly damped, lightweight, stiff panels of small surface area. At frequencies approaching f_c, σ increases very rapidly towards unity and τ_r approaches τ_∞.

It is an experimentally observed fact that the *subcritical* sound reduction indices of many simple homogeneous partitions, as measured when they are inserted between reverberation rooms, approximate fairly closely to that given by the infinite-partition field incidence formula, Eq. (5.52). It is clear, therefore, that a model based exclusively upon transmission by the mechanism of excitation and radiation of modes *at resonance* is not adequate. Further convincing evidence for the dominance of a non-resonant sub-critical transmission mechanism is provided by the observation that the subcritical sound reduction indices of many partitions are not significantly altered by a large increase in their total damping. One may infer from this that a form of response and radiation that is insensitive to the action of damping mechanisms is responsible for the major part of the sound transmission process in such cases.

There are two forms of qualitative explanation of the origin of this behaviour. As the basis of the first, we observe that the vibrational response of a uniform bounded plate to an external force field may be considered to be the sum of the response to that field of an unbounded plate having the same dynamic properties, plus the response of the unbounded plate to the action of boundary force fields that are conceived as forcing it to satisfy its actual boundary conditions: the result of the action of these forces is seen in boundary wave reflection. A similar concept was introduced in Chapter 2 in connection with the impedance analysis of a bounded beam, in which the incidence of outgoing bending waves upon the beam boundaries causes reflected waves to be generated. Such a decomposition of response is not straightforward in the case of fluid-loaded plates because regions of the imaginary unbounded plate outside the boundaries can communicate acoustically with regions within the boundary, a process that is not physically valid. However, the influence of fluid loading on the vibration of most practical structures transmitting *airborne* sound is small.

Hence, we may visualise the response of a bounded plate to an incident plane sound wave as comprising two components: (i) the infinite-plate response component, which is 'forced' to travel at the trace wave speed $c/\sin\phi$ of the incident wave; (ii) the waves caused by the incidence of this forced wave on the actual boundaries. The latter waves, which are *free* bending waves travelling at their natural, or free, wave speeds, are multiply reflected by the various boundaries, and those components having frequencies equal to the natural frequencies of the bounded plate modes interfere constructively to create resonant motion in these modes. We may, at least qualitatively, consider the transmission processes associated with free- and forced-wave components to coexist independently, one controlled by damping and the other not. As already shown by Eq. (5.56), the forced-wave process (corresponding to τ_∞) tends to transmit more energy than the free-wave process (corresponding to τ_r), in agreement with experimental results.

In Eq. (5.56), the panel dimensions factor $(A/P\lambda_c)$ appears because it controls subcritical modal radiation efficiencies, as explained in Section 3.9. There is an additional geometric effect associated with the ratio of panel dimensions to the wavelength of *forced* waves. Equation (5.56) is based upon a model assumed that an infinitely extended forced bending wave exists, having a radiation efficiency equal to sec ϕ. In fact, forced-wave motion actually exists physically only within the boundaries of the plate, and therefore the number of forced wavelengths between boundaries is limited. This 'windowing' effect, already met in connection with bounded panel radiation, spreads the wavenumber spectrum of forced vibration around the line value $k\sin\phi$ of the equivalent infinitely extended wave train. Some of these spectral components are subsonic and do not radiate. Those forced-wave components produced by near grazing incidence ($\phi \rightarrow \pi/2$), which dominate forced transmission, leak most strongly into the subsonic wavenumber range ($k_x > k$). The effect is akin to that which allows the subsonic free-wave components to radiate, as already seen in Chapter 3; however, it acts in a reverse manner. We see, therefore, that the effect of reducing the size of a panel is to enhance free-wave, resonant transmission, and to reduce forced-wave transmission. Since the latter normally dominates in large panels, the low-frequency sound reduction index of small panels is likely to be greater and more affected by damping than that of larger panels of the same material.

An alternative explanation of the dominance at subcritical frequency of non-resonant transmission mechanisms is based upon the fact that the radiation efficiency of a given mode, which has a subcritical natural frequency, increases with the frequency of modal vibration. It transpires from detailed analysis that the greatest subcritical contribution of each mode (except the fundamental) to sound energy transmission occurs at frequencies far above its resonance frequency, in which case its response, and the resulting transmission, is mass controlled.

The effect of panel size on forced-wave (non-resonant) transmission is quantified in an analysis by Sewell (1970). His expression for the transmission coefficient of a panel of area A, at a frequency corresponding to acoustic wavenumber k, may be expressed in terms of a transmission loss for subcritical frequencies ($\omega < \omega_c$) as

$$TL_{nr} \approx TL(0) - 10\log_{10}[\ln(kA^{1/2})] + 20\log_{10}[1 - (\omega/\omega_c)^2] \quad \text{dB} \quad (5.57)$$

where the subscript 'nr' indicates non-resonant. Sewell's original formula contains a panel shape factor and another area factor, neither of which is usually significant. This expression has also been compared by Quirt (1982) with the field incidence formula and found to be superior at low frequencies (Fig. 5.9). Note that the low frequency transmission loss exceeds the normal incidence mass law. The increase in measured sound reduction index at low frequencies with a reduction of partition area has been observed in many transmission loss measurements, for example by Kurra and Arditi (2001). Sewell's result has been refined by Leppington *et al.* (1987). Villot *et al.* (2001) apply a specialised holographic technique, called 'phonoscopy', to the study of the transmission of sound through plain panels that separate rectangular rooms. The source room is reverberant but the wall of the receiver room parallel to the plane of the test panel is made totally sound absorbent. The vibration field of the panel and the near-field sound pressure field on the transmission side are decomposed into their wavenumber components. From these, the wavenumber spectrum of the radiated power is determined. This technique accounts for the effect of the finite extent of a panel on the wavenumber spectrum of the excitation field and on the radiation efficiency. It is related to the model analysed in Section 6.3. The efficacy of this technique is illustrated by application to a plain aluminium plate, a double glazing panel and a multilayer composite plate, of which the first is shown in Fig. 5.10. The differences between the sound reduction indices of the bounded and infinite panel are clearly shown.

5.5 Transmission of Sound through a Partition between Two Rooms

In Section 5.4, it was assumed that the panel was located in an aperture in an infinite, rigid, plane baffle and that the incident sound field was diffuse. It is of considerable practical interest to know how the transmission loss formulas obtained using this model compare with those of the sound reduction index derived on the basis of a model of a rectangular partition that forms the complete dividing wall between rectangular rooms. A comprehensive analysis of this problem has been presented by Josse and Lamure (1964). The essential difference between their model and the diffuse-field model is that the sound field is represented in terms of a series of room acoustic modes. The response of the

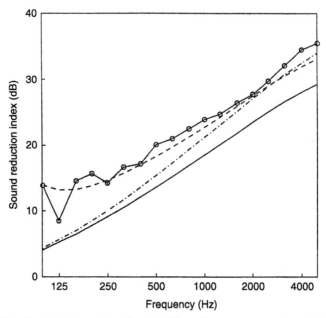

Fig. 5.10. Sound reduction index of an aluminium plate; ⊖-⊖-⊖ measured, —— infinite system, - - - spatially windowed system, –·–·–· infinite system with plane wave incidence angle limited to the range 0–78° (Villot *et al.*, 2001).

panel and the transmission of energy into the receiving room depends upon two coupling factors: the first is a function of the closeness of the natural frequencies of the acoustic modes of the rooms and of the panel vibration modes; the second depends upon the degree of spatial matching between the acoustic mode pressure distributions over the panel and the distributions of panel mode displacements. The analysis is not based upon a fully coupled system formulation, since the exciting pressure field on the surface of the panel is assumed to be that which would exist in the absence of panel motion; this assumption is not likely to lead to significant errors in most practical problems of airborne sound transmission through load-bearing partitions in buildings.

The reader interested in the details of the analysis is referred to the cited reference; the principal results for a panel measuring $a \times b$ follow.

(i) Subcritical frequencies; $\omega < \omega_c$:

$$R = R(0) - 10 \log_{10} \left\{ \left[\frac{3}{2} + \ln \left(\frac{2\omega}{\Delta\omega} \right) \right] + \frac{16c^2}{\eta\omega_c} \frac{1}{(\omega\omega_c)^{1/2}} \left(\frac{a^2 + b^2}{a^2 b^2} \right) \right.$$
$$\left. \times \left[1 + \frac{2\omega}{\omega_c} + 3 \left(\frac{\omega}{\omega_c} \right)^2 \right] \right\} \quad \text{dB} \qquad (5.58)$$

The term in the braces which contains η represents the resonant contribution to transmission; it is seen to increase in significance as the partition size decreases, a characteristic observed in the case of a baffled panel. A valuable feature of Eq. (5.58) is that it offers a means of estimating the minimum value of η that will significantly affect sound transmission. The general expression for the sound reduction index indicates that resonance-controlled transmission is relatively more important (3–6 dB) in this case of separation of two rooms than in the case of a panel in a rigid baffle separating two semi-infinite volumes of fluid.

The non-resonant component of sound reduction index is

$$R = \begin{cases} R(0) - 5.6\,\text{dB} & (1/3 \quad \text{octave} \quad \text{band}) \\ R(0) - 4.0\,\text{dB} & (1/1 \quad \text{octave} \quad \text{band}) \end{cases}$$

These expressions are close to the empirical field incidence formula.

(ii) Supercritical frequencies, $\omega > \omega_c$:

$$R = R(0) + 10\log_{10}[(2\eta/\pi)(\omega/\omega_c - 1)] \quad \text{dB} \tag{5.59}$$

This formula agrees with Cremer's [Eq. (5.53)] and indicates a 9 dB per octave increase with frequency.

(iii) Frequency band centred on ω_c, $\Delta\omega/\omega_c > (k_c a)^{-1}$, $(k_c b)^{-1}$:

$$R = R(0) + 10\log_{10}(2\eta/\pi)(\Delta\omega/\omega_c) \quad \text{dB} \tag{5.60}$$

Experimental verification of Eqs. (5.58)–(5.60) is provided by the results shown in Fig. 5.11 of measurement on a 5 cm-thick brick wall plastered on both sides to a thickness of 1.5 cm. Note the low frequency rise at 100 Hz.

Application of the Phonoscopy technique by Villot *et al.* (1992) to the study of sound transmission through a 60 mm thick plaster block wall separating a reverberant source room and an anechoically terminated receiver room reveals the distribution of the structural wavenumber components that dominate the transmission of sound. The critical frequency of the wall is 606 Hz and k_p is the free plate bending wavenumber. Figure 5.12(a) shows that only the wavenumber components less than the acoustic wavenumber k radiate (transmit) effectively. Figure 5.12(b) shows how the wavenumber vector components having magnitudes just below k dominate transmission at the critical frequency, as predicted theoretically by Anderson and Bratos-Anderson (2005). Figure 5.12(c) shows that the radiating wavenumbers cluster close to k at supercritical frequencies.

Theoretical and experimental measurements of sound transmission through walls between reverberant rooms at very low frequencies (<100 Hz) when the

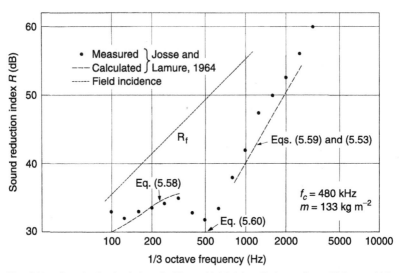

Fig. 5.11. Sound reduction index of a 50 mm thick brick wall plastered to a thickness of 15 mm. (Josse and Lamure, 1964).

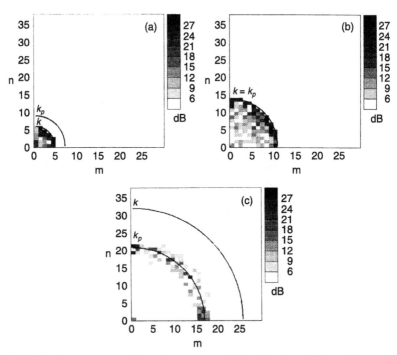

Fig. 5.12. k-space spectra of sound power transmission through a plain partition revealing the wavenumber components principally responsible as measured by phonoscopy: (a) subcritical frequency; (b) close to the critical frequency; (c) supercritical frequency (Villot *et al.*, 1992).

acoustic wavelengths are of the same order of magnitude as the room dimensions have shown that the results are highly dependent on the particular geometry of the system, the effectiveness and location of the principal absorbing surfaces and the location(s) of the source. This is not surprising since the sound power transmission coefficient depends upon the coupling between a rather small number of acoustic and panel modes. The concept of the sound reduction index as a quantity that characterises the transmission properties of a partition, and which is largely independent of its environment, is not really tenable at such low frequencies, although an International Standard for its measurement exists. This issue is discussed by Kropp *et al.* (1991) and Hopkins and Turner (2005).

5.6 Double-Leaf Partitions

Theoretical and experimental analyses of sound transmission through single-leaf partitions show that the sound reduction index at a given frequency generally increases by 5–6 dB per doubling of mass, provided that no significant flanking or coincidence-controlled transmission occurs. In practice, it is often not only necessary for structures to have low weight, but also to provide high transmission loss: examples include the walls of aircraft fuselages, partition walls in tall buildings and movable walls between television studios. This requirement can clearly not be met by single-leaf partitions.

The most common solution to this problem is to employ constructions comprising two leaves separated by an air space or cavity. It would be convenient if the sound reduction index of the combination were to equal the sum of those of the two leaves when used as single-leaf partitions. Unfortunately, the air in the cavity dynamically couples the two leaves, with the result that the sound reduction index of the combination may fall below this ideal value, sometimes by a large amount. In the following sections, various idealised models are theoretically analysed in order to illustrate the general sound transmission characteristics of double-leaf partitions and the dependence of these characteristics on the physical parameters of the systems. It is clear from even a superficial review of the available literature that theoretical analysis of the sound transmission behaviour of double-leaf partitions is far less well developed than that of single-leaf partitions, and that consequently greater reliance must be placed upon empirical information. The reason is not hard to find; the complexity of construction and the correspondingly larger number of parameters, some of which are difficult to evaluate, militate against the refinement of theoretical treatments. In particular, it is difficult to include the effects of mechanical connections between leaves, and of non-uniformly distributed mechanical damping mechanisms, in mathematical models. The following analyses are offered, therefore, more as vehicles for the discussion of the general physical mechanisms involved, than as means of accurate quantitative assessments of the performance of practical structures. These are followed by a brief survey of a number of experimental studies of the problem.

5.7   Transmission of Normally Incident Plane Waves through an Unbounded Double-Leaf Partition

The idealised model is shown in Fig. 5.13. Uniform, non-flexible partitions of mass per unit area m_1 and m_2, separated by a distance d, are mounted upon viscously damped, elastic suspensions, having stiffness and damping coefficients per unit area of s_1, s_2 and r_1, r_2, respectively.

It is assumed initially that the same fluid exists in the cavity, as on both sides of the partition. A plane wave of frequency ω is incident normally upon leaf 1.

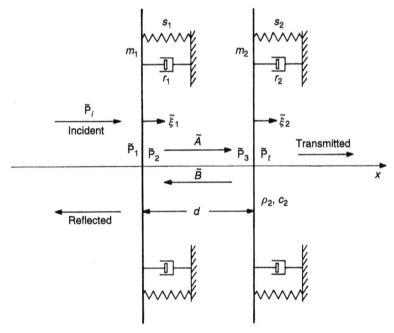

Fig. 5.13. Idealised model of normal incidence sound transmission through a double-leaf partition (Fahy, 1987).

The cavity wave coefficients $\tilde{A}$ and $\tilde{B}$ are related to the leaf displacements $\tilde{\xi}_1$ and $\tilde{\xi}_2$ and cavity pressures $\tilde{p}_2$ and $\tilde{p}_3$ as follows:

$$\tilde{p}_2 = \tilde{A} + \tilde{B} \tag{5.61}$$

$$\tilde{p}_3 = \tilde{A}\exp(-jkd) + \tilde{B}\exp(jkd) \tag{5.62}$$

$$j\omega\tilde{\xi}_1 = (\tilde{A} - \tilde{B})/\rho_0 c \tag{5.63}$$

$$j\omega\tilde{\xi}_2 = [\tilde{A}\exp(-jkd) - \tilde{B}\exp(jkd)]/\rho_0 c \tag{5.64}$$

The equations of motion of the leaves are

$$j\omega\tilde{\xi}_1\tilde{z}_1 = \tilde{p}_1 - \tilde{p}_2 \tag{5.65}$$

$$j\omega\tilde{\xi}_2\tilde{z}_2 = \tilde{p}_3 - \tilde{p}_t \tag{5.66}$$

in which the specific mechanical impedances of the leaves are given by

$$\tilde{z}_1 = j\omega m_1 + r_1 - js_1/\omega = m_1(j\omega + \eta_1\omega_1) - js_1/\omega \tag{5.67}$$

$$\tilde{z}_2 = j\omega m_2 + r_2 - js_2/\omega = m_2(j\omega + \eta_2\omega_2) - js_2/\omega \tag{5.68}$$

where η_1 and η_2 are the respective mechanical loss factors and ω_1 and ω_2 are the *in-vacuo* natural frequencies of the two leaves. The pressure $\tilde{p}_1$ is related to the pressure $\tilde{p}_i$ of the incident wave by

$$\tilde{p}_1 = 2\tilde{p}_i - j\omega\rho_0 c\tilde{\xi}_1 \tag{5.69}$$

and the transmitted wave has pressure $\tilde{p}_t$ given by

$$\tilde{p}_t = j\omega\rho_0 c\tilde{\xi}_2 \tag{5.70}$$

Let us assume first that the cavity width is very small compared with an acoustic wavelength, in which case $kd \ll 1$. Equations (5.61) and (5.62) indicate that, in this case, $\tilde{p}_3 \approx \tilde{p}_2 = \tilde{p}_c$; in other words, we may assume that the cavity pressure is uniform. Equations (5.63) and (5.64) may be combined to give

$$j\omega\rho_0 c\tilde{\xi}_2 = j\omega\rho_0 c\tilde{\xi}_1 - jkd(\tilde{A} + \tilde{B})$$

or

$$(\rho_0 c^2/d)(\tilde{\xi}_1 - \tilde{\xi}_2) = \tilde{p}_c \tag{5.71}$$

which indicates that the air acts as a spring of stiffness per unit area $s' = \rho_0 c^2/d$. Equations (5.65)–(5.71) may be combined to yield the leaf displacement ratio

$$\tilde{\xi}_1/\tilde{\xi}_2 = [j\omega(\tilde{z}_2 + \rho_0 c) + \rho_0 c^2/d]/(\rho_0 c^2/d) \qquad (5.72)$$

and the pressure amplitude transmission coefficient

$$\frac{\tilde{p}_t}{\tilde{p}_i} = -\frac{2j(\rho_0 c)^2/kd}{[\tilde{z}_2 + \rho_0 c - j\rho_0 c/kd][\tilde{z}_1 + \rho_0 c - j\rho_0 c/kd] + (\rho_0 c/kd)^2} \qquad (5.73)$$

Comparison of the terms in square brackets in the denominator of Eq. (5.73) with Eqs. (5.67) and (5.68) shows that the impedance of each leaf is combined with an acoustic radiation (damping) term $\rho_0 c$ and an acoustic stiffness term $\rho_0 c/kd$. The mechanical stiffness s_1 of leaf 1 may be equated to $\omega_1^2 m_1$ where ω_1 is the *in-vacuo*, undamped resonance frequency of leaf 1 on its mounting. Hence the ratio of mechanical to acoustic stiffness is

$$\delta_1 = \frac{s_1/\omega_1}{\rho_0 c/k_1 d} = \frac{\omega_1 m_1 k_1 d}{\rho_0 c} \qquad (5.74)$$

The same form of relationship can be written for leaf 2. If the model is considered to represent an approximation to normal incidence sound transmission through a bounded panel, ω_1 and ω_2 can be taken as the fundamental *in-vacuo* natural frequencies of each panel. The products $\omega_1 m_1$ and $\omega_2 m_2$ are proportional to the square of the ratios of the panel thicknesses to the typical panel dimension, and it turns out that for many lightweight double-leaf partitions of practical dimensions δ_1 is less than unity, so that the acoustic stiffness predominates. (This is not necessarily true in cases where the partitions are connected by mechanical links.)

If acoustic damping and mechanical damping and stiffness are neglected, the maximum transmission coefficient $\tau = |\tilde{p}_t/\tilde{p}_i|^2$ occurs at a frequency such that

$$(-\omega^2 m_1 + \rho_0 c^2/d)(-\omega^2 m_2 + \rho_0 c^2/d) = (\rho_0 c^2/d)^2$$

The solution is

$$\omega_0 = \left[\left(\frac{\rho_0 c^2}{d}\right)\left(\frac{m_1 + m_2}{m_1 m_2}\right)\right]^{1/2} \qquad (5.75)$$

This is termed as the 'mass-air-mass resonance frequency', which is seen to decrease with increase of the leaf separation d. This frequency is a minimum when $m_1 = m_2$: we shall denote it by ω_{0m}.

The mass-air-mass resonance phenomenon (which we shall see later also exists also for any angle of incidence, the frequency increasing as $\sec\phi$ where ϕ is the angle of incidence) crucially affects the performance of all double-leaf partitions. The overall performance improves as the mass-air-mass resonance frequency is decreased. Consequently, it is necessary to maximise the cavity width d, and/or the weights of the leaves, if high TL is required in the low-and mid-frequency range. Because of its deleterious influence, it is better in some cases of predominantly low frequency excitation to employ single-leaf partitions instead.

At low frequencies, such that $kd \ll 1$, the transmission behaviour may be classified as follows:

(i) Frequencies below the mass-air-mass resonance frequency, $\omega < \omega_0$: In this case, $\omega^2 m_2 m_1 < (m_1 + m_2)(\rho_0 c^2/d)$, the damping terms have negligible influence, and Eq. (5.73) becomes

$$\tilde{p}_t/\tilde{p}_i \approx -2j\rho_0 c/\omega(m_1 + m_2) \qquad (5.76)$$

Hence

$$\tau \approx (2\rho_0 c/\omega m_t)^2$$

and

$$TL = TL(0, m_t) \quad \text{dB} \qquad (5.77)$$

where $m_t = m_1 + m_2$. Comparison with Eq. (5.20) shows that the partition behaves like a single-leaf partition having a mass equal to the sum of the masses of the two leaves: damping has negligible effect.

(ii) Frequencies close to the mass-air-mass resonance frequency, $\omega \approx \omega_0$: In this case, the pressure transmission coefficient is

$$\tilde{p}_t/\tilde{p}_i \approx -2\rho_0 c/(\eta_1 \omega_1 m_2 + \eta_2 \omega_2 m_1 + \mu\rho_0 c) \qquad (5.78)$$

where the factor μ equals $[(m_1/m_2)+(m_2/m_1)]$. This result suggests that, if mechanical damping is low, it is preferable to minimise transmission at resonance. This can be done by maximising μ either by making m_1/m_2 or $m_2/m_1 \gg 1$: in these cases $\omega_0 > \omega_{0m}$. However, later analysis shows that benefit near ω_0 is gained at the expense of performance at higher frequencies. In the special case of leaves of equal mass m, *in-vacuo* fundamental natural frequency ω', and loss factor η,

$$\frac{\tilde{p}_t}{\tilde{p}_i} \approx -\frac{2}{2\eta(m\omega'/\rho_0 c) + 2} \qquad (5.79)$$

If, in addition, the mechanical damping is sufficiently large to make η much greater than $\rho_0 c / \omega' m$, which is generally greater than $\rho_0 c / \omega_0 m$, then

$$\tilde{p}_t / \tilde{p}_i \approx -\rho_0 c / m \omega' \eta \tag{5.80}$$

The transmission loss is hence

$$TL = TL(0, m_t, \omega') + 20 \log_{10} \eta \quad \text{dB} \tag{5.81}$$

where $TL(0, m_t, \omega')$ is based upon the total mass and the *in-vacuo* fundamental natural frequency of the leaves: the transmission is damping controlled. If $\eta \ll \rho_0 c / \omega' m$, then τ is close to unity, and virtually all the incident sound energy is transmitted. As already stated, the transmission peak caused by resonance is made less severe by using leaves of different weight.

(iii) Frequencies above the mass-air-mass resonance frequency, $\omega > \omega_0$: In this case, $\omega^2 m_2 m_1 > (m_1 + m_2)(\rho_0 c^2 / d)$ and

$$\frac{\tilde{p}_t}{\tilde{p}_i} \approx \frac{2j(\rho_0 c)^2 / kd}{\omega^2 m_1 m_2} \tag{5.82}$$

Substitution from Eq. (5.75) for $\rho_0 c^2 / d$ yields

$$\frac{\tilde{p}_t}{\tilde{p}_i} \approx \frac{2j\rho_0 c}{\omega(m_1 + m_2)} \left(\frac{\omega_0}{\omega} \right)^2$$

Hence

$$TL \approx TL(0, m_t) + 40 \log_{10}(\omega / \omega_0) \quad \text{dB} \tag{5.83a}$$

which may also be expressed as

$$TL \approx TL(0, m_1) + TL(0, m_2) + 20 \log_{10}(2kd) \quad \text{dB} \tag{5.83b}$$

where $TL(0, m_t)$ is based upon the total mass of the partition. The transmission loss therefore rises at 18 dB/octave from the value it would have at the resonance frequency if simply controlled by total mass. The great improvement over the performance below the resonance frequency, as indicated by the term $40 \log_{10}(\omega / \omega_0)$, is typical of transmission through inertial layers coupled by a resilient layer. The physical explanation is that leaf 2 acts as a mass driven through a spring by the motion of leaf 1, above the system resonance frequency: the air acts as in a vibration isolation spring. The transmission characteristics in the low frequency range, for which $kd \ll 1$, are presented in graphical form in Fig. 5.14.

The behaviour of the system at higher frequencies, for which it may not be assumed that $kd \ll 1$, may be analysed by solving Eqs. (5.61)–(5.70) for arbitrary kd. The general solution for the ratio of transmitted to incident pressures is

$$\frac{\tilde{p}_t}{\tilde{p}_i} = -\frac{2j\rho_0^2 c^2 \sin kd}{\tilde{z}_1' \tilde{z}_2' \sin^2 kd + \rho_0^2 c^2} \qquad (5.84)$$

where $\tilde{z}' = \tilde{z} + \rho_0 c(1 - j\cot kd)$. Note that Eq. (5.84) reduces to Eq. (5.73) if $kd \ll 1$. The variation of this ratio with frequency is complicated; it varies between minima, which correspond to acoustic anti-resonances of the cavity, when $kd = (2n - 1)\pi/2$, and maxima, at resonances, when $kd = n\pi$, n being any non-zero positive integer. At the anti-resonance frequencies the ratio takes the approximate form

$$\frac{\tilde{p}_t}{\tilde{p}_i} = -\frac{2j\rho_0^2 c^2}{\omega^2 m_1 m_2} \qquad (5.85a)$$

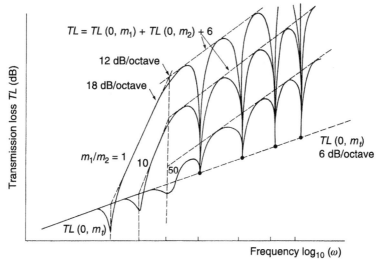

Fig. 5.14. Illustration of the theoretical effect on normal incidence *TL* of varying leaf mass ratios while keeping the total mass constant (Fahy, 1987).

which gives

$$TL \approx TL(0, m_1) + TL(0, m_2) + 6 \quad \text{dB} \qquad (5.85\text{b})$$

which, unlike the case at ω_0, is maximised for a given m_t by making $m_1 = m_2$. This is greater than the sum of the mass-controlled sound reduction indices of the two leaves considered as single partitions. At the resonance, we may find a solution by considering the acoustic impedance imposed on the first leaf by the combination of the fluid in the cavity and second leaf. The general expression for the specific acoustic impedance of a column of fluid terminated by a specific acoustic impedance $\tilde{z}_1$ is

$$\tilde{z}_0 = \rho_0 c \frac{\tilde{z}_1 + j\rho_0 c \tan kl}{\rho_0 c + j\tilde{z}_1 \tan kl} \qquad (5.86)$$

(Prove.) At resonance, $\tan kl = 0$ and

$$\tilde{z}_0 = \tilde{z}_1 \qquad (5.87)$$

Hence the loading on leaf 1 is the same as if leaf 2 were directly attached to it. The corresponding transmission loss is

$$TL = TL(0, m_t) \quad \text{dB} \qquad (5.88)$$

which is that given by a single leaf of mass $m_t = m_1 + m_2$. The general variation of the transmission loss with frequency is shown in Fig. 5.14. The asymptotic frequency average value of TL is approximately equal to the sum of $TL(0, m_1)$ and $TL(0, m_2)$ which increases at 12 dB/octave. If this line is extrapolated to low frequencies it will intersect the line given by Eq. (5.83), at a frequency given by $kd = 1/2$ which corresponds approximately to one-sixth of the lowest cavity acoustic resonance frequency.

5.8 The Theoretical Effect of Cavity Sound Absorption on Normal Incidence Transmission Loss

Sound absorbing materials are placed in the cavities of double-leaf building constructions and form interlayers between the external shells of vehicle cabins and the trim sheets that are used to line their inner surfaces to provide sound insulation. The types of sound absorbent materials most commonly used in for this purpose in building constructions are mineral wool and glasswool; their principal function is to suppress acoustic resonances of the cavities that would otherwise strongly couple the two leaves. Various other forms of fibrous materials such as felt and shoddy (shredded fibre of old woollen cloth) and porous plastic foam are used in road vehicles. Their principal function is to decouple the trim sheet from the vibration field induced in the shell by various acoustic and mechanical sources. They are generally too thin to be effective as sound absorbent elements over most of the important frequency range. They reduce noise transmission and radiation by physically separating the outer (shell) and inner (trim sheet) elements so that, in principle, they are coupled only through the air contained in the decoupler. However, their influence on internal noise levels is very different in the cases of excitation by externally incident sound and excitation by structure-borne waves travelling in the shell structure. At the typical fibre densities employed, fibrous materials act as effective vibration decouplers. However, some forms of plastic foam transmit low frequency vibration fairly effectively through their skeleton structures as well as through the air, even though they feel 'squashy' to the touch. The rather thin sheets of decoupler that are practicable in car trim do little to attenuate the mass-air-mass resonance dip in *TL* which appears typically at about 200 Hz. Examples of the performance of car trim are presented in Section 5.12.

The acoustic properties of a porous/fibrous material that has an effectively *rigid skeleton* may be characterised by three principal parameters: flow resistivity *r* porosity *h* and structure factor *s* (Fahy, 2001). Flow resistivity is defined as the pressure difference across unit thickness of material due to unit steady volume velocity per unit cross-sectional area. Porosity is defined as the ratio of the volume of voids to the total volume. The various influences of geometric form of the skeleton on the effective fluid density and compressibility are lumped together into the structure factor. The flow resistivity quantifies the viscous mechanism responsible for dissipating into heat the energy of sound waves travelling through the air within the material. The associated wave attenuation requires the acoustic wavenumber to be complex, thus:

$$k' = \beta - j\alpha \approx \omega(\rho'h/\kappa)^{1/2} \qquad (5.89)$$

in which the complex density is given by

$$\rho' = s\rho_0/h - jr/\omega$$

and the bulk modulus κ is given by $K\rho_0 c^2$ where K typically lies between 0.7 and 1.0. The real component β is known as the 'propagation constant', and the imaginary component α is known as the 'attenuation constant'. Except at frequencies below about 100 Hz they may be approximated by the following expressions: $\alpha \approx (1/2)(rh/\rho_0 c)(K/s)^{-1/2}$ and $\beta \approx (\omega/c)(s/K)^{1/2}$ which we shall use in the following numerical example of the effect of cavity absorbent on the normal incidence *TL* of a double-leaf partition.

Adapting the equations of Section 5.7, the expression for the pressure field in the cavity becomes

$$\tilde{p}(x) = \tilde{A} \exp(-jk'x) + \tilde{B} \exp(jk'x) \tag{5.90}$$

The particle velocities at the surfaces of the two leaves are given by

$$j\omega\tilde{\xi}_1 = (\tilde{A} - \tilde{B})/\tilde{z}_c \tag{5.91}$$

and

$$j\omega\tilde{\xi}_2 = [\tilde{A} \exp(-jk'd) - \tilde{B} \exp(jk'd)]/\tilde{z}_c \tag{5.92}$$

where $\tilde{z}_c$ is the characteristic specific acoustic impedance associated with plane sound waves propagating within the material. The specific impedance ratio may be approximated by

$$\tilde{z}'_c = \tilde{z}_c/\rho_0 c \approx (Ks/h^2)^{1/2} - j(r/2\omega\rho_0)(K/s)^{1/2} \tag{5.93}$$

The general solution for the ratio of transmitted to incident sound pressure is (Fahy, 2000)

$$\frac{\tilde{p}_t}{\tilde{p}_i} = -\frac{2\tilde{z}'_c}{j[\tilde{z}'_2\tilde{z}'_1 + \tilde{z}'_2 + \tilde{z}'_1 + (\tilde{z}'_c)^2]\sin(k'd) + \tilde{z}'_c[\tilde{z}'_2 + \tilde{z}'_1 + 2]\cos(k'd)} \tag{5.94}$$

in which $\tilde{z}'_1 = \tilde{z}_1/\rho_0 c$ and $\tilde{z}'_2 = \tilde{z}_2/\rho_0 c$. This equation reduces to Eq. (5.84) when $\alpha = 0$ and $\beta = k$.

Because it is not straightforward to discern the effect of the absorber parameters by means of parametric approximations (as was done for the untreated cavity) the results of a numerical study are presented in Fig. 5.15. The absorbent has little effect at frequencies well below the first cavity resonance at 860 Hz ($kd = \pi$). (The low frequency rise in *TL* in the highest resistivity case is due to the assumption of the approximate formula for $\tilde{z}'_c$ that is not valid at such low frequencies.)

The mass-air-mass resonance frequency may be significantly changed by the presence of a porous absorbent material in the cavity because the acoustic bulk modulus, together with effective density and phase speed can be very different from those in free air. On the basis of the assumption that $\alpha d < 1$, which is

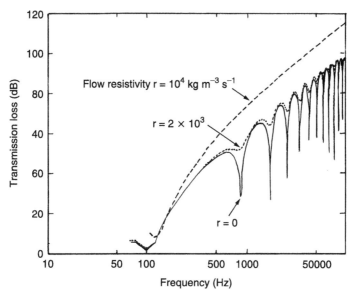

Fig. 5.15.   Theoretical normal incidence transmission loss of a double-leaf partition as a function of the flow resistivity of the cavity absorbent: $m_1 = m_2 = 3\,\mathrm{kg m}^{-2}$; $d = 0.2\,\mathrm{m}$; $s = h = K = 1$.

generally true for typical values of flow resistance at frequencies close to the mass-air-mass resonance frequency, and also that $\beta d \ll 1$ (since β is of the order of k), this model predicts that cavity absorption has a measurable effect on the mass-air-mass TL minimum of a partition comprising two leaves of equal mass per unit area m only if $r > \rho_0^2 c/m$. At the higher frequencies of cavity resonance, αd may well exceed unity, which ensures that the resonances do not significantly degrade the TL. In the case to which Fig. 5.15 applies, values of r of 2000 and 10000 $\mathrm{kg m}^{-3}\mathrm{s}^{-1}$ correspond to values of αd at 100 Hz of 0.49 and 2.42, respectively.

Although this model predicts that increasing the flow resistivity of a cavity absorbent substantially increases the normal incidence TL, measured data presented later in this chapter show that, under *diffuse field* excitation, its thickness is a more potent agent in this respect. It is also observed from measurements that the initial introduction of a relatively thin layer of absorbent produces the bulk of the improvement. The reasons appear to be that plane sound waves transmitted *obliquely* into the cavity suffer increased attenuation because of their increased path length between reflections, and that resonances associated with the finite extent of a cavity parallel to the plane of the partition are strongly attenuated by even a thin sheet of absorbent.

5.9   Transmission of Obliquely Incident Plane Waves through an Unbounded Double-Leaf Partition

So far, only normally incident sound has been considered. When a plane wave is incident at an oblique angle ϕ upon leaf 1 it sets up a bending wave travelling in the plane of the leaf with the trace wavenumber $k_z = k \sin \phi$. Provided that the leaves and cavity are unbounded and uniform, waves travelling with the same wavenumber vector component parallel to the plane of leaf 1 are set up in the fluid in the cavity, in leaf 2, and in the fluid external to the partition, as shown in Fig. 5.16. In satisfaction of the acoustic wave equation, and in the absence of cavity absorption, the wavenumber vector components of the cavity wave in the direction normal to the planes of the leaves have magnitudes given by $k_x = k(1 - \sin^2 \phi)^{1/2} = k \cos \phi$. Hence the pressure wave system in the cavity takes the form

$$\tilde{p}(x, z) = [\tilde{A} \exp(-jk_x x) + \tilde{B} \exp(jk_x x)] \exp(-jk_z z) \qquad (5.95)$$

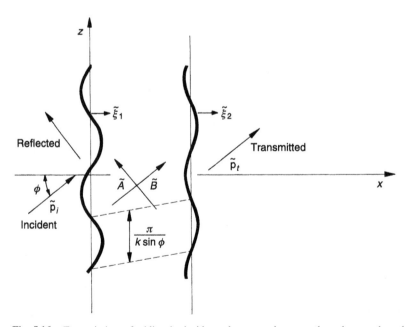

Fig. 5.16. Transmission of obliquely incident plane sound waves through an unbounded double-leaf partition.

and the corresponding particle velocity normal to the planes of the leaves is

$$\tilde{u}_x(x, z) = \frac{\cos\phi}{\rho_0 c}[\tilde{A}\exp(-jk_x x) - \tilde{B}\exp(jk_x x)]\exp(-jk_z z) \tag{5.96}$$

The physical interpretation of Eqs. (5.95) and (5.96) is that the acoustic impedance of the cavity is increased by a factor $\sec\phi$ compared with the normal incidence value. The fluid-loading (radiation) impedance produced by the fluid external to the partition is increased by the same factor to $\rho_0 c\sec\phi$.

The relevant impedances of the leaves are the wave impedances corresponding to a progressive bending wave of wavenumber $k\sin\phi$ as given by Eq. (5.30) with $\phi_1 = \phi$. It is assumed in the following analysis that the fluids inside the cavity and outside the partition are the same, so that no refraction occurs. We have already seen in the analysis of single-leaf transmission that at frequencies well below the critical frequency of a leaf the inertial component of wave impedance greatly exceeds the stiffness component, and therefore in this frequency range the latter may be neglected. Equations (5.61)–(5.64) become

$$\tilde{p}_2 = \tilde{A} + \tilde{B} \tag{5.97}$$

$$\tilde{p}_3 = \tilde{A}\exp(-jkd\cos\phi) + \tilde{B}\exp(jkd\cos\phi) \tag{5.98}$$

$$j\omega\tilde{\xi}_1 = (\tilde{A} - \tilde{B})/(\rho_0 c\sec\phi) \tag{5.99}$$

$$j\omega\tilde{\xi}_2 = [\tilde{A}\exp(-jkd\cos\phi) - \tilde{B}\exp(jkd\cos\phi)]/(\rho_0 c\sec\phi) \tag{5.100}$$

Equations (5.67)–(5.70) become

$$\tilde{z}_1 = j\omega m_1 + r_1 \tag{5.101}$$

$$\tilde{z}_2 = j\omega m_2 + r_2 \tag{5.102}$$

and

$$\tilde{p}_1 = 2\tilde{p}_i - j(\omega\rho_0 c\sec\phi)\tilde{\xi}_1 \tag{5.103}$$

$$\tilde{p}_t = j(\omega\rho_0 c\sec\phi)\tilde{\xi}_2 \tag{5.104}$$

The general solution for the ratio of transmitted to incident pressures is

$$\frac{\tilde{p}_t}{\tilde{p}_i} = -\frac{2j\rho_0^2 c^2\sec^2\phi\sin(kd\cos\phi)}{\tilde{z}_1'\tilde{z}_2'\sin^2(kd\cos\phi) + \rho_0^2 c^2\sec\phi} \tag{5.105}$$

where $\tilde{z}' = \tilde{z} + \rho_0 c \sec \phi [1 - j \cot(kd \cos \phi)]$. When $kd \cos \phi \ll 1$, Eq. (5.105) reduces to

$$\frac{\tilde{p}_t}{\tilde{p}_i} = -\frac{2j\rho_0^2 c^2 \sec^2 \phi / (kd \cos \phi)}{[\tilde{z}_1 + \rho_0 c \sec \phi - j\rho_0 c / (kd \cos^2 \phi)][\tilde{z}_2 + \rho_0 c \sec \phi - j\rho_0 c / (kd \cos^2 \phi)] +}$$
$$\cdots + [\rho_0 c / (kd \cos^2 \phi)]^2$$
$$(5.106)$$

Comparison with the equivalent normal-incidence, low-frequency result, Eq. (5.73), shows that the effective stiffness of the cavity has increased by a factor $\sec^2 \phi$. Hence the oblique-incidence mass-air-mass resonance frequency is greater than the normal-incidence value by a factor $\sec \phi$. (As explained in Section 5.7, the mass-air-mass resonance phenomenon has a crucial effect on the overall performance of a double-leaf partition. In the case of diffuse field excitation, the factor $\sec \phi$ spreads the phenomenon over a theoretically infinite frequency range. However, cavity absorption suppresses the resonances at high angles of incidence and the effect is manifested only in a small frequency range above ω_{0m}.)

The low-frequency transmission behaviour may be classified as follows:

(i) Frequencies below the oblique incidence mass-air-mass resonance frequency, $\omega < \omega_0 \sec \phi$: The pressure transmission coefficient is

$$\tilde{p}_t / \tilde{p}_i \approx -2j\rho_0 c / \omega (m_1 + m_2) \cos \phi \qquad (5.107a)$$

which is the same as the oblique-incidence expression for a single leaf of mass m_t. Hence

$$TL(\phi) = TL(\phi, m_t) \quad \text{dB} \qquad (5.107b)$$

(ii) Frequencies close to the oblique-incidence mass-air-mass resonance frequency, $\omega \approx \omega_0 \sec \phi$:
In this case, we must take mechanical damping of the leaves into account; the result is

$$\tilde{p}_t / \tilde{p}_i \approx -2\rho_0 c \sec \phi / (\eta_1 \omega_1 m_2 + \eta_2 \omega_2 m_1 + \mu \rho_0 c \sec \phi) \qquad (5.108)$$

where $\mu = [(m_1/m_2) + (m_2/m_1)]$, as in the normal-incidence case. The influence of mechanical damping, in comparison with that of different leaf weights, is seen to decrease with increasing angle of incidence. It is also seen that mass-air-mass resonance can take place at all frequencies above ω_0, the frequency increasing with ϕ.

(iii) Frequencies above the oblique-incidence mass-air-mass resonance frequency, $\omega > \omega_0 \sec \phi$: In this case, leaf inertia dominates and the result takes the same form as for normal incidence:

$$\frac{\tilde{p}_t}{\tilde{p}_i} \approx -\frac{2j\rho_0 c}{\omega(m_1 + m_2) \cos \phi} \left(\frac{\omega_0 \sec \phi}{\omega}\right)^2 \qquad (5.109a)$$

and

$$TL(\phi) \approx TL(\phi, m_t) + 40 \log_{10}[(\omega/\omega_0) \cos \phi] \quad \text{dB} \qquad (5.109b)$$

or, alternatively,

$$TL(\phi) \approx TL(\phi, m_1) + TL(\phi, m_2) + 20 \log_{10}(2kd \cos \phi) \quad \text{dB} \quad (5.110)$$

The behaviour at higher frequencies, for which it may not be assumed that $kd \ll 1$, may be analysed by solving Eq. (5.105) for arbitrary kd. As in the normal-incidence case, transmission maxima produced by acoustic resonance of the cavity alternate with transmission minima caused by anti-resonance. These frequencies are higher than the corresponding values for normal incidence by the factor $\sec \phi$. At the anti-resonance frequencies given by $kd \cos \phi = (2n-1)\pi/2$, the pressure transmission coefficient minimum is

$$\tilde{p}_t/\tilde{p}_i \approx 2j\rho_0^2 c^2/\omega^2 m_1 m_2 \cos^2 \phi \qquad (5.111)$$

and

$$TL(\phi) = TL(\phi, m_1) + TL(\phi, m_2) + 6 \quad \text{dB} \qquad (5.112)$$

Because the anti-resonance frequencies increase in proportion to $\sec\phi$, the sound reduction index maxima for any particular value of n are actually independent of angle of incidence and are given by Eqs. (5.85).

At the acoustic resonance frequencies, the panels move as one and the sound reduction index minimum corresponds to Eq. (5.46):

$$TL(\phi) = TL(\phi, m_t) \quad \text{dB} \qquad (5.113)$$

The value is the same for all angles of incidence because $\omega \cos \phi$ is a constant. This conclusion is extremely important because at every frequency above the lowest cavity resonance frequency $\omega = \pi c/d$, there is an angle of incidence for which resonant transmission occurs: the same is true for mass-air-mass resonance above ω_0. Hence, in a diffuse field, acoustic resonance phenomena effectively control the maximum achieved sound reduction index. Therefore, it is vital to suppress these resonances, by inserting absorbent material into the cavity and/or by dividing up the cavity in order to suppress lateral wave motion. A generalised form of the oblique incidence *TL* curve for a double-leaf partition that contains no absorbent is presented in Fig. 5.17. Note that the maxima and minima are independent of ϕ.

The general analysis of transmission of obliquely incident plane sound waves through an infinite double-leaf partition containing absorbent material is complicated by the refraction effects caused by the difference of phase velocities of waves in the free air and in the absorbent: it is too involved to be presented in detail here, but is discussed later in Section 5.12 in relation to vehicle trim. However, if the cavity is filled with absorbent material of appropriate flow resistivity, acoustic wave propagation parallel to the leaves is strongly inhibited and the behaviour approximates to that for normal incidence. The reason is that, while the phase change undergone by the acoustic wave during its return journey across the cavity relative to that undergone by the leaf between transmission and return points, is equal to $\beta d \cos \phi$, the attenuation of the pressure amplitude is equal to $\exp(-\alpha d \sec \phi)$, which increases with angle of incidence. As mentioned in the previous section, it is found that the insertion into a cavity of sheets of absorbent material considerably thinner than the cavity width can produce substantial improvements in the performance of a lightweight, double-leaf partition.

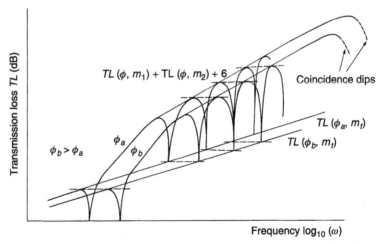

Fig. 5.17. Oblique incidence *TL* curves of a double-leaf partition at two angles of incidence: no cavity absorption (Fahy, 1987).

The mechanism is probably the attenuation of waves in the cavity having rel-atively large wavenumber vector components parallel to the leaves (caused by highly oblique incidence). However, such minimal treatment does not signifi-cantly reduce the adverse influence of the mass-air-mass resonance phenomenon. In the case of double glazing, absorption can only be provided in the reveals at the edges of the cavity, which does little to influence mass-air-mass resonance effects.

The individual leaves of a double-leaf partition naturally exhibit coincidence effects in response to the imposed sound fields. In order to derive expressions for the *TL* in the frequency range around and above the critical frequency, the leaf impedance terms in Eqs. (5.101) and (5.102) would have been modified to include bending stiffness and appropriate damping terms. However, it is intuitively obvi-ous that coincidence effects in partitions consisting of nominally identical leaves are likely to be rather more severe in a frequency range around the critical fre-quency than those in partitions having dissimilar leaves; empirical data shows this to be the case. Where the two leaves are effectively decoupled by cavity absorbent, the decrease in R below the mass-controlled value caused by coinci-dence effects in the two leaves can be approximated by the arithmetic sum of the individual coincidence dips in R of the two leaves when tested in isolation (Sharp, 1978). As with single-leaf coincidence, mechanical damping of the leaves largely controls the severity of coincidence effects.

In a theoretical study of 'low' frequency sound transmission through double-leaf partitions, with particular emphasis on the relative location of the mass-air-mass and critical frequencies, Kropp and Rebillard (1999) contend that, in cases where the cavity contains an absorbent having high flow resistance so that sound wave propagation parallel to the partitions is strongly inhibited, the air coupling may be represented by an interlayer comprising a uniform array of independent, undamped, springs. (Note: this model is valid only below the resonance frequency of the lowest acoustic 'thickness mode' of the cavity.) They derive the dispersion equations for wave propagation in the coupled plate waveg-uide. The solutions comprise both propagating and non-propagating modes, as also shown in the analogous coupled plate waveguide analysis in Chapter 7. The study reveals the benefits and disadvantages of employing partitions of different thickness. For partitions of the same material, a thickness ratio of about 1:2 is found to maximise the *TL* around the critical frequency.

Although double-leaf walls in buildings invariably incorporate stiffening frames, known as 'studs', it is appropriate here to present the results of sound transmission measurements made on partitions lacking studs because some sound insulation systems such as double-glazed windows and vehicle wall trim lack such connections. Also, the results form a baseline reference for the study of the influence of studs, which is the subject of the following section. Sharp (1978) presents expressions for the diffuse field *TL* of double-leaf partitions containing

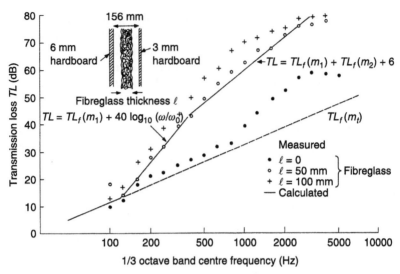

Fig. 5.18. Performance of a mechanically isolated double-leaf partition (Sharp, 1978).

efficient absorbent material that correspond to Eqs. (5.77), (5.83b) and (5.85b), but with $TL(0)$ replaced by TL_f. Figure 5.18 compares these expressions with some of his measured results.

According to Sharp, the change from a slope of 18 dB/octave above the mass-air-mass resonance to 12 dB/octave occurs at a 'limiting' frequency f_1 given by $c/2d$, which is 366 Hz for the case shown. This is one-third the resonance frequency of the fundamental acoustic 'thickness' mode of the cavity. (Note that many other lower frequency 'room' modes of a bounded cavity exist, but can be effectively suppressed by cavity absorption.)

Many other studies of sound transmission through double-leaf partitions have been performed, of which only a small selection can be discussed here. A valuable review of literature relating to the sound transmission behaviour of double-leaf partitions is presented by Hongisto *et al.* (2002) as a precursor to their experimental study of the parametric dependence of the transmission behaviour of lightweight double-leaf partitions incorporating studs and cavity sound absorbent, installed between two reverberation rooms, and tested according to ISO 140–3:1995. The partition leaves had dimensions of 1105 mm × 2250 mm and were made of 2 mm thick steel having a critical frequency of 6200 Hz. The principal conclusions relating to partitions lacking mechanical connections between the leaves are as follows. In the range between the lowest mass-air-mass resonance and the critical frequency, the sound

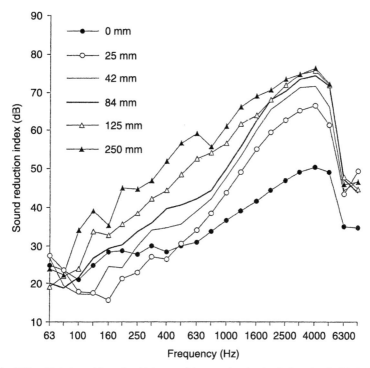

Fig. 5.19. Variation with cavity thickness of the sound reduction index of a double-leaf panel with no cavity absorbent (Hongisto *et al.*, 2002).

reduction index *R* increases with *empty* cavity width, the rate of rise with frequency averaging between 10 and 12 dB per octave. (Fig. 5.19): this dependence contrasts with Sharp's experimental curve in Fig. 5.18 for no absorbent. Nearly filling the cavity with sound absorbent generally almost doubles the rate of increase of *R* with frequency above the mass-air-mass resonance up to a 'plateau', dependent upon cavity width, for which no satisfactory explanation was found (Fig. 5.20). Unlike the empty cavity case, the presence of absorbent caused *R* to increase with cavity width in the vicinity of the critical frequency.

In agreement with many other studies, it was found that the introduction of an absorbent sheet having a thickness of only one-quarter to one-third of the cavity width was almost as effective as a nearly complete fill over the entire frequency range. However, a significant benefit of about 5 dB was produced in the mid-frequency range by an increase in thickness from 0.24*d* to 0.88*d* (Fig. 5.21). Increasing the resistivity of a nearly full thickness absorbent over a range from

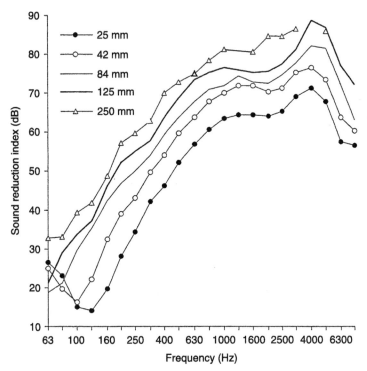

Fig. 5.20. Variation with cavity thickness of the sound reduction index of a double-leaf panel with cavity absorbent filling between 60 and 88% of the cavity width (Hongisto *et al.*, 2002).

8 to $300 \, \mathrm{kPasm}^{-2}$ had little effect below f_1, and only a marginal, and non-systematic, effect up to the critical frequency also with stud frames (Fig. 5.22), again in agreement with other studies. The authors conclude that, for given leaf weights, the cavity width and the presence of a relatively thin sheet of absorbent are the most important factors in determining the performance of a double-leaf partition, the flow resistivity of the absorbent having far less influence.

The principal physical difference between the double-leaf constructions discussed above and double-glazing systems is that sound absorbent material can only be installed in the reveals (inner boundaries of the frame) in the latter. This is fairly effective in suppressing acoustic resonances involving wave propagation parallel to the glass, but does not affect the 'thickness' resonances of the cavity. Quirt (1983) assesses double-glazing formulas. Brekke (1981) shows that triple-leaf partitions outperform double-leaf partitions constructed in the same form of panel above the frequency of the fundamental acoustic 'thickness' mode of the wider of the two cavities.

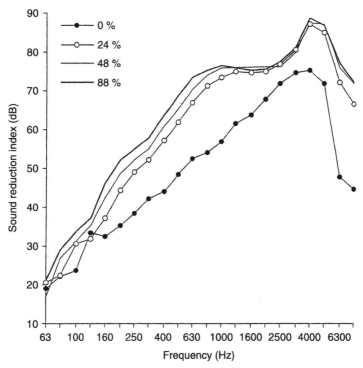

Fig. 5.21. Variation of the sound reduction index of a double-leaf partition with ratio of absorbent to cavity thickness (Hongisto *et al.*, 2002).

5.10 Mechanical Stiffening and Coupling of Double Partition Leaves

In the double-leaf system that has been discussed in the previous section, the two leaves are coupled only by the fluid in the cavity. In practice, it is necessary to mount the leaves on a system of frames (studs). In building practice, the studs take the form either of timber beams of rectangular cross section or thin-wall metal beams having a wide variety of dimensions and cross-sectional forms. In some cases, the studs directly connect the two leaves and in others each leaf carries a separate set of studs (staggered stud construction) so that no direct connections exists. Connections may take the forms of screws, nails or adhesive bonds. Timber studs may be connected to the leaves via thin flexible strips to minimise vibration transmission. The incorporation of studs introduces a very large number of geometric and material parameters that pose a challenging problem for theoretical analysis. In addition, the dynamic properties and behaviours

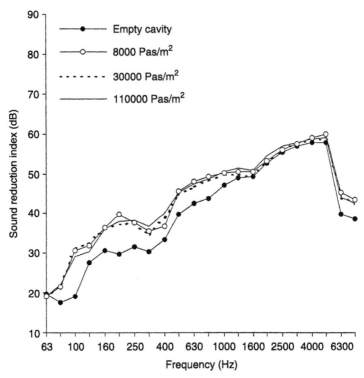

Fig. 5.22. Variation of the sound reduction index of a double-leaf partition with the flow resistivity of the absorbent: cavity width 84 mm, wooden stud pitch 1100 mm, screw pitch 170 mm (Hongisto *et al.*, 2002).

of the various connections between the studs and the leaves are not amenable to simple idealisation, not least because they are subject to considerable uncertainty and variability. Unfortunately, they have considerable influence on partition performance.

Faced with such a complex problem, it is not practicable in this book to present a comprehensive theoretical analysis of sound transmission through such structures. A highly idealised theoretical model and analysis of sound transmission produced by coupling between leaves and timber studs is followed by a presentation of a selection of experimental data. In the simplest form of construction used in buildings, sheets of gypsum (plaster) board are attached to both sides of a common timber frame, which effectively constrains the two leaves to undergo equal transverse displacements at the positions of the studs. In order to approximately evaluate the influence of such connections on sound transmission, we may take advantage of some earlier results relating to scattering of waves in plates by discontinuities and to sound radiation generated by line forces. It is assumed that the dynamic behaviour of each stud is independent of any other, which is reasonable on a frequency-average basis, and that the studs simply translate as rigid bodies in their own planes. By these assumptions we restrict our attention to bending waves incident normally on the stud-leaf connection lines, and we neglect the bending and torsional stiffnesses of the studs. These limitations are fairly severe, but they allow some of the more important features of the leaf-coupling process to be understood. We also assume that the air path is suppressed by cavity absorption, so that the mechanical path dominates: but see Fig. 5.22.

An incident plane wave having a frequency below the critical frequency of the leaves drives a bending wave in leaf 1 which is incident normally upon a stud. The impedance discontinuity scatters the bending wave and leaf 2 is driven by the stud. The sound radiated from leaf 2 due to this excitation is considered to be the transmitted field, the acoustic path through the cavity being assumed inactive. The total mechanical impedance of a stud, as 'seen' by leaf 1, is equal to the inertial impedance of the stud itself plus the line force impedance of leaf 2. The incident bending-wave velocity amplitude is approximately given by

$$j\omega\tilde{\xi}_1 \approx 2\tilde{p}_i/j\omega m \qquad (5.114)$$

in which fluid-loading effects have been neglected. The velocity of the stud is given by

$$j\omega\tilde{\xi}_s = j\omega\tilde{\xi}_1\tilde{z}_F/(2\tilde{z}_F + j\omega m_s) \qquad (5.115)$$

where $\tilde{z}_F$ is the line force impedance of each leaf and m_s the mass per unit length of the stud. The effective line force per unit length applied to the second leaf is

$$\tilde{F}_2 = j\omega\tilde{\xi}_s\tilde{z}_F \qquad (5.116)$$

and the sound power radiated per unit length of the stud is, for frequencies much less than the critical frequency of the leaves, given by Eq. (3.140) as

$$\overline{P} = |\tilde{F}_2|^2 \rho_0/4m^2\omega \qquad (5.117)$$

If there are n studs per unit width of the partition, the ratio of transmitted to incident sound power per unit area, for a plane sound wave incident at angle ϕ is

$$\tau = 2n\overline{P}\rho_0 c/|\tilde{p}_i|^2 \cos\phi \qquad (5.118)$$

The line force impedance of an infinite thin plate is, by analogy with the beam impedance equation,

$$\tilde{z}_F = 2D^{1/4}\omega^{1/2}m^{3/4}(1+j) \tag{5.119}$$

Below about 1 kHz, the line force impedance of 13 mm gypsum board exceeds half the inertial impedance of a 100 mm × 50 mm timber stud; hence the stud impedance term may be neglected in the denominator of Eq. (5.115). Subsequent substitution of the plate impedance expression (5.119) into Eqs. (5.115), (5.117) and (5.118) yields the following expression for sound power transmission coefficient associated with the bending near fields induced by the stud motion:

$$\tau_s = \frac{2n\rho_0^2 cc_l'}{\sqrt{3}\omega^2 \rho_s^2 h \cos\phi} \tag{5.120}$$

in which ρ_s, c_l' and h are the density, quasi-longitudinal wave speed and thickness of the plate material, respectively. It is clearly of interest to compare this expression with that for transmission through a double wall by the mechanism of cavity acoustic coupling between the leaves. However, it is difficult to express the parametric dependence of the ratio of transmission coefficients in simply interpreted terms. Therefore, we derive an expression for the ratio of the stud-induced transmission coefficient to that corresponding to the oblique incidence mass law based upon the total mass of the partition. The result is

$$\frac{\tau_s}{\tau_a} = \frac{\tau_s}{(\rho_0 c/\omega m \cos\phi)^2} = \frac{2nc_l' h \cos\phi}{\sqrt{3}c} = \frac{0.7nc \cos\phi}{f_c} \tag{5.121}$$

This ratio is independent of frequency. Substitution of $\phi = 45°$ and values of n and f_c typical of double-leaf constructions used in building yields values of this ratio of the order of 0.1 or -10 dB. Superimposition of a line 10 dB above the m_t mass law line in Fig. 5.18 suggests that the connection by timber degrades and limits the performance of a double-leaf partition containing cavity absorbent above a frequency given by $40 \log_{10}(f/f_0) = 10$ or $f/f_0 = 1.8$. The result of Hongisto *et al.* shown in Fig. 5.23 generally confirms the conclusion that, not far above f_0, timber studs degrade R to a mass-controlled line rising at only 6 dB per octave; in this case, f/f_0 is estimated to be 2.0. However, R lies only 4 dB above R_f of a single partition of mass per unit area m_t, whereas Eq. (5.121) suggests 12 dB. A plausible reason for this discrepancy is offered below.

The presence of f_c in the denominator of Eq. (5.121) indicates that transmission via timber studs is more important for partitions having low critical frequencies. This is confirmed by the results shown in Fig. 5.24, in which the

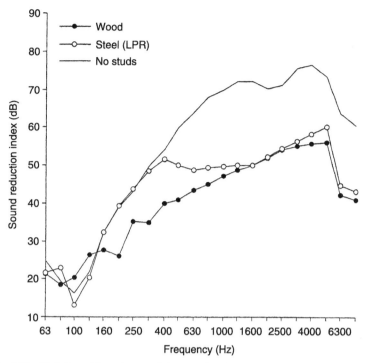

Fig. 5.23. Variation with stud type of the sound reduction index of a double-leaf partition containing sound absorbent: cavity width 42/45 mm, stud pitch 550 mm. (Hongisto *et al.*, 2002).

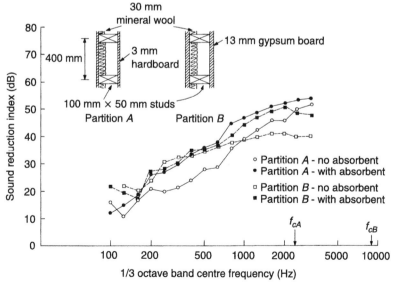

Fig. 5.24. Improvement made by cavity absorption with leaves of different critical frequency. (Ingemansson and Kihlman, 1959).

improvement in sound reduction index afforded by the introduction of cavity absorbent is shown for two constructions in which the leaf critical frequencies are very different (Ingemansson and Kihlman, 1959). The mid-frequency performance of partition A is considerably improved by cavity absorption, showing that the unattenuated air path transmits more energy than the mechanical vibration path, unlike the mid-frequency behaviour of partition B.

The above analysis neglects the contribution to radiated sound power of the reverberant field generated in leaf 2. The ratio of sound power radiated by the reverberant field induced by a single line velocity input to that radiated by the bending near field is given by Eq. (3.144) as $1/2(\omega_c/\omega)^{1/2}(\sigma/\eta)$. The sound power transmission coefficient τ_r associated with the reverberant field is obtained by multiplying the expression in Eq. (5.120) by this factor. There is no simple expression for the radiation efficiency of a periodically beam-stiffened plate, so that a result cannot be explicitly derived. However, the measurements of Drotleff (1996) and Fig. 3.36 suggests that, well below the critical frequency $\sigma \approx 2(\omega/\omega_c)^2$ and Hongisto's measurements give $\eta \approx 0.8\omega^{-1/2}$. These estimates suggest that reverberant transmission could exceed near-field transmission by 4 and 12 dB at 630 and 1600 Hz, respectively. Although not quantitatively accurate, these estimates suggest that reverberant transmission may be dominant over much of the frequency range.

The simple analysis leading to Eq. (5.121) gives results very similar to those of Sharp (1978). He elaborates by giving expressions for the frequency-independent exceedance of the total mass-controlled TL as function of the form of attachments between panels and studs. The results of a complex theoretical analysis by Lin and Garrelick (1977) of normal incidence sound transmission through a timber stud partition treated as a periodic structure indicate that resonances of the inter-stud leaf panels can also reduce the minimum sound reduction index to that given by the mass. Guigo-Carter and Villot (2003) model the process of sound transmission through a double-leaf panel in which the leaves are stiffened by independent sets of parallel beams. Inter-stiffener panel resonances decrease the mid-frequency TL by up to 10 dB. These results support the conclusion from the foregoing discussion of reverberant field radiation that damping of the leaves is almost certainly important, even well below the critical frequency.

Our elementary analysis of transmission due to rigid line connection between the leaves has yielded a result from which conclusions of practical importance can be drawn. Since the insertion of absorbent into a cavity raises the frequency-average sound reduction index well above the minimum values associated with cavity thickness resonance, the presence of rigid line connections will normally short circuit the absorbent and hence severely degrade the partition performance, except at frequencies near the mass-air-mass resonance frequency. We may also infer that reducing the impedance of the connections will improve the TL, at least in the frequency range between the mass-air-mass resonance frequency and the onset of the lowest acoustic thickness mode of the cavity. Damping of the radiating leaf is likely to have a beneficial effect. The short-circuiting effect is less serious in the case of leaves with a high critical frequency, as indicated by Eq. (5.121).

Experimental measurements on partitions incorporating timber studs show that the *TL* is greatly influenced by the spacing of the attachment screws. When the screw spacing is less than half a bending wavelength in the leaves, the stud acts as a continuous line exciter; at greater spacings, each connection acts as a point exciter. Application of Eqs. (3.145) and (3.146) indicates that the ratio of the sound power radiated per point connection to that radiated by line connection over screw spacing length *L* as

$$\overline{P}_p/\overline{P}_L = (4c/\pi^2)(1/f_c L) \qquad (5.122)$$

indicating that it is desirable to use the largest practicable screw spacing and that the benefit of point connection increases with leaf critical frequency.

If a very flexible connection is used between the studs and one of the leaves, the behaviour of the radiating leaf in response to stud movement can be modelled approximately by a base-excited mass-spring system. Elementary theory shows that above the resonance frequency the ratio of mass to base velocities decreases in proportion to the square of frequency. The sound power transmission coefficient is inversely proportional to the square of this velocity ratio and thus varies inversely as the fourth power of frequency, as it does above the mass-air-mass resonance frequency of partitions coupled by air stiffness. It is therefore essential that short circuiting of the absorbent cavity by mechanical connections should be reduced by the use of flexible stud connections or by making the studs themselves flexible. For example, it is common in modern stud wall constructions in buildings to use open section channels formed from 0.5-mm-thick steel sheet, having cross sections so designed as to provide the necessary static stiffness and stability, but to provide a high degree of dynamic flexibility at audio frequencies.

Hongisto *et al.* compare the performance of a double-leaf partition based on 2 mm thick steel panels without studs with those in which the leaves are connected by timber or commercial steel studs spaced at 550 mm: cavity absorption was in place. Fig. 5.25 shows that that the steel studs outperform the timber studs in the mid-frequency range. The compressional and bending stiffnesses of the AWS studs were about 7 and 40%, respectively, of those of the LR and TC studs. Improvements to the mid- and high-frequency sound insulation performance of existing masonry walls, which is based upon similar principles, can be effected by attaching thin-sheet panels via flexible-point or -line connections (Cremer *et al.*, 1988). As shown by Eqs. (5.121) and (5.122), it is important that only very flexible sheets having high critical frequencies should be used in order

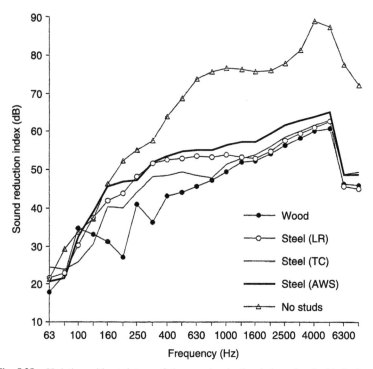

Fig. 5.25. Variation with stud type of the sound reduction index of a double-leaf partition containing sound absorbent: cavity width 125 mm, stud pitch 550 mm (Hongisto *et al.*, 2002).

to minimise radiation from the connection points or lines; also, the number of connections per unit width should be reduced as far as possible.

5.11   Close-Fitting Enclosures

A common method of reducing sound radiation from machinery or industrial plant is partially or fully to cover the radiating surfaces with a sheet of impervious material; such covering is sometimes known as 'cladding'. The cavity formed between the surface and its enclosure is usually relatively shallow compared with an acoustic wavelength over a substantial fraction of the audio frequency range, and it normally contains sound-absorbent material. Theoretical predictions of the performance of such enclosures have not been conspicuously successful to date, and designers still rely heavily on empirical data. The reasons are threefold: (i) the enclosure and source surfaces are strongly coupled by the intervening fluid, so that the radiation impedance of the source is affected by the dynamic behaviour of the enclosure; (ii) the geometries of sources are often such that the cavity wavefields are very complex in form and difficult to model deterministically; (iii) the dimensions of the cavities are not sufficiently large for statistical models of the cavity sound fields to be applied with confidence.

In view of the lack of reliable theoretical treatments of the problem, and the emphasis of the book on vibroacoustic phenomena, we shall confine detailed analysis to a simple one-dimensional model that exhibits some but not all of the mechanisms that operate in practical cases. The major difference between this model and that of a double partition is that the motion of the primary source surface is assumed to be inexorable, i.e., unaffected by the presence of the enclosure. This assumption is reasonable because the internal impedance of a machinery structure is generally much greater than that of its enclosure.

The model is shown in Fig. 5.26. The complex amplitude of pressure at the surface of the source is given by

$$\tilde{p}_0 = \tilde{A} + \tilde{B} \tag{5.123}$$

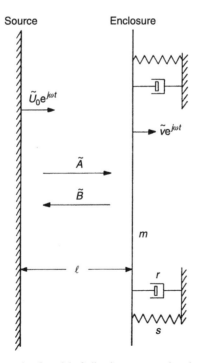

Fig. 5.26. One-dimensional model of vibrating source and enclosure (Fahy, 1987).

and the associated particle velocity is given by

$$\rho_0 c \tilde{u}_0 = \tilde{A} - \tilde{B} \tag{5.124}$$

The pressure in the cavity that drives the panel is

$$\tilde{p}_1 = \tilde{A} \exp(-jkl) + \tilde{B} \exp(jkl) \tag{5.125}$$

and the associated particle velocity, which equals the panel velocity, is given by

$$\rho_0 c \tilde{v} = \tilde{A} \exp(-jkl) - \tilde{B} \exp(jkl) \tag{5.126}$$

Let the specific impedance of the panel that represents the enclosure be represented by

$$\tilde{z}_p = j(\omega m - s/\omega) + r \tag{5.127}$$

to which must be added in series the specific radiation impedance, which we assume to be equal to $\rho_0 c$; we denote the total impedance by $\tilde{z}_t$. The equation of motion of the panel is hence

$$\tilde{z}_t \tilde{v} = \tilde{p}_1 \tag{5.128}$$

Substituting for $\tilde{B}$ from Eq. (5.124) and using Eq. (5.125), Eq. (5.128) becomes

$$\tilde{z}_t \tilde{v} = 2\tilde{A} \cos kl - \rho_0 c \tilde{u}_0 \exp(jkl) \tag{5.129}$$

Equations (5.124) and (5.126) allow us to obtain a second equation relating $\tilde{A}$ and $\tilde{v}$:

$$\rho_0 c \tilde{v} = -2j\tilde{A} \sin kl + \rho_0 c \tilde{u}_0 \exp(jkl) \tag{5.130}$$

Hence we can eliminate $\tilde{A}$ in order to relate $\tilde{v}$ and $\tilde{u}_0$:

$$\frac{\tilde{z}_t \tilde{v} + \rho_0 c \tilde{u}_0 \exp(jkl)}{2 \cos kl} = \frac{\rho_0 c \tilde{u}_0 \exp(jkl) - \rho_0 c \tilde{v}}{2j \sin kl} \tag{5.131}$$

of which the solution is

$$\frac{\tilde{v}}{\tilde{u}_0} = \frac{1}{\cos kl + j(\tilde{z}_t/\rho_0 c)\sin kl} \tag{5.132}$$

The ratio of sound power radiated by the enclosure panel to that radiated in the absence of the enclosure is

$$\frac{\overline{P}_e}{\overline{P}} = \left|\frac{\tilde{v}}{\tilde{u}_0}\right|^2 = \left\{ \left[\cos kl - \frac{(m\omega - s/\omega)\sin kl}{\rho_0 c} \right]^2 + \sin^2 kl \left[1 + \frac{r}{\rho_0 c} \right]^2 \right\}^{-1} \tag{5.133}$$

The 'insertion loss (IL)' is actually a logarithmic measure of the difference of radiated sound pressure levels with and without the enclosure. In this one-dimensional case

$$IL = 10\log_{10}(\overline{P}/\overline{P}_e) \quad \text{dB}$$

It may immediately be seen that the insertion loss is zero whenever $\sin kl = 0$, irrespective of the mechanical damping of the enclosure. This occurs at frequencies when the cavity width is equal to an integer number of half-wavelengths; the impedance at the source surface then equals the impedance of the panel plus the radiation impedance, and the panel velocity equals the source surface velocity. This situation is similar to that of the double partition at normal incidence, when the impedances of the two partitions simply add. The difference here is that, according to our assumption, the source surface motion is inexorable, which is equivalent to assuming that the load impedance is very much less than the internal impedance of the source. Hence the panel velocity $\tilde{v}$ equals the source velocity $\tilde{u}_0$, and the presence of the panel has no effect.

The insertion loss also takes minimum values when

$$\tan kl = \frac{\rho_0 c}{m\omega - s/\omega} \tag{5.134}$$

In practice the lowest frequency at which this occurs is normally such that $kl \ll 1$ and $\tan kl \approx kl$. Hence

$$\omega_1^2 \approx \rho_0 c^2 / ml + \omega_0^2 \tag{5.135}$$

where $\omega_0^2 = s/m$: the fluid cavity bulk stiffness $\rho_0 c^2 / l$ is added to the mechanical stiffness, as in the case of mass-air-mass resonance of the double partition. In this case the insertion loss becomes

$$IL = 20\log_{10}(1 + r/\rho_0 c) + 10\log_{10}(\rho_0 l/m + k_0^2 l^2) \, \text{dB} \qquad (5.136)$$

where $k_0 = \omega_0 c$, and $k_0 l \ll 1$ because $kl \ll 1$.

It is clearly beneficial to make the *in-vacuo* natural frequency of the panel as high as possible because the second term will normally be negative. The specific panel damping factor $r/\rho_0 c$ may be written $\eta \omega_0 m/\rho_0 c$, and the damping is therefore only significant if $\eta \gg \rho_0 c/\omega_0 m$. If the mechanical damping is rather low, the minimum insertion loss is normally negative; more power is radiated from the enclosure at this resonance frequency than from the unenclosed source! How can this be? It has nothing to do with the surface area of the enclosure in comparison with that of the source; they are equal. The answer is revealed by recalling that the basic expression for acoustic power radiation per unit area of vibrating surface is $\overline{P'} = \overline{pv}$, which, for a surface vibrating at a single frequency, is $\overline{P'} = 1/2\mathrm{Re}\{\tilde{p}\tilde{v}^*\} = 1/2\,|\tilde{v}|^2\,\mathrm{Re}\{\tilde{z}_r\}$, where $\tilde{z}_r$ is the specific acoustic impedance 'seen' by the surface. Since the source vibration is inexorable, the real part of the impedance presented by the fluid plus enclosure must, in this case, exceed that for the unbounded fluid. Reference to Eq. (5.86) indicates that $\mathrm{Re}\{\tilde{z}_r\}$ for the source is maximised when Eq. (5.134) is satisfied. The resonant behaviour of the enclosure/airspace combination creates high acoustic pressures in the air space. However, not all this power is radiated from the enclosure; some is dissipated by enclosure motion, which is why the enclosure damping is an important factor in controlling the minimum insertion loss.

Equation (5.136) clearly indicates that a combination of high stiffness, high damping and low mass is required for good *low-frequency* enclosure performance. These requirements are very different from those for good performance of a single-leaf partition at frequencies below the critical frequency, although the maxima in insertion loss correspond to the normal incidence sound reduction index for a panel of twice the mass per unit area. The mechanical stiffness of the enclosure is only significant, however, if it exceeds the acoustic stiffness of the fluid, i.e., $\omega_0^2 m > \rho_0 c^2/l$. Although increasing the cavity width l will reduce the severity of this insertion loss minimum, it decreases the frequencies of standing-wave resonance in the cavity, and therefore may not always be beneficial, at least in theory.

Higher-frequency minima in insertion loss occur whenever Eq. (5.134) is satisfied, but the values of these minima are greater than that at the lowest resonance frequency, as can be shown by substituting the corresponding values of $\sin kl$ in Eq. (5.133). Let $\tan kl = \alpha$, then

$$\sin^2 kl = \alpha^2/(1 + \alpha^2)$$

Assuming that these higher resonances occur well above the *in-vacuo* natural frequency of the enclosure, then $\alpha \approx \rho_0 c / \omega m$, and the insertion loss minima are given by

$$IL = 20\log_{10}(\rho_0 c/\omega m) - 10\log_{10}[1+(\rho_0 c/\omega m)^2] + 20\log_{10}(1+r/\rho_0 c) \quad \text{dB} \tag{5.137}$$

which can be negative. In fact, the frequencies at which the minima occur are very close to those at which sin kl and therefore IL are zero. The presence of sound-absorbent material in the cavity will improve the insertion loss at these higher resonances but is not likely to be very effective at the lowest resonance frequency given by Eq. (5.135). A theoretical insertion loss curve is shown in Fig. 5.27.

The foregoing analysis is based upon a very simplistic model of a source and enclosure. Byrne *et al.* (1988) present a somewhat more realistic model. The effect of multiple poroelastic interlayers between the source and the cover has been analysed by Au and Byrne (1987, 1990). A number of attempts have been made to develop more realistic analytical models, particularly with respect to the three-dimensional nature of acoustic fields in real enclosure cavities, with modest success (Tweed and Tree, 1978). Low-frequency mass-cavity resonance dips appear in most published results, although the frequencies are often well

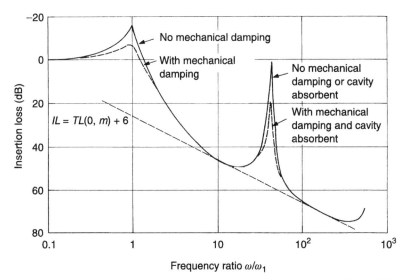

Fig. 5.27.   Generalised theoretical insertion loss performance of an enclosure (Fahy, 1987).

below 100 Hz and therefore may not be of great consequence. In the case of sources that are small in comparison with the enclosure dimensions, there is evidence of resonance due to the acoustic coupling of opposite walls, which is given by Eq. (5.135) with l as the enclosure half-width dimension. The lower-order cavity standing-wave resonances can be discerned in some results also. In the absence of absorbent material in the cavity, the insertion loss of an enclosure above the lowest standing-wave resonance usually rises by about 3 dB per octave, and falls substantially below the sound reduction index of the material when used as a partition. When even a moderate thickness of absorbent material (e.g., 50 mm) is introduced, the insertion loss above the lowest standing-wave resonance frequency generally increases to values close to the field incidence sound reduction index, or even to the normal incidence value in the case of small cavities and piston-like source motion. In the frequency range encompassing the mass-cavity resonance, and the lowest standing-wave resonance, the one-dimensional theory appears to give fairly accurate predictions of enclosure behaviour.

The insertion loss of an enclosure is severely degraded by stiff connection to the vibrating source. Expressions for the sound powers radiated by the bending near fields generated by the transmission of vibration from source to cover by discrete rigid connectors may be derived by the use of expressions derived in Section 3.10. Excitation of a cover plate by a 'point' velocity v_0 produced by rigid connection to a vibrating source generates two vibration field components: direct and reverberant. The sound power radiated by *each* direct field is given by Eq. (3.145) in terms of the associated mean square velocity as

$$\overline{P}_d = \frac{16\rho_0 c^3 \, |\tilde{v}_0|^2}{\pi \omega_c^2} = \frac{16\rho_0 D \, |\tilde{v}_0|^2}{\pi \, cm} \tag{5.138}$$

The ratio of power radiated by the direct fields to that radiated by the associated reverberant field is given by Eq. (3.135) as

$$\overline{P}_d / \overline{P}_p = (4/\pi)(\omega/\omega_c)(\eta/\sigma) \tag{5.139}$$

This will tend to be of the order of unity unless the cover sheet is highly damped when it will exceed unity. Since σ tends to vary as $(\omega/\omega_c)^{1/2}$ at frequencies far below ω_c, and η tends to vary as $\omega^{-1/2}$, this ratio tends to be independent of frequency and to decrease with increase in ω_c. The total radiated power per connection

$$\overline{P}_t = \overline{P}_d + \overline{P}_p = \frac{16\rho_0 D \, |\tilde{v}_0|^2}{\pi \, cm} \left[1 + \left(\frac{\pi}{4} \right) \left(\frac{\omega_c}{\omega} \right) \left(\frac{\sigma}{\eta} \right) \right] \tag{5.140}$$

The design implications for plane cover sheets rigidly supported at discrete points, which are incompatible with those for low frequencies are as follows:

(i)   cover sheet mass per unit area should be as large as possible
(ii) cover sheet stiffness should be as small as possible
(iii) cover sheet damping should be as large as possible
(iv) minimise the number of connections.

Naturally, it is advisable to support the cover sheet via flexible connections. However, it must be understood that these will be ineffective unless their differential impedances are substantially smaller than the point impedance of the cover sheet (see Section 2.7).

Holes in an enclosure are usually necessary but degrade insertion loss, especially at high frequencies. Thin slits efficiently transmit sound having a wavelength less than the slit length. A simple analysis of the performance of enclosures based upon energetic considerations can be found in Fahy and Walker (1998).

5.12 Transmission of Sound through Stiffened, Composite, Multilayer and Non-Uniform Panels

For reasons of weight, static stiffness, stiffness-to-weight ratio or noise control requirements, many forms of partition employed in practice differ from plain, uniform, homogeneous plates. Examples include lightweight corrugated factory walls, beam-stiffened floors of buildings, floors of road vehicles and trains, and aircraft cabin structures. The transmission loss of load-bearing (base) structures is often enhanced by the application of non-load-bearing secondary layers of materials such as fibrous mats and plastic foam sheets, usually covered by very flexible cover sheets: these secondary layers are known by vehicle manufacturers as 'trim'. The range of forms and structures and materials is huge, and only a small selection of archetypical configurations can be treated here. The principal aim of this section is to explain the effects of the distinctive features of each class on sound transmission behaviour. A brief discussion of the effects of trim on the reduction of sound radiation caused by mechanical excitation of base structures is also included. The exposition is largely qualitative and is illustrated by a selection of experimental data.

In this section, 'composite' panels are defined as those consisting of layers of different materials that are bonded together and in which all the layers are designed to contribute substantially to the overall bending, shear or compressional stiffness of the structures. 'Multilayer' panels are defined as those comprising a number of different layers of which only some have a specific structural function, the others acting as means to increase sound insulation or vibration isolation, as explained above. The section begins with an explanation of the effect on sound transmission of the attachment of line stiffeners, or ribs, to an otherwise plain panel: the effect on radiation efficiency was considered in Chapter 3. The effect of corrugation is then discussed. A brief review of the vibrational wave and sound transmission characteristics of three-layer composite panels is followed by consideration of the behaviour of various forms of multilayer panel.

For reasons of weight and static stiffness, structural panels are often stiffened by ribs or corrugations. In considering the transmission of diffuse incidence sound through a bounded panel in Section 5.4, a ratio of the mass-controlled, subcritical, transmission coefficient τ_∞ to resonant-mode transmission coefficient τ_r was introduced: the ratio τ_r / τ_∞ is proportional to the square of the radiation efficiency σ and inversely proportional to the loss factor η. For most practical, uniform, homogeneous panels below the critical frequency, the loss factor and radiation efficiencies are such as to make the ratio τ_∞ / τ_r much greater than unity. However, if the radiation efficiency is increased by modification of the plain panel, the ratio may decrease, resonance-controlled transmission may predominate, and the sound reduction index will then fall below the mass-controlled value. A consequence of major practical importance is that the panel damping will, to a greater or lesser extent, control the overall sound reduction index, in contrast to the purely mass-controlled case.

The effects of support constraints and impedance discontinuities on the radiation efficiency of a panel have been briefly discussed in Chapters 3 and 4. One way of qualitatively understanding the effect of attaching line stiffeners to a panel is to consider the panel to consist of an assemblage of smaller panels, each bounded by the adjacent ribs. The radiation efficiency of small panels is shown in Chapter 3 to exceed that of larger panels of the same material; hence, the radiation efficiency of the assemblage exceeds that of the unstiffened panel. In order for this argument to be fully justified, the panel vibrations on the two sides of a rib should be uncorrelated. In practice, vibrational waves are transmitted across such features, and therefore this assumption cannot strictly be correct. An alternative approach, which allows for such transmission, is based upon the derivation of dispersion relationships for bending waves that interact with the ribs. As explained in Chapter 3, the wave reflections produced by the ribs alter the dispersion relationship in such a way that free waves having wavenumber vector components of supersonic phase velocity can propagate at frequencies below the uniform-plate critical frequency (Mead, 1975). These components increase the subcritical radiation efficiency and may cause the panel to be excited in a coincident manner by incident sound waves at frequencies below critical. In practice, the effect on sound transmission is as though the critical frequency had been lowered by one

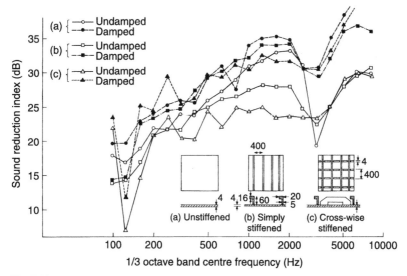

Fig. 5.28. Effects of stiffening and damping on the sound reduction index of aluminium plates
(von Venzke *et al.*, 1973).

or two octaves, the degree of change being dependent upon the spacing and stiffnesses (bending and torsional) of the ribs. Figure 5.28 shows sound reduction
index curves of stiffened panels and their unstiffened counterpart, in which the
influence of damping can clearly be seen to be much greater for the former than
for the latter (von Venzke *et al.*, 1973). The implication for the designers of stiffened panel structures is clear. Guigou-Carter and Villot (2003) model the effect
of a set of parallel, periodically spaced, line stiffeners on the sound transmission
of a 4 mm thick aluminium plate and demonstrate the deleterious effect on *TL*
produced by close stiffener spacing *L*, as shown in Fig. 5.29.

A common means of increasing the static bending stiffness and buckling resistance of thin panels is to corrugate them. This makes them orthotropic, giving
very great differences of bending stiffness in the two orthogonal directions. We
note in passing that waves in deeply convoluted plates and shells (e.g., flexible bellows) are not governed by the ordinary plate bending-wave equation.
Assuming that a 'smeared', orthotropic, bending model is appropriate, in which
the corrugation stiffness is spread uniformly over the panel surface, it is immediately clear that the critical frequency for bending waves travelling in the direction
parallel to the corrugations is greatly decreased in comparison to that of a uniform flat plate of equal thickness. It is therefore not surprising that the effect of
corrugations appears to spread the effect of coincidence on sound transmission
over a wider frequency range.

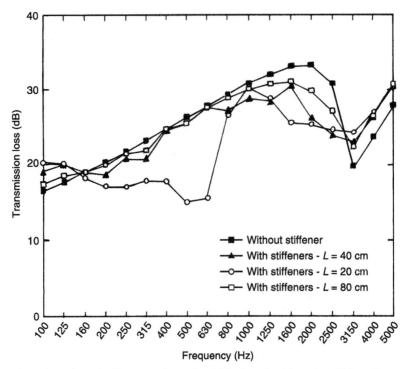

Fig. 5.29. Effect of stiffeners on the predicted *TL* of an aluminium plate (Guigou-Carter and Villot, 2003).

An analysis of sound transmission through orthotropic plates (Heckl, 1960) yielded the following expression for the diffuse-field transmission coefficient:

$$
\tau_d \approx
\begin{cases}
\dfrac{\rho_0 c}{\pi \omega m} \dfrac{f_{c_1}}{f} \left[\ln \left(\dfrac{4f}{f_{c_1}} \right) \right]^2, & f_{c_1} < f < f_{c_2}, & \text{(5.141a)} \\[3ex]
\dfrac{\pi \rho_0 c}{\omega m} \dfrac{(f_{c_1} f_{c_2})^{1/2}}{f}, & f > f_{c_2} & \text{(5.141b)}
\end{cases}
$$

where f_{c_1} and f_{c_2} are the critical frequencies based upon the maximum and minimum bending stiffnesses. Figure 5.30 compares Eq. (5.141a) with some results of sound transmission tests on corrugated panels made by Cederfeldt (1974) in which f_{c_1} and f_{c_2} are 230 Hz and 18.1 kHz respectively. Drotleff (1996) shows a similar degree of agreement between Heckl's theory and experiment. A comprehensive study of sound transmission through corrugated plates typical of those employed as outer walls of industrial premises is presented by Lam and Windle (1995a,b) and Lam (1995). Their studies were specifically aimed at explaining the

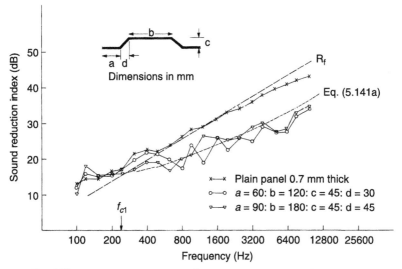

Fig. 5.30.   Sound reduction indices of two corrugated plates (Cederfeldt, 1974).

substantial 'dips' observed in measured *TL* curves. They conclude that the 'dips' are not acoustic but structural in origin, are associated with resonances of individual flat strips of the structure and are not some form of coincidence phenomenon. Damping therefore plays a vital role in controlling sound transmission.

Composite panels take the form of sandwich constructions, especially where a combination of high static strength, stability and low weight is necessary. Honeycomb panels that consist of a resin-impregnated paper, plastic or metal foil cellular core sandwiched between very thin metal or fibre-reinforced plastic face plates are widely used in aerospace structures. The cores are so stiff in compression that the distance between the face plates may be assumed not to vary during vibration. The form of the transverse wave dispersion curve is determined partly by the bending stiffness of the whole construction and partly by the shear stiffness of the core, except at very low frequencies, where the section bending stiffness predominates, and at very high frequencies, where the individual face plate bending stiffnesses become significant. A generalised dispersion curve is shown in Fig. 3.40.

The relationship between the wavenumber of transverse waves in a honeycomb panel and the wavenumbers corresponding to pure bending and to pure shear is (Kurtze and Watters, 1959)

$$1 + \left(\frac{k_s}{k_b}\right)^2 \left(\frac{k}{k_b}\right)^2 - \left(\frac{k}{k_b}\right)^4 - \left(\frac{k_b}{k_{bf}}\right)^4 \left(\frac{k_s}{k_b}\right)^2 \left(\frac{k}{k_b}\right)^6 = 0 \qquad (5.142)$$

where k_s is the shear wavenumber in the absence of transverse bending forces, k_b the overall cross-section bending wavenumber in the absence of shear distortion, and k_{bf} the wavenumber for faceplate bending alone. With core depth d, shear modulus G, total mass per unit area m, face plate thickness $h(\ll d)$, and elastic modulus E, these wavenumbers are given by

$$k_s^2 = m\omega^2/Gd \quad \text{(non-dispersive)} \tag{5.143}$$

$$k_b^4 = m\omega^2/D_1 \quad \text{(dispersive)} \tag{5.144}$$

$$k_{bf}^4 = m\omega^2/2D_2 \quad \text{(dispersive)} \tag{5.145}$$

where

$$D_1 = Ed^2h/2(1 - v^2) \quad \text{and} \quad D_2 = Eh^3/12(1 - v^2)$$

Because $h \ll d$, $D_1 \gg D_2$ and $k_b^4 \ll k_{bf}^4$ for most practical constructions, the last term on the left-hand side of Eq. (5.142) can be neglected, since k_s/k_b and k/k_b are generally of the order of unity in the frequency range of interest. The frequency at which k_s equals k_b, which is shown in Fig. 3.40, is then given by

$$\omega_{bs}^2 = G^2d^2/mD_1 \tag{5.146}$$

at which frequency

$$k \approx 1.25k_b = 1.25k_s \tag{5.147}$$

Below this frequency $k_b > k_s$, and the solution of Eq. (5.142) tends to $k = k_b$, so that shear distortion has little influence. At higher frequencies, $k_b < k_s$, and the solution tends to $k = k_s$, so that shear distortion dominates.

The free wavenumber in the shear-controlled region is approximately $k_s = \omega(m/Gd)^{1/2}$. Hence, if the phase speed $(Gd/m)^{1/2}$ is less than the speed of sound, coincidence should not occur until the very high frequencies at which face plate bending becomes significant and the wave becomes dispersive. Hence a honeycomb panel having this property might be expected to exhibit mass-controlled transmission behaviour over the whole of the frequency range of practical interest; in addition, it would be expected to form a good low-frequency enclosure structure because of its combination of low mass and high transverse stiffness at low frequencies.

In practice, a number of other factors combine to affect adversely the sound reduction index of many honeycomb panel constructions. If the shear wave phase-peed is not very much lower than the speed of sound, panel boundary effects can significantly raise the radiation efficiency, especially since the inherent damping of the type of structure is normally quite low. Studies of the transmission characteristics of the types of honeycomb structure used in aircraft and helicopters have shown that the construction must be carefully optimised to obtain acceptable performance (see Davis, 1999). Figure 5.31(a) and (b) shows the effect on wave speed and transmission loss of the core shear stiffness of a honeycomb panel comprising carbon-fibre reinforced laminated sheets bonded to a flexible Nomex core (an aramid fibre paper dipped in phenolic resin) typical of a composite aircraft fuselage (Kim and Bolton, 2004). It is seen that the transmission loss does not vary monotonically with core shear stiffness, a conclusion previously drawn by Heron (1979). Performance data for a variety of sandwich panels containing rigid plastic foam or paper honeycomb cores are presented by Jones (1981).

Honeycomb panels usually contain cores that are very stiff in compression normal to the plane of the panel. Sandwich panels containing cores of more compressible materials, such as polystyrene, exhibit a transverse dilatational resonance between the masses of the face plates and the compressional stiffness of the core. This phenomenon is analogous to the mass-air-mass resonance phenomenon of double-leaf partitions, but tends to occur at relatively higher frequencies in practice. The effect is to produce a sound reduction index curve below the resonance frequency that rises very little with frequency and lies well below the mass law line. A typical performance curve, is shown in Fig. 5.32. The adverse effects of resonance do not seem to be greatly mitigated by the use of a highly damped core, and such constructions are generally to be avoided. A form of sandwich construction that is widely used for partitions in ships takes the form of steel face plates, a few millimetres thick, separated by a core of 'cross-cut' mineral wool having the fibres oriented normally to the plane of the plate. The compressional stiffness is sufficiently high to avoid adverse dilatational effects in the frequency range of interest, but the shear stiffness is low and the damping is moderately high. There is, of course, an upper limit to the core thickness, for which the dilatational effects are not of concern. Einarsson and Soderquist (1982) describe a means of combining a honeycomb sandwich with a plain sheet to provide good insulation performance with minimum thickness.

Multilayer sandwich constructions in which the core offers little or no contribution to the mechanical stiffness of the assembly are commonly used to increase the transmission loss and to reduce the radiation efficiency of the structural shells of vehicle compartments. They typically consist of an impervious sheet of plastic or rubber-based material separated from the load-bearing structural shell by a layer of fibrous or poroelastic material. In some cases, impervious septa are incorporated within the core material. In most cases, the structural shell is treated

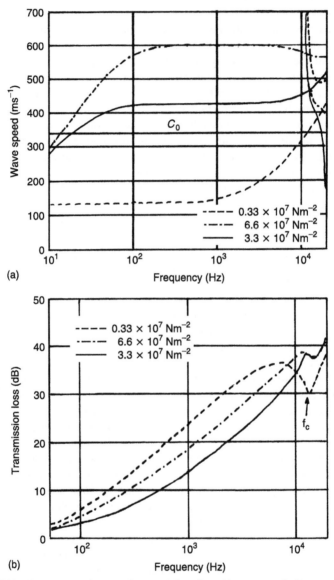

Fig. 5.31. Parameter study: core shear modulus effect. (a) wave speed; (b) transmission loss (Kim and Bolton, 2004).

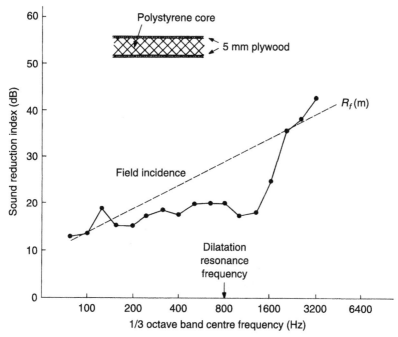

Fig. 5.32.   The effect of dilatation resonance on sound reduction index.

with viscoelastic damping material. The sound transmission characteristics of thin metal plates covered with multilayer trim typically installed in road vehicle are different from those of the double-leaf partitions employed in buildings, for the following reasons: (i) the core layers are considerably thinner, the thicknesses typically lying in the range 15–35 mm; (ii) the core material is in physical contact with both enclosing sheets, may be bonded to them, and may offer some degree of stiffness and damping; (iii) the dynamic properties of the impervious cover sheets and of the base structure are very different, the former possessing almost zero bending stiffness; (iv) the critical frequencies of the base structure and cover sheet are usually above 10 kHz and therefore of no consequence. The resonance that corresponds to the mass-air-mass resonance of double-leaf partitions involves anti-phase motion of the two outer sheets, the stiffness being provided mainly by the entrapped air. In road vehicles, the thinness of the core causes the resonance frequency to lie typically in the range 180–250 Hz.

Theoretical analysis of sound transmission through multilayer partitions is conventionally based upon the derivation of an impedance matrix for a layer of core material that expresses the normal velocity wave impedance on one face to a general terminal normal velocity wave impedance on the other. It is assumed that the partition comprises layers of infinitely extended, uniform, sheets, and that one of the outer sheets is exposed to an incident plane sound wave. The waves excited in the whole assembly consequently all share the trace wavenumber of the incident wave. The equations of continuity of normal velocity at the interfaces allow the impedance matrix to be expressed in terms of the complex amplitudes of the various wave types in the core. It is assumed that the face plates do not bear shear waves, so that velocities tangential to the interface are set to zero.

An assumption must be made about the acoustical and mechanical properties of the core. The most general poroelastic model, based upon the Biot formulation (or derivatives thereof) (Allard, 1993), demonstrates that three forms of wave may exist: (i) a longitudinal 'airborne' wave that is influenced by the fluid–acoustical properties of the core (e.g., flow resisitivity and structure factor) and that travels at a speed generally somewhat lower than a sound wave in free air; (ii) a longitudinal 'frame' wave that travels at a lower speed than the airborne wave and is determined in part by the bulk elastic stiffness of the core (iii) a shear 'frame' wave of which the speed is generally very much slower than that of the airborne wave and determined primarily by the shear moduli of the skeleton. In many practical cases, the shear wave can be neglected and if, as with most fibrous materials, the core does not support a longitudinal frame wave, an 'equivalent fluid' model may used to represent the airborne wave. The impedance matrix is derived by assuming a temporally and spatially harmonic distribution of normal velocity over the 'input' face, expressed in terms of complex amplitudes of the various wave types. Once the impedance matrix has been defined, equations of motion of the cover sheets can be written in terms of the excitation pressures and stresses at the interfaces. A good overview of the physics and various models of poroelastic materials is given by Bolton (2005).

The process of modelling a core material and deriving the matrix equation relating incident and transmitted plane wave pressures is too involved to be reproduced here, but it is exposed in detail by Bolton et al. (1996). Their results illustrate the substantial influence on TL of the elastic properties of the core of a multilayer partition consisting of a 27 mm thick sheet of partially reticulated polyurethane sandwiched between two thin aluminium plates. Figure 5.33 compares the measured performance of the configuration in which the core was bonded to both face plates (BB) with that in which a small air gap was introduced between them (UU): in case (PP), the core was lightly squashed between the face plates but not bonded. The stiffness added by the core raises the low frequency TL of case BB slightly above that of case UU. The latter is considerably superior at mid and high frequencies because the face plates do not mechanically excite the frame waves in the core. The PP result is rather surprising. The measured TLs compared very favourably with the theoretical predictions.

Measurements have been given by Drotleff (1996) of the diffuse field sound reduction indices of a flat, 0.91 mm thick, aluminium plate ($m = 2.5\,\text{kg}\cdot\text{m}^{-2}$)

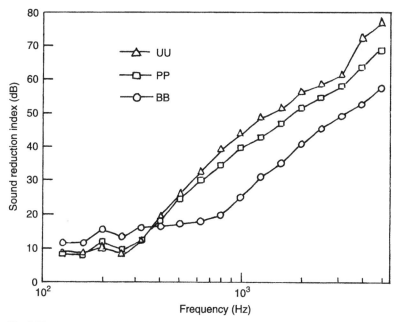

Fig. 5.33.   Measured effects of the contact condition between a polyurethane core and aluminium face plates on the sound reduction index of a sandwich panel (Bolton *et al.*, 1996).

with and without parallel stiffening beams, and of a corrugated aluminium plate of the same thickness [see Fig. 3.39(a)], with and without a 'trim' covering consisting of a 20 mm thick layer of open cell plastic foam ($m = 0.9 \, \text{kg} \cdot \text{m}^{-2}$) covered by a sheet of linoleum (lino) ($m = 1.1 \, \text{kg} \cdot \text{m}^{-2}$). The added mass theoretically increases the diffuse field R of the assembly by 5 dB relative to that of the uncovered plate. The Rs of the uncovered and covered plates are shown in Figs. 5.34 and 5.35 (against a *linear* frequency scale). The R of the covered flat plate increases at about 10 dB per octave for two octaves above the resonance frequency of approximately 400 Hz and then 'turns' to oscillate about a mean value of about 43 dB. The 'turn', 'oscillation' and 'dip' (here at about 3.5 kHz) were also observed in the results of another, quite different, method of test of the 'Insertion Loss' (*IL*) of a similar form of trim, and appear to be characteristic of this type of core material. Since the speed of the airborne wave in the foam was about 75% of that in free air, the lowest acoustic 'thickness' resonance frequency was about 6.4 kHz. The frequencies of thickness resonances associated with the longitudinal frame wave are much lower than those of the fast wave. At these frequencies, the effective foam stiffness is high, thereby 'short-circuiting' the air path. It is of practical significance that *fibrous* materials do not support a frame

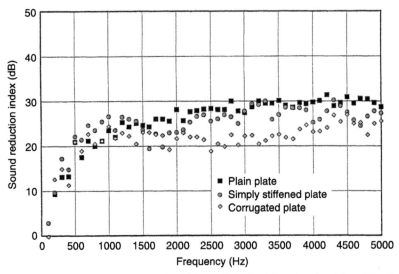

Fig. 5.34. Sound reduction index of a 0.91 mm thick aluminium sheet in plain, stiffened and corrugated configurations (Drotleff, 1996).

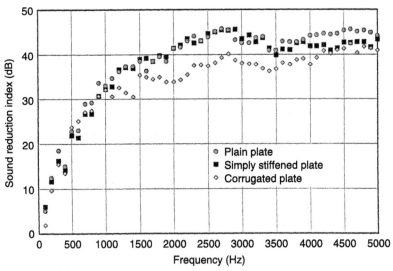

Fig. 5.35. Sound reduction index of the aluminium sheet structures when covered with a 'trim' layer of 20 mm thick open cell plastic foam and linoleum cover sheet (Drotleff, 1996).

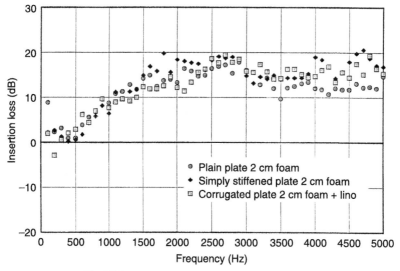

Fig. 5.36.   Insertion loss of the 'trim' (Drotleff, 1996).

wave and are therefore likely to give superior mid- and high frequency insertion loss performance.

The insertion losses of the 'trim' are shown in Fig. 5.36 in which the effect of the mass-air-mass resonance is seen in the vicinity of 400 Hz. It clearly undermines the 5 dB increase expected from the greater total mass per unit area. The effectiveness of the trim apparently varies with the form of supporting structure, but the irregularity of the curves, combined with the uncertainty of the measured data, precludes any definitive quantitative statement in this respect.

In a recent development in the optimisation of trim design for automotive vehicles, it has been found that it is beneficial in terms of trim weight not to utilise an impermeable cover sheet, but to give it a finite flow resistance (Godano *et al.*, 2000). The trim therefore acts both as a sound insulation 'barrier' and as an effective sound absorber: the practical realisation of this principle is optimised to minimise the sound pressure levels in the passenger compartment. An example of this principle is presented by Lai (2001).

Vehicle trim must also be designed to minimise the radiation of sound by vibration of the structural shell generated either by local mechanical excitation, such as suspension units, or excited by structure-borne waves generated elsewhere in the structure. The vibrational waves in the shell propagate freely and therefore have quite different wavenumber spectra from those forced by externally incident airborne plane waves. This type of problem does not seem to have been specifically addressed in the literature. Drotleff (1996) measured the insertion losses of the trim described above applied to the three forms of structure shown in Fig. 3.39(a) excited by three uncorrelated point forces. Two forms of measure of the effect of the trim on sound radiation were employed: one evaluated the radiation efficiencies with and without trim based upon the space-averaged mean square velocity of the base structure with and without trim; the other evaluated the change in radiated power per unit mean square applied force.

The radiation efficiencies of the structures covered by a 10 mm thick layer of foam and the cover sheet are presented in Fig. 5.37. Comparison of these with those of the uncovered plates (Fig. 3.39(b)) demonstrates the very substantial changes caused by the trim, especially for plain and corrugated plates. Above 1 kHz, the radiation efficiency of the latter is reduced on average by about 10 dB, whereas that of the former is greatly increased between 3.0 and 4.5 kHz. It must be understood that a change in radiation efficiency may be brought about by combination of changes in both radiated sound power and plate kinetic energy. Application of only a 10 mm thick, uncovered, layer of foam increased the average loss factor of the plain panel from 0.02 to 0.27 in the frequency range up to 2 kHz. Application of the cover sheet increased it still

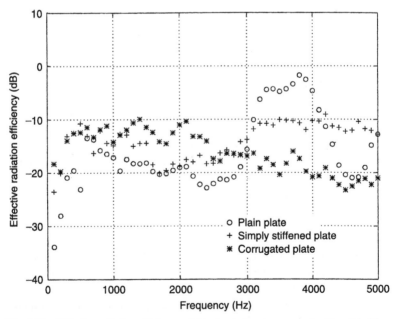

Fig. 5.37. Effective radiation efficiency of the structures when covered with a 'trim' layer of 10 mm thick open cell plastic foam and linoleum cover sheet (Drotleff, 1996).

further to an average of 0.60. However, the additional damping was minimal around 3.5 kHz. The physical origin of the damping enhancement is most probably that attributed by Cummings (2001) to the near field generated in the foam by subcritical bending wavenumbers in the plate (see also Section 4.9). The minimal effect around 3.5 kHz suggests that an acoustic near field was not present in the foam. The additional damping was considerably less in the case of the stiffened plate and negligible in the case of the corrugated plate.

Given that an increase in damping altered the relative amplitudes of the local and reverberant bending wave fields in the base structures, a more relevant quantity is the radiated sound power per unit applied force. An example of the effect of trim on this measure (a form of insertion loss) is shown in Fig. 5.38. The marked transition from almost linear frequency dependence of *IL* up to about 2 kHz to the quite different behaviour at higher frequencies cannot be attributed simply to foam thickness resonance, since the transition and forms of the curves were very similar for both 10 and 20 mm thick foam layers. It is believed that the regression of the plain plate/trim *IL* to almost zero dB in the frequency range 3–4 kHz is a manifestation of a coincidence between the plate bending wave and one of the foam waves. The inferred wave speed is about 160 m·s^{-1} which is much lower than the measured airborne wave speed in the foam of about 260 m·s^{-1} and probably relates to one of the frame waves. This is not inconsistent with the 'dip' seen in the plain plate diffuse field *IL* curve and the additional damping minimum around 3.5 kHz referred to above.

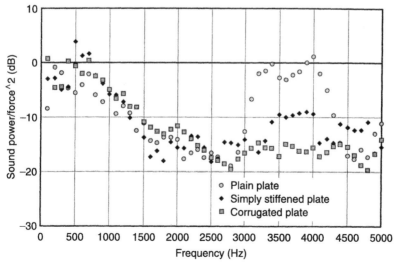

Fig. 5.38.   Effect on radiated sound power per unit mean square force of a 'trim' layer of 10 mm thick open cell plastic foam and linoleum cover sheet (Drottleff, 1996).

5.13   Transmission of Sound through Circular Cylindrical Shells

The mechanisms of sound transmission through thin-walled circular cylindrical shells are of practical interest in relation to two rather different classes of structure. One class contains large-diameter, thin-walled shells, characteristic of aircraft fuselages and rocket launchers; the other includes generally smaller diameter shells, commonly described as pipes, that are widely used in industrial plant and in fluid distribution networks. The ratio of length to diameter of cylinders in the first class is generally much smaller than those of the second class; this difference is significant in relation to the natural frequency distribution and influence of the acoustic modes of the contained fluid on sound transmission. Most commonly, it is of practical concern to keep sound energy from entering the first class of structures and from escaping from those in the second class. Large diameter heating and ventilation ducts carrying low-pressure airflows constitute an exception to this distinction. We shall henceforth refer to all circular cylindrical shells as 'cylinders'.

The transmission of plane wave sound energy through an unbounded, uniform, flat plate is governed by seven independent non-dimensional parameters. If the type and condition of the fluid on each side of the structure is the same, the number of parameters is reduced by two. The transmission of externally incident plane wave sound into an *infinitely long*, uniform, cylinder is controlled by nine (or seven) independent non-dimensional parameters. In cases of cylinders of finite length, the problem is compounded by an extra non-dimensional length parameter, together with associated resonant structural and acoustic modal behaviour. A major consequence is that it is not possible to identify any simple parametric dependences for cylinders of the type represented by the mass law. The theoretical modelling and analysis of the process are too complex to be presented in complete detail in this book. Some basic theory will be presented, together with largely qualitative interpretations of the principal features and dependences of the process, supported by some examples of experimental data.

The four principal features that cause the sound transmission characteristics of cylinders to differ from those of flat plates are (i) the finite diametral dimension; (ii) the curvature of the shell; (iii) the constraint imposed upon shell waves by circumferential continuity; (iv) the constraint imposed by the cylinder walls on the sound field in the contained fluid. The first of these features causes the circumferential distribution of the *blocked* sound field on the surface of a cylinder insonified by an externally incident plane wave to contain many wavenumber components quite different from that of the incident wave. The cylindrical coordinate system natural to geometry of a cylinder does not 'match' plane acoustic wave fronts (although the latter can be mathematically decomposed into a sum of cylindrical waves). The consequent diffraction allows any one wavenumber component of an incident field to excite shell waves of many different circumferential orders. Examples of the ratio of the amplitude of the nth circumferential order of blocked pressure to the amplitude of the incident plane wave, known as the 'scattering coefficient' are presented in Section 5.17.

The effect of curvature on vibrational waves in a cylinder wall has been explained in Section 1.9. Whereas small-amplitude flexural waves in a uniform flat plate of infinite extent propagate independently of in-plane longitudinal and shear waves, the curvature of the walls of a cylinder couples the radial, axial and tangential motions so that the three equations of motion in these directions contain contributions from all three displacements or their spatial derivatives. The solutions of these equations represent three classes of propagating wave, of which the one in which the radial displacement is dominant, labelled 'flexural', is of most interest in the audio-frequency vibroacoustics of cylinders because an inviscid fluid can only exchange energy with a shell via radial motion. Also, the wave impedances of the flexural class are generally best matched to those of sound waves in the audio-frequency range. We have seen in Section 1.9 that the shell thickness parameter $\beta = h/(12)^{1/2}a$ is an indicator of the proportion of strain energy of flexural waves associated with bending strain and with 'membrane' strain of the median surface of a cylinder wall: it is much smaller in the first class of cylinders mentioned above than in the second. It controls the shape of the flexural wave dispersion curves presented in Fig. 1.22 as a function of the ratio of frequency to the ring frequency of the cylinder, denoted by Ω. As we shall see, this shape is crucial in controlling the coupling between the shell waveguide modes and both the external and internal sound fields.

The third and fourth features produce waveguide behaviour in the form of acoustic and structural modes in which the associated physical quantities are constrained to be harmonic functions of the angular polar coordinate. The effect of this constraint in producing selectivity in coupling between acoustic and structural modes is explained in the following section

5.14 Coupling between Shell Modes and Acoustic Modes of a Contained Fluid

A cylindrical shell clearly represents a waveguide for waves in both the solid and fluid media. In principle, only coupled waves can propagate in a fluid-filled cylinder (see Chapter 7). But in cases of metal cylinders filled with gas, these waves in general correspond closely to the uncoupled *in-vacuo* shell waves and to the rigid-wall duct acoustic waves (Fuller and Fahy, 1981). We shall assume that this condition obtains in the analysis that follows. The acoustic coupling between the fluid contained in a cylinder and the shell is strongly dependent upon the relative axial phase speeds (or axial wavenumbers) of the waveguide modes in the two media. Acoustic duct theory (Pierce, 1989) shows that the modes proper to a rigid-walled, uniform, cylindrical waveguide of infinite length take the form

$$p_{np}(r, \phi, z) = \tilde{p}_{np} \, {}^{\cos}_{\sin}(n\phi) J_n(k_r r) \exp(-jk_z z) \tag{5.148}$$

where J_n is a Bessel function and the radial wavenumber k_r is determined by the zero normal-particle wall boundary condition as solutions of the equation $[J'_n(k_r r)]_{r=a} = 0$. The characteristic solutions of the equation are multi-valued for given n and therefore they are superscripted as k_r^{np}, in which n indicates the number of diametral pressure nodes and p the number of concentric circular pressure nodes (Fig. 5.39).

The radial and axial wavenumbers satisfy the acoustic wave equation

$$k_z^2 + (k_r^{np})^2 = k^2 \tag{5.149}$$

The cut-off frequencies below which a particular mode cannot propagate freely and carry energy in an infinitely long duct are given by

$$k_z^2 = k^2 - (k_r^{np})^2 = 0 \tag{5.150a}$$

or

$$ka = k_r^{np} a \tag{5.150b}$$

Values of $k_r^{np} a$ are given in Table 5.2. In terms of the ring frequency of the shell, Eq. (5.150b) may be expressed as

$$\Omega_{np} = (k_r^{np} a)(c_i/c_l'') \tag{5.151}$$

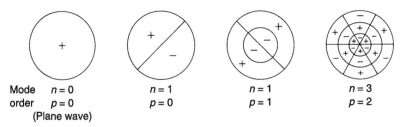

Mode	$n = 0$	$n = 1$	$n = 1$	$n = 3$
order	$p = 0$	$p = 0$	$p = 1$	$p = 2$
	(Plane wave)			

Fig. 5.39. Cross-sectional distribution of pressure phase and nodal surfaces of the acoustic modes of a circular cylindrical waveguide.

TABLE 5.2

Cut-off Frequencies of Acoustic Modes of Hard-Wall
Ducts of Circular Cross Section

n	p				
	0	1	2	3	4
0	0	3.83	7.02	10.17	13.32
1	1.84	5.33	8.53	11.71	14.86
2	3.05	6.71	9.97	13.17	16.35
3	4.20	8.02	11.35	14.59	17.79
4	5.32	9.28	12.68	15.96	19.20
5	6.42	10.52	13.99	17.31	20.58
6	7.50	11.73	15.27	18.64	21.93
7	8.58	12.93	16.53	19.94	23.27
8	9.65	14.12	17.77	21.23	24.59

[a] Values of $k_r^{np} a$ are tabulated: see Eq. (5.150b).
For steel or aluminium and atmospheric air $ka = 15.5 \, \Omega$.

where c_i is the speed of sound in the contained fluid. For instance the lowest cut-off frequency is given by

$$\Omega_{10} = 1.84 c_i / c_l'' \tag{5.152}$$

which for steel or aluminium alloy, and air at room temperature becomes

$$\Omega_{10} = 0.125 \tag{5.153}$$

The dispersion relation [Eq. (5.149)] may be represented graphically in the same form as Fig. 1.22 for the shell modes (Fig. 5.40). The appropriate values of β and c_i / c_l'', must be employed, because there is no universal form of combined structural and acoustic wavenumber diagram.

As mentioned above, the acoustic modes and the modes of a cylindrical shell do not exist independently in a fluid-filled duct. But, in cases of coupling between metal cylindrical shells and gases (except at extremely high pressure), the coupled modes resemble closely their uncoupled components and, for the purposes of approximate analysis and qualitative understanding of the coupling characteristics they may be assumed to retain their uncoupled characteristics. Hence, the dispersion diagrams for the shell waves and the fluid waves may be superimposed as shown in Fig. 5.41(a) for a typical industrial steel pipe. In this cases, it

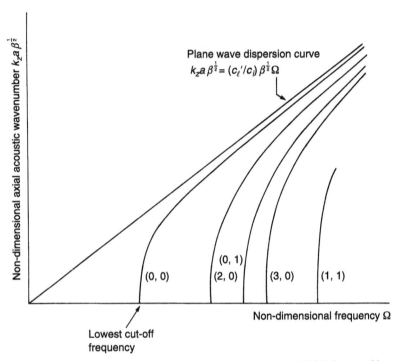

Fig. 5.40. Dispersion curves of the acoustic modes of a circular cylindrical waveguide.

is assumed that the speed of sound is 343 m·s^{-1} in the contained fluid. Note that only waves of equal circumferential order n may couple. At the intersection(s) of the dispersion curves of shell mode (solid line) of circumferential order n with those of acoustic modes (faint lines) of the same order, equality of frequency and of both axial and circumferential wavenumber vector components of the fluid and shell waves occurs: that is, equality of k_{cs}. This is a condition of coincidence.

In the cases typical of industrial pipes containing a non-flowing fluid, each acoustic mode is involved in, at most, only one coincidence, but a shell mode may be involved in more than one. However, in cases of large diameter aerospace structures, an individual acoustic mode may also be involved in more than one. The presence of a mean flow of Mach number M reduces the cut-off frequencies of modes propagating with the flow by a factor $(1 - M^2)^{1/2}$ (Norton and Karczub, 2003). This can have a substantial, and apparently non-systematic, effect on pipe response and radiation, as observed by Holmer and Heymann (1980). This complication will be avoided here by the assumption of zero flow.

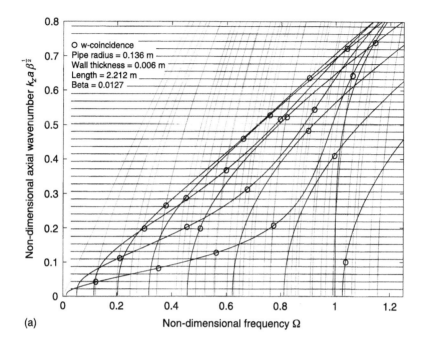

(a)

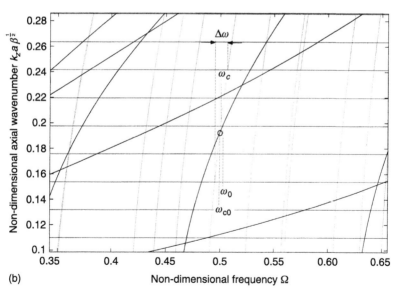

(b)

Fig. 5.41. (a) w-coincidence plot typical of an industrial pipe; (b) magnified coincidence between the $n = 6$ shell mode and the $(n = 6, p = 0)$ acoustic mode (refer to Fig. 5.44).

5.15   Vibrational Response of Pipes to Internal Acoustic Excitation

Many large industrial process and production plants incorporate thousands of metres of pipework much of which often contains high pressure gas. The gas flow is controlled by valves that produce large amounts of sound energy in the contained fluid. This not only generates sound in the surrounding air, sometimes at an environmentally unacceptable level, but can also induce very high levels of pipe wall vibration. Structural failure due to acoustically induced fatigue of pipe connections is not at all uncommon.

Straight lengths of pipes are usually terminated by flanged joints that connect them to bends, branches, valves and other components. They are also supported at intervals along their lengths. For simplicity of analysis, we shall consider a cylinder model that is constrained only at each end where it is simply supported. The natural frequencies of a pipe of length L occur where the non-dimensional shell mode dispersion curves in Fig. 1.22 intersect the horizontal lines of constant wavenumber spaced at intervals of non-dimensional axial wavenumber equal to $\pi a \beta^{1/2}/L$, as shown in Fig. 5.41(a). Since the principal wavenumber of a structural mode of axial order m is $m\pi/L$ (*independent of frequency*), resonance r-coincidence only occurs if the acoustic and structural dispersion curves of the same circumferential order intersect precisely on one of the horizontal lines. This is rare and would be very sensitive to small variations of pipe parameters, boundary conditions and mean fluid flow speed. We refer to non-resonant wavenumber coincidence as 'w-coincidence'. Figure 5.41(a) illustrates the distribution of w-coincidences typical of an industrial pipe (strong lines) containing air at N.T.P. Figure 5.41(b) shows a magnified region of the coincidence between the $n = 6$ shell mode and the ($n = 6$, $p = 0$) acoustic mode (refer to Fig. 5.44). Sound waves propagating in the contained fluid may also be reflected to some extent by features such as bends and junctions; but because each case will differ in this respect, and a degree of dissipative attenuation is produced by pipe flow, it is assumed that propagation occurs only in the positive z-direction.

Spectra of acoustically induced pipe wall vibration measured in both small scale experiments and full scale operation are commonly found to be characterised by three regions of frequency. Below Ω_{10}, the response is rather weak. The region between Ω_{10} and the ring frequency is dominated by coincidence peaks that are close to the cut-off frequencies of acoustic modes having non-zero n. These features are illustrated by the example shown in Fig. 5.42 measured in a laboratory experiment in which a 2.92 m long steel pipe having $\beta = 0.007$ and $a/L = 0.0124$ was excited by flow through an upstream 90-degree mitred bend. (Bull and Norton, 1982). Except very close to the bend, the acoustic pressures exceeded the turbulent wall pressures. The cut-off frequencies of the low order $n \neq 0$ acoustic modes are marked. (The Mach number is 0.22 and the 2.5% flow

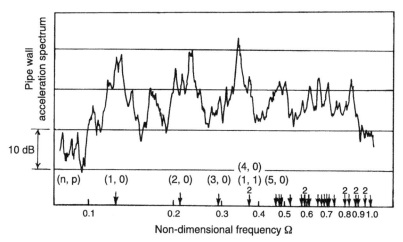

Fig. 5.42. Radial acceleration spectrum of the wall of a pipe exited by sound in the contained air (Bull and Norton, 1982).

correction to the cut-off frequency has been neglected.) Above the ring frequency, the wall response falls rapidly with increasing frequency.

Although the coincidence-related components are the dominant in the response and radiation of steel pipes having values of β of the order of 10^{-2}, coincidence becomes increasingly rare as β is increased above this value, as indicated by Fig. 5.43. The response per unit internal sound pressure falls and the response spectrum is no longer clearly separable into three distinct regions. Consequently, the interactions of many more non-coincident acoustic and structural modes must be modelled. Practical problems raised by acoustically induced wall response and radiation are confined largely to pipes having values of β less than 0.02. Figure 5.43 provides pipe designers with a guide to selecting pipe wall thickness to reduce vibrational response and radiation. Coincidence excitation is not a dominant contributor to either the vibration of pipes such as hydraulic lines or to that of liquid distribution systems. The following analysis deals exclusively with coincidence-related response.

We now turn to the mathematical modelling of the interaction between propagating acoustic waveguide modes and standing structural modes. The complex amplitude of generalised force produced by an acoustic mode of order (n, p), axial wavenumber k_z and complex harmonic pressure amplitude $\widetilde{P}_{np}$ acting on a shell mode of order (m, n) is given by

$$\widetilde{F}_{mnp} = \widetilde{P}_{np} \int_0^L \int_0^{2\pi} \sin(m\pi z/L) \exp(-jk_z z) \cos^2(n\phi) a\, d\phi\, dz \qquad (5.154)$$

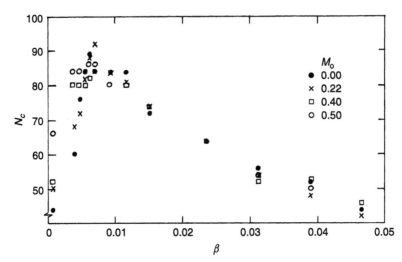

Fig. 5.43. Variation with wall thickness parameter β of the total number of w-coincidences in air up to a non-dimensional frequency $\Omega = 0.88$ as a function of flow Mach number (Norton and Karczub, 2003).

The result of the integration over z has the same form as Eq. (3.92). The square of the modulus of this term is given by Eq. (3.101) with a replaced by L, p replaced by m and x replaced by z. The integration over ϕ yields $a\pi$. Hence,

$$\left| \widetilde{F}_{mnp}(k_z) \right|^2 = \left| \widetilde{P}_{np} \right|^2 \left[\frac{2\pi p/L}{k_x^2 - (p\pi/L)^2} \right]^2 \sin\left(\frac{k_x L - p\pi}{2} \right) \pi^2 a^2 \qquad (5.155)$$

The form of $\left| \widetilde{F}_{mnp}(k_z) \right|^2$ a function of k_z is shown in Fig. 3.20. The wavenumber 'width' of the base of central lobe is $4\pi/L$. If the acoustic mode is excited by a broadband source, the expression for the *single-sided* wavenumber spectral density of generalised force $G_F^{mnp}(k_z)$ is given by Eq. (5.155) with $|P_{np}|^2$ replaced by $G_p^{np}(k_z)$. The maximum value of $G_F^{mnp}(k_z)$ at w-coincidence, when $k_z = m\pi/L$, is $(\pi^2 a^2 L^2/4) G_p^{np}(m\pi/L)$. Assuming that $(k_z + m\pi/L)$ may be approximated by $2m\pi/L$ within the span of the central lobe,

$$G_F^{mnp}(k_z) \approx G_p^{np}(k_z)(\pi^2 a^2 L^2/4)\text{sinc}^2[(k_z L - m\pi)/2] \qquad (5.156)$$

In the majority of cases, the frequency of w-coincidence does not coincide precisely with the resonance frequency of the structural mode. If the separation of the resonance and w-coincidence frequencies is substantially greater than the half-power bandwidth of the structural mode, the modal response spectrum exhibits

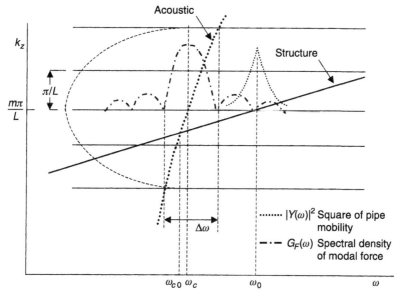

Fig. 5.44.   Spectral contributions of r- and w-coincidences to pipe mode response (diagrammatic).

two local 'peaks', as shown by Fig. 5.44. One sharp peak is centred on its resonance frequency ω_0, and another broader 'peak' is centred on the frequency
of the associated w-coincidence ω_c. The relative magnitudes and proximities
of the two are extremely sensitive to the separation of these two frequencies.
The separation can be theoretically calculated for an ideal model, but it cannot
be precisely estimated for physical systems. Prediction could, in principle, proceed on the basis of a probabilistic model of the distribution of the frequency
separation in conjunction with a generalised description of the topology of the
two sets of dispersion curves. However, the generally rather small population of
w-coincidences presents an obstacle to reliable statistical estimates of response
and radiation. Numerical evaluation of the contributions of coincidence to pipe
response has been implemented, principally by M. Bull and M.P. Norton (for
a list of references see Norton and Karczub, 2003). In cases with many coincidences, this method is reliable, but it becomes increasingly unreliable as the ratio
of wall thickness to radius increases and/or the length of the pipe decreases.

For the sake of illustration of the contributions of the two coincidence 'peaks'
to shell response, approximate estimates may be made as follows. If we assume
that it is most likely that difference between the w-coincidence frequency ω_c
and associated structural mode resonance frequency ω_0 substantially exceeds the
half-power bandwidth of the structural mode, we may treat the local generalised

force spectrum $G_F^{mnp}(\omega_0)$ as being uniform and effectively broadband. Random vibration theory gives the mean square velocity of a single-degree-of-freedom oscillator of mass M, natural frequency ω_0 and damping loss factor η subject to a broadband force of uniform spectral density G_F as

$$\overline{v^2} = G_F \pi / 2M^2 \omega_0 \eta \tag{5.157}$$

For all shell modes of non-zero circumferential order, the modal mass M_m of the pipe is equal to one quarter of its physical mass. If we assume that $G_F^{mnp}(\omega)$ varies little over the half-power bandwidth $\omega_{mn}\eta_{mn}$ of the structural mode, we may assume it to be uniform with a value given by the assumption that $k_z(\omega_0)$ is the axial wavenumber of the participating acoustic mode at the *resonance frequency* of the participating shell mode. Now, $G_F^{mnp}(k_z)/G_p^{np}(k_z) = G_F^{mnp}(\omega)/G_p^{np}(\omega)$, and according to the geometric construction shown in Fig. 5.44, $(k_z(\omega_0)L - m\pi)/2 \approx 2\pi(\omega_0 - \omega_c)/\Delta\omega$. Thus, from Eq. (5.157), the space-average mean square velocity of the shell associated with mode resonance may be written as

$$\left\langle \overline{v^2} \right\rangle_{\omega_0} \approx \pi G_p^{np}(\omega_0)\mathrm{sinc}^2[2\pi(\omega_0 - \omega_c)/\Delta\omega)]/8\rho_s^2 h^2 \omega_0 \eta \tag{5.158}$$

The 'peak' in the generalised force spectrum associated with w-coincidence occurs at ω_c. The resulting mean square velocity of the shell mode should strictly be obtained by integrating the quotient of the spectral density of the force and the square of the modulus of the impedance of the shell mode over frequency. In order to obtain an approximate algebraic expression for the response velocity, the average impedance over the frequency range of the coincidence 'peak' is approximated by its value at ω_c. Since this represents non-resonant response, damping has little influence, and the corresponding square of the modulus of the modal impedance is

$$|\tilde{Z}|^2 = \omega_c^2 M_m^2[(1 - (\omega_0/\omega_c)^2)^2] \approx 2M_m^2(\omega_0 - \omega_c)^2 \tag{5.159}$$

The mean square generalised force is given by the integral of $G_F^{mnp}(k_z)$, as expressed by Eq. (5.156), over the range of k_z within the w-coincidence peak. As we shall see later, it is reasonable to assume $G_p^{np}(k_z)$ to be uniform over this range. The integral of $\mathrm{sinc}^2[(k_z L - m\pi)/2]$ between the limits $(m + 2)\pi/L$ and $(m - 2)\pi/L$ equals $5.67/L$. Conversion of $G_p^{np}(k_z)$ to $G_p^{np}(\omega)$ gives the mean square generalised force on the shell mode as

$$\overline{F^2} = 0.35\pi G_p^{np}(\omega_c)a^2 L^2(\omega_0 - \omega_c) \tag{5.160}$$

and the resulting space-average mean square shell velocity as

$$\left\langle \overline{v^2} \right\rangle_{\omega_c} = 0.06 \Delta \omega \, G_p^{np}(\omega_c) / \pi \rho_s^2 h^2 (\omega_0 - \omega_c)^2 \tag{5.161}$$

The response increases with increase of $\Delta \omega$ (decrease in the slope of the acoustic dispersion curve at coincidence) and with decrease of $(\omega_0 - \omega_c)$. The forms of the dispersion curves in Fig. 5.41(a) suggests that these factors would favour coincidences in the upper part of the frequency range between Ω_{10} and the ring frequency. Although this is true for $\langle \overline{v^2} \rangle_{\omega_c}$, their influences on $\langle \overline{v^2} \rangle_{\omega_0}$ are opposed by the presence of ω_0 in the denominator of Eq. (5.158). Theoretical and experimental studies suggest that pipe response is only weakly dependent upon pipe damping, which suggests that w-coincidence tends to be more influential than resonant modal response.

A much stronger influence on pipe response is the non-dimensional spacing between the modal lines of constant axial wavenumber in Fig. 5.41. This is proportional to $(ha)^{1/2}/L$, which is proportional to the ratio of the square root of the cross-sectional area of the pipe *wall* to its length. As this parameter decreases, the average value of $(\omega_0 - \omega_c)$ decreases and the probability of r-coincidence increases. Since neither $\langle \overline{v^2} \rangle_{\omega_c}$ nor $\langle \overline{v^2} \rangle_{\omega_0}$ are explicitly dependent upon L, one must conclude that the influences of coincidence and damping on pipe response increase with increase of uninterrupted length. This factor has not apparently been researched systematically, but may partly account for discrepancies between the results of published studies.

The next challenge is to model the relation between the values of $G_p^{np}(\omega)$ of the acoustic modes associated with coincidence and the spectral density of the total sound pressure field in the band of interest. It is here where a model of the sound source is required. It is suspected that this is another major source of the substantial discrepancies between various published theoretical predictions of pipe response and transmission loss. This vexed question cannot be comprehensively addressed in this book and therefore only one source model is adopted for transmission loss calculations. A spatially uncorrelated axial dipole source appears to provide a reasonable model of sources created by obstructions to flow, provided that they do not generate a static pressure ratio of more than two to one, so that the flow remains subsonic. Joseph *et al.* (2003) show that such a source excites all higher order acoustic modes to equal pressure amplitude. At frequencies above $2f_{10}$, the number of acoustic modes propagating in a pipe increases nearly linearly with frequency. Hence, one can express the spectral density of the total sound pressure field in terms of the spectral density per mode as $G_p(\omega) = N_p G_p^{np}(\omega)$ where $N_p = (\gamma \omega + \delta)$ and $\gamma \gg \delta$.

An alternative acoustic source model that generates equal *sound power* per mode leads to an *inverse* dependence of mean square modal pressure amplitude upon axial wavenumber. Since acoustic modes coincident at frequencies much lower than the ring frequency have much smaller coincident axial wavenumbers than those in the upper part of this range, they would have proportionately greater pressure amplitudes. The two source models produce substantially different forms of pipe wall response spectra.

5.16   Transmission of Internally Generated Sound through Pipe Walls

Holmer and Heymann (1980) define a pipe wall transmission coefficient as the ratio of the sound power radiated per area of the pipe equal to its cross-sectional area to the sound power carried by the internal fluid. Shell modes excited by coincident acoustic modes have radiation efficiencies very close to unity. The dipole source model poses a problem for the estimation of the internal sound power because the power per mode is proportional to the axial wavenumber of the mode which, at any frequency, ranges from zero to k. However, since the dipole model generates uncorrelated acoustic modes of equal amplitude, we may assume as an approximation that the multi-mode internal sound field is hemi-diffuse. In this case, the sound power in a frequency band of width $\Delta\omega_b$ is given by $\pi a^2 G_p(\omega) \Delta\omega_b / 4\rho_0 c$. In the case where the contained and external fluids are identical, the sound power transmission coefficient is

$$\tau = 4(\rho_0^2 c^2 / N_p \Delta\omega_b) \langle \overline{v^2} \rangle / G_p(\omega) \qquad (5.162)$$

where $\langle \overline{v^2} \rangle = \langle \overline{v^2} \rangle_{\omega_c} + \langle \overline{v^2} \rangle_{\omega_0}$

A calculation of τ in ten frequency bands of width $\Delta\Omega = 0.1$ in the range $\Omega = 0.2$–1 has been made on the basis of Eqs. (5.158), (5.161) and (5.162) for a steel pipe of 0.149 m diameter, 3.17 mm thickness and 4.1 m length with still air inside and outside at atmospheric pressure and 20°C. This corresponds to pipe 'E' of Holmer and Heymann (1980) for which $\beta = 0.012$. A comparison between measured and calculated normalised transmission loss of this pipe is presented in Fig. 5.45. The differences are quite large in some bands and the measured minimum of TL around Ω_{10} is not reproduced. This is partly because the slope of the shell dispersion curve near coincidence is such that $(\omega_0 - \omega_c)/\Delta\omega$ is, on average, large and the probability of r–coincidence is very small. It is initially surprising, therefore, that most measurements of pipe response clearly exhibit a maximum around Ω_{10}. However, measurements of low order acoustic modal pressure amplitudes in a pipe excited by flow at a Mach number of 0.4 through

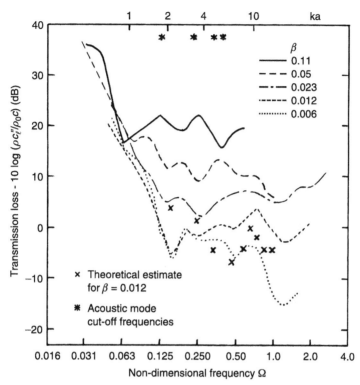

Fig. 5.45.   Normalised transmission loss for no flow. Experimental data (Holmer and Heymann, 1980); theoretical estimate for wall thickness parameter $\beta = 0.012$.

an orifice plate show the pressure amplitude of the plane wave mode at the pipe wall to be 13–17 dB below those of modes (1,0) and (2,0) just above their cut-off frequencies (Kerschen and Johnston, 1980). This large difference undermines the assumption of equal modal pressure amplitudes assumed above and explains why the predicted *TL*s in the bands two and three are overestimated. It must be emphasised that the detailed form of the calculated *TL* curve is also very sensitive to variation of all the system dimensions and properties except the fluid density. In particular, the theoretical presence or absence of coincidence greatly affects the *TL*. The probability of occurrence of r-coincidence increases with pipe length because the interval between the values of the axial wavenumbers of pipe modes is π/L. The presence of flow, which alters the positions of the acoustic dispersion curves, is also likely to strongly influence the calculated *TL*. These sensitivities, plus the influence of the type of source, may well explain why no 'universal' parametric dependences have been established and no reliable theoretical scaling rules fully developed. Fagerlund (1980) provides some useful proposals in these respects.

Pipe *TLs* at frequencies below f_{10} are of practical interest in many cases. In air at N.T.P. this frequency ranges from 400 to 2000 Hz for pipe diameters in the range 500–100 mm. In principle, a plane wave in a perfectly axisymmetric pipe can excite only the $n = 0$ peristaltic wave controlled by the 'hoop' stiffness of the pipe. This would produce a *TL* that varies at a rate of -9 dB per octave. Holmer and Heymann (1980) observed an average rate of -18 dB per octave for their test pipes. This figure is difficult to explain. However, it is likely in practice that a combination of non-axisymmetry, weight sag and the scattering effect of bends, junctions and supports of industrial pipe installations, together with excitation by turbulence and acoustic near fields local to valves, induces primarily $n = 1$ bending vibration. This would bring the figure to -15 dB per octave. The actual figure will depend upon the particular form of pipework geometry and support. Reliable theoretical predictions are therefore not possible in this range.

In the absence of a fully reliable theoretical model for any portion of the frequency range, a number of empirical formulae have been developed. In Europe, perhaps the most widely used method is that presented in the Guideline Document VDI 3733 (VDI, 1996).

5.17   Transmission of Externally Incident Sound through Large-Diameter, Thin-Walled Cylinders

The transmission of sound into the interiors of large-diameter, thin-walled cylinders is of interest principally in aerospace technology. The generation of cabin noise by jet engines and propellers and the acoustically induced vibration of payloads by the noise of the launcher rocket are two prime examples. The three features that distinguish these problems from that of sound radiation by internally excited pipes are (i) that the external sound field is unconstrained and non-modal; (ii) it is incident from a limited range of solid angle; (iii) the interior space is fully bounded and exhibits resonant modal behaviour. The very large ratio of cylinder radius to shell thickness, together with a typical value of ka of the order of 100 at 4 kHz, produces such a high density of structural and acoustic modes that it is not practicable to base a general parametric analysis of the transmission of incident plane wave sound into the interior over this entire frequency range, although the application of a computationally intensive numerical analysis of the interaction between shell modes and acoustic modes of the interior volume is possible in ideal examples (Gardonio *et al.*, 2001). However, the phenomenon of coincidence is of consequence for the transmission of plane wave sound energy *into* large diameter, thin-walled, shells in the frequency range below the ring frequency, which is typically in the range 100–400 Hz for rocket launch vehicles and transport aircraft fuselages. Theoretical aspects of coincidence related transmission will therefore be briefly discussed and illustrated. Higher frequency transmission characteristics will be illustrated by examples of experimental results.

A shell dispersion curve diagram for an idealised model of the fuselage of a medium range civil transport (assumed, for simplicity, to be uniform and circular in cross section), for which $\beta = 2.9 \times 10^{-4}$, is presented in Fig. 5.46(a). It is distinguished from that of a typical pipe in Fig. 5.41 by the much greater influence of membrane stresses on the axial wave speed of the shell waveguide modes of low circumferential order and by the fact that all the shell modes have axial wavenumbers that exceed the acoustic wavenumber over some portions of their dispersion curves below the non-dimensional critical frequency of the cylinder wall which is of the order of 10 compared with 0.5 for a typical industrial pipe. Figure 5.46(b) presents the dispersion curves of the lower shell order modes upon which are superimposed the dispersion lines of plane waves incident upon the cylinder at a range of angles to the normal to the cylinder axis. It is seen that there is near coincidence of axial wavenumber over a substantial fraction of this frequency range between each shell mode of low circumferential order and plane waves incident within a small range of angles. Since the horizontal lines corresponding to the axial wavenumbers of modes of a shell of finite length are typically spaced at intervals of non-dimensional axial wavenumber of the order of only 0.05, multiple coincidence-driven resonances are highly probable. This feature dominates sound transmission below the ring frequency and must be carefully considered in predictions of shell vibration and the resulting noise in the interior volume.

We shall return to the coincidence phenomenon after considering the coupling between plane waves incident upon a cylindrical shell and the shell modes in terms of ka and circumferential wavenumber. For this purpose, we initially consider a model of a cylinder of radius a and length L mounted between two semi-infinite, concentric, rigid baffles. A harmonic plane wave is incident at an angle ϕ_i between the wavenumber vector and *normal* to the cylinder axis (Fig. 5.47). We shall assume that the pressure field on the external surface of the cylinder is that which would be present on a infinitely long rigid shell of radius a and that it is not affected by shell response. (Note: This assumption may not be valid close to lightly damped shell resonance frequencies of real structures: it is certainly not so for external liquids.)

The wavenumber vector of an incident plane wave has an axial component $k_z = k \sin \phi_i$ and a radial component $k_r = k \cos \phi_i$. Only the latter component produces scattering of the incident wave into a series of cylindrical waves of circumferential order n. The first stage of the analysis is to calculate the scattering coefficients $\sigma_n(ka, \phi_i)$ that are defined as the ratios of the amplitudes of the scattered wave components to the amplitude of the incident wave. As an example,

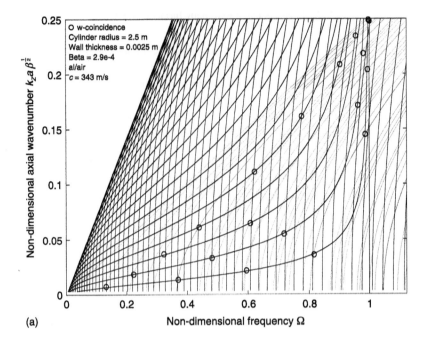

(a)

Non-dimensional axial wavenumber $k_z a \beta^{\frac{1}{2}}$

Non-dimensional frequency Ω

O w-coincidence
Cylinder radius = 2.5 m
Wall thickness = 0.0025 m
Beta = 2.9e-4
al/air
c = 343 m/s

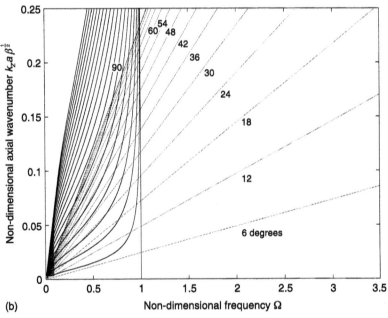

(b)

Non-dimensional axial wavenumber $k_z a \beta^{\frac{1}{2}}$

Non-dimensional frequency Ω

54
60 48
42
36
30
90
24
18
12
6 degrees

Fig. 5.46. (a) Typical w-coincidence plot for a large-diameter thin-walled cylinder; (b) superposition of dispersion curves of shell modes and externally incident plane waves.

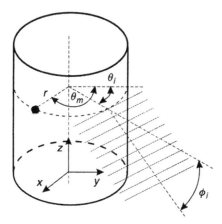

Fig. 5.47.   Coordinates for the incidence of a plane sound wave upon a cylinder.

Fig. 5.48 shows the modulus of σ_n as a functions of ka and circumferential order n (1 to 8) for $\phi_i = \pi/4$ and a frequency range that corresponds to a range of ka of 0 to 10. (Gardonio *et al.*, 2001). (Note: ϕ_i is defined relative to the cylinder axis in the cited reference.)

We see that a plane wave incident at any angle can couple with all orders of shell mode, each component of the scattered field of order n coupling only with the shell waveguide mode of the same order. The expression for the generalised harmonic force on a shell mode is the same as that of Eq. (5.155) with the pressure amplitude $\tilde{P}_{np}$ replaced by the scattered field amplitude $\sigma_n(ka, \phi_i)\tilde{P}_i$. In the particular case to which Fig. 5.46(b) applies, the scattered field components of plane waves incident at less than about 30° to the normal can be near-coincident in axial wavenumber and coincident in circumferential wavenumber with low order shell modes over a substantial portion of the frequency range below the ring frequency. The calculation of the response of the acoustic modes of the interior volume would proceed by the evaluation of their mode shapes, natural frequencies and loss factors followed by application of Modal-Interaction Analysis as set out in Section 7.6. This would yield the 'Noise Reduction (*NR*) as a logarithmic measure of the ratio of the mean square pressure of the incident wave to the space-average mean square pressure in the volume. The evaluation of the *TL* can be made by equating the rate of energy dissipation of the acoustic modes by sound absorption in the volume to the transmitted sound power. If the near-coincident excitation of modal resonances dominates the transmission, damping of the cylinder would be a very effective control measure.

Figure 5.49 presents examples of measured values of the *NR* of a thin-wall aluminium cylinder of 350 mm diameter and 0.5 mm wall thickness ($\beta = 8.0 \times 10^{-4}$) insonified by plane waves at angles ϕ_i of 30°, 60° and 90° (Ljunggren, 1987).

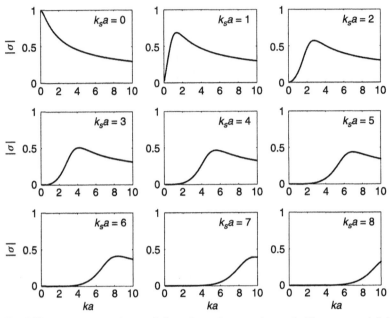

Fig. 5.48. Pressure scattering coefficients for a plane sound wave incident upon an infinitely long rigid cylinder (Gardonio *et al.*, 2000).

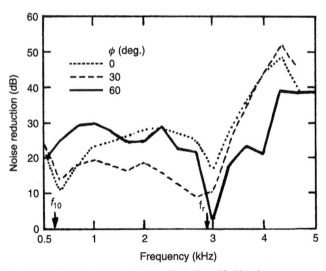

Fig. 5.49. Noise reduction of a thin-walled cylinder insonified by plane waves at various angles of incidence (Ljunggren, 1987).

The interior of the cylinder was made rather sound absorbent. As expected from the above discussion, the average *NR* of the cylinder in the frequency range between f_{10} (574 Hz) and f_r (4730 Hz) excited at 30° was the lowest of the three. The plane wave dispersion line for $\phi_i = 60°$ passes through a dense cluster of coincidences just below the ring frequency. The effect is evident in Fig. 5.49. At 0°, the phase of the incident plane wave is uniform over the length of the cylinder and w-coincidence occurs at the cut-off frequency of each structural mode. The modes of a finite cylinder that have significant wavenumber spectral components at $k_z = 0$ are of axial order $m = 1$. Consequently, resonance coincidence is not possible. The result is that the *NR* is close to the normal incidence mass law of a flat plate over the range 1–2.3 kHz and the ring frequency dip is not so severe as at the other two angles.

Because of the enormous number and high density of acoustic and structural modes, calculation of the *TL* and *NR* of large-diameter thin-wall cylinders subject to *diffuse field* acoustic excitation is best achieved on the basis of a probabilistic model as employed in Statistical Energy Analysis. This approach also has the advantage over deterministic numerical computations of providing the possibility of identifying parametric dependences and providing associated physical insights. There is no space here for setting out the details of the analysis, for which reference should be made to Szechenyi (1971), in which a comparison between theoretical calculation and experimental measurement of *NR* of a cylinder ($\beta = 5.3 \times 10^{-4}$) is presented (Fig. 5.50). The curve shows some resemblance to

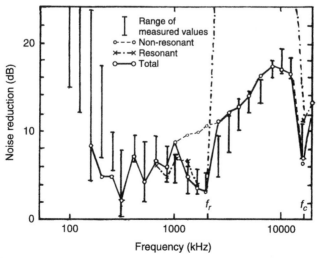

Fig. 5.50. Comparison between theoretical and measured noise reduction of a large diameter, thin-walled, circular cylinder subject to diffuse acoustic field excitation (Szechenyi, 1971).

Ljunggren's result at 60°, but is much lower in value, no doubt because the cylinder did not contain sound absorbent material. It is clear that it is dangerous to apply measurements of TL made under diffuse field excitation to cases in which the sound waves are incident upon a cylinder from within a small range of solid angle, especially below the ring frequency.

Problems

5.1 A piston of 100-mm diameter and 0.02-kg mass is elastically mounted in a rigid co-planar baffle so that it has an *in-vacuo* natural frequency of 50 Hz. By considering the normal incidence of plane waves upon the piston, together with the scattered and transmitted fields (fields radiated due to piston vibration), derive an equation of motion for the piston. Hence evaluate the sound power transmission coefficient in air as a function of frequency between 10 and 500 Hz. Compare the sound reduction index of the piston with the mass law and explain the physical reasons for the differences. Hint: Use small ka approximations for fluid loading and radiated field. How will τ change with angle of incidence of the excitation field?

5.2 Determine the magnitude of the loss factor necessary to reduce the resonant transmission coefficient of the system described in problem 5.1 by a factor of ten.

5.3 The sound of an over-flying aircraft is assumed to fall as plane waves on the surface of a deep lake. Determine the sound-pressure level, relative to the incident-field sound-pressure level, produced at a water depth of 500 mm by the 1-kHz component of the incident sound, when the aircraft is at an angle of elevation of 72°. Is it appropriate to evaluate an intensity transmission coefficient τ in this case?

5.4 A plane sound wave in air is obliquely incident upon a large horizontal metal sheet. The temperature of the sheet is raised by some means to higher than ambient temperature. Qualitatively investigate the influence of increase of temperature on the sound transmission coefficient (a) at subcritical frequencies and (b) at supercritical frequencies.

5.5 A simply supported, baffled, rectangular plate has an elastic modulus of $10^9 \, \text{Nm}^{-2}$, a density of $700 \, \text{kg·m}^{-3}$ and a thickness of 4 mm. The dimensions of the plate are 1 m × 1.2 m. Determine the value of the plate loss factor that makes the resonance-controlled transmission coefficient for 500-Hz octave band diffuse incidence sound in air approximately equal to the mass-controlled transmission coefficent. Equation (5.56) applies. Also try Eq. (5.58).

5.6 Evaluate δ_1 in Eq. (5.74) for a double-glazed window construction consisting of two panes of 6 mm thick glass mounted in a 1500 mm × 700 mm frame at a separation distance of 19 mm. Also evaluate the mass-air-mass resonance frequency of this construction.

5.7 Using the model of a double-leaf partition illustrated in Fig. 5.13, derive an expression for the ratio of leaf displacements ξ_1/ξ_2 at the mass-air-mass resonance frequency in terms of the leaf mass ratio m_1/m_2.

5.8 By reference to the simple enclosure model illustrated in Fig. 5.26 and the associated analysis, find an expression for the ratio of power radiated by the enclosed source to power radiated by the enclosure, in terms only of the specific damping ratio of the enclosure. Hint: Estimate the power dissipated by the enclosure damping. Show that, for practical values of enclosure damping, the maximum value of high-frequency *IL* is equal to the normal incidence sound reduction index for a panel of twice the mass per unit area of the enclosure.

6 | Acoustically Induced Vibration of Structures

6.1 Practical Aspects of Acoustically Induced Vibration

Any solid structure exposed to a sound field in a contiguous fluid medium will respond to some degree to the fluctuating pressures acting at the interface between the two media. It is common experience that heavy road traffic can produce visually observable low-frequency displacements in window panes of houses adjacent to a highway; concert goers are particularly thrilled when they can actually feel, through their seats, the response of the auditorium structure to a fortissimo passage; viewers at air displays frequently experience strong chest cavity vibration when a pilot turns on the afterburners; and heavy rock devotees love the body vibrations induced by the very high powered loudspeaker systems used by performing groups.

The process of transmission of airborne sound energy through partitions involves vibrational response to sound pressures acting on one side, together with the consequent radiation of sound from the other side. This process, which has been described and analysed in Chapter 5, generally involves extremely small transverse displacement amplitudes of the partitions, which therefore do not constitute a threat to the integrity of the structures. There are, however, cases of considerable practical significance in which acoustically induced vibration levels are of such a magnitude as to require an engineer to assess the likely effects on the structural and operational integrity of the system concerned. Acoustically induced fatigue failure first came to prominence in the 1950s owing to the high levels of jet noise to which aircraft structures were exposed. In the 1960s aerospace engineers became concerned with the high levels of vibration induced in spacecraft structures and on-board equipment packages due to rocket noise at launch. Acoustically induced fatigue failure of industrial plant and gas-cooled nuclear reactors components came to light at about the same time.

Other examples of acoustically induced vibration of practical significance include the degradation of receiving sonar array performance by the response of the surrounding hull structure to the incoming sound field; low-frequency, airborne sound-induced vibration of complete building structures by sources such as large reciprocating compressors; vibration of the tank tops (engine support structures) of ships by airborne noise radiated by the underside of the engines; and the process of absorption of sound by structures in buildings, such as lightweight ceilings, glazing and panelling. The mechanisms of structural vibration caused by explosive blasts and by sonic booms are qualitatively similar, but these airborne disturbances differ from those considered in this book because they are of finite amplitude and are therefore non-linear in nature.

It will be shown in this chapter that the characteristics of response of a structure to incident sound are closely linked to its sound radiation characteristics. This feature is the result of an extremely important fundamental property of sound sources and sound fields, which is expressed in the *principle of reciprocity*, together with its various ramifications. This principle will be discussed in this chapter.

6.2   Decomposition of a Sound Field

Before entering upon any mathematical analysis of the response of structures to sound, it is helpful to discuss certain fundamental principles of the interaction process in qualitative terms. Consider a *rigid* body of arbitrary shape upon which a sound field is incident. At all points on the surface of the body the sound field particle velocity normal to the surface must be zero. This condition may be satisfied by imagining the surface of the body to undergo a form of motion such that its normal surface velocity is equal and opposite to that which would exist in the sound field in that normal direction, but in the absence of the body. The *unobstructed* incident sound field and that 'radiated' by the body in its imagined motion are then superimposed: the latter field component is correctly termed the 'scattered field'. This form of thinking is helpful because we already have some physical feel for the nature of sound fields radiated by vibrating bodies.

In the simplest case, imagine a plane sound wave normally incident upon an *infinitely extended* rigid plane surface. We know that the pressure at the surface of a uniformly vibrating, infinite, plane surface, on the side to which positive velocity is directed, is $p(t) = \rho_0 c v(t)$. Hence the total pressure on the rigid surface is

$$p_{bl}(t) = p_i(t) + \rho_0 c[p_i(t)/\rho_0 c] = 2p_i(t) \qquad (6.1)$$

in which p_{bl} symbolises the total surface pressure in the obstructed or blocked, sound field, and p_i the pressure at the position of the surface in the unobstructed sound field. Now imagine that a simple harmonic plane sound wave is incident at an angle ϕ to the normal to a rigid surface. In the absence of the surface, the incident-wave particle velocity component normal to position of the absent surface takes the form

$$u_n(x, t) = (\tilde{p}_i/\rho_0 c) \cos \phi \exp[j(\omega t - kx \sin \phi)] \qquad (6.2)$$

in which x is the coordinate in the plane of the surface. We now imagine the surface to be present and vibrating with normal surface velocity $(-u_n)$. Analysis in Chapters 3 and 4 of the radiation from a vibrating-plane surface shows that the wave impedance presented by a fluid to such a travelling wave is

$$\tilde{z}_{wf} = \rho_0 c/[1 - (k \sin \phi/k)^2]^{1/2} = \rho_0 c \sec \phi \qquad (6.3)$$

Hence the pressure on the rigid surface, which equals $p_i + u_n z_{wf}$ is

$$p_{bl}(x, t) = 2\tilde{p}_i \exp[j(\omega t - kx \sin \phi)] = 2p_i(x, t) \qquad (6.4)$$

We see therefore that 'pressure doubling' occurs for all angles of incidence, except for 90° when the incident wave is not obstructed.

Imagine now that the plane surface is not rigid but has *uniform* elastic and inertial properties. In this special case, the spatial distribution of response of the surface must necessarily match that of the incident pressure field, i.e., it has a surface (trace) wavenumber $k_x = k \sin \phi$. The dynamic properties of the surface may be expressed in terms of a structural wave impedance $\tilde{z}_{ws}(k_x, \omega)$. The surface pressure cannot now be equal to $2p_i(x, t)$, because the surface moves under the action of surface pressure and therefore produces an additional scattered field, which we call the radiated field. Now, the blocked pressure field corresponds to zero motion of the surface, and hence it is valid to superimpose the blocked field and the radiated field i.e., that produced by the absence of motion plus that produced by motion. As far as the structure is concerned, it moves in response to the total surface pressure that arises from the sum of these two field components. Hence we may write a surface response equation as

$$\tilde{v}(k_x, \omega)\tilde{z}_{ws}(k_x, \omega) = -\tilde{p}_{bl}(k_x, \omega) - \tilde{p}_{rad}(k_x, \omega) \qquad (6.5)$$

in which positive v is directed out from the structure into the fluid, and the fluid is assumed to exist only on one side of the structure.

As in the case of our imagined surface motion, $\tilde{p}_{rad}(k_x, \omega)$ may be expressed in terms of the surface velocity and the fluid wave impedance, as

$$\tilde{v}(k_x, \omega)\tilde{z}_{ws}(k_x, \omega) = -\tilde{p}_{bl}(k_x, \omega) - \tilde{v}(k_x, \omega)\tilde{z}_{wf}(k_x, \omega) \qquad (6.6)$$

The structural and fluid wave impedances are seen to add to form a fluid-loaded structure wave impedance, introduced in Chapter 4 [Eq. (4.43)]. The response equation may be rewritten as that of the fluid-loaded structure to the blocked-pressure field

$$\tilde{v}(k_x, \omega)[\tilde{z}_{ws}(k_x, \omega) + \tilde{z}_{wf}(k_x, \omega)] = -\tilde{p}_{bl}(k_x, \omega) \qquad (6.7)$$

Although this equation has been developed here specifically in relation to an infinite plane surface of uniform properties exposed to an incident plane wave, in which case both the scattered and radiated fields have a unique wavenumber component k_x, the principles involved apply to any flexible surface of any geometric form. However, it is often difficult to evaluate the appropriate forms of $\tilde{z}_{ws}$ and $\tilde{p}_{bl}$ in cases where the structure is of irregular geometry and non-uniform dynamic properties. In such cases the scattered and vibrational fields do not possess a unique wavenumber even if the blocked field does. In cases of planar structure, the response may be evaluated by including the contributions of all possible values of k_x, by inverse Fourier transformation of Eq. (6.7) as illustrated in Section 6.3. The numerical approach to more general scattering problems is explained in Chapter 8.

One valuable conclusion that can be drawn from our consideration of the components of the total pressure field acting on the surface of an elastic structure is that, if we can reasonably accurately assess the influence of a contiguous fluid on the natural frequencies and mode shapes of a structure, then we may apply the estimated blocked-pressure field to the fluid-loaded structure. The influence of fluid loading on structural vibration characteristics has been discussed in Chapter 4. In many practical cases the fluid-loaded structural mode shapes are altered little from their *in-vacuo* forms, even in the presence of a dense fluid such as water, and reasonable estimates can be made of the reduction of natural frequencies by inertial fluid loading.

It should not be assumed that the blocked pressure on the surface of a structure is always close to twice the incident pressure in the absence of the body. This is particularly not the case for bodies that have one or two typical dimensions small when compared with an acoustic wavelength, as consideration of the equivalent radiated fields readily shows. For example, consider a long rigid circular cylinder of radius a exposed to plane sound waves incident at 90° to the cylinder axis. If it is assumed that the acoustic wavelength greatly exceeds the cylinder circumference ($ka \ll 1$), then the scattered field can be equated approximately to that radiated by the cylinder in undergoing uniform translational vibration with a velocity amplitude $\tilde{u}$ equal and opposite to that in the incident wave. (Why is the small-ka condition necessary for this assumption?) The scattered surface pressure is given, correct to first order in ka, by Eq. (4.30) as

$$(\tilde{p}_s)_{r=a} = j\rho_0 cka\tilde{u} \cos \theta \tag{6.8}$$

where θ is the azimuthal coordinate of the cylinder measured relative to the plane of motion. Now $\tilde{u} = -\tilde{p}_i/\rho_0 c$. Therefore, the ratio of amplitudes of scattered surface pressure to incident pressure is given approximately by

$$|\tilde{p}_s|_{r=a}/|\tilde{p}_i| \approx ka \cos \theta \ll 1 \tag{6.9}$$

The pressure in the incident wave at the location of the cylinder surface, but in the absence of the cylinder, is given approximately by

$$(\tilde{p}_i)_{r=a} \approx \tilde{p}_i(1 + jka \cos \theta) \tag{6.10}$$

Thus, although the scattered pressure is small in comparison with the incident pressure, the total force per unit length due to the blocked pressure is actually double that which is produced by the integration of the unobstructed incident field component around the circumference of the cylinder. (Check this conclusion yourself.)

6.3   Response of a Baffled Plate to Plane Sound Waves

The following analysis of the response of a finite plate in a rigid baffle to incident plane sound waves serves to illustrate the phenomenon of scattering of sound by reflection from a surface of non-uniform specific acoustic impedance, as well as the modal-analysis approach to the problem of estimating the response of flexible structures to sound. It is not always realised that scattering of incident waves caused by spatial variations of surface impedance is at least equal in practical importance to that due to irregularity of surface geometry, particularly when the surface can respond in strongly resonant modes. A practical example of the exploitation of this phenomenon is in the installation of low-frequency resonant panel (membrane) absorbers in recording and broadcasting studios. Not only do these panels absorb low-frequency energy and thereby control undesirable room acoustic resonances, but they also scatter incident sound energy into many directions and improve the diffusion of the sound fields in a way that cannot be achieved by shaping the room surfaces because of their small dimensions in comparison with the wavelength of the sound to be controlled.

The analysis is of a two-dimensional problem because the complexity of the expression is somewhat less than that for a three-dimensional system; however, the principles are the same. Consider an infinitely long, uniform, thin, elastic plate of width a set between two semi-infinite rigid-plane baffles: the whole system lies in one common plane (Fig. 6.1). The edges of the plate are assumed to be simply supported in order to simplify the modal expressions; this is not, however, a necessary condition. A simple harmonic plane wave, of frequency ω and wavenumber k, is assumed to be incident on the plane at an angle ϕ to the normal.

As we have seen, the blocked pressure is equal to twice the incident pressure at the surface of the plane. The transverse normal velocity v of the plate is expressed in terms of a summation over its *in-vacuo* normal modes: the primary objective

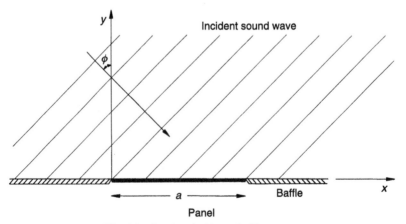

Fig. 6.1.   Panel excited by an incident wave.

of the analysis is to find the velocity amplitudes of these modes. Let

$$v_n(x, t) = \sum_p \tilde{v}_p \sin(p\pi x/a) \exp(j\omega t) \qquad (6.11)$$

In order to derive an expression for the field radiated by the plate we may express the transverse velocity of the whole plane surface in terms of an infinite set of simple harmonic travelling waves of wavenumber k_x by the Fourier integral transform pair, as in Chapters 3 and 4:

$$\tilde{V}(k_x) = \int_0^a \sum_p \tilde{v}_p \sin(p\pi x/a) \exp(-jk_x x) \, dx \qquad (6.12a)$$

$$v_n(x) = \frac{1}{2\pi} \int_{-\infty}^{\infty} \tilde{V}(k_x) \exp(jk_x x) \, dk_x \qquad (6.12b)$$

Now the radiated field can be obtained by using the relationship between pressure and surface velocity in the field radiated by a single travelling-wave component [Eq. (3.94)]:

$$\tilde{P}(k_x, y) = \tilde{V}(k_x)(\omega\rho_0/k_y) \exp(-jk_y y) \qquad (6.13)$$

in which $k_y = \pm(k^2 - k_x^2)^{1/2}$, the choice of sign being decided upon physical grounds, as explained in Section 3.8. The radiated-field component expressed by Eq. (6.13) corresponds physically to a plane wave radiated at angle $\phi(k_x) = \cos^{-1}(k_y/k)$, for surface wavenumbers satisfying the condition $k_x < k$, and to a near field, which decays exponentially with distance from the plane, for $k_x > k$. The reason that k_x takes an infinite range of values is that the plate modes exist over only a finite region of the plane. The solution to Eq. (6.12a) for $\tilde{V}_p(k_x)$ corresponding to vibration in a single mode p is

$$\tilde{V}_p(k_x) = \tilde{v}_p \frac{(p\pi/a)[(-1)^p \exp(-jk_x a) - 1]}{k_x^2 - (p\pi/a)^2} \qquad (6.14)$$

Figure 3.20 illustrates the dependence of $|\tilde{V}_p(k_x)|^2$ on $k_x/(p\pi/a)$. Equations (6.13) and (6.14) effectively express the scattering behaviour of one vibration mode of the plate; a wave of single $k_x = k \sin\phi$ is incident, and plane waves having k_x within the range $-k < k_x < k$ are radiated at all angles, in the direction opposite to the x-wise component of the incident wavenumber vector as well as in the same direction. It should not, however, be concluded that sound energy is necessarily radiated symmetrically about the normal, because

the incident field will force the relative phases of the modal responses to take a
certain form that will influence the directional characteristics of the field radiated
by all modes when vibrating simultaneously. The plate responds to the total
acoustic-pressure field on its surface, which is the sum of the blocked pressure
(twice the incident pressure) and the radiated field. From Eqs. (6.12a), (6.13) and
(6.14) the wavenumber spectrum of the radiated field at the surface is given by

$$\tilde{P}(k_x)_{y=0} = \sum_p \tilde{V}_p(k_x)(\omega\rho_0/k_y) \tag{6.15a}$$

and the surface pressure distribution due to modal motion is given by the inverse
transform [Eq. (4.53)]

$$\tilde{p}(x,0) = \frac{1}{2\pi} \int_{-\infty}^{\infty} \sum_p \tilde{V}_p(k_x)(\omega\rho_0/k_y) \exp(jk_x x)\, dk_x \tag{6.15b}$$

The total surface pressure is equal to the blocked pressure plus the radiated
pressure given by Eq. (6.15b). Hence the generalised force per unit length on
mode m is given by

$$\tilde{F}_m = \int_0^a 2\tilde{p}_i \sin(m\pi x/a) \exp(-jkx \sin\phi)\, dx$$

$$+ \frac{1}{2\pi} \int_0^a \left[\int_{-\infty}^{\infty} \sum_p \tilde{V}_p(k_x)(\omega\rho_0/k_y) \exp(jk_x x)\, dk_x \right] \sin(m\pi x/a)\, dx \tag{6.16}$$

The result of the first integral, which gives the blocked modal force, is similar in
form to Eq. (6.14) with p replaced by m and k_x replaced by $k\sin\phi$. However,
unlike k_x, which varies between $-\infty$ and $+\infty$, $k\sin\phi$ is limited to the range
between $-k$ and $+k$, the limits corresponding to grazing incidence and zero
corresponding to normal incidence.

Hence, the blocked force can only achieve the peak value at $k\sin\phi = \pm p\pi/a$
if $|k| > p\pi/a$, which for *resonant* vibration means that the modal natural frequency
exceeds the critical frequency (Fig. 6.2). The blocked modal force can peak for
lower-order modes at frequencies above their natural frequencies; it requires only
a sufficiently high excitation frequency to satisfy the above condition.

This condition is reminiscent of that encountered in Section 3.5 in connection with radiation from plate modes: only the components of $\tilde{V}_p(k_x)$ with
$-k < k_x < k$ can radiate sound energy to the far field, and only those same components can be directly excited by incident plane waves. Resonant vibration of all

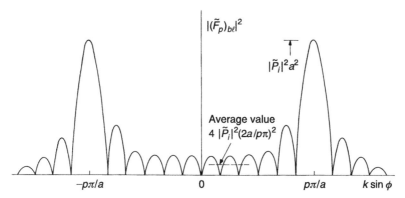

Fig. 6.2. Squared modulus of the wavenumber spectrum of the blocked modal force produced on a baffled panel by an incident plane sound wave.

modes having natural frequencies above the critical frequency is most strongly excited by waves incident at angles corresponding to the angles of maximum energy radiation by those modes at resonance. Resonant vibration of all other lower-order modes is excited most strongly by waves close to grazing incidence, with the exception of the fundamental mode, for which the blocked force peaks at $k \sin \phi = 0$.

The panel mode shape $\sin(m\pi x/a)$ in Eq. (6.16) may be expressed by the Fourier integral transform pair

$$\sin(m\pi x/a) = \frac{1}{2\pi} \int_{-\infty}^{\infty} \tilde{A}_m(k'_x) \exp(jk'_x x)\, dk'_x$$

$$\tilde{A}_m(k'_x) = \int_0^a \sin(m\pi x/a) \exp(-jk'_x x)\, dx$$

Thus, Eq. (6.16) may be written as

$$\tilde{F}_m = 2\tilde{p}_i \tilde{A}_m(k \sin \phi) + \frac{1}{4\pi^2} \int_{-\infty}^{\infty} \left\{ \left[\int_{-\infty}^{\infty} \sum_p \tilde{V}_p(k_x)(\omega\rho_0/k_y)\exp(jk_x x)\, dk_x \right] \right.$$

$$\left. \times \left[\int_{-\infty}^{\infty} \tilde{A}_m(k'_x)\exp(jk'_x x)\, dk'_x \right] \right\} dx \quad (6.17)$$

in which the range of integration over x has been extended from 0 to a to $-\infty$ to $+\infty$, because the form of $\tilde{A}_m(k'_x)$ ensures that the modal displacement does not extend outside the panel. Now, since the modal displacement function is real,

$\tilde{A}_m(-k'_x) = \tilde{A}_m^*(k'_x)$, and use of the identity

$$\int_{-\infty}^{\infty} \exp[j(k_x - k'_x)x]\,dx = 2\pi\delta(k_x - k'_x)$$

allows Eq. (6.17) to be reduced to

$$\tilde{F}_m = 2\tilde{p}_i\tilde{A}_m(k\sin\phi) + \frac{1}{2\pi}\int_{-\infty}^{\infty}\sum_p \tilde{V}_p(k_x)\tilde{A}_m^*(k_x)(\omega\rho_0/k_y)\,dk_x \qquad (6.18)$$

in which

$$\tilde{A}_n(\alpha) = \frac{(n\pi/a)[(-1)^n\exp(-j\alpha a) - 1]}{\alpha^2 - (n\pi/a)^2}$$

Evaluation of the integral in Eq. (6.18) is left as an exercise for advanced students. This result indicates that each mode m is subject to pressures generated by vibration in all the other modes p through the summation term, which arises because every mode makes a contribution to each k_x component of $\tilde{P}(k_x)_{y=0}$. Physically this represents cross-modal coupling between modes through the scattered pressure field. The cross coupling term will normally represent a mixture of reactive and resistive loading through the different forms of k_y in the ranges $-k \leq k_x \leq k$, $k_x < -k$ and $k_x > k$ (see Section 3.8). The total response of the structure is formally given by

$$v_n(x, t) = \exp(j\omega t)\sum_m \tilde{F}_m\sin(m\pi x/a)/\tilde{Z}_m$$

where $\tilde{Z}_m$ is the *in-vacuo* modal impedance.

The general solution for the multi-mode response and scattering by a plate is clearly rather complicated to evaluate, even in this relatively simple case. However, if the modal density of the structure is rather low, it may be reasonable to neglect inter-modal coupling and remove the summation sign from Eq. (6.18), letting $m = p$. The second term in Eq. (6.18) then represents the combination of modal radiation damping and reactive fluid loading appropriate to each mode at its resonance frequency. Fortunately, reactive fluid loading generally alters mode shapes little from their *in-vacuo* forms, even if substantially lowering modal natural frequencies. An example of a general scattering analysis for a thin rod is presented by Liamshev (1958).

The degree of excitation of any mode by an incident plane wave, and the consequent scattering of the wave energy into angles different from the specular direction (rigid-plane surface reflection), is clearly strongly related to the radiation characteristics of the mode. The actual magnitude of resonant modal vibrations is controlled not only by the blocked-pressure force, but by the modal mass and total damping, the latter including radiation damping. The strength of resonant modal scattering increases as the total damping *decreases*.

The two-dimensional analysis can, of course, be generalised to any arrangement of structure and fluid. However, the blocked-pressure distribution and modal-radiation characteristics may be difficult to evaluate in cases of complex geometry and it may be necessary to employ numerical analyses for the purpose.

6.4   The Principle of Vibroacoustic Reciprocity

Our preceding discussions of structural response to incident sound, and its consequent scattering by radiation, have hinted at the existence of a fundamental relation between the acoustically induced response and radiation characteristics of structures. In the following sections we investigate the origin and nature of these relations. As a preliminary to dealing with flexible bodies immersed in fluids it is helpful to consider further the generation of blocked acoustic pressure distributions on rigid bodies by incident sound. The general principle of point-to-point acoustic reciprocity is presented in detail in 'Theory of Sound' by Lord Rayleigh (1896). In essence, the principle states that the acoustic pressure produced at one point A in a fluid by a simple harmonic, omni-directional, point source of volume velocity at another point B in the fluid is the same as the pressure produced at B by the same source located by point A, irrespective of the presence of arbitrary, *dynamically linear*, boundaries to the fluid. We now imagine one of the points to be adjacent to the rigid surface of a body in the fluid. A point volume velocity source located immediately adjacent to the body surface will produce an acoustic pressure at an observation point in the fluid. On interchanging the source and observation points, the same (blocked) pressure will be produced on the body surface.

The elementary source of volume velocity at the body surface can be created equally well by the vibration of a very small element of the body surface. Now we know that the total acoustic pressure produced in a fluid by the vibration of the surface of a body can, by the principle of superposition, be evaluated by integrating over the surface of the body the product of the strength of elemental sources on the surface times an appropriate transfer function between the source points and the observation point. In an anechoic environment the transfer function will depend upon the shape of the (motionless) body, as well as the

source-observation point distance, because the waves emitted by any one elemental surface source will be scattered by the presence of the rest of the (motionless)
body (Cremer, 1981). This fact is expressed mathematically by the surface pressure term in the Kirchhoff–Helmholtz integral equation (see Section 3.13). If the
environment is reflective, the integral must be extended over all surfaces exposed
to the fluid. Unless the surface of the body takes the form of an infinitely extended
plane surface, the transfer function does not take the simple form $C \exp(-jkR)/R$,
which is appropriate only to a point source in a fluid free of boundaries (free field)
and to the special planar source case to which the Rayleigh equation applies
[see Eq. (3.4)]. However, the principle of reciprocity can be invoked to provide
a means of bypassing the problem of evaluating the transfer function directly,
as explained in the next section based upon analysis by Smith (1962). Another
form of reciprocity that is of considerable practical importance in experimental
vibroacoustics is that which relates to a point force that acts on a *linear* elastic
structure and the sound pressure generated in a contiguous fluid by the action of
that force. It is explained and illustrated in Section 6.6.

6.5   Modal Reciprocity: Radiation and Response

Imagine that we are in a position to make a reasonable assumption about the
form of motion generated in a body by acoustic excitation, as in the case of a
flexibly mounted but otherwise rigid body, or in cases of vibration dominated by
response in an individual mode. Let points on the body surface be defined by the
position vector $\mathbf{r}_s$ and points in the fluid be defined by $\mathbf{r}$. As we have already
seen, the response of a fluid-loaded mode is obtained by evaluating the modal
force caused by the blocked pressure field on the surface of the structure. This
blocked force is defined by

$$\widetilde{F}_{bl}^{m} = \int_{S} \psi_m(\mathbf{r}_s)\tilde{p}_{bl}(\mathbf{r}_s)\, dS \tag{6.19}$$

where $\psi_m(\mathbf{r}_s)$ describes the assumed mode shape. A point volume velocity source
$\widetilde{Q}(\mathbf{r}_0)\exp(j\omega t)$ in the fluid volume at $\mathbf{r}_0$ is assumed to generate a simple harmonic
excitation field. A reciprocal relationship may be written between the blocked
pressure produced on the *rigid* surface at $\mathbf{r}_s$ by source Q, and the pressure $\tilde{p}(\mathbf{r}_0)$
produced at $\mathbf{r}_0$ by an elemental surface source of volume velocity $\tilde{v}_m\psi_m(\mathbf{r}_s)\delta S$,
where $\tilde{v}_m$ is the modal coordinate velocity amplitude:

$$\tilde{p}(\mathbf{r}_0)/\tilde{v}_m\psi_m(\mathbf{r}_s)\delta S = \tilde{p}_{bl}(\mathbf{r}_s)/\widetilde{Q}(\mathbf{r}_0) \tag{6.20}$$

This relationship is true irrespective of the shape of the body. Hence, from Eq. (6.19) the elemental contribution to $\widetilde{F}_{bl}$ may be written as

$$\delta \widetilde{F}_{bl}^{m} = \widetilde{Q}(\mathbf{r}_0)\tilde{p}(\mathbf{r}_0)/\tilde{v}_m \tag{6.21}$$

By superposition, the total modal blocked force may be written as

$$\widetilde{F}_{bl}^{m} = (\tilde{p}(\mathbf{r}_0)/\tilde{v}_m)\widetilde{Q}(\mathbf{r}_0) \tag{6.22}$$

where $p(\mathbf{r}_0)$ is the total pressure produced at $\mathbf{r}_0$ by modal vibration of the whole surface with velocity amplitude $\tilde{v}_m$. Although this equation is valid for any position of point source relative to that of the structure, and for any (linear) condition of fluid boundaries, it is convenient initially to assume that apart from the presence of the structure, the fluid is unbounded and that the source is in the far acoustic field of the vibrating body. The reason for making these assumptions is that pressure can then be simply related to sound intensity, and also a reference incident pressure can be unambiguously defined.

Modal vibration can be attributed with a radiation resistance, thus

$$\overline{P}_{rad}^{m} = \frac{1}{2}|\tilde{v}_m|^2 R_{rad}^{m} = \frac{1}{2}\rho_0 c\sigma_m |\tilde{v}_m|^2 \int_S \psi_m(\mathbf{r}_s)^2 dS$$
$$= \frac{1}{2}\left(\eta_{rad}^{m}\,\omega_m M_m |\tilde{v}_m|^2\right) \tag{6.23}$$

where σ_m is the modal radiation efficiency and η_{rad}^{m} is the modal radiation loss factor. In the far radiation field, the radial sound intensity is given by

$$I = \frac{1}{2}|\tilde{p}|^2/\rho_0 c \tag{6.24}$$

A source directivity factor $D(\theta, \phi)$ can be defined as the ratio of sound intensity at radius r in the direction (θ, ϕ) to the sound intensity I_0 produced at the same radius by a uniformly directional source of the same power, where $4\pi r^2 I_0 = \overline{P}_{rad}$. Hence

$$D(\theta, \phi) = \frac{I(\theta, \phi)}{I_0} = \frac{2\pi r^2 |\tilde{p}(\theta, \phi)|^2}{\rho_0 c\overline{P}_{rad}} = \frac{4\pi r^2 |\tilde{p}(\theta, \phi)|^2}{\rho_0 c R_{rad}^{m} |\tilde{v}_m|^2} \tag{6.25}$$

Thus, if position $\mathbf{r}_0$ has coordinates (r, θ, ϕ), Eq. (6.25) relates the modulus $|\tilde{p}(\mathbf{r}_0)/\tilde{v}_m|$ of the complex ratio appearing in Eq. (6.22) to the radiation directivity factor of the modal source.

A fluid-loaded modal impedance $\tilde{Z}_m$ may be defined by

$$\tilde{Z}_m = \tilde{F}_{bl}^m / \tilde{v}_m \tag{6.26}$$

When the fluid-loaded mode is excited at resonance, the imaginary component of $\tilde{Z}_m$ disappears and only the resistive component R_m remains to control the response amplitude. Hence

$$\tilde{v}_m = \tilde{F}_{bl}^m / R_m = \tilde{F}_{bl}^m / (R_{int}^m + R_{rad}^m) \tag{6.27}$$

in which the total modal resistance has been separated into that associated with energy dissipation within the structure (or radiation of mechanical energy to connected systems) and that associated with acoustic energy radiation. Normally only the mean-square modal velocity, and the associated modal energy, is of interest, the phase of the response being of less concern. Hence Eqs. (6.22), (6.24) and (6.25) are combined to give

$$|\tilde{v}_m|^2 = \frac{\rho_0 c D(\theta, \phi) |\tilde{Q}(\mathbf{r}_0)|^2}{4\pi r_0^2} \frac{R_{rad}^m}{(R_{int}^m + R_{rad}^m)^2} \tag{6.28}$$

It is convenient to replace $\tilde{Q}$ by an expression involving the pressure in the incident field at the location of the structure, but in its absence, this is the pressure $\tilde{p}_0$ produced by $\tilde{Q}$ at distance r. Equation (3.2) gives

$$|\tilde{p}_0|^2 = \omega^2 \rho_0^2 |\tilde{Q}|^2 / 16\pi^2 r^2 \tag{6.29}$$

Substitution into Eq. (6.28) yields

$$\frac{|\tilde{v}_m|^2}{|\tilde{p}_0|^2} = \frac{4\pi D(\theta, \phi)}{\rho_0 c k^2} \frac{R_{rad}^m}{(R_{int}^m + R_{rad}^m)^2} \tag{6.30}$$

The significance of this result is that it shows that the resonant response of a structure in a single, well-defined mode to excitation by the field of a point source, located in the direction (θ, ϕ), can be estimated from the knowledge of the modal radiation resistance and its directional radiation characteristics: the direct scattering problem has been avoided by the use of the reciprocity principle.

Spectral analysis may be used to show that if the source is not of a single frequency, but has a broad random character, the equivalent form of response equation is

$$\frac{\overline{v_m^2}}{G_{p_0}(\omega)} = \frac{2\pi^2 c D(\theta, \phi)}{\rho_0 \omega_m^2 M_m} \frac{R_{rad}^m}{R_{int}^m + R_{rad}^m} \tag{6.31}$$

where M_m is the modal generalised mass, $G_{p_0}(\omega)$ is the single-sided mean-square spectral density of the incident pressure, which is assumed to be effectively uniform over the bandwidth of the mode, and ω_m is the natural frequency of the fluid-loaded mode. Note that $M_m \overline{v_m^2}$ is the time-average total energy of vibration of the mode.

In practice it is often likely that one wishes to estimate modal response to a multi-directional sound field because high sound levels that can produce stresses capable of threatening the integrity of exposed structures are more easily generated within enclosed spaces because of reverberant build-up. Provided that equivalent point source locations can be identified, the reciprocity Eq. (6.22) applies equally well to this more complex case. However, it is then not possible to use Eq. (6.24), nor is the incident-sound pressure associated with each source readily identified. Nevertheless, it is possible to derive an approximate response formula for the case of excitation by a *diffuse* field, which is an idealisation in which plane waves are incident upon a structure from all directions with equal probability and random phase.

An incident diffuse field may be modelled as that produced by a uniform distribution over a spherical surface of large radius in free field of point sources of uniform strength and random phase. If the mean-square source strength per unit area is $\bar{q}^2$, an element of the source surface of area δS produces an elemental modal response $\delta |\tilde{v}_m|^2$ given by Eq. (6.28), with $|\tilde{Q}|^2 = 2\bar{q}^2 \delta S$. Because all the elemental sources are uncorrelated, the total mean-square pressure produced at the location of the absent structure by the whole source surface is given by Eq. (6.29) with $|\tilde{Q}|^2 = 4\pi r^2 \bar{q}^2$. Integration of Eq. (6.28) over the source surface to obtain the total modal response gives

$$\overline{v_m^2} = \frac{\rho_0 c}{4\pi r^2} \frac{R_{rad}^m}{(R_{int}^m + R_{rad}^m)^2} \overline{q^2} \int_{4\pi} D(\theta, \phi) r^2 d\theta d\phi \tag{6.32}$$

where δS has been replaced by $r^2 d\theta d\phi$. By definition of the directivity factor

$$\int_{4\pi} D(\theta, \phi) r^2 d\theta d\phi = 4\pi \tag{6.33}$$

and since

$$\overline{p_0^2} = \omega \rho_0^2 \overline{q^2} / 4\pi \tag{6.34}$$

the normalised modal response is

$$\frac{\overline{v_m^2}}{p_0^2} = \frac{4\pi}{\rho_0 ck^2} \frac{R_{rad}^m}{(R_{int}^m + R_{rad}^m)^2} \tag{6.35}$$

The response to a broadband random diffuse field is, by analogy with Eq. (6.31),

$$\frac{\overline{v_m^2}}{G_{p0}(\omega)} = \frac{2\pi^2 c}{\rho_0 \omega_m^2 M_m} \frac{R_{rad}^m}{R_{int}^m + R_{rad}^m} \tag{6.36}$$

where again $M_m \overline{v_m^2}$ is the total energy of modal vibration and $G_{p0}(\omega)$ is assumed to be uniform.

If a number of modes of a structure have their natural frequencies within the bandwidth of a random diffuse field, the total energy of vibration is exactly equal to the sum of the modal energies, provided that the modes form an orthogonal set. In this case the total time-average energy of structural vibration is related to the mean-square pressure in a bandwidth centered on ω_c by

$$\frac{\overline{E}_s}{p_0^2} = \frac{2\pi^2 cn_s(\omega)}{\rho_0 \omega_c^2} \left\langle \frac{R_{rad}^m}{R_{int}^m + R_{rad}^m} \right\rangle_m \tag{6.37}$$

where $n_s(\omega)$ is the average modal density of the structure, which is the average number of natural frequencies per rad s^{-1}, and the brackets denote an average over the modes resonant in the bandwidth. Modal-average radiation efficiency, which is discussed in Chapter 3, is related to modal-average radiation resistance, through Eq. (6.23) by

$$\langle \sigma \rangle_m = \frac{\langle R_{rad}^m / S_m \rangle_m}{\rho_0 c} \tag{6.38}$$

where $S_m = \int_S \psi_m^2 \, dS$.

It has been shown by alternative analysis techniques that Eqs. (6.36) and (6.37) apply to any case of diffuse field excitation, irrespective of the actual forms of the acoustic sources producing the field. In enclosures having dimensions much greater than an acoustic wavelength, the acoustic modal density is generally so high that even a rather small excitation bandwidth $\Delta\omega$ is sufficient to produce a close approximation to a diffuse field. The acoustic-field energy density is $\overline{p_0^2}/\rho_0 c^2$ and the modal density $n_s(\omega)$ is $V\omega^2/2\pi^2 c^3$, where V is the enclosure volume. Hence the average energy per mode is $2\pi^2 c \overline{p_0^2}/\rho_0 \omega^2 \Delta\omega$. Equation (6.37) may therefore be written

$$\frac{\overline{E}_s}{\overline{E}_a} = \frac{n_s(\omega)}{n_a(\omega)} \left\langle \frac{1}{1 + R_{int}^m / R_{rad}^m} \right\rangle_m = \frac{n_s(\omega)}{n_a(\omega)} \left\langle \frac{1}{1 + \eta_{int}^m / \eta_{rad}^m} \right\rangle \quad (6.39)$$

The physical interpretation of this equation is that the average ratio of structural energy per mode to acoustic energy per mode is determined by the ratio of modal radiation and internal resistances or loss factors. If $R_{int}^m / R_{rad}^m \gg 1$ the energy ratio is proportional to R_{rad}^m / R_{int}^m, but if $R_{int}^m / R_{rad}^m \ll 1$ the energy ratio is nearly independent of R_{rad}^m and R_{int}^m. This condition represents an *upper limit* on structural response to excitation by a broadband diffuse sound field in an enclosure; it is known as 'equipartition of modal energy' in the terminology of Statistical Energy Analysis. A similar conclusion regarding dependence on the resistance ratios may be drawn from broadband response Eqs. (6.31) and (6.36). The influence of the resistance ratio on modal response *at resonance* is rather different as indicated by Eqs. (6.32) and (6.35). The maximum response is produced by a resistance ratio of unity. We shall discover later how to make use of this result in designing low-frequency resonant acoustic absorbers.

6.6 Radiation Due to Point Forces and Response to Point Sources

There exists a fundamental relationship between the radiation field of a plate or shell structure subjected to point force excitation, and its response to the acoustic field produced by a point source in a contiguous fluid (Liamshev, 1960). It relates the acoustic pressure at an observation point A in a contiguous fluid produced by the vibration of any (linear) plate or shell structure subject to the action of a single harmonic point mechanical force at point B, and the vibration velocity produced in the structure at point B, in the absence of the mechanical force, by a point source located in the fluid at point A. The direction of the velocity is defined to be the same as that of the force (Fig. 6.3):

$$\tilde{p}(\mathbf{r}_0)/\tilde{F}(\mathbf{r}_s) = -\tilde{v}(\mathbf{r}_s)/\tilde{Q}(\mathbf{r}_0) \quad (6.40)$$

where $\mathbf{r}_0$ and $\mathbf{r}_s$ are the position vectors of points A and B, respectively.

The following analysis illustrates a particular example of the general relationship. In Section 4.2, the response of a small, fluid-loaded, baffled piston to a

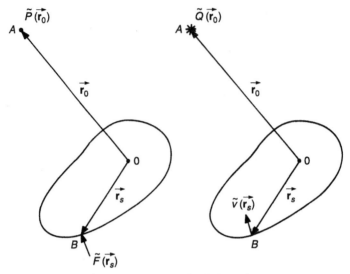

Fig. 6.3. Reciprocal cases of radiation and response.

mechanically applied, simple harmonic force was examined. Equations (4.7)–(4.10) give

$$\frac{\tilde{v}}{\tilde{F}} = \frac{1}{\tilde{Z}_{tot}} = \left\{ j\left[\omega M - S/\omega + (8/3)\rho_0 a^3 \omega\right] + \left[B + \tfrac{1}{2}\rho_0 c\pi a^2/(ka)^2\right]\right\}^{-1}$$

(6.41)

in which $(\omega M - S/\omega)$ is the reactive mechanical impedance, to which is added the mechanical equivalent of the reactive fluid-loading impedance, and B is the mechanical damping coefficient, to which is added the equivalent acoustic radiation resistance: equivalence is necessary because of the different definitions of mechanical and acoustic impedance. The effect of fluid density on the natural and resonance frequencies of the system is clearly indicated by Eq. (6.41). We know that the response of an elastic system to acoustic excitation may be evaluated by decomposing the total acoustic field on the surface of the system into its blocked and radiated components, and that this leads to the concept of the fluid-loaded system responding to the blocked pressure component alone. This approach is most useful when the dominant spatial form of system response can be assumed *a priori*: the piston is an example of such a system. The mechanical force $\tilde{F}$ may be replaced by the blocked-pressure force due to any form of incident field.

We are currently most interested in the particular type of incident field to which the force-source relationship of Eq. (6.40) applies, namely that produced by an omni-directional 'point' source. The blocked pressure on an infinite baffle may be evaluated most easily by imagining an identical image source to exist on the other side of the baffle in the mirror-image position (Fig. 6.4). The particle velocity normal to the plane of the (now removed) baffle is zero, and the blocked-pressure amplitude is twice that at the plane due to the real source alone. It is given by

$$\tilde{p}_{bl}(\mathbf{r}_s) = [j\omega\rho_0\tilde{Q}(\mathbf{r}_0)/2\pi r]\exp(-jkr) \tag{6.42}$$

where $r = |\mathbf{r}_0 - \mathbf{r}_s|$. Since the piston is small ($ka \ll 1$) it may be assumed that the blocked pressure is uniform. With the centre of the piston at the origin of the coordinate system,

$$\tilde{F}_{bl} = -[j\omega\rho_0\pi a^2\tilde{Q}/2\pi r]\exp(-jkr) \tag{6.43}$$

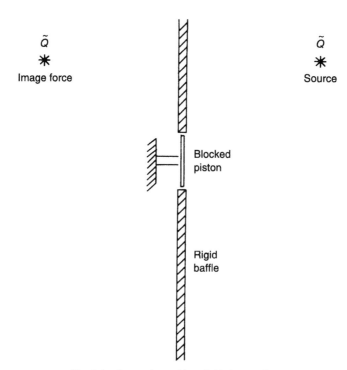

Fig. 6.4.   Source imaged in a rigid plane surface.

The piston response to acoustic excitation is

$$\tilde{v}/\tilde{Q} = -\,(j\omega\rho_0 a^2/2r\tilde{Z}_{tot})\exp(-jkr) \tag{6.44}$$

When the piston is vibrated by the mechanical force $\tilde{F}$ the radiated field is given by that of a point source in a baffle, because it has been assumed that $ka \ll 1$. Hence, from Eq. (3.3),

$$\tilde{p}/\pi a^2 \tilde{v} = (j\omega\rho_0/2\pi r)\exp(-jkr) \tag{6.45}$$

However, $\tilde{v} = \tilde{F}/\tilde{Z}_{tot}$ and therefore

$$\tilde{p}/\tilde{F} = (j\omega\rho_0 a^2/2r\tilde{Z}_{tot})\exp(-jkr) = -\tilde{v}/\tilde{Q} \tag{6.46}$$

which confirms Eq. (6.40) in the elementary case.

The general relationship expressed by Eq. (6.40), which holds for any configuration and boundary conditions of structure and fluid, subject to the condition of linearity of dynamic behaviour, has considerable practical importance. It is often necessary to investigate the dependence on mechanical forces applied to a structure of the sound field radiated by that structure, but it is frequently very difficult and costly, or even impossible, to replace the forces operating in practice by measurable, controlled forces. This is particularly difficult in cases where couples as well as forces are applied to a structure. However, if the reciprocal experiment is performed, in which acoustic sources, placed in the fluid at positions of interest, are used to excite vibrations in the structural system, the resulting linear and rotational velocities at the points of force operation can be measured relatively easily by small transducers. In principle, the point of operation of the force should be completely free of external constraints during the reciprocal measurement of response. However, in cases where the forces (couples) of concern are caused by strains of resilient vibration isolator elements installed to reduce force transmission, the constraints are usually sufficiently weak to allow them to remain in place without causing unacceptable error in the transfer function(s) inferred from the measurement.

This technique was pioneered by the TNO, Delft, The Netherlands. It has been extensively applied to measurements on industrial machinery, ships, aircraft, trains and road vehicles. A good example is that of the diagnosis and evaluation of the sources of structure-borne noise in car interiors. In the simplest implementation, an omni-directional acoustic source is placed at the position of the passengers' heads and accelerations are measured on the engine mountings, exhaust suspension points, etc. A more sophisticated arrangement replaces the omni-directional source by an anthropomorphic dummy containing acoustic

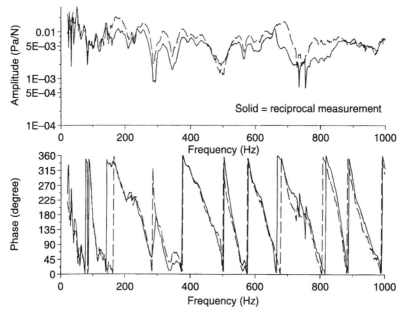

Fig. 6.5. Comparison of directly and reciprocally measured transfer functions between the sound pressure at a car driver's right ear and the force applied to the gearbox mounting point.

sources in the head that radiate from orifices located at the ear positions thus replicating the diffraction effect of the body (Pankoke, 1993). An example of the results so obtained as presented in Fig. 6.5. A further interesting potential use of the relationship is in determining the optimum positions for the attachment of vibrating machinery via resilient isolators to associated structures so as to minimise sound radiation. This is done by finding positions of minimum vibration velocity under acoustic excitation. This technique is not applicable in cases where the source component acts as a velocity source. Comprehensive surveys of these and other applications are presented by Fahy (1995a, 2004). Reciprocal relationships of the forms discussed above are not limited to simple harmonic time dependence, since they hold for any Fourier components of complex time histories, provided the systems are linear.

The literature of sound radiation from plates and shells contains many expressions for the far fields generated by point-excited structures. The responses of the structures to acoustic excitation by point sources in the far field may be evaluated from the same expressions: response to plane-wave excitation may also be evaluated using Liamshev's extension of the point-to-point relation. Important examples of these relations are the expression for the response of a uniform plate to point source and to plane-wave excitation: the fluid is considered to exist on one side of the plate only.

In the former case

$$\frac{\tilde{v}}{\tilde{Q}} = \frac{jk\exp(-jkr)}{2\pi r} \frac{\cos\theta}{1 - jkh(\rho_s/\rho_0)(\cos\theta)[1 - (\omega/\omega_c)^3\sin^4\theta]} \tag{6.47}$$

where θ is the angle between the normal to the plate and the line joining the point of observation of v to the source point, $h\rho_s$ is the plate mass per unit area, and ω_c is the critical frequency of the plate. At low frequencies, $\omega/\omega_c \ll 1$ and Eq. (6.47) reduces to

$$\frac{\tilde{v}}{\tilde{Q}} = \frac{jk\exp(-jkr)}{2\pi r} \frac{\cos\theta}{1 - jkh(\rho_s/\rho_0)(\cos\theta)} \tag{6.48}$$

Liamshev's plane-wave response theory gives, for $\omega/\omega_c \ll 1$ and incidence angle ϕ,

$$\frac{\tilde{v}}{\tilde{p}_i} = -\frac{(2/\rho_0 c)\cos\phi}{1 - jkh(\rho_s/\rho_0)\cos\phi} \tag{6.49}$$

where $\tilde{p}_i$ is the incident plane-wave amplitude. In the limit of very heavy fluid loading $(kh\rho_s/\rho_0 \ll 1)$, or high angle of incidence, the response tends to the equivalent free surface particle velocity $-2(\tilde{p}_i/\rho_0 c)\cos\phi$, indicating that the plate impedance is insignificant compared with the fluid impedance.

6.7   An Application of Response Theory to Building Acoustics

The principal applications of the theories of acoustically induced response are to problems of estimating the vibrational responses of aircraft, spacecraft, industrial plant and pipework to excitation by intense sound fields to which they are exposed in operation. They may also be applied to problems of flanking-sound transmission in buildings, in which acoustically induced vibration is transmitted along the building structure, thereby bypassing partition structures such as walls and floors. One application in building acoustics that has received little attention is that of the absorption of sound in auditoria by non-load-bearing structures exposed to the sound field; these include lightweight ceilings, gypsum board soffits, stage structures, reflectors, wall panelling, ductwork and even seats. These are all known to act as low-frequency absorbers, controlling low-frequency reverberation time, but little quantitative information has been published to provide a guide for acoustic designers. In auditoria, sound fields can be considered to be quasi-diffuse, and most sounds of interest are transient and have energy distributed over rather broad frequency bands. However, the decay rates of sound fields in auditoria tend to be far less than those of such structures, and therefore the steady-state, broadband, response equations may be used as a reasonable approximation to the short-time average behaviour.

The rate at which energy is dissipated and/or transmitted by acoustically excited structures to supporting structures is

$$\overline{P_d} = \sum_m \eta_{int}^m \omega_m \overline{E}_m \tag{6.50}$$

in which $\eta_{int}^m = R_{int}^m / \omega_m M_m$. The vibrational energy of an individual mode in response to broadband, diffuse-field acoustic excitation is given by Eq. (6.36). Multi-mode response energy is given by Eq. (6.37). It is generally the case for all but very lightweight sheets excited at frequencies below critical in air that $R_{rad}^m \ll R_{int}^m$. Using this approximation, together with Eq. (6.50), gives the power dissipated in acoustically induced, multi-mode, vibration as

$$\overline{P_d} = [2\pi^2 c G_{p0}(\omega)/\rho_0] \sum_m \eta_{rad}^m / \omega_m \tag{6.51}$$

This equation is interesting because it shows the dissipated energy to be dependent only on the radiation loss factors and *not on the internal loss factors* (damping) of the structure. The modal-radiation loss factor is related to the modal-radiation efficiency by

$$\eta_{rad}^m = \rho_0 c A \sigma_m / \omega_m M \tag{6.52}$$

Note the appearance of the surface area A and the total structural mass M as opposed to the modal mass. Thus

$$\overline{P_d} = [2\pi^2 c^2 A G_{p0}(\omega)/M] \sum_m \sigma_m / \omega_m^2 \tag{6.53}$$

In a frequency band of relatively small percentage bandwidth and centre frequency ω_0 this equation may be further simplified to

$$\overline{P_d}/\overline{p_0^2} = [2\pi^2 c^2 A n_s(\omega)/M\omega_0^2] \langle \sigma_m \rangle_m \tag{6.54}$$

where $\langle \sigma_m \rangle_m$ is the modal-average radiation efficiency appropriate to the band.

In architectural acoustics the measure of the capacity of a surface to dissipate sound energy is the absorption coefficient α, which is defined as the ratio of absorbed to incident sound power. Clearly, its value for any particular surface depends upon the spatial form of the incident field, as well as on frequency and the mechanical and geometric properties of the surface. It is conventional to evaluate the diffuse incidence, or statistical, absorption coefficient, since in practice the actual spatial characteristics of fields are not known. In a diffuse field the intensity incident upon a plane from one side is

$$I = \overline{p_0^2}/4\rho_0 c \tag{6.55}$$

Therefore,

$$\alpha = \frac{\overline{P_d}/A}{\overline{p_0^2}/4\rho_0 c} = \frac{8\pi^2\rho_0 c^3 n_s(\omega)}{M\omega_0^2}\langle\sigma_m\rangle_m \tag{6.56}$$

Since the radiation efficiency of baffled, uniform panels at frequencies well below the critical frequency tends to vary approximately as $\omega^{1/2}$, and the average modal density of flat panels is independent of frequency, it is clear from Eq. (6.56) that panel vibration constitutes primarily a low-frequency absorption mechanism. An approximate expression for the low-frequency radiation efficiency of a baffled rectangular panel is

$$\langle\sigma_m\rangle_m = B(1.8hc_l')^{3/2}\omega_0^{1/2}/c^2\pi^2 A(2\pi)^{1/2} \tag{6.57}$$

where B, A and h are respectively, the perimeter, area and thickness of the panel, and c_l' is the speed of quasi-longitudinal waves in the plate. The modal density is

$$n_s(\omega) = \sqrt{3}A/2\pi h c_l' \tag{6.58}$$

and hence the absorption coefficient becomes

$$\alpha(\omega_0) = 3.3\rho_0 Bc_l'^{1/2}c/A\rho_s h^{1/2}\omega_0^{3/2} \tag{6.59}$$

This result indicates that one large panel will absorb less than a number of smaller panels making up the same total area, and that the absorption coefficient decreases by a factor of approximately three per octave: absorption by a panel of a given material increases as the thickness is decreased. The value of α for a $2\,\text{m} \times 2\,\text{m} \times 13\,\text{mm}$ gypsum board panel at 100 Hz is approximately 0.9. These conclusions are consistent with observations on buildings. It should be carefully noted that these specific expressions do not apply to unbaffled panels for which the radiation efficiency is far lower and increases much more rapidly with frequency.

Equation (6.59) relates to broadband absorption by multi-modal vibration. Another form of panel absorber commonly used in buildings, particularly in recording studios, is the so-called membrane absorber. This consists of a thin panel, typically about 1 m square, which is fixed to a wall with a shallow air cavity between it and the wall. The function of this type of absorber is selectively to absorb low-frequency energy by resonance in its fundamental mode; the air cavity bulk stiffness, rather than the panel bending stiffness, usually determines this frequency. The equation relevant to single-frequency diffuse-field excitation is the single-mode Eq. (6.35). The power dissipated by vibration at resonance in the fundamental mode is

$$\overline{P_d} = \eta_{int}^1 \omega_1 \overline{E}_1 \tag{6.60}$$

$$= \frac{4\pi \overline{p_0^2} c}{\rho_0 \omega_1^2} \frac{R_{rad}^1 R_{int}^1}{(R_{rad}^1 + R_{int}^1)^2} \tag{6.61}$$

and

$$\alpha = \frac{\overline{P_d}/A}{\overline{p_0^2}/4\rho_0 c} = \frac{16\pi c^2}{A\omega_1^2} \frac{R_{rad}^1 R_{int}^1}{(R_{rad}^1 + R_{int}^1)^2} \tag{6.62}$$

The absorption coefficient is *maximised* by making $R_{rad}^1 = R_{int}^1$; then

$$\alpha_{max} = 4\pi c^2 / A\omega_1^2 = \lambda_1^2 / \pi A \tag{6.63}$$

This result is analogous to that for a Helmholtz resonator in a wall in that the maximum absorption is obtained by matching the radiation and internal resistances; also the maximum absorption $(\alpha_{max} A)$ m^2 is independent of panel or neck area. It is the necessity of matching the two resistances to achieve optimum performance that gives membrane absorbers their characteristic single low-frequency peak, corresponding to fundamental-mode resonance. As discussed in Chapter 3, the fundamental mode of an edge supported panel has a much higher radiation efficiency than any of the higher order modes within the next few octaves above f_1 because no volume cancellation occurs. The radiation resistance corresponds to that of a baffled piston of equivalent volume velocity at low ka and approximates to

$$R_{rad}^1 = \rho_0 A^2 \omega_1^2 / 8\pi c \tag{6.64}$$

Optimum absorption at resonance is obtained by equating R^1_{int} to this value. The corresponding mechanical loss factor is given by $\eta^1_{int} = \rho_0 A^2 \omega_1 / 2\pi c M$.

Problems

6.1 Why is the field scattered by a rigid circular cylinder located in a plane wave sound field propagating in a direction normal to the cylinder axis not exactly equal to that produced by uniform transverse oscillation of the velocity equal to that of the plane wave particle velocity at the position of the axis of the cylinder, but in its absence? Why does the parameter ka influence the error incurred by equating the two fields?

6.2 Under what conditions would you expect the acoustically induced response of the fundamental mode of a flat panel to be rather insensitive to the direction of incidence of a plane sound wave? Why is the fundamental mode unique in this respect?

6.3 Would you expect an array of small, independently mounted, flexible panels to scatter an incident plane wave more 'diffusely' than a single panel of the same area. Explain the physical basis of your answer.

6.4 A flat plate steel structure is excited into vibration in a 200 m^3 reverberation chamber by random mechanical excitation in the one-third octave band centred on 500 Hz. The spatial-average mean-square acceleration is measured to be 10^6 m^{2}s^{-4}, the average sound pressure level in the chamber is 106 dB and its reverberation time in the 500 Hz band is 4 s. The area of the plate is 2 m^2 and its thickness is 3 mm. Estimate the average mean-square velocity of response of the structure to a 500-Hz one-third-octave band random sound field of 100 dB average sound-pressure level produced in the chamber by loudspeakers. Assume that the average internal loss factor of the structure in this band is 5×10^{-3}. Equations (3.25) and (6.37) apply.

6.5 It is desired to make an estimate of the rate at which a lightweight sheet roof of a factory absorbs sound when a large number of different types of machine operate simultaneously, so that an effective roof absorption coefficient can be approximately determined. Assuming that you have access to the roof via scaffolding, suggest how this objective might be achieved. Hint: In addition to the appropriate transducers and measurement equipment, you are provided with a large hammer and information concerning the material and thickness of the roof.

6.6 A nuclear reactor designer is concerned that the vibrational stresses in the steel casing of a heat exchanger may be unacceptably high because of acoustic excitation by the sound field generated by a circulator that drives the coolant gas through the exchanger. The coolant is helium gas, which is at an average temperature of 800°C, and a static pressure of 20 bar in the heat exchanger. Tests are made on a heat exchanger–circulator combination in atmospheric air at an average pressure of 1.3 bar and an average temperature of 45°C. Measurements are made of sound pressures in the heat exchanger and vibration strains in the casing.

Assuming that the sound fields in both cases can be considered to be broadband and diffuse, use the appropriate response equation to assess whether the results of such a test could reasonably be extrapolated to operating conditions. Indicate specifically, in qualitative terms, where the major uncertainties are expected to exist.

6.7 Assuming that the radiation loss factor of a structure is much smaller than the internal loss factor, show that the sound power dissipated by the structure when excited by a broadband acoustic field is nearly independent of the actual internal loss factor. Is this true for pure tone excitation at the resonance frequency of one of the structural modes? What are the implications for the design of plate-like sound absorbers for application in building acoustics?

7 Acoustic Coupling between Structures and Enclosed Volumes of Fluid

7.1 Practical Importance of the Problem

In our earlier considerations of sound radiation by vibrating structures and the associated fluid-loading effects, we made the implicit assumption that the geometry of the boundaries of the fluid were such that sound waves could be radiated away to infinity and were not reflected back on the surface of the structure to produce added fluid loading. There are, however, many systems of practical interest in which a structure is in contact with a fluid that is contained within a finite volume by physical boundaries that may wholly or only partly comprise the surface of the structure under consideration. The most significant difference between the acoustic behaviour of fluid contained within physical boundaries and that of unconstrained fluid is the existence in the former of natural modes of vibration and associated natural frequencies. As in solid structures, acoustic modes arise from interference between intersecting waves, and the natural frequencies are associated with the correspondence between the spatial characteristics of the interference pattern and the geometry of the physical boundaries of the fluid volume. An enclosed fluid exhibits resonant acoustic behaviour, the effects of which are to produce a strongly frequency-dependent response to vibration by contiguous structures, together with fluid-loading effects that also exhibit strong dependence upon frequency. Practical examples of interest include machinery noise control enclosures, double-leaf partitions, vehicle cabin spaces, tympanic and bodied musical instruments, loudspeaker enclosures and fluid transport ducts. The skin drum is an example of a system in which solid–fluid interaction is fundamental to its vibrational behaviour.

Where the influence of the fluid loading has a significant effect on the vibration of the enclosing structure, the problem of analysis of the vibrational behaviour of the resulting coupled system is rather involved, except in certain rather idealised cases, such as the one-dimensional case analysed in the following section. It is, however, by no means invariably the case that the presence of an enclosed fluid greatly changes the structural vibration characteristics: the 'art of the game' is to be able to judge whether a fully coupled vibration analysis is necessary or not. To some extent the necessity for such analysis depends upon the objective sought. For example, it is known that the sound power radiated by a vibrating structure into an enclosed volume of gas is often largely insensitive to the detailed form of fluid behaviour and its influence on structural vibration, provided that the bandwidth of excitation is broad and that the power is evaluated in frequency bands encompassing more than about five natural frequencies of the coupled system. In fact, it may often be assumed that the fluid exhibits no resonant behaviour at all. However, if discrete frequency response of the structure and/or fluid is required, it will probably be necessary to account for resonant behaviour of the coupled system. Clearly, enclosed volumes of liquid exert greater reaction forces on containment structures than gases.

The chapter opens with an analysis of a very simple one-dimensional system that exhibits the characteristic features of the interaction between elastic structures and enclosed volume of fluids. This is followed by a section that examines the application of the Kirchhoff–Helmholtz integral equation, introduced in Section 3.13, to the 'interior' problem in which a volume of fluid is enclosed within boundaries of various types. Analytical expressions for fluid–structure interaction are then derived and various physical features are explained. The next section models the interaction process in energetic terms and introduces the Statistical Energy Analysis formulation. The last two sections analyse the interaction between structures and fluid volumes that take the form of waveguides.

7.2 A Simple Case of Fluid–Structure Interaction

The free-vibration equations of strongly coupled structure-bounded fluid systems are not usually amenable to analytic solution in closed form, and numerical methods are now commonly applied to such problems, for which further details

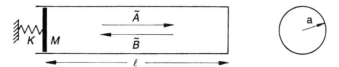

Fig. 7.1.   Spring-mounted piston in a closed tube.

will be found in Chapter 8. However, it is possible to solve the one-dimensional coupled problem rather easily, and the solution reveals features common to more complex systems. We consider the one-dimensional model of a spring-mounted piston sliding in a tube terminated by a rigid closure (Fig. 7.1).

The acoustic impedance (pressure/volume velocity) at the right-hand face of the piston is obtained by considering acoustic waves travelling to the right and left with amplitudes $\tilde{A}$ and $\tilde{B}$, respectively. A force $\tilde{F}e^{j\omega t}$ acts on the piston, and fluid-loading effects on the left-hand face of the piston are ignored. The acoustic impedance of the column of fluid is given by

$$\tilde{Z}_a = \frac{\rho_0 c}{\pi a^2}\left[\frac{\tilde{A} + \tilde{B}}{\tilde{A} - \tilde{B}}\right] \tag{7.1}$$

Application of the boundary conditions at the closed end of the tube gives

$$\tilde{B} = \tilde{A}\exp(-2jkl) \tag{7.2}$$

Hence the acoustic impedance presented to the piston is given by

$$\tilde{Z}_a = -(j\rho_0 c/\pi a^2)\cot kl \tag{7.3}$$

The corresponding form of impedance of the *in-vacuo* piston is

$$\tilde{Z}_p = j(\omega M - K/\omega)/(\pi a^2)^2 \tag{7.4}$$

Note that the acoustic impedance can represent either inertial or elastic fluid loading according to the sign of cot kl. When $kl = n\pi$, $\tilde{Z}_a$ is infinite, and the piston cannot move. When $kl = (2n-1)/2$, $\tilde{Z}_a$ is zero, and the fluid does not influence the motion of the piston. The former conditions correspond to the acoustic natural frequencies of the fluid in a tube closed at both ends. The latter correspond to the acoustic natural frequencies of the fluid in a tube closed at one end and having a pressure release condition at the other: they are also the acoustic resonance frequencies of the fluid in the tube closed at one end and driven by the piston at the other. The impedances $\tilde{Z}_a$ and $\tilde{Z}_p$ act in series because the two components share the same volume velocity; hence they add. The equation of simple harmonic motion of the piston may now be written in terms of its displacement ξ as

$$\widetilde{F} = j\omega\widetilde{\xi}(\pi a^2)^2(\widetilde{Z}_a + \widetilde{Z}_p) \qquad (7.5)$$

The natural frequencies ω_n of the coupled system correspond to the condition $\widetilde{F} = 0$: this corresponds to zero total impedance. Hence

$$\cot[(\omega_n/\omega_0)k_0 l] = (M\omega_0/\rho_0 c\pi a^2)(\omega_n/\omega_0)[1 - (\omega_0/\omega_n)^2] \qquad (7.6)$$

where $\omega_0^2 = K/M$ and $k_0 = \omega_0/c$. The term $\rho_0 c\pi a^2/M\omega_0$ may be considered to be a form of non-dimensional fluid-loading parameter, since it represents the ratio of the characteristic specific acoustic impedance of the fluid to the magnitude of the impedance of the piston mass per unit area at the *in-vacuo* natural frequency: we shall denote it by β. The non-dimensional tube length is represented by the parameter $k_0 l$.

Solutions to Eq. (7.6) may be obtained numerically or by graphical means. In Fig. 7.2, $\cot(kl)$ and $\beta^{-1}(kl/k_0 l)[1 - (k_0 l/kl)^2]$ are plotted against kl for various ranges of $k_0 l$ and β. The intersections represent coupled system natural frequencies given by $k_n l = \omega_n l/c$. A number of special cases may be identified. Suppose first that the *in-vacuo* natural frequency of the mass-spring system is well below that at which $kl = \pi/2$, or $l = \lambda/4$, the lowest acoustic resonance frequency of the fluid when driven by the piston. In this case $\omega_0 \ll \pi c/2l$, and $\cot kl$ can be approximated by $(kl)^{-l}$. Then

$$\omega_1^2 \approx \omega_0^2(1 + \beta/k_0 l) \qquad (7.7)$$

and the effective stiffness of the piston system has been increased by a factor $1 + \beta/k_0 l$.

The physical explanation of this result is the fact that at frequencies where the acoustic wavelength is much greater than the tube length, the fluid reaction is produced by almost pure bulk compression. The pressure change produced by a volume displacement $\xi\pi a^2$ is equal to $\rho_0 c^2\xi\pi a^2/\pi a^2 l = \rho_0 c^2\xi/l$. Hence the effective fluid stiffness is $\pi a^2\rho_0 c^2/l$. The term $\beta/k_0 l$ can be written as $\rho_0 c^2\pi a^2/M\omega_0^2 l = \rho_0 c^2\pi a^2/Kl$, giving a fractional increase of stiffness as indicated. The effect of the fluid on the natural frequency clearly increases as the piston mass per unit area decreases, as the characteristic fluid impedance increases and as the length of the tube decreases.

As another special case we may assume that a coupled system natural frequency lies far above ω_0, and that β is considerably less than unity.

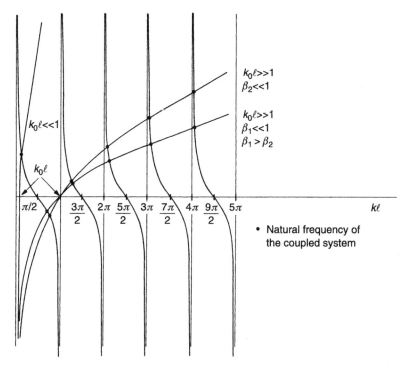

Fig. 7.2.   Plot of $\cot kl$ and $\beta^{-1}(kl/k_0 l)[1 - (k_0 l/kl)^2]$ vs. kl.

In this case

$$\tan[(\omega_n/\omega_0)k_0 l] \approx (\omega_0/\omega_n)\beta \ll 1$$

and

$$(\omega_n/\omega_0)k_0 l \to n\pi \tag{7.8}$$

where n is any integer. Hence

$$\omega_n \approx n\pi c/l \tag{7.9}$$

These frequencies correspond to the acoustic natural frequencies of the tube when closed at both ends. This makes physical sense since the impedance of the piston is dominated by the inertial term $j\omega M$, which is large at high frequencies and can combine with the high negative acoustic impedance (stiffness) of the fluid in the tube to produce zero total impedance. The single natural frequency of the *in-vacuo* system has been joined by an infinite host of colleagues due to the extended form of the elastic-fluid system.

Intermediate cases have to be evaluated for specific combinations of β and $k_0 l$. However, this example shows that fluid in a bounded space can have a profound influence on the values and *numbers* of natural frequencies of a coupled structural system. The corresponding three-dimensional case is analysed in Sections 7.5 and 7.6. Such dramatic fluid stiffness effects as we have encountered here are not very common in practice, but have been shown to influence the resonances of windows coupled to small rooms and are, of course, important in low-frequency cabinet loudspeaker performance. They are also at work in double-leaf partitions, as we have seen in Chapter 5, and under floating floors in buildings. (As an experiment, try measuring the acoustic impedance of a loudspeaker fitted to one end of a tube that is open or closed, at the other end.)

If the tube in our example were terminated by an impedance that contained a real (resistive) term, the free vibration of the piston would be damped, even if the piston suspension were not itself damped. Analysis of the power radiated by a piston into an acoustically 'lossy' tube shows that it is very much influenced by the resonant behaviour of the system. The influence of waves reflected from a tube termination, or impedance discontinuity, can have a profound influence on the acoustic source and is central to the operation of many musical instruments, including pipe organs and wind instruments. In these cases, analysis of the coupled source-tube system is far more complicated than in the case of the piston, because the aerodynamic source characteristics are essentially non-linear.

7.3   Harmonic Sound Fields in an Enclosed Volume of Fluid

The Kirchhoff–Helmholtz integral equation for harmonic time-dependence that relates sound pressure to conditions on closed and open surfaces in an unbounded fluid was introduced in Section 3.13. This section presents a general analysis of the acoustic response of a fluid within a closed cavity of arbitrary shape to acoustic sources distributed within the volume and to the enclosure boundary conditions. As shown in Fig. 7.3, the enclosure can be defined by three different types of boundary surface. On surface S_v the velocity component $\tilde{v}_v(\mathbf{r}_v)$ normal to the surface is prescribed so that the fluid particle velocity in direction normal to the radiating surface is given by $\tilde{v}_n(\mathbf{r}_v) = \tilde{v}_v(\mathbf{r}_v)$ (Neumann boundary condition). Surface S_s represents that part of the radiating object which is modelled as a flexible structure whose transverse vibration $\tilde{v}_s(\mathbf{r}_s)$ is coupled to the acoustic field so that $\tilde{v}_n(\mathbf{r}_s) = \tilde{v}_s(\mathbf{r}_s)$. On surface S_p the pressure $\tilde{p}_p(\mathbf{r}_p)$ is prescribed so that

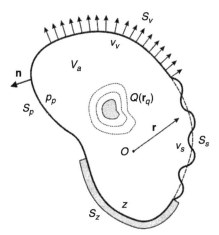

Fig. 7.3.   Interior acoustic system with volume-distributed source and three types of boundary condition.

$\tilde{p}(\mathbf{r}_p) = \tilde{p}_p(\mathbf{r}_p)$ (Dirichlet boundary condition). Surface S_z is characterised by normal specific acoustic impedance $\tilde{z}(\mathbf{r}_z)$, so that $\tilde{p}(\mathbf{r}_z)/\tilde{v}_n(\mathbf{r}_z) = \tilde{z}(\mathbf{r}_z)$. The fluid momentum equation relates the normal particle velocity to the normal gradient of sound pressure so that on the boundary surface $\tilde{v}_n(\mathbf{r}_{S_a}) = j/\omega\rho_0(\partial\tilde{p}/\partial n)_{\mathbf{r}=\mathbf{r}_{S_a}}$, where $\mathbf{r}_{S_a}$ gives the positions on the enclosure surface S_a and n is the local coordinate normal to S_a. Note that n is directed outwards.

When distributed harmonic acoustic sources represented by the complex function $\widetilde{Q}(\mathbf{r}_s)$ operate within an enclosure, the complex amplitude of the pressure field $\tilde{p}(\mathbf{r})$ in the cavity may be considered as the superposition of the direct sound pressure field $\tilde{p}_d(\mathbf{r})$ that is due to the distributed acoustic sources radiating into *free field* and the sound pressure field $\tilde{p}_f(\mathbf{r})$ that is imposed by the boundary conditions on S_a, as illustrated in Fig. 7.4:

$$\tilde{p}(\mathbf{r}) = \tilde{p}_d(\mathbf{r}) + \tilde{p}_f(\mathbf{r}) \tag{7.10}$$

The sound radiation into free field by distributed harmonic sources is expressed by the integral of the response to point sources $\widetilde{Q}(\mathbf{r}_q)$

$$\tilde{p}_d(\mathbf{r}) = \int_{V_q} \widetilde{G}(\mathbf{r}, \mathbf{r}_q)\widetilde{Q}(\mathbf{r}_q)\, dV_q \tag{7.11}$$

where, according to the formulation introduced in Section 3.13, $\widetilde{Q}(\mathbf{r}) = j\rho_0\omega\tilde{q}(\mathbf{r}) - \nabla\tilde{\mathbf{f}}(\mathbf{r})$ represents arbitrary distributions within the fluid of rates of

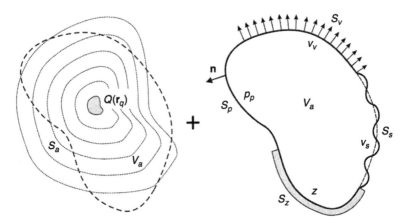

Fig. 7.4. Breakdown of the bounded acoustic problem into that of free-field radiation by the volume-distributed sources plus the field imposed by the boundary conditions.

volume displacement q per unit volume (monopole) and distributions of external forces per unit volume $\mathbf{f}$ (dipole). $\widetilde{G}(\mathbf{r}, \mathbf{r}_0)$ is the free-space Green's function

$$\widetilde{G}(\mathbf{r}, \mathbf{r}_0, \omega) = \frac{e^{-jk|\mathbf{r}-\mathbf{r}_0|}}{4\pi|\mathbf{r} - \mathbf{r}_0|} \tag{7.12}$$

which satisfies the inhomogeneous Helmholtz equation (3.164) with a source of unit strength per unit volume concentrated at $\mathbf{r}_0$

$$\nabla^2\widetilde{G}(\mathbf{r}, \mathbf{r}_0, \omega) + k^2\widetilde{G}(\mathbf{r}, \mathbf{r}_0, \omega) = -\delta(\mathbf{r} - \mathbf{r}_0) \tag{7.13}$$

together with the Sommerfeld radiation condition in Eq. (3.167) in the form

$$\lim_{|\mathbf{r}-\mathbf{r}_0|\to\infty} |\mathbf{r} - \mathbf{r}_0| \left(\frac{\partial\widetilde{G}(\mathbf{r}, \mathbf{r}_0)}{\partial|\mathbf{r} - \mathbf{r}_0|} + jk\widetilde{G}(\mathbf{r}, \mathbf{r}_0) \right) = 0 \tag{7.14}$$

The component of the sound pressure field $p_f(\mathbf{r})$ that is produced by the presence of the fluid boundaries is then derived from the homogeneous Helmholtz equation

$$\nabla^2\tilde{p}_f(\mathbf{r}) + k^2\tilde{p}_f(\mathbf{r}) = 0 \tag{7.15}$$

subject to the appropriate boundary conditions.

As for the unbounded problem considered in Section 3.13, this problem can be solved by the application of Green's second identity given by Eq. (3.170) by assuming ψ is the free-space Green's function $\widetilde{G}(\mathbf{r}, \mathbf{r}_a)$ in Eq. (7.12) and φ is the sound pressure function $p(\mathbf{r}_a)$ which satisfies the homogeneous Helmholtz Eq. (7.15) and boundary conditions. In this case the acoustic volume V_a is enclosed by the surface $S_a = S_v \cup S_s \cup S_p \cup S_z$. Also, as in the formulation presented in Section 3.13 for the unbounded problem, the position $\mathbf{r}$ in Fig. 7.5(a) is bounded by a spherical surface S_{R1} of radius R_1 in order to avoid the singularity problem at $\mathbf{r} = \mathbf{r}_a$ of the Green's function $\widetilde{G}(\mathbf{r}, \mathbf{r}_a, \omega)$ in Eq. (7.12). Application of Green's second identity to the volume V_a bounded by the surfaces S_a and S_{R1} leads to

$$\int_{S_a + S_{R1}} \left(\tilde{p}(\mathbf{r}_a) \frac{\partial \widetilde{G}(\mathbf{r}, \mathbf{r}_a)}{\partial n} - \widetilde{G}(\mathbf{r}, \mathbf{r}_a) \frac{\partial \tilde{p}(\mathbf{r}_a)}{\partial n} \right) dS = 0 \qquad (7.16)$$

The volume integral on the right hand side of Eq. (3.170) is equal to zero since both $\varphi = \tilde{p}(\mathbf{r}_a)$ and $\psi = \widetilde{G}(\mathbf{r}, \mathbf{r}_a)$ satisfy Eq. (7.15) in V_a and therefore both $\varphi \nabla^2 \psi$ and $\psi \nabla^2 \varphi$ are equal to $-k^2 \widetilde{G}(\mathbf{r}, \mathbf{r}_a) \tilde{p}(\mathbf{r}_a)$.

Consider first the case shown in Fig. 7.5(a), with the closed surface S_a and one spherical surface, S_{R1} with radius R_1 and centre position defined by $\mathbf{r}$. As shown by Eqs. (3.172)–(3.176), the integration over the surfaces S_a and S_{R1} gives

$$\int_{S_a} \left(\tilde{p}(\mathbf{r}_a) \frac{\partial \widetilde{G}(\mathbf{r}, \mathbf{r}_a)}{\partial n} - \widetilde{G}(\mathbf{r}, \mathbf{r}_a) \frac{\partial p(\mathbf{r}_a)}{\partial n} \right) dS_a = -\tilde{p}(\mathbf{r}) \qquad (7.17)$$

Consider, now, the second case where, as shown in Fig. 7.5(b), if the point at position $\mathbf{r}$ is located on the closed surface S_a so that, provided the surface is smooth, as $R_1 \to 0$ the surface S_{R1} becomes a hemisphere. Therefore, the

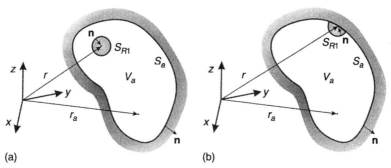

Fig. 7.5.   Schematic of an acoustic domain V_a bounded by an outer surface S_a and a spherical surface S_{R1} (a) placed inside and (b) bisected by the surface S_a.

normal direction to S_a is uniquely defined at all positions on the surface and, as shown by Eqs. (3.177) and (3.178), the integral in Eq. (7.16) gives

$$\int_{S_a} \left(\tilde{p}(\mathbf{r}_a) \frac{\partial \tilde{G}(\mathbf{r}, \mathbf{r}_a)}{\partial n} - \tilde{G}(\mathbf{r}, \mathbf{r}_a) \frac{\partial p(\mathbf{r}_a)}{\partial n} \right) dS_a = -\frac{1}{2} \tilde{p}(\mathbf{r}) \qquad (7.18)$$

For the third case, which is shown in Fig. 7.6(a), the point at position $\mathbf{r}$ is not located in the acoustic domain V_a and thus, in the limiting case for $R_1 \rightarrow 0$, the spherical surface S_{R1} is also located outside the acoustic domain V_a and $\tilde{p}(\mathbf{r}) = 0$ for any position $\mathbf{r}$ outside the acoustic domain V_a. Thus

$$\int_{S_a} \left(\tilde{p}(\mathbf{r}_a) \frac{\partial \tilde{G}(\mathbf{r}, \mathbf{r}_a)}{\partial n} - \tilde{G}(\mathbf{r}, \mathbf{r}_a) \frac{\partial p(\mathbf{r}_a)}{\partial n} \right) dS_a = 0 \qquad (7.19)$$

Now, as done for the unbounded problem in Section 3.13, the three cases discussed above can be summarised in one formula which, using Eq. (3.163), gives the Kirchhoff–Helmholtz integral equation:

$$c(\mathbf{r}) \tilde{p}(\mathbf{r}) = \int_{S_a} \left(\tilde{p}(\mathbf{r}_a) \frac{\partial \tilde{G}(\mathbf{r}, \mathbf{r}_a)}{\partial n} + j\rho_0 \omega \tilde{G}(\mathbf{r}, \mathbf{r}_a) \tilde{v}_n(\mathbf{r}_a) \right) dS_a \qquad (7.20a)$$

where

$$c(\mathbf{r}) = \begin{cases} -1 & \mathbf{r} \in V_a \\ 0 & \mathbf{r} \notin V_a \\ -1/2 & \mathbf{r} \in S_a \end{cases} \qquad (7.20b)$$

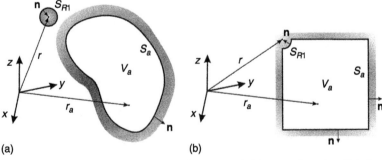

Fig. 7.6. Schematic of an acoustic domain V_a bounded by a surface S_a which is either (a) smooth or (b) with protrusions.

If, as shown in Fig. 7.6(b), the point defined by $\mathbf{r}$ is located at a protrusion of the closed surface S_a so that the normal to the surface itself cannot be unambiguously defined, then

$$c(\mathbf{r}) = 1 + \frac{1}{4\pi} \int_{S_a} \frac{\partial}{\partial n} \left(\frac{1}{|\mathbf{r} - \mathbf{r}_a|} \right) dS_a \qquad (7.20c)$$

In summary, the direct boundary integral formulation for time-harmonic interior problems relates the acoustic pressure at any point in an enclosed fluid that satisfies the homogeneous Helmholtz equation to the distribution of pressure and normal velocity on the boundary surface of the acoustic domain. The more general equation for arbitrary time dependence involves the surface distributions of pressure and normal acceleration (Pierce, 1989). When acoustic sources operate within the volume of fluid, the total pressure field is given by the superposition of the free-field pressure, which is the response to the acoustic sources in the absence of the boundary surface, and the scattered field, which is governed by the corresponding Helmholtz equation. The total sound pressure field is thus given by the integral equation

$$\tilde{p}(\mathbf{r}) = \frac{1}{c(\mathbf{r})} \int_{S_a} \left(\tilde{p}(\mathbf{r}_a) \frac{\partial \tilde{G}(\mathbf{r}, \mathbf{r}_a)}{\partial n} + j\rho_0 \omega \tilde{G}(\mathbf{r}, \mathbf{r}_a) \tilde{v}_n(\mathbf{r}_a) \right) dS_a$$
$$+ \int_{V_q} \tilde{Q}(\mathbf{r}_q) \tilde{G}(\mathbf{r}, \mathbf{r}_q) dV_q \qquad (7.21)$$

which is known as the 'direct boundary integral' formulation. The solution of this equation requires a specification of the sound pressure and normal particle velocity on the bounding surface. This can be obtained by a two step procedure. First, Eq. (7.21) is evaluated on the boundary surface assuming $c(\mathbf{r}) = -1/2$ and the given sound pressure or normal particle velocity on the boundary surface; second, Eq. (7.21) is evaluated within the volume V_a assuming $c(\mathbf{r}) = -1$ and substituting in the integral the sound pressure or normal particle velocity derived in the first step. As discussed in Section 3.13, the first step integral analysis is affected by a singularity problem when the Green's function $\tilde{G}(\mathbf{r}, \mathbf{r}_a, \omega)$ is estimated at collocated positions. However, for smooth surfaces the singularity is weak and thus the Kirchhoff–Helmholtz integral equation converges (Wu, 2000). If the enclosure has flexible walls, a fully coupled structural-acoustic analysis is carried out in which the compatibility condition is imposed between the transverse velocity of the flexible structure and normal particle velocity of the acoustic field on the boundary surface.

7.4   Sound Field in a Closed Space with Rigid Surfaces

In a volume of fluid completely bounded by rigid surfaces it is relatively easy to appreciate the physical interpretation of the Green's function. If a small source of harmonically fluctuating volume, a point monopole, is placed in a fluid that is contained within boundaries, then, as we have seen, the pressure is a solution to Eq. (7.13) with a suitable form of mass fluctuation source term on the right-hand side. If the boundaries are rigid, then $\partial \tilde{p}/\partial n = 0$ on the boundaries, and therefore a form of Green's function that satisfies (7.13) and the boundary condition $\partial \tilde{G}/\partial n$ on the walls is acceptable. In the absence of sources in the fluid, the acoustic pressure satisfies the Helmholtz Eq. (7.15) subject to the rigid-wall boundary condition. Let a solution to Eq. (7.15) be written as

$$p(\mathbf{r})\exp(j\omega_n t) = \tilde{A}_n \psi_n(\mathbf{r})\exp(j\omega_n t) \tag{7.22}$$

where ψ_n is the acoustic pressure mode shape corresponding to the natural frequency ω_n of the rigid-walled space. Hence

$$\nabla^2 \psi_n(\mathbf{r}) + k^2 \psi_n(\mathbf{r}) = 0 \tag{7.23}$$

where $k_n = \omega_n/c$. These functions ψ_n satisfy the condition $\partial \psi_n/\partial n = 0$ on the walls and therefore are candidates for incorporation in a Green's function satisfying the same condition. Hence it is reasonable to try to express $\tilde{G}$ as

$$\tilde{G}(\mathbf{r}, \mathbf{r}_0, \omega) = \sum_{n=0}^{\infty} \tilde{B}_n \psi_n(\mathbf{r}) \tag{7.24}$$

and

$$\nabla^2 \tilde{G} = \sum_{n=0}^{\infty} \tilde{B}_n \nabla^2 \psi_n(\mathbf{r}) \tag{7.25}$$

From Eq. (7.23)

$$\nabla^2 \psi_n(\mathbf{r}) = -k_n^2 \psi_n(\mathbf{r}) \tag{7.26}$$

giving

$$\nabla^2 \tilde{G}(\mathbf{r}, \mathbf{r}_0, \omega) = -\sum_n k_n^2 \tilde{B}_n \psi_n(\mathbf{r}) \tag{7.27}$$

Equation (7.13) can now be written as

$$-\sum_n k_n^2 \tilde{B}_n \psi_n(\mathbf{r}) + k^2 \sum_n \tilde{B}_n \psi_n(\mathbf{r}) = -\delta(\mathbf{r} - \mathbf{r}_0) \qquad (7.28)$$

It is a condition of natural modes of closed, non-dissipative, elastic systems that are mutually orthogonal. Since the mean fluid density is assumed to be uniform, and we are assuming a lossless fluid,

$$\int_V \psi_m(\mathbf{r})\psi_n(\mathbf{r})dV = \begin{cases} 0, & m \neq n \\ \Lambda_n, & m = n \end{cases}$$

where $\Lambda_n = \int_V \psi_n^2(\mathbf{r})dV$. If Eq. (7.28) is multiplied by $\psi_m(\mathbf{r})$ and integrated over the fluid volume, the summations disappear by virtue of the orthogonality condition to yield

$$\tilde{B}_n = \psi_n(\mathbf{r}_0)/\Lambda_n(k^2 - k_n^2) \qquad (7.29)$$

Hence

$$\tilde{G}(\mathbf{r}, \mathbf{r}_0, \omega) = \sum \frac{\psi_n(\mathbf{r})\psi_n(\mathbf{r}_0)}{\Lambda_n(k_n^2 - k^2)} \qquad (7.30)$$

This function is reciprocal in $\mathbf{r}$ and $\mathbf{r}_0$ as it should be, and $\partial \tilde{G}/\partial n = 0$ on the walls because $\partial \psi_n/\partial n = 0$ there also.

Now, if it is desired to evaluate the pressure field created in a fluid enclosed in a rigid boundary by a distribution of sources of volume acceleration within the volume, the volume integral of Eq. (7.21) may be used, with $\tilde{G}$ expressed by Eq. (7.30): the surface integral disappears. In the not so common case of a volume enclosed by boundaries on which the pressure vanishes, Eq. (7.30) takes the same form, but the mode shape functions ψ_n and frequencies ω_n are those appropriate to this boundary condition.

7.5 Interaction Analysis by Green's Function

We are primarily concerned with fluid–structure interaction, and therefore we now proceed to apply Eqs. (7.21) and (7.30) to the case of a fluid volume bounded by a thin-plate or shell structure that is excited by mechanical forces.

The equation of motion of the structure is

$$L[w(\mathbf{r}_s)] + m(\mathbf{r}_s)\partial^2 w(\mathbf{r}_s)/\partial t^2 = f(\mathbf{r}_s) + p(\mathbf{r}_s) \qquad (7.31)$$

where w is the normal surface displacement of the structure, L is the operator governing the elastic forces in the structure, f is the distribution of mechanically applied force per unit area and p is the distribution of surface pressures: f and w are directed outwards from the fluid volume, and fluid pressures external to the enclosed volume are ignored for simplicity. The surface pressure distribution is given by Eq. (7.18) for harmonic excitation,

$$L[\tilde{w}(\mathbf{r}_s)] - \omega^2 m(\mathbf{r}_s)\tilde{w}(\mathbf{r}_s) = \tilde{f}(\mathbf{r}_s) + 2\omega^2 \rho_0 \int_{S'} \tilde{w}(\mathbf{r}'_s)\tilde{G}(\mathbf{r}_s, \mathbf{r}'_s, \omega)dS' \quad (7.32)$$

where $\tilde{G}(\mathbf{r}_s, \mathbf{r}'_s, \omega) = \sum_n [\psi_n(\mathbf{r}_s)\psi_n(\mathbf{r}'_s)/\Lambda_n(k_n^2 - k^2)]$. Hence, if the structural motion is expressed as a summation over the displacements in the *in-vacuo* natural modes as $\tilde{w}(\mathbf{r}_s) = \sum_p \tilde{w}_p \phi_p(\mathbf{r}_s)$, in which each mode satisfies

$$L[\phi_p(\mathbf{r}_s)] - \omega_p^2 m(\mathbf{r}_s)\phi_p(\mathbf{r}_s) = 0 \qquad (7.33)$$

Equation (7.32) becomes

$$m(\mathbf{r}_s)\left[\sum_p \omega_p^2 \tilde{w}_p \phi_p(\mathbf{r}_s) - \omega^2 \sum_p \tilde{w}_p \phi_p(\mathbf{r}_s)\right]$$

$$= f(\mathbf{r}_s) + 2\omega^2 \rho_0 \int_{S'} \left[\sum_q \tilde{w}_q \phi_q(\mathbf{r}'_s)\sum_n [\psi_n(\mathbf{r}_s)\psi_n(\mathbf{r}'_s)/\Lambda_n(k_n^2 - k^2)]\right]dS'$$

$$(7.34)$$

Multiplication of Eq. (7.34) by $\phi_r(\mathbf{r}_s)$ and integration over the surface of the structure yields

$$\tilde{w}_p \Lambda_p(\omega_p^2 - \omega^2) - 2\omega^2 \rho_0 \tilde{w}_p \sum_n \alpha_{ppn}$$

$$= \int_S f(\mathbf{r}_s)\phi_p(\mathbf{r}_s)dS + 2\omega^2 \rho_0 \sum_{q\neq p}\sum_n \tilde{w}_q \alpha_{pqn}$$

$$(7.35)$$

where

$$\alpha_{ppn} = \frac{\left[\int_S \psi_n(\mathbf{r}_s)\phi_p(\mathbf{r}_s)dS\right]^2}{\Lambda_n(k_n^2 - k^2)},$$

$$\alpha_{pqn} = \frac{\left[\int_S \psi_n(\mathbf{r}_s)\phi_p(\mathbf{r}_s)dS\right]\left[\int_{S'} \psi_n(\mathbf{r}_s')\phi_q(\mathbf{r}_s')dS'\right]}{\Lambda_n(k_n^2 - k^2)}$$

and $\Lambda_p = \int_S \phi_p^2(\mathbf{r}_s)m(\mathbf{r}_s)dS$, which is the generalised modal mass. In Eq. (7.35) the modal self (direct)-fluid-loading terms α_{ppn} have been separated from the modal mutual (cross)-fluid-loading terms α_{pqn}. It is seen from the presence of all the modal displacements $\tilde{w}_{q \neq p}$ in the equation for $\tilde{w}_p$ that the action of the fluid is to couple the *in-vacuo* structural modes. In fact, provided that the *in-vacuo* structural modes form an orthonormal set, such a closed elastic system does possess orthogonal natural modes and associated natural frequencies, but they are different from those of the uncoupled fluid and structural systems.

It is clear from the form of the denominators of α_{ppn} and α_{pqn} that the resonant behaviour of the fluid in the closed volume has a significant influence on the fluid loading. The physical nature of α_{ppn} can be appreciated by considering the term $-\omega^2 \rho_0 \tilde{w}_p / (k_n^2 - k^2)$. At frequencies well *below* the acoustic mode natural frequency ω_n, $k \ll k_n$, and this term represents *inertial* fluid loading, proportional to ω^2. At frequencies well *above* ω_n, $k \gg k_n$, and the term represents frequency-independent elastic *stiffness*, the numerator containing the fluid adiabatic bulk modulus $\rho_0 c^2$. The physical influence of α_{pqn} depends upon the sign of α_{pqn}, which varies with p, q and n; therefore the corresponding fluid loading may be mass-like or stiffness-like.

Equation (7.35) suggests infinite fluid loading (and zero structural response) at the acoustic natural frequencies of the rigid-walled enclosure ($k = k_n$). This condition corresponds to the complete rigidity of the fluid column discussed in Section 7.2 at frequencies where the length of the fluid column equals an integer number of half-wavelengths; that is, at the acoustic natural frequencies of the duct when terminated at each end by a rigid plug. In practice, the fluid loading is limited by dissipative mechanisms in the fluid and at its boundaries; this can be modelled in the Green's function of Eq. (7.30) by an *ad hoc* complex fluid bulk modulus which leads to a small imaginary term in the denominator. Similarly, it is customary to incorporate damping into the structural model by means of a complex elastic modulus. Although these dissipative mechanisms limit the fluid-loading effects, they do not change their basic characteristics. The effect of very high fluid loading on a structural mode due to resonance of a coupled acoustic mode is similar to that by a vibration neutraliser on structural vibration: modal motion at the acoustic resonance frequency is largely suppressed. If the structural and acoustic mode resonance frequencies coincide, the resulting coupled modes have one resonance frequency above the uncoupled value and one below: this is termed 'frequency splitting'. Near a natural frequency of the fluid volume with pressure release ($p = 0$) boundary conditions, the fluid loading produced by the associated acoustic mode disappears, but of course the contributions of other acoustic modes still exist.

As already mentioned, evaluation of the modes and responses of fully coupled system requires the application of numerical analysis procedures. Although these provide powerful means of solving specific problems, the results do not necessarily reveal the physics underlying the system behaviour. In generic studies of fluid–structure coupling that are based upon models having simple geometries and forms of structure, it is common to employ approximations: the most common one is to neglect the mutual (cross)-fluid-loading terms α_{pqn}. In many cases of reasonably large volumes of a gas, such as air, at static pressure close to atmospheric, it is generally unnecessary to account for fluid loading at all. Analysis proceeds in two separate stages: first, the *in-vacuo* structural response to mechanical excitation is evaluated; then Eq. (7.21) is used to estimate the pressure field produced by that given vibration. Fairly frequently only a relatively small proportion of the surface integral terms of the form $\int_S \psi_n(\mathbf{r}_s)\phi_p(\mathbf{r}_s)dS$ are significant, and further simplifications can be made. This feature is particularly useful in statistical analyses of fluid–structure interaction (see, for example, Fahy, 1970).

7.6 Modal-Interaction Model

There is an alternative formulation of the fluid–structure coupling problem to the integro–differential equation formulation of the interaction between structures and enclosed fluids based upon the Green's functions of the fluid. The essential difference from the foregoing analysis is that the fluid field is expressed directly in terms of its uncoupled modes and its behaviour is described by means of a differential equation for each mode. The fluid pressure on the surface of the structure is the agent by which the fluid influences structural motion, and the normal surface acceleration of the structure is the agent by which the structure influences the fluid field. Naturally, the response solutions correspond to those obtained by Green's function representation, but the form of the governing equations corresponds more directly to the coupled oscillator model that forms the basis of the probabilistic modelling of coupled systems embodied in Statistical Energy Analysis (SEA) that is introduced in Section 7.8.

We may write the inhomogeneous wave equation for pressure as

$$\nabla^2 p - (1/c^2)\partial^2 p/\partial t^2 = -\rho_0 \partial q/\partial t \tag{7.36}$$

where q represents the distribution of source volume velocity per unit volume. As discussed earlier, the right-hand side represents a rate of change of mass flux per unit volume. The vibration of a bounding surface may be represented as a volume velocity distribution q_b confined to an infinitesimally thin layer situated on a surface S_0 just inside a rigid boundary, by the use of a one-dimensional Dirac delta function in a coordinate normal to surface. Since we have already assumed the normal structural displacement w to be directed outwards from the fluid volume, we may write

$$q_b(\mathbf{r}) = -2(\partial w(\mathbf{r}_s)/\partial t)\delta(\xi - \xi_0) \tag{7.37}$$

Since δ is one dimensional, it has the dimensions L^{-1} and therefore $q(\mathbf{r})$ has the correct dimensions of volume velocity per unit volume. Hence

$$\nabla^2 p - (1/c^2)\partial^2 p/\partial t^2 = 2\rho_0[\partial^2 w(\mathbf{r}_s)/\partial t^2]\delta(\xi - \xi_0) - \rho_0 \partial q(\mathbf{r}_0)/\partial t \tag{7.38}$$

where $q(\mathbf{r}_0)$ is the distribution of sources other than those representing boundary motion. We now express the acoustic pressure as a sum of the pressure distributions in the acoustic modes of the fluid volume with rigid boundaries:

$$p(\mathbf{r}, t) = \sum_{n=0}^{\infty} p_n(t)\psi_n(\mathbf{r}) \tag{7.39}$$

Using Eqs. (7.23) and (7.39) in Eq. (7.38) we obtain

$$\sum_n -k_n^2 p_n(t)\psi_n(\mathbf{r}) - (1/c^2)\sum_n \ddot{p}_n(t)\psi_n(\mathbf{r})$$
$$= 2\rho_0[\partial^2 w(\mathbf{r}_s)/\partial t^2]\delta(\xi - \xi_0) - \rho_0 \partial q(\mathbf{r}_0)/\partial t \tag{7.40}$$

Multiplying by $\psi_m(\mathbf{r})$, integrating over the fluid volume and applying the orthogonality condition, Eq. (7.40) yields a differential equation for the modal-pressure coordinate $p_n(t)$:

$$\ddot{p}_n + \omega_n^2 p_n = -\frac{\rho_0 c^2}{\Lambda_n} \int_S \psi_n(\mathbf{r}_{s0})[\partial^2 w(\mathbf{r}_s)/\partial t^2]dS + (\rho_0 c^2/\Lambda_n)\dot{Q}_n \tag{7.41}$$

where Q_n is a generalised volume velocity source strength given by $\int_V q(\mathbf{r}_0)\psi_n(\mathbf{r}_0)dV$ and $\mathbf{r}_{s0}$ refers to the surface at an infinitesimally small distance from the surface of the structure: ω_n is the natural frequency of the acoustic mode with rigid boundaries. It should be noted that, because of a mathematical condition associated with infinite sums, termed Gibb's phenomenon, the summation of rigid-wall acoustic modes *does not converge to the correct boundary normal velocity*, but does converge correctly to the surface pressure, which is all that is needed for a correct formulation of the coupled equations.

Substitution into Eq. (7.31) of an expression for the structural displacement in terms of a summation over the *in-vacuo* normal modes,

$$w(\mathbf{r}_s t) = \sum_{p=1}^{\infty} w_p(t)\phi_p(\mathbf{r}_s) \tag{7.42}$$

together with the application of Eq. (7.33), followed by multiplication by $\phi_q(\mathbf{r}_s)$, and integration over the structural surface, yields the coupled modal equations of motion for the structure:

$$\ddot{w}_p + \omega_p^2 w_p = \frac{S}{\Lambda_p} \sum_p^{\infty} p_n C_{np} + F_p/\Lambda_p \tag{7.43}$$

where F_n is a generalised force given by $\int_S f(\mathbf{r}_s)\phi_p(\mathbf{r}_s)dS$ and $\Lambda_p = \int_S m(\mathbf{r}_s)\phi_p^2(\mathbf{r}_s)dS$. (Check the derivation of the equation yourself.) For the fluid,

$$\ddot{p}_n + \omega_n^2 p_n = -(\rho_0 c^2 S/\Lambda_n) \sum_p \ddot{w}_p C_{np} + (\rho_0 c^2/\Lambda_n)\dot{Q}_n \tag{7.44}$$

C_{np} is a dimensionless coupling coefficient given by the integral of the product of the structural and acoustic mode shape functions over the surface of the structure:

$$C_{np} = \frac{1}{S}\int_S \psi_n(\mathbf{r}_s)\phi_p(\mathbf{r}_s)dS \tag{7.45}$$

where S is the total surface area of the structure. Note that Sp_n has the dimensions of force and $S\ddot{w}_p$ the dimensions of volume acceleration, as comparison with the applied generalised force F_n and source strength Q_n shows they should. Equations (7.43) and (7.44) together actually represent a doubly infinite set of simultaneous differential equations; these sets have to be limited in order to obtain solutions.

These coupled equations of motion have been developed in terms of the acoustic pressure p because it is the dynamical quantity most commonly measured and most familiar to readers. However, the equations take on a more obvious and aesthetically satisfying symmetry if pressure is replaced by the acoustic-field velocity potential Φ. This is a scalar quantity in terms of which an irrotational fluid field can be completely described by its spatial distribution. Books on fluid mechanics, such as that by Duncan *et al.* (1970), present complete treatments of the velocity potential. It suffices here to state that the pressure and Cartesian components of fluid particle velocity are related to it by the following equations:

$$p = -\rho_0 \partial\Phi/\partial t$$

$$u = \partial\Phi/\partial x, \quad v = \partial\Phi/\partial y, \quad w = \partial\Phi/\partial z$$

The acoustic pressure mode shape functions ψ_n and frequencies ω_n are also appropriate to Φ. Equations (7.43) and (7.44) can now be written in terms of the structural displacement w and Φ as

$$\ddot{w}_p + \omega_p^2 w_p = -(\rho_0 S/\Lambda_p) \sum_n \dot{\Phi}_n C_{np} + F_p/\Lambda_p \tag{7.46}$$

$$\ddot{\Phi}_n + \omega_n^2 \Phi_n = (c^2 S/\Lambda_n) \sum_p \dot{w}_p C_{np} - c^2 Q_n/\Lambda_n \tag{7.47}$$

Note that Q_n and not $\dot{Q}_n$ now appears as the acoustic source term, the sign of the boundary velocity term being opposite because $\dot{w}_p$ is directed out of the volume and Q_n is directed into the volume. The aforementioned symmetry is now apparent in the dependence of the coupling terms on $\dot{\Phi}_n$ and $\dot{w}_p$. This particular form of coupling (gyrostatic) does not involve energy dissipation, unlike viscous coupling, which is also proportional to the first time derivative of displacement.

Up to this point no allowance has been made for internal dissipation mechanisms in the structure or the fluid. The most practical means of representing dissipation is to employ *ad hoc* viscous damping terms in the modal equations; these usually take the form $\beta_p \dot{w}_p$ and $\beta_n \dot{\Phi}_n$, where β_p and β_n are generalised modal damping coefficients, normally based upon empirical data. The corresponding modal loss factors can only strictly be related to β_p and β_n in terms of the energies of modal vibration at the corresponding resonance frequencies

$$\omega_p^2 \Lambda_p \beta_p \overline{w_p^2} = \eta_p \omega_p \overline{E}_p = \eta_p \omega_p^3 \Lambda_p \overline{w_p^2}$$

or

$$\eta_p = \beta_p/\omega_p \tag{7.48a}$$

Similarly,

$$\eta_n = \beta_n / \omega_n \tag{7.48b}$$

In using these models it is implicitly assumed that damping coupling between modes is neglected. This is generally a reasonable assumption and is adopted so as to simplify modal analysis.

7.7   Solutions of the Modal-Interaction Model

The natural frequencies and modes of a coupled system are different from those of the individual uncoupled systems. The total energy of vibration of a mode of a coupled system is divided between the structure and the fluid. In many cases the greater proportion of energy resides either in the fluid or in the structure; this fact leads us to talk loosely of 'fluid'- and 'structure'- (dominated) modes. Although the coupled modes differ from the uncoupled modes, the motions and energies of the two (or more) coupled components may still be expressed in terms of combinations of the uncoupled modes, as the previous analysis shows. In this case these modes may simply be thought of as convenient basis functions rather than having physical significance. In particular, the energy of vibration of the system is equal to the sum of the energies of the uncoupled modes, these energies being computed according to the proportional contribution of each uncoupled mode to the coupled mode under consideration. This useful fact arises from the orthogonality of the uncoupled modes. The coupled modes are orthogonal over the whole fluid–structure system, but not over any one subsystem.

It is revealing to analyse the vibration of a coupled system modelled in terms of just one uncoupled fluid mode and one uncoupled structural mode. Consider Eqs. (7.46) and (7.47) in this case, with simplified coefficients and *ad hoc* damping terms:

$$\ddot{w}_p + \beta_p \dot{w}_p + \omega_p^2 w_p = -K_{np} \dot{\Phi}_n + K_p F_p \tag{7.49}$$

$$\ddot{\Phi}_n + \beta_n \dot{\Phi}_n + \omega_n^2 \Phi_n = G_{np} \dot{w}_p - G_n Q_n \tag{7.50}$$

Let us first consider free vibration. We wish to solve for the natural frequencies ω:

$$\left(-\omega^2 + j\omega\beta_p + \omega_p^2\right)\tilde{w}_p = -j\omega K_{np} \tilde{\Phi}_n \tag{7.51}$$

and

$$(-\omega^2 + j\omega\beta_n + \omega_n^2)\tilde{\Phi}_n = j\omega G_{np}\tilde{w}_p \quad (7.52)$$

yielding

$$[(-\omega^2 + j\omega\beta_p + \omega_p^2)(-\omega^2 + j\omega\beta_n + \omega_n^2) - \omega^2 K_{np}G_{np}]\tilde{\Phi}_n = 0 \quad (7.53)$$

The expression in square brackets equals zero in the non-trivial case.
  In the undamped case

$$2\omega^2 = \omega_n^2 + \omega_p^2 + K_{np}G_{np} \pm [(\omega_n^2 - \omega_p^2)^2 + 2K_{np}G_{np}(\omega_n^2 + \omega_p^2) + K_{np}^2 G_{np}^2]^{1/2} \quad (7.54)$$

The product $K_{np}G_{np}$ is always positive because it is proportional to C_{np}^2. Suppose $\omega_p = \omega_n$; then

$$2\omega^2 = 2\omega_n^2 + K_{np}G_{np} \pm [4K_{np}G_{np}\omega_n^2 + K_{np}^2 G_{np}^2]^{1/2} \quad (7.55)$$

Since the value of the term in brackets exceeds $K_{np}G_{np}$, one of the new natural frequencies is greater than ω_n and one is less than ω_n. This is an example of the well-known frequency-splitting phenomenon exhibited by coupled oscillators of the same uncoupled natural frequencies. The interaction of an uncoupled fluid mode with an uncoupled structural mode can be likened to the influence of an auxiliary-mass-spring system on a primary oscillator, a phenomenon utilised in the design of dynamic absorbers or vibration neutralisers. Suppose $\omega_n \gg \omega_p$: then

$$2\omega^2 \approx \omega_n^2 + K_{np}G_{np} \pm (\omega_n^2 + K_{np}G_{np}) \quad (7.56)$$

and the two natural frequencies ω_1 and ω_2 of the coupled system satisfy the conditions $\omega_1^2 \ll \omega_2^2$, $\omega_2^2 \approx \omega_n^2 + K_{np}G_{np}$.
  The magnitudes of the frequency changes from the uncoupled state clearly depend upon the magnitude of $K_{np}G_{np}$. Reference to Eqs. (7.49) and (7.50) show this term to be given by

$$K_{np}G_{np} = \rho_0 c^2 S^2 C_{np}^2 / \Lambda_n \Lambda_p \quad (7.57)$$

in which $0 \le C_{np} \le 1, 1/8V \le \Lambda_n \le V$ and $1/4M \le \Lambda_p \le 1/2M$ assuming sinusoidal structural modes and cosinusoidal fluid modes in a rectangular box geometry. Hence

$$\left(K_{np}G_{np}\right)_{max} = 32\rho_0 c^2 S^2/MV = 32\rho_0 c^2 S/V \langle m\rangle \qquad (7.58)$$

where $\langle m\rangle$ is an average structural mass per unit area. As indicated by the ratio S/V, this term is greatest for fluid volumes having one dimension considerably less than the other two: these are termed 'disproportionate volumes' in room acoustics. For a rectangular volume of side lengths d, e and f, with a uniform flexible plate forming one face of dimensions e and f, the coupling term becomes

$$(K_{np}G_{np})_{max} = 32\rho_0 c^2/md = 32\omega_0^2 \qquad (7.59)$$

where ω_0 is the fundamental natural frequency of a plate of mass per unit area m on a fluid spring formed by a cavity of depth d. The bulk modulus of the fluid $\rho_0 c^2$ is, as expected, a controlling parameter. The effect of coupling in altering natural frequencies from their uncoupled values is clearly a function of the ratios ω_n/ω_0 and ω_p/ω_0. When both ratios greatly exceed unity, Eq. (7.54) indicates that little change is expected.

In fact, the maximum value of the coupling term, given in Eq. (7.58), is extremely unlikely to be achieved in practice because of the characteristics of the coupling coefficient C_{np}. The value of this coefficient is a measure of the degree of spatial matching of structural and acoustic mode shapes at the fluid–structure interface. Modes are standing waves formed by the interference between multiply reflected travelling waves, and it is more useful for the present purpose of discussing C_{np} to consider the natural wavenumbers in the structure and fluid. For simplicity, consider the rectangular panel-box system mentioned above: the panel is located in the x-y plane.

In the panel

$$k_{x_p}^2 + k_{y_p}^2 = k_b^2 \qquad (7.60)$$

and in the fluid

$$k_{x_f}^2 + k_{y_f}^2 + k_{z_f}^2 = k^2 \qquad (7.61)$$

For the best matching to occur, $k_{x_p} \approx k_{x_f}$ and $k_{y_p} \approx k_{y_f}$. Hence

$$k_b^2 + k_{z_f}^2 \approx k^2 \qquad (7.62)$$

Now, at frequencies below the critical frequency of the panel, $k_b > k$. Hence, for fluid modes in which k_{zf} is real, the best matching condition cannot occur, the nearest approach corresponding to acoustic modes with $k_{zf} = 0$; that is to say, two-dimensional acoustic modes that have component wavenumber vectors parallel to the plate. Best coupling for a flat uniform panel can only occur at frequencies above the critical frequency, in which case $\omega_n, \omega_p \gg \omega_0$, except for extremely shallow cavities. The fundamental mode of a panel is the sole exception to this general form of behaviour because it can couple efficiently with the zero-order, bulk-compression mode of the fluid for which $n = 0$ and $\omega_n = 0$. In this case, assuming a simply supported panel,

$$C_{np} = \frac{1}{S} \int_0^e \int_0^f \sin\left(\frac{\pi x}{e}\right) \sin\left(\frac{\pi b}{f}\right) 1 dS = \frac{4}{\pi^2}$$

$$K_{np} G_{np} = (64/\pi^4)\omega_0^2$$

which is far less than $(K_{np} G_{np})_{max}$. In this case, the mode shape of the panel in the fundamental coupled mode can be significantly altered by the influence of the fluid, as shown in Fig. 7.7 after Pretlove (1965). Note: the solution for mode shape has not converged with only six terms included.

The above argument based upon wavennumber matching must be modified for curved-shell structures because membrane stress effects can significantly increase flexural-wave phase velocities, and hence reduce flexural wavenumbers at given frequencies. This phenomenon has already been extensively

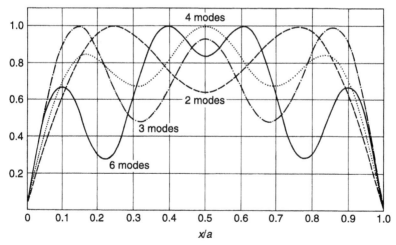

Fig. 7.7. Successive approximations to the fundamental modal deflection of a panel using an increasing number of modes in the synthesis (Pretlove, 1965).

discussed in Chapter 1. Therefore natural flexural wavenumbers less than acoustic wavenumbers can exist in a circular cylinder below the critical frequency of a shell wall considered as a flat plate. Hence Eq. (7.62) can be satisfied in terms of the appropriate coordinate system, and good matching can occur, as indicated by the coincidence behaviour described in Chapter 5.

We now turn to the analysis of forced simple harmonic vibration of the coupled system. As already stated, if the coupling is fairly weak between the modes of interest (ω_n/ω_0, $\omega_p/\omega_0 \gg 1$), and direct excitation is applied either only to the fluid or only to the structure, then the relevant uncoupled modal equation [(7.50) or (7.49), respectively] is solved for the uncoupled modal response: *ad hoc* modal damping must be employed to limit the resonant responses. Then the other equation of the pair is solved with excitation arising only from the coupling term into which the uncoupled response of the other component is substituted.

Suppose the structure is directly excited by applied forces, Eq. (7.49) becomes, with coupling neglected,

$$\left(-\omega^2 + j\omega\omega_p\eta_p + \omega_p^2\right)\tilde{w}_p = \tilde{F}_p/\Lambda_p \tag{7.63}$$

Substituting for $\dot{w}_p = j\omega w_p$ in Eq. (7.50), modified to include acoustic-mode damping, and setting $Q_n = 0$, gives

$$\tilde{\Phi}_n = \frac{j\omega c^2 S}{\Lambda_n\left(-\omega^2 + j\omega\omega_n\eta_n + \omega_n^2\right)} \sum_p \frac{\tilde{F}_p C_{np}}{\Lambda_p\left(-\omega^2 + j\omega\omega_p\eta_p + \omega_p^2\right)} \tag{7.64}$$

The total acoustic field is given by

$$\tilde{\Phi}(\mathbf{r}_s, t) = \exp(j\omega t) \sum_n \tilde{\Phi}_n \phi_n(\mathbf{r}_s) \tag{7.65}$$

The relative contributions to the acoustic field from the doubly infinite sum of modal terms depend upon the ratios ω/ω_p, ω/ω_n the coupling coefficients C_{np}, and the generalised forces $\tilde{F}_p$. Maxima in the acoustic-frequency response would be expected when the excitation frequency is close to ω_p or ω_n, or to a pair of each, the magnitude of the maxima depending upon the relevant loss factors and the magnitude of the appropriate coupling coefficient.

The fact that this approximate method of response estimation yields maxima near the uncoupled-mode resonance frequencies is simply a result of ignoring the interaction between the two systems in Eq. (7.63). Of course, the maxima in the frequency response of the coupled system should occur at frequencies near the resonance frequencies of the coupled system, obtained by solving the simple harmonic forms of Eqs. (7.49) and (7.50), with $F_p = Q_n = 0$. Conventional numerical techniques may be used to obtain solutions for the natural frequencies and mode shapes, once the infinite sets of equations (or numbers of uncoupled acoustic and structural modes) are limited according to the frequency range of interest.

In practice, increasing numbers of modes are included in the computation until an appropriate convergence criterion is satisfied. In cases where the mode shapes and natural frequencies of the uncoupled systems can be determined analytically, a process of selection ('triage') is employed in which the computation is restricted to those mode pairs (one acoustic and one structural) having the highest spatial coupling coefficients and/or close natural frequencies, in order to limit the size of the task (computation time). An example of this procedure may be found in a paper that addresses the problem of minimising the transmission of sound into a circular cylindrical shell by attaching a set of concentrating masses to the shell in order to disturb the spatial matches between dominant mode pairs (Gardonio *et al.*, 2001a).

In many practical applications, the geometric forms of the structures and fluid enclosures are irregular. Although the physical effects of coupling are the same in kind as those described in this section, simple algebraic expressions for the uncoupled natural frequencies, mode shapes and coupling coefficients are not available. Numerical procedures in the form of Finite Element Analysis and Boundary Element Analysis must be employed. These are explained in Chapter 8. They are, of course, applicable to any form of system. But, for the sake of comparison with the conclusions drawn from the analysis presented in this section, these methods of analysis are illustrated by application to a coupled rectangular panel-box system.

7.8   Power Flow and Statistical Energy Analysis

As an alternative to solving the integro–differential or modal equations of motion directly, a technique of estimating the time-average energy flux (or 'power flow') between coupled multimode subsystems called 'Statistical Energy Analysis' (SEA) has been under development during the past forty years. It is most suitable for cases of broadband excitation over a bandwidth encompassing many uncoupled-system natural frequencies. The reasons are twofold: (i) the basic relationship between power flow and energy difference between individual pairs of oscillators (modes) takes a very simple linear algebraic form provided that the excitation spectrum extends over a frequency range encompassing the resonance frequencies of the uncoupled modes involved; (ii) statistical estimates of the coupling parameter, termed the 'coupling loss factor' (CLF), which is the factor of proportionality relating time-average power flow between subsystems and their time-average vibrational/acoustic energies, are fairly easily obtained provided that a reasonably large population of modes is involved. However, it must be noted that SEA in its present form is most readily applicable to 'weakly coupled' systems in which the modes and natural frequencies of the coupled system are not greatly different from those of the uncoupled systems (having appropriate boundary conditions). SEA is also valid for strongly coupled systems, but it is not straightforward to determine appropriate expressions for the coupling coefficients and subsystem energies.

In SEA, a system comprising an assembly of components of different material properties and geometric forms is modelled as a network of coupled subsystems. Each of the subsystems is conceptually modelled as a set of oscillators corresponding to its uncoupled modes: the modal parameters are represented as members of random populations having quantifiable expected values. The application of SEA produces a response estimate that applies not to any specific system but represents an estimate of the average response of an ensemble of many systems that are grossly similar but differ in detail, such as cars coming off a production line. The vibroacoustic behaviour of any one sample at frequencies well above its fundamental natural frequency is unpredictable because it is sensitivity to detail that is unpredictable; but SEA is capable of estimating the average behaviour of a population. Recent research is beginning to produce predictions of the probability distribution about that average (Fahy and Langley, 2004).

As we have seen, the interaction between the subsystems can be represented exactly by coupled differential equations of motion in these modes, even though they do not form the actual orthogonal set of modes of the coupled system. The advantages of using the uncoupled modes are twofold: (i) knowledge of the average spatial or wavenumber characteristics, and of the frequency-average density of natural frequencies, is more readily available than the equivalent information about the coupled system; (ii) the total time-average energies of each component subsystem are exactly given by the sum of the energies expressed in terms of the uncoupled modes of that system. General introductions to SEA may be found in the following publications: Lyon and DeJong (1995), Keane and Price (1997), Craik (1996) and Fahy and Langley (2004). Tutorial material may be found at www.seanet.be. We concentrate here on coupling between enclosed volumes of fluid and structures that enclose them.

The time-average power flow from a structure to a contiguous fluid may be expressed exactly as

$$
\begin{aligned}
\overline{P}_{12} &= \frac{1}{T} \int_0^T \left[\int_S \dot{w}(\mathbf{r}_s) p(\mathbf{r}_s) dS \right] dt \\
&= \frac{\rho_0}{T} \int_0^T \left[\int_S \sum_p \dot{w}_p \phi_p(\mathbf{r}_s) \sum_n \dot{\Phi}_n \psi_n(\mathbf{r}_s) dS \right] dt
\end{aligned}
\tag{7.66}
$$

where T is chosen to suit the frequency spectrum of the vibration. The space and time integrals are independent, and therefore

$$\overline{P}_{12} = \rho_0 \sum_p \sum_n \overline{\dot{w}_p \dot{\Phi}_n} \int_S \psi_p(\mathbf{r}_s)\phi_n(\mathbf{r}_s)dS = \rho_0 S \sum_p \sum_n C_{np}\overline{\dot{w}_p \dot{\Phi}_n} \quad (7.67)$$

in which the overbar indicates time average and C_{np} is defined by Eq. (7.45). Referring back to Eq. (7.46), we can check Eq. (7.67) for energy conservation. Multiply Eq. (7.46) (modified by a viscous damping term $\beta_p \dot{w}_p$) by $\dot{w}_p$ and consider the time average:

$$\Lambda_p \left(\overline{\ddot{w}_p \dot{w}_p} + \beta_p \overline{(\dot{w}_p)^2} + \omega_p^2 \overline{w_p \dot{w}_p} \right) = -\rho_0 S \sum_n C_{np} \overline{\dot{w}_p \dot{\Phi}_n} + \overline{F_p \dot{w}_p} \quad (7.68)$$

If the vibration is time-stationary, the terms $\overline{\ddot{w}_p \dot{w}_p}$ and $\overline{w_p \dot{w}_p}$ are zero. (Prove.) The term $\Lambda_p \beta_p \dot{w}_p^2$ is the rate of dissipation of energy attributable to vibration in mode p, and the term $\overline{F_p \dot{w}_p}$ is the rate of work done by external forces attributable to motion in mode p. Note that the phrases 'work done on mode p' and 'energy dissipated by mode p' are avoided in order to reinforce understanding of the fact that these are not members of the set of actual orthogonal modes of the whole coupled system, but are mathematically useful base functions in terms of which the motion of the structural portion of the coupled system may be expressed. Summation of the remaining terms in Eq. (7.68) over all p shows correctly that the total power injected into the structure is equal to the sum of that dissipated internally by the structure and that transferred to the fluid, and dissipated therein.

Of course, the power flow attributable to coupling between any two modes of the two uncoupled systems, expressed by Eq. (7.67) as $\rho_0 S \overline{\dot{w}_p \dot{\Phi}_n}$ cannot strictly be evaluated simply by considering the equations of motion of those two modes in isolation, since in principle all structural modes are coupled to all acoustic modes, as Eqs. (7.46) and (7.47) indicate. However, engineering analysis is largely about making reasonable assumptions that simplify the process of calculation and increase the ability of the analyst or user to understand the influence of the basic parameters, without reducing the accuracy of the results to an extent that renders them misleading or useless. Consequently, a commonly used simplifying assumption is that the total power flow between the structural and fluid systems is approximately equal to the sum of the power flows attributable to coupling between isolated pairs of modes. This is known as the 'weak coupling' assumption. It is justified if the energies of the coupled modes reside principally in one or the other of the fluid and structural subsystems. This is generally true in cases where the fluid is atmospheric air, and often, but not always, in cases where a liquid is involved.

Detailed analysis (Scharton and Lyon, 1968) shows that the time-average power flow between a pair of randomly excited, gyrostatically coupled, isolated, linear oscillators is

$$\bar{P}_{12} = \eta_{12}\omega_c(\bar{E}_1 - \bar{E}_2) \tag{7.69}$$

where $\bar{E}_1$ and $\bar{E}_2$ are the oscillator energies, and the coupling loss factor η_{12} is given by

$$\eta_{12} = \frac{2\rho_0 c^2 S^2 C_{12}^2}{\Lambda_1 \Lambda_2} \left[\frac{\left(\beta_2\omega_1^2 + \beta_1\omega_2^2\right)(\omega_1 + \omega_2)^{-1}}{\left(\omega_1^2 - \omega_2^2\right)^2 + \left(\beta_2\omega_1^2 + \beta_1\omega_2^2\right)(\beta_1 + \beta_2)} \right] \tag{7.70}$$

in which the excitation band centre frequency ω_c has been replaced by the arithmetic mean of the modal natural frequencies, $(\omega_n + \omega_p)/2$: C_{12}, Λ_1 and Λ_2 are equivalent to C_{np}, Λ_n and Λ_p, respectively. It is significant that the coupling loss factor is symmetric in subscripts 1 and 2.

Equation (7.70) shows that the CLF is proportional to the square of the coupling coefficient and is extremely sensitive to the difference between the natural frequencies of the two uncoupled oscillators. We may obtain approximate expressions for η_{12} that correspond either to conditions of close, or proximate, natural frequencies or to conditions of distant, or remote, natural frequencies. In the former case we write $\omega_1 \approx \omega_2 \approx (\omega_1 + \omega_2)/2$. If, in addition, $2|\omega_1 - \omega_2| < \beta_1 + \beta_2$, then the first term in the denominator is small compared with the second and

$$\eta_{12} \approx \frac{2\rho_0 c^2 S^2 C_{12}^2}{\Lambda_1 \Lambda_2} \left[\frac{1}{(\omega_1 + \omega_2)(\beta_1 + \beta_2)} \right] \tag{7.71}$$

In this case the power flow produced by a given coupling coefficient between oscillators of given energy difference is *inversely proportional* to the sum of the damping loss factors of the oscillators.

In the case of remote natural frequencies, for which $2|\omega_1 - \omega_2| \gg \beta_1 + \beta_2$, the first term in the denominator of Eq. (7.70) is dominant. Since $\beta_1 = \eta_1\omega_1$ and $\beta_2 = \eta_2\omega_2$ are normally much less than ω_1 and ω_2, respectively, the CLF is, in this case, very small compared with that produced by proximate modal frequencies, given similar values of the coupling coefficient. Also, the CLF is *proportional* to the average damping loss factors of the oscillators. If we consider a population of oscillator pairs for which the frequencies ω_1 and ω_2 are random and assumed to lie anywhere within the analysis band with uniform probability, the ensemble-average CLF is *independent of damping* (Lyon and DeJong, 1995).

If it is assumed, as a first approximation, that the total power flow between a structure and a volume of fluid is given by the sum of the power flows attributable to coupling between isolated pairs of modes of the uncoupled components, then it is only necessary in practice to identify mode pairs that are well matched spatially and that have proximate natural frequencies; the other mode pairs can be discounted. This is a form of 'triage'.

This approach was used by Fahy (1969b, 1970) to evaluate the power flow between a rectangular flat plate and fluid in a box of which the plate formed one face, and also between a uniform, circular, cylindrical shell and a contained fluid. It was found that the modal-average coupling loss factors so evaluated corresponded to those between the structures and the unbounded fluid external to the system, provided that the average difference between the natural frequencies of the modes of the two uncoupled components was not too large. This external radiation coupling loss factor is related to the radiation efficiency by $\eta_{rad} = \rho_0 c \sigma / \omega_c m$. The important conclusion to be drawn from these analyses is that, in cases in which the modal density of each component is sufficiently high, the broadband power exchanged between a structure and an enclosed volume of fluid can be evaluated from knowledge of the modal average free-space radiation characteristics. This convenient result does not apply to discrete-frequency or narrow-band excitation.

Although this discrete oscillator pair model is useful for the purposes of research studies of systems involving spatially uniform subsystems and simple geometries, for which analytical expressions for mode pair CLFs are available, it is not of much utility for the majority of practical vibroacoustic problems in the fields of automotive, aerospace and marine engineering, except as in providing conceptual guidance and insight into the coupling phenomenon. Fortunately, there is an alternative approach to the problem of estimating the frequency-average CLF between two, weakly coupled, multimode subsystems.

The weak coupling assumption allows the power flow-energy difference relation for an individual mode pair [(Eq. (7.69)] to be extended to an analogous relation involving many mode pairs. (The underlying reasoning may be found in the general SEA literature of which four references are cited above.) The relation is

$$P_{12} = \eta_{12}\omega_c N_1 \left(\frac{E_1}{N_1} - \frac{E_2}{N_2} \right) \tag{7.72}$$

in which P_{12} is the time-average, band-limited, power flow between subsystems 1 and 2, E_1 and E_2 are the time-average energies stored in the subsystems, N_1 and N_2 are the statistically estimated number of modes of the uncoupled subsystems having resonance frequencies in the band, ω_c is the centre frequency of the band and η_{12} is the average CLF between all pairs of modes that are resonant in the band. The important distinction between Eqs. (7.69) and (7.72) is that the power flow between multimode subsystems is proportional to the difference of the average energies *per (uncoupled) subsystem mode* (known as 'modal energies'). [Note: In the SEA literature, both the numerator and denominator of Eq. (7.72) are conventionally divided by the analysis bandwidth, so that N becomes n, denoting the average modal density of the subsystem in the band.] In an analogy with heat flow, the modal energies correspond to 'temperatures' and the factor of proportionality $\eta_{12}\omega_c N_1$ corresponds to the coefficient of thermal conductivity. As with heat, vibrational energy flows from a 'hot' to a 'cold' subsystem in proportion to the 'temperature' difference. It may be shown that $n_1\eta_{12} = n_2\eta_{21}$, the so-called 'consistency relation'.

It has been shown by numerous studies that the average CLF between two weakly coupled, multimode subsystems in a frequency band encompassing a number of modal natural frequencies (typically at least five) may be estimated by means of an analysis of the transmission of travelling wave energy across the interface between coupled subsystems. The essential assumption is that the waves transmitted 'forward' through an interface are uncorrelated with those returning to the interface from the 'receiving' subsystem. Hence, the net power flow is the sum of the 'forward' and 'backward' power flows. This is the essential condition that the 'weak coupling' assumption imposes. It is not valid in cases where the coupled subsystems exhibit 'global mode' behaviour in which the energies of modes of the total system are not resident predominantly in one or the other subsystem. In the cases of 'non-weak' coupling, the CLF so estimated is generally an overestimate of the true value; it is then also proportional to subsystem damping.

In the case of weak vibroacoustic coupling between a structure (subsystem 1) and a fluid (subsystem 2), η_{12} is given by

$$\eta_{12} = \rho c \sigma / \omega_c \langle m \rangle \tag{7.73}$$

in which σ is the modal- (or frequency-) average radiation efficiency of the structure in the analysis band and $\langle m \rangle$ is the spatial-average structural mass per unit area. Hence, calculation or measurement of σ allows estimates to be made of the power flow from a directly excited structure into a contiguous volume of fluid, and of the acoustically induced response to sound in a confined volume of fluid. The SEA approach to vibroacoustic coupling complements that presented in Section 7.5 and confirms that *efficient radiators are efficient responders*.

7.9 Wave Propagation in Plates Loaded by Confined Fluid Layers

So far in this chapter we have assumed that the fluid is completely bounded, and consequently we have constructed modal models of the coupled structure–fluid system. As we have seen before, the modes of a system are formed by constructive interference between progressive waves, and consequently it is equally valid to formulate a wave propagation model for a coupled system. It transpires that the results of analyses of such models are particularly useful in cases where a plate or shell structure is coupled to a volume of fluid confined within boundaries essentially parallel to the structural surface, in which case the fluid can greatly modify the free-wave characteristics of the structure, and *vice versa*. Practical examples include fluid transport ducts, flooded sonar compartments in ships, double-leaf partitions and the basilar membrane in the human auditory system. The latter, being a biological system, does not really come within the scope of this book, which is confined to linear, constant-parameter structures; however, like blood vessels, it shares some of the features of the systems dealt with below. This section presents an analysis of wave propagation in a two-dimensional model comprising two parallel, unbounded, uniform, flat plates separated by a uniform layer of fluid. The following section presents a brief, largely qualitative, introduction to the subject of wave propagation in fluid-filled, flexible tubes that shares some of the phenomenological features of the flat-plate 'sandwich' system.

The parallel plate system is shown in Fig. 7.8. A presence of fluid outside the sandwich would considerably complicate the wave propagation behaviour. Therefore it is excluded, although it would be mathematically straightforward to include. (Its inclusion would constitute a suitable assignment for advanced students.) The acoustic pressure in the fluid satisfies the homogeneous wave equation subject to normal particle velocity boundary conditions on the surfaces of the plates. We shall seek a relationship between an assumed propagation

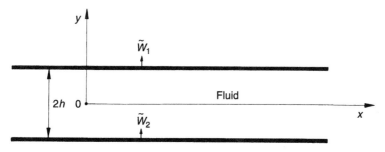

Fig. 7.8. Coupled plate–fluid layer system.

wavenumber in the x-direction and frequency; in other words, we wish to establish the dispersion characteristics of this system. Let the assumed wavenumber be k_x. The two-dimensional acoustic wave equation for the simple harmonic pressure $\tilde{p}(y) \exp[j(\omega t - k_x x)]$ yields

$$\partial^2 \tilde{p}(y)/\partial y^2 + (k^2 - k_x^2)\tilde{p}(y) = 0 \tag{7.74}$$

The solution for $p(y)$ is

$$\tilde{p}(y) = \tilde{A} \exp(-jk_y y) + \tilde{B} \exp(jk_y y) \tag{7.75}$$

with $k_y = (k^2 - k_x^2)^{1/2}$.

The equation of motion of the upper plate, neglecting external fluid, gives

$$(Dk_x^4 - \omega^2 m)\tilde{w}_1 = (\tilde{p})_{y=h} \tag{7.76}$$

and for the lower plate gives

$$(Dk_x^4 - \omega^2 m)\tilde{w}_2 = -(\tilde{p})_{y=-h} \tag{7.77}$$

The boundary conditions on the acoustic field are

$$[\partial \tilde{p}(y)/\partial y]_{y=h} = \rho_0 \omega^2 \tilde{w}_1 \tag{7.78}$$

$$[\partial \tilde{p}(y)/\partial y]_{y=-h} = \rho_0 \omega^2 \tilde{w}_2 \tag{7.79}$$

It is clear from considerations of symmetry that a solution must exist with $\tilde{w}_1 = -\tilde{w}_2 = \tilde{w}$. This model also represents a single plate separated from a parallel rigid plane by a layer of fluid of thickness h. In this case, Eqs. (7.75), (7.78) and (7.79) yield $\tilde{A} = \tilde{B}$ and

$$\tilde{p}(y) = 2\tilde{A} \cos(k_y y) \tag{7.80}$$

giving a pressure amplitude $2\tilde{A}$ on the axis of symmetry. The solution for $\tilde{A}$ is

$$\tilde{A} = -\omega^2 \rho_0 \tilde{w}/2k_y \sin(k_y h) \tag{7.81}$$

Hence Eq. (7.76) becomes

$$D(k_x^4 - k_b^4)\tilde{w} = -\omega^2 \rho_0 \tilde{w}(k^2 - k_x^2)^{-1/2} \cot\left[(k^2 - k_x^2)^{1/2}h\right] \tag{7.82}$$

where k_b is the *in-vacuo*, free flexural wavenumber of the plate. This equation may be expressed in a non-dimensional form in terms of the ratio of axial to acoustic wavenumber k_x/k, the ratio of acoustic wavenumber to the acoustic (and flexural) wavenumber at the critical frequency $k/k_c = \omega/\omega_c$, a fluid-layer thickness parameter $k_c h$ and a plate mass parameter $m/\rho_0 h$, thus:

$$(k_c h)(m/\rho_0 h)(k/k_c)^3\left[1-(k_x/k)^2\right]^{1/2}\left[(k_x/k)^4-(k_c/k)^2\right]$$
$$= -\cot\left[(k/k_c)(k_c h)\left(1-(k_x/k)^2\right)^{1/2}\right] \tag{7.83}$$

This is the dispersion relationship for symmetric waves, which is, of course, independent of $\tilde{w}$ for the assumed linear system. It is effectively a fifth-order equation in k_x^2, but because of the presence of the trigonometric term it is an implicit equation in k_x/k, which must be solved by graphical or numerical techniques. Although numerical solutions are complete and accurate, the physical interpretation of equations such as (7.83) is often more clearly revealed by seeking approximate analytical solutions for various special cases. We do not know *a priori* whether k_x will be greater or less than k; that is to say, whether the axial phase velocity is smaller or greater than the speed of sound in the fluid. However, it is clear from examination of the equation that the forms of solution will be rather different in the two cases because of the terms $[1 - (k_x/k)^2]^{1/2}$.

We proceed by first considering cases with $k_x < k$.
(1) $k_x < k$: $k_x < k_b$. If $k_x < k_b$ then $(k_x/k)^4 \ll (k_c/k)^2$ and Eq. (7.83) maybe approximated by

$$(m/\rho_0 h)(k/k_c)(k_c h)\left[1-(k_x/k)^2\right]^{1/2} \approx \cot\left[(k/k_c)(k_c h)(1-(k_x/k)^2)^{1/2}\right] \tag{7.84}$$

or

$$(m/\rho_0 h)\beta \approx \cot\beta \tag{7.85}$$

where $\beta = (k/k_c)(k_c h)[1 - (k_x/k)^2]^{1/2}$. The two sides of Eq. (7.85) are plotted against β in Fig. 7.9.

When $m/\rho_0 h \gg 1$, the first intersection occurs at a value of β much less than unity. In this case, $\cot\beta \to (1 - \frac{1}{2}\beta^2)/\beta$ and

$$\beta^2 \approx \rho_0 h/m$$

or

$$(k_x/k)^2 \approx 1 - (\omega_0/\omega)^2 \tag{7.86}$$

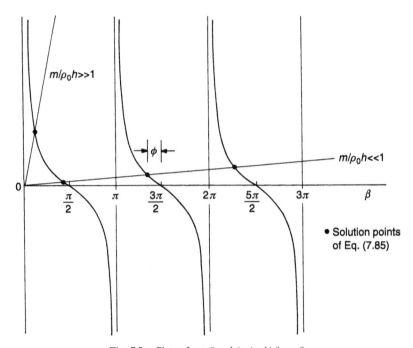

Fig. 7.9. Plots of cot β and $(m/\rho_0 h)\beta$ vs. β.

where $\omega_0 = (\rho_0 c^2/mh)^{1/2}$ is the frequency of resonance of mass per unit area m on a rigidly terminated column of fluid of length h. The ratio $(\omega_0/\omega_c)^2 = (1/12)(c_l'/c)^2(t/h)(\rho_0/\rho_s)$, where ρ_s is the density of the plate material and t is the plate thickness. Since it was assumed initially that $k_x < k_b$, this solution is only valid if $k^2[1 - (\omega_0/\omega)^2] \ll kk_c$. This condition corresponds to a condition on frequency of $2(\omega/\omega_c) \ll 1 + [1 + (2\omega_0/\omega_c)^2]^{1/2}$. Propagating solutions corresponding to this first intersection (branch) do not exist for frequencies below ω_0 because k_x is then imaginary. This observation suggests that the cut-off frequency for propagation of waves in this category may be maximised by minimising the mass per unit area of the plates, a measure that might offer a means of attenuating low frequency 'plane' wave propagation in thin-walled ducts of rectangular cross section, such as air conditioning, heating and ventilation ducts. When $\omega/\omega_0 \gg 1$, the wave corresponding to this branch propagates with an axial phase speed close to the speed of sound in the fluid.

Further intersections will occur at values of β close to $n\pi$. In these cases,

$$(k_x/k)^2 \approx 1 - (n\pi/kh)^2, \quad \text{integer } n \geq 1 \tag{7.87}$$

The condition $k_x < k_b$ is satisfied for $2(\omega/\omega_c) \ll 1 + [1 + 2n\pi/k_ch)^2]^{1/2}$ and propagation does not occur at frequencies ω less than $n\pi c/h$. The latter are the frequencies corresponding to n acoustic wavelengths between the plates. Figure 7.10(a) shows the behaviour of these various branches.

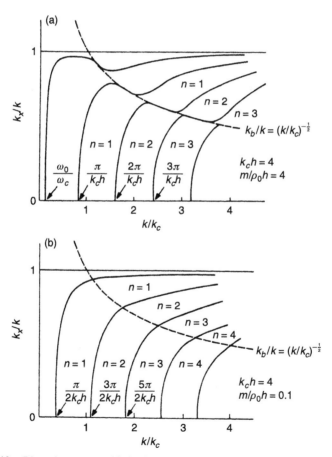

Fig. 7.10. Dispersion curves with $k_x/k < 1$: (a) $k_ch = 4$, $m/(\rho)_0h = 4$; (b) $k_ch = 4$, $m/\rho_0h = 0.1$.

On the other hand, if $m/\rho_0 h \ll 1$ [(Fig. 7.10(b)], intersections of the curves in Fig. 7.9 occur at values of β less than $(2n-1)\pi/2$ by a small value ϕ, which is much less than unity. Now $\cot[(2n-1)\pi/2 - \phi] \approx \phi$. Hence intersection occurs when

$$\beta(1 + m/\rho_0 h) \approx (2n-1)\pi/2, \quad n \geq 1$$

or

$$\left(\frac{k_x}{k}\right)^2 \approx 1 - \left[\frac{(2n-1)\pi}{2kh(1 + m/\rho_0 h)}\right]^2, \quad \text{integer } n \geq 1 \qquad (7.88)$$

This wave does not propagate at frequencies below that for which

$$kh = \frac{(2n-1)\pi}{2(1 + m/\rho_0 h)} \qquad (7.89)$$

In the limit $m/\rho_0 h \to 0$, these frequencies correspond to n half-wave-lengths between the 'plates', which, of course, are the cut-off frequencies of acoustic modes in a two-dimensional waveguide with pressure-release ($p = 0$) boundaries. The assumed condition $k_x < k_b$ holds at frequencies satisfying

$$2(\omega/\omega_c) \ll 1 + \left\{1 + \left[(2n-1)\pi/k_c h(1 + m/\rho_0 h)\right]^2\right\}^{1/2} \qquad (7.90)$$

The transition of the wave characteristics from a similarity to those in a duct with rigid boundaries ($m/\rho_0 h \to \infty$), to a similarity to those in a duct with pressure-release boundaries ($m/\rho_0 h \to 0$), is seen clearly from Fig. 7.9 by following the intersection of straight lines of slope varying from infinity to zero with any one cotangent curve.

We now consider another special case:
(2) $k_x < k$: $k_x = k_b$. Equation (7.82) indicates that this will occur when $(k^2 - k_b^2)^{1/2}h = (2n-1)\pi/2$. Solving for frequency yields

$$2(\omega/\omega_c) = 1 + \left\{1 + \left[(2n-1)\pi/k_c h\right]^2\right\}^{1/2} \qquad (7.91)$$

Hence, equality of the axial phase speed of the coupled wave with that of the *in-vacuo* flexural wave in a plate can only occur above the critical frequency. At these frequencies the impedance of the fluid layer, as seen by the plate, is zero and therefore has no influence on the plate. Figure 7.10 shows how the branches that reveal above for $k_x < k$ cross the k_b curve at frequencies given

by Eq. (7.91). Although Eq. (7.91) indicates that the value of $\omega/\omega_0 = k/k_c$ at which $k_x = k_b$, for a given n, is independent of $m/\rho_0 h$, it is clear from Fig. 7.10 that the tendency for the dispersion curves in this region to follow that of the *in-vacuo* bending wavenumber k_b, increases as $m/\rho_0 h$ increases. We now seek branches representing waves travelling with axial phase speeds greater than that of sound in the fluid, but less than that of the *in-vacuo* flexural waves in the plate. This can obviously occur only at frequencies greater than the critical frequency.
(3) $k_x < k$: $k_x > k_b$. Equation (7.83) becomes

$$(k_c h)(m/\rho_0 h)(k/k_c)^3 (k_x/k)^4 \left[1 - (k_x/k)^2\right]^{1/2}$$

$$= -\cot\left[(k/k_c)(k_c h)(1 - (k_x/k)^2)^{1/2}\right]$$

or

$$(m/\rho_0 h)(k_x/k)^4 (k/k_c)^2 \beta \approx -\cot\beta \tag{7.92}$$

It is difficult to generalise about this equation because, although k_x/k is less than unity, k/k_c, which must be greater than unity, may be indefinitely large. When $m/\rho_0 h \to 0$, solutions correspond to $\beta \to (2n-1)\pi/2$, and when $m/\rho_0 h \gg 1$, solutions correspond to $\beta \to n\pi$.

Consequently it is necessary to resort to numerical solution, the form of which is shown in Fig. 7.10. Of course, if $m/\rho_0 h$ is very small, the branches correspond closely with pressure-release boundary modes except at very high frequencies, when $(k/k_c)^2$ is very large and there is a transition to rigid-wall duct behaviour. If $m/\rho_0 h$ is much greater than unity, the modes correspond closely to rigid-walled duct modes except close to the frequencies of intersection with k_b, given by Eq. (7.90).

So far we have assumed that k_x/k is less than unity. We now relax this condition, but in doing so, we should return to the original derivation of Eqs. (7.82) and (7.83) to check the physics.
(4) $k_x > k$: $k_x < k_b$. Here we seek waves that have axial phase speeds less than the speed of sound in the fluid but greater than *in-vacuo* flexural waves in the plates. Returning to Eq. (7.75) we must instead write

$$\tilde{p}(y) = \tilde{A}\exp(-k_y y) + \tilde{B}\exp(k_y y) \tag{7.93}$$

where $k_y = (k_x^2 - k^2)^{1/2}$. Application of the boundary conditions that the plate displacements be in anti-phase yields $\tilde{A} = \tilde{B}$ and

$$\tilde{p}(y) = 2\tilde{A}\cosh(k_y y) \tag{7.94}$$

The solution for $\tilde{A}$ is

$$\tilde{A} = \omega^2 \rho_0 \tilde{w} / 2k_y \sinh(k_y y) \tag{7.95}$$

Hence Eq. (7.76) becomes

$$D(k_x^4 - k_b^4)\tilde{w} = \omega^2 \rho_0 \tilde{w}(k_x^2 - k^2)^{-1/2} \coth\left[(k_x^2 - k^2)^{1/2}h\right]$$

or

$$(k_c h)(m/\rho_0 h)(k/k_c)^3 \left[(k_x^2/k)^2 - 1\right]^{1/2}\left[(k_x/k)^4 - (k_c/k)^2\right]$$
$$= \coth\left\{(k_c h)(k/k_c)[(k_x/k)^2 - 1]^{1/2}\right\} \tag{7.96}$$

Applying the condition $k_x < k_b$ gives an equation equivalent to (7.85):

$$(m/\rho_0 h)\gamma \approx -\coth\gamma \tag{7.97}$$

where $\gamma = (k_c h)(k/k_c)[(k_x/k)^2 - 1]^{1/2}$. The two sides of this equation are plotted against γ for values of $m/\rho_0 h$ greater than, equal to and less than unity in Fig. 7.11. It is seen that no solution exists for this combination of conditions. Let us now see if waves can exist that have axial phase speeds less than both the speed of sound in the fluid and *in-vacuo* flexural waves in the plates.
(5) $k_x > k$: $k_x > k_b$. Equation (7.96) reduces to

$$(k/k_c)^2(m/\rho_0 h)(k_x/k)^4 \gamma \approx \coth\gamma$$

This may also be written

$$(k/k_c)^{-2}(m/\rho_0 h)(k_c h)^{-4}\alpha^5 \approx \coth\alpha \tag{7.98}$$

where $\alpha = k_x h$ and it has been assumed that $[(k_x/k)^2 - 1]^{1/2} \approx k_x/k$. The positive sign on the right-hand side allows this equation to possess a solution for any chosen value of the parameters k/k_c, $(m/\rho_0 h)$, and $k_c h$, as Fig. 7.12 shows. However, there is only one intersection, only one branch, and therefore only one type of wave. Numerical analysis shows that this branch corresponds to flexural plate waves that are inertially loaded by the evanescent fields in the fluid, and hence have phase speeds below their *in-vacuo* values. The fact that no waves having $k_x < k_b$ and $k_x > k$ exist indicates that the fluid does not exert a stiffness-type loading on this branch. Above the critical frequency, this branch

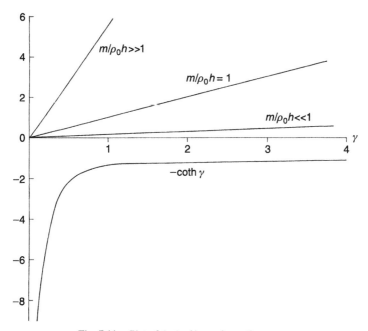

Fig. 7.11. Plot of $(m/\rho_0 h)\gamma$ and $-\coth\gamma$ vs. γ.

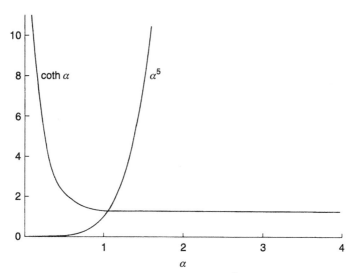

Fig. 7.12. Plots of $\coth\alpha$ and α^5 vs. α.

asymptotes to a wave with a constant-phase velocity just below that of sound in the fluid, as shown in Fig. 7.13 in which the broken curve of k_b/k is shown for reference.

[The analyses of anti-symmetric wave motion, in which $\tilde{w}_1 = \tilde{w}_2$, and of axially evanescent fields, are left as exercises for the (keen) student.]

The model analysed above is two-dimensional and infinitely extended. The more realistic model of the radiation of sound into a fluid layer contained within two parallel, rigid, planes by a vibrating rectangular panel set into one of the planes is analysed by Schroter and Fahy (1981). It is shown that, well below the plate critical frequency, the modal radiation efficiency is inversely proportional to the layer thickness.

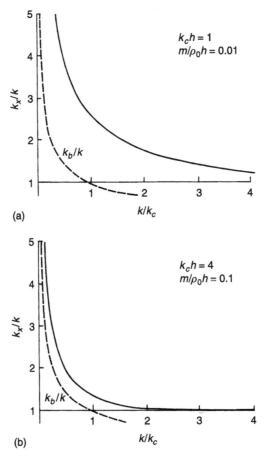

Fig. 7.13. Dispersion curves with $kx/k > 1$: (a) $k_c h = 1$, $m/\rho_0 h = 0.01$; (b) $k_c h = 4$, $m/\rho_0 h = 0.1$.

7.10   Wave Propagation in Fluid-Filled Tubes of Circular Cross Section

The propagation of acoustic disturbances in fluids contained in tubes (the word 'pipe' will be used synonymously) is of interest in many fields including those concerned with air, water, oil and gas distribution systems, electrical power generation equipment, hydraulic systems and blood flow. While the thin flat plate modelled in the previous section can resist applied transverse forces only by the action of the transverse shear forces associated with bending deformation (assuming that the deflection is sufficiently small to make contributions from in-plane strain negligible and the response linear), the curvature and closure of the cross-section of a circular cylindrical shell allows it to resist radial applied forces by means of tangential and longitudinal direct stresses. Analysis of the wave propagation characteristics of fluid-filled, thin-walled, circular cylindrical shells reveals them to be substantially more complex than those of the two-dimensional flat-plate-cavity system analysed in the previous section (Fuller and Fahy, 1981). One of the complicating factors is the existence of complex axial wavenumbers, even in the undamped *in-vacuo* shell.

In this section, we shall restrict our attention to structural waveguide modes involving only circumferential mode orders $n = 0$ ('breathing') and $n = 1$ ('beam bending') on the basis that they tend to dominate the energy transport in the majority of practical cases of interest. In doing so, we are implicitly assuming that the upper frequency of interest is less than the cut-off frequency of the $n = 2$ *in-vacuo* shell mode, given by $0.12hc'_l/a^2$, where h and a are the tube wall thickness and radius, respectively, and c'_l is the quasi-longitudinal wave speed in the wall material. The corresponding ratio of this frequency to the ring frequency, denoted by Ω_2, is equal to $0.77(h/a)$. The ratio of the cut-off frequency of the $n = 1$ acoustic mode of the fluid contained in a rigid tube to the ring frequency is given by $\Omega_1 = 2.4(c/c'_l)(h/a)$. The ratio Ω_2/Ω_1 is given by $0.32(c'_l/c)$, which is close to unity for a steel tube filled with fresh water.

We start with the simplest model of zero-order ($n = 0$) wave propagation in a flexible tube filled with a compressible fluid, illustrated by Fig. 7.14(a). This is known as the 'ring' model because the structure is considered to consist of a stack of short, independent rings that may displace radially without constraint from adjacent rings: the ring axial length δx is assumed not to change. The associated circumferential strain gives rise to the circumferentially directed stress that is assumed to produce the only structural elastic force on an elemental arc of the shell. In a continuous shell, *axial* variation of radial displacement produces shell flexure and associated radially directed shear force. However, for values of Ω much less than unity, and wall thickness parameter h/a less than 0.1,

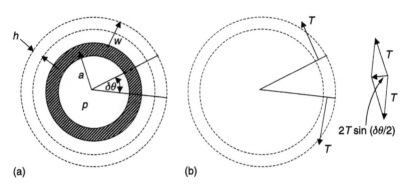

Fig. 7.14. (a) Two-dimensional ring model of a section of fluid-filled tube. (b) Circumferential forces on a strained segment of the pipe wall.

the shear force is very much less than the radial force due to circumferential strain, and may be neglected.

As shown in Fig. 7.14(a), the uniform radial displacement of a ring element is denoted by w. The associated circumferential strain is equal to w/a. Since δx is assumed constant, the circumferential force per unit length is given by $T = (E/(1 - v^2))(h/a)w$. The radial force on the element of arc of length $a\delta\theta$ is equal to $T\delta\theta$. The excess (acoustic) pressure in the fluid is assumed to be uniform over the cross section. This is clearly an approximation, since the associated radial pressure gradient at the wall is zero, whereas it should be equal to $-\rho_0$ times the radial acceleration of the wall. The radial distribution of sound pressure in a circular cylindrical duct is given by Eq. (5.148) in terms of Bessel Functions $J_n(k_r r)$ of order n, where k_r equals $(k^2 - k_z^2)^{1/2}$ and k_z is the axial wavenumber. The forms of J_0, J_1 and J_2 are illustrated by Fig. 7.15.

If the pipe wall undergoes radial acceleration, it is clear that $k_z \neq k$ because $J_0(0)$ is a constant. The equation of motion of the element of arc is

$$\rho_s h a \frac{\partial^2 w}{\partial t^2} = pa - (E/(1 - v^2))(h/a)w \qquad (7.99)$$

where ρ_s is the density of the wall material.

The volumetric strain of the element of fluid of unstrained length δx is given by

$$\frac{\delta V}{V} = \frac{\partial \xi}{\partial x} + 2\frac{w}{a} \qquad (7.100)$$

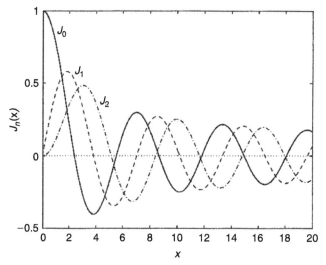

Fig. 7.15.   Bessel functions J_0, J_1 and J_2.

in which ξ is the axial displacement of the fluid. The associated acoustic pressure is given by

$$p = -\rho_0 c^2 \frac{\delta V}{V} \qquad (7.101)$$

and the axial pressure gradient is given by

$$\frac{\partial p}{\partial x} = -\rho_0 c^2 \left[\frac{\partial^2 \xi}{\partial x^2} + \frac{2}{a} \frac{\partial w}{\partial x} \right] \qquad (7.102)$$

The resulting linearised equation of motion of the fluid element is

$$\frac{\partial^2 \xi}{\partial t^2} = c^2 \left[\frac{\partial^2 \xi}{\partial x^2} + \frac{2}{a} \frac{\partial w}{\partial x} \right] \qquad (7.103)$$

Substitution of the expression for p from Eq. (7.101) into Eq. (7.99) yields the equation of motion of the shell

$$\rho_s h a \frac{\partial^2 w}{\partial t^2} = -\rho_0 c^2 \left[\frac{\partial \xi}{\partial x} + 2\frac{w}{a} \right] - \left(\frac{E}{(1 - \nu^2)} \right) \left(\frac{h}{a} \right) w \qquad (7.104)$$

We now assume that the radial shell displacement and axial fluid displacement take the form of travelling harmonic waves of frequency ω having the same axial wavenumber k_w. Substitution of these forms into Eqs. (7.103) and (7.104) yields the coupled simultaneous algebraic equations for the complex amplitudes of displacements

$$\tilde{\xi}\left[-c^2 k_w^2 + \omega^2\right] = j\left(\frac{2c^2}{a}\right)k_w \tilde{w} \tag{7.105}$$

$$\tilde{w}\left[-\omega^2 \rho_s h a + 2\rho_0 c^2 + \left(\frac{E}{(1-v^2)}\right)\left(\frac{h}{a}\right)\right] = j\rho_0 c^2 k_w \tilde{\xi} \tag{7.106}$$

The solution for the ratio of k_w to the acoustic wavenumber k is

$$\left(\frac{k_w}{k}\right)^2 = \frac{(1+\alpha-\Omega^2)}{(1-\Omega^2)} \tag{7.107}$$

where $\alpha = 2\rho_0 c^2/[(h/a)(E/(1-v^2))]$. For $\Omega < 1$ this ratio is always greater than unity, so that k_w is real. The wave propagates with a phase speed less than that of the acoustic wave in a rigid tube. For steel pipes containing water, $\alpha = 0.019(h/a)$. According to this simple model, wall flexibility significantly reduces the low frequency wave speed for values of (h/a) even as high as 0.1. As either the wall thickness parameter (h/a), or the elastic modulus parameter $(E/\rho_0 c^2)$, tend to zero, the phase speed tends to zero: blood vessels represent the latter case, typical phase speeds being in the range 10–$14\,\text{ms}^{-1}$. The ratio of the effective acoustic impedance of the fluid in a flexible tube to that in a rigid tube equals k/k_w, which is less than unity. This fact may be exploited to attenuate the passage of pressure pulsations along liquid-filled, stiff-walled, piping systems by introducing a highly flexible insert.

It is intuitively obvious that straight tubes *in-vacuo* can also support torsional and quasi-longitudinal waves of zero circumferential order. The purely circumferential shell displacement and shear strain of the former do not couple to a contained fluid. In quasi-longitudinal wave motion the shell is coupled to a contained fluid through the thickness strain of the pipe wall due to the Poisson phenomenon. A more rigorous analysis of both the 'breathing' and quasi-longitudinal wave motions produces the following expressions for the associated axial wavenumbers (Pinnington and Briscoe, 1994).

$$\left(\frac{k_w}{k}\right)^2 = \frac{\left(1+\alpha-\Omega^2-v^2\right)}{\left(1-\Omega^2-v^2\right)} \tag{7.108}$$

$$\left(\frac{k_z}{k_L}\right)^2 = \frac{\left(1+\alpha-\Omega^2\right)}{\left(1+\alpha-\Omega^2-v^2\right)} \tag{7.109}$$

where $k_L = \omega/c_l'$. Poisson's ratio ν cannot exceed 0.5 and in cases where tube flexibility significantly influences 'breathing' wave speed, α is of the order of unity or greater. The expression in Eq. (7.108) yields values close to that derived on the basis of the 'ring' model [(Eq. (7.107)] except at values of Ω^2 less than ν^2. In this range of non-dimensional frequency, the quasi-longitudinal wave speed in the structure is slightly affected by the presence of a contained fluid.

The simplest model of the effect of contained fluid on the wave speed of the $n = 1$ 'beam bending' mode of a tube assumes that the mass per unit length of the fluid, considered as a lumped (rigid) element, is simply added to that of the tube. The effect is to increase the bending wavenumber by a factor of $\left[1 + (\rho_0 \rho_s 2h/a)\right]^{1/4}$. This assumption is valid at frequencies well below that of the non-dimensional cut-off frequency Ω_1 of the $n = 1$ acoustic mode in the rigid tube. As the frequency increases toward Ω_1 the effective mass of the fluid increases and the bending wavenumber increasingly exceeds that of the empty tube. In addition to the beam bending mode there exists an $n = 1$ 'acoustic' mode, the inverted commas indicating that it also involves tube bending motion, since an $n = 1$ acoustic field imposes a transverse force on the tube. This mode does not propagate below Ω_1. At higher frequencies, it propagates with a phase speed between that of fluid in a rigid tube and a column of fluid having pressure release ($p = 0$) boundaries, depending upon the material and geometric parameters of the tube.

Further reading on the subject of wave motion in straight pipes is provided by Variyart and Brennan (2002), who present a simplified analysis of wave motion in *in-vacuo* pipes. Fuller and Fahy (1981) and Finnveden (1997) present comprehensive analyses of wave motion in fluid-filled pipes, and Muggelton *et al.* (2002) consider the effects of a surrounding fluid in relation to the detection of leaks in buried water pipes. The subject of vibrational energy flow in fluid-filled circular cylindrical shells is treated by Xu and Zhang (2000).

Problems

7.1 Obtain explicit expressions for the modal fluid-loading terms a_{ppn} and a_{pqn} in Eq. (7.35) in the case of a rectangular, simply supported uniform panel that forms one wall of an otherwise rigid rectangular box. Hence demonstrate that inter-modal coupling may vary widely in magnitude and in dynamic form, i.e., analogous to stiffness or mass.

7.2 Why are the acoustic-pressure mode shapes and natural frequencies of an enclosure also appropriate to the velocity potential?

7.3 Check Eqs. (7.43) and (7.44) for dimensional consistency.

7.4 Prove that gyrostatic coupling of the form appearing in Eqs. (7.46) and (7.47) represents energy exchange between fluid and structural modes, and not energy dissipation in the coupling, unlike mode coupling through a viscous dashpot. Hint: Write the equations of motion of two mass-spring systems coupled through a dashpot.

7.5 Evaluate the maximum change from the *in-vacuo* natural frequency of the fundamental panel mode of a rectangular panel on a rectangular box according to Eq. (7.54), when the box measures $300\,mm \times 400\,mm \times 500\,mm$, the panel is 1 mm thick, measures $500\,mm \times 400\,mm$, and is made of aluminium alloy. Assume the fluid medium is air at atmospheric pressure and $20°C$. How would this frequency change if the temperature of the air were increased to $200°C$, assuming that the pressure was maintained constant? What reasons would you have for doubting the validity of this model?

7.6 A thin square panel is simply supported across one end of a square section duct, the other end of which is open to the atmosphere. Assuming that the resulting fundamental natural frequency of the panel, loaded by the fluid in the duct, is sufficiently low for only plane-wave propagation to occur in the duct, and that the small-*ka* expressions for circular piston radiation impedance provide reasonable approximations to the open end impedance, derive an equation of free vibration of the panel. Evaluate the damped natural frequency and loss factor of the lowest 'panel' mode for the following conditions: aluminium alloy panel of dimensions $300\,mm \times 300\,mm \times 2\,mm$; duct length 3 m; air at N.T.P. in the duct. Numerical or graphical solutions may be employed. Obtain an equivalent piston radius by equating areas.

7.7 Derive an expression for power radiated by a piston vibrating with frequency-independent velocity amplitude into a uniform tube of length l and diameter d that is terminated by a plug of acoustic impedance Z_l. Determine by numerical or graphical means the lowest frequencies of maximum and minimum power radiation in air when $d = 50\,mm$, $l = 500\,mm$ and $Z_l = 2e-5 \times (5-10j)$ $kgm^{-4}s^{-1}$. Explain your results in physical terms.

7.8 A piston slides freely in one end of a tube that is closed at the other end by a rigid plug pierced by a small hole. By assuming that the fluid velocity through the hole is proportional to the pressure difference across the hole, appropriately modify the equation in Section 7.2 to investigate the influence of the hole on the impedance characteristics of the piston–tube system. Neglect inertial loading of the fluid in the hole. Under what conditions is such neglect justified?

8 | Introduction to Numerically Based Analyses of Fluid–Structure Interaction

8.1 The Role of Numerical Analysis

So far we have mainly considered analytical techniques for solving the equations that express relations between the quantities we chose for our mathematical models of the physical world. The biggest advantage of analytic solutions is that they often display the dependence of the model behaviour on its non-dimensional parameters in an explicit fashion, and they lend themselves well to 'large-parameter' and 'small-parameter' asymptotic approximations, so that physical interpretation of the solutions may readily be achieved. The other side of the coin is that it is very often not possible to obtain analytic solutions to the model equations because of the geometric, material, dynamic and/or kinematic complexity of the system modelled. The prime example of this problem is that of fluid turbulence: the governing equations are well known, but a general analytic solution cannot be found. As seen in the previous chapter, the vibrational behaviour of bounded fluid and solid systems may generally be expressed in terms of series solutions to the governing equations; however, it is generally not possible to derive analytic expressions for the terms in the series, unless the systems and their boundaries are particularly simple.

As an alternative to seeking approximate analytical means of solving equations, we may assume various forms of distribution of the field quantities that are present in the governing equation of motion, together with sets of unknown coefficients that specify the contributions of these distributions to the total field. By entering expressions for these contributions into equations that satisfy the mechanical principles that govern the vibration of elastic systems, various computational techniques are used to produce estimates of the unknown coefficients. These procedures fall into the general category of 'numerical analysis'. Many systematised routines for performing numerical analysis have been introduced and the development of efficient computational techniques for this implementation is an important area of research.

It cannot be emphasised too strongly that the first priority in the mathematical analysis of the behaviour of physical systems is to generate a mathematical model appropriate to the desired objective of the analysis and to the likely degrees of influence and ranges of variation of the physical quantities involved. In other words, a model must incorporate representations of all the physical quantities thought to exert significant influences on the behaviour of the primary quantities of interest, and the selection of the quantities must be made on the basis of a reasonably thorough qualitative understanding of the physics of a problem, and of the physical inter-relations between the quantities involved. The desired degree of correspondence between the behaviour of the mathematical model and that of the physical system modelled necessarily influences the detail to which the modelling process extends. The reader may like to consider the latter point in relation to the choice of reference coordinate system for the analysis of dynamic behaviour of systems on the earth's surface. When and why should account be taken of the rotation of the earth? Closer to home, the question of how to represent the boundaries of a distributed system, which is part of a much more extensive system, is a problem frequently encountered in the field of vibration and acoustic analysis. The reason for emphasising the paramount importance of giving very careful thought to the process of generating a mathematical model is that, however sophisticated the numerical analysis techniques used, an inadequate model yields inadequate equations, and the subsequent solution is of little value. The technique of solution must never take pride of place over the model to which it is applied.

Numerical analysis can, of course, be performed to any degree of precision desired, subject to the limitations of the precision to which numbers can be held and manipulated in the computer. Although this is generally very high, significant errors can accumulate as a result of successive arithmetic operations, and vigilance in this respect should always be maintained. Although very precise numerical solutions can usually be obtained, the influence of individual system parameters which is vital in design development, may not be easy to evaluate from a welter of computer output, and much tedious interpretive work may often be avoided by employing a judicious combination of approximate analytical techniques and sensible parametric studies on the computer. Naturally the acid test is whether the theoretical results compare favourably with observed physical behaviour, and a healthy scepticism regarding the results of theoretical studies is the hallmark of the mature analyst.

Having noted the traps awaiting the unwary, let us now proceed with caution to consider various approaches to the numerical analysis of vibrating systems.

8.2   Numerical Analysis of Vibration in Solids and Fluids

We concentrate here on both vibration fields in solid structures and sound fields in fluids with the aim of providing tools for the vibroacoustic analysis of systems in which the two forms of field are coupled (i.e. they influence each other). The problems of practical interest divide into three main categories: analysis of the vibrational behaviour of fluid in volumes enclosed by structures such as shell or folded plates; analysis of the interaction between flexible shell or plate structures and sound fields incident upon them; and analysis of the acoustic fields generated in essentially unbounded regions of fluid by vibrating bodies or shell and plate structures.

We have already seen that fluid-loading effects may sometimes significantly affect structural motion, in which case a fully coupled analysis is necessary. Examples of such cases include underwater sonar transducers and liquid containment vessels such as fuel tanks and nuclear reactor pressure vessels. In the two most commonly used forms of numerical vibroacoustic analysis, field quantities that are actually continuous throughout a region of an elastic continuum are represented by their numerical values at discrete points in space, together, in most cases, with assumed interpolation functions, which approximate the distribution of these quantities between the points. Both, a vibration field in a structure and a sound field in a fluid have to satisfy the relevant governing wave equations, together with certain conditions on bounding edges or surfaces and, in the case of transient problems, certain initial conditions. In the case of steady-state radiation into a fluid region that is infinitely extended, certain conditions at infinity, which ensure that the field is physically valid, also have to be applied.

The analysis of vibrations in structures and sound fields in closed or nearly closed volumes is most commonly accomplished by using 'Finite Element Analysis' (FEA), in which the structure or fluid space are theoretically divided into contiguous elements of linear dimension substantially smaller than a structural or acoustic wavelength at the highest frequency of interest. This requirement is analogous to the Shannon criterion for the digital representation of signals, in which the sampling rate must exceed twice the highest frequency present in a signal. Field-variable distributions are assumed, and expressions for potential and kinetic energies, and work done by non-conservative forces, are derived in terms of variables assigned as degrees of freedom at nodal points on element boundaries. Then, Hamilton's variational principle (see Eq. (8.1)) is implemented and solved for the degrees of freedom that define the field (Meirovitch, 1967; Reddy, 1984a). This technique has been successfully implemented for all types of structure but it is more useful for enclosed volumes than for unbounded volumes because the necessary number of degrees of freedom is limited. It has been used most successfully for the analysis of sound fields in enclosures of arbitrary boundary geometry and for studying the behaviour of sound in ducts.

It is, in principle, possible to apply 'Finite Difference Analysis' (FDA) to sound fields. Again, the fluid region is divided by a line grid and field values are assigned to the grid intersection points. In the rectangular Cartesian coordinate system the grid is necessarily a square, whereas FEA can handle diverse geometric forms. The reason is that in FDA the derivatives in the partial differential wave equation are represented by finite-difference approximations, those pertaining to any given point involving values of surrounding grid points. An initial set of field values is assumed, and systematic techniques exist for iterating the values to a stable solution. The main practical problem with FDA in application to sound fields in volumes of arbitrary geometry is that it does not readily accommodate boundaries that do not conform to the grid line pattern. In addition, it is much more sensitive to local errors of field representation than FEA, in which intermediate energy expressions, which are the results of integration over the whole region, are formed.

The evaluation of the acoustic field generated by a vibrating surface in an infinitely extended volume of fluid is commonly accomplished by the application of 'Boundary Element Analysis' (BEA) to the evaluation of the field on the surface of the radiator, through the approximate solution of the Kirchhoff–Helmholtz integral [Eq. (3.180)]: the radiated field can easily be determined once the surface field is known. Numerical solutions are usually required in cases where the surface geometry of the radiator is not one of the small family for which the Green's function can be expressed in closed, or series form and satisfies either zero pressure or zero normal particle velocity on the surface. Hence it is usual to use the free-space form of the Green's function which is given by Eq. (3.169). The radiating surface is divided into discrete elements and the aim of the analysis is to obtain an estimate of the surface pressure distribution that corresponds to the assumed normal velocity distribution. As we shall see, there are certain fundamental analytical problems to overcome before numerical analysis can be successfully applied to this problem. In the fully coupled case, where the surface is driven by some form of applied forces, or is subject to an incident sound field, it is of course necessary to solve simultaneously for the surface field and for the motion, which both generate, and is affected by, the fluid loading. In such cases, it is possible to combine FEA and BEA.

The various computational methods employed to implement FEA and BEA are known collectively as FEM and BEM. In the following sections the principles of various numerical techniques are introduced, examples of applications are presented and some of the practical aspects of their application are discussed. In recent years, alternative techniques for estimating the sound field radiated by a vibrating solid body have been under development. In the 'Infinite Element' method, the non-reflecting boundary condition at infinity is replaced by a single layer of elements which have infinite radial extent. The infinite elements are constructed with radial wave functions which automatically satisfy the Sommerfeld condition at infinity. This chapter will not present the details of this method which can be found in the comprehensive summary written by Thompson (2006). In the 'Equivalent Source' method, distributions of virtual point monopole and dipole sources are assumed to exist within the structural envelope and the positions, amplitudes and phases are adjusted to minimise the error is reproducing the actual surface distribution of normal acceleration (Ochmann and Mechel, 2002).

8.3   Finite Element Analysis

The vibrations of distributed solid or fluid elastic systems, or combination thereof, are governed by equations of motion which must be satisfied at every point of the system and are subject to boundary and initial conditions. The equations of motion involve spatial derivatives of the field variables of order $2p$ while boundary conditions imply spatial derivatives of the field variables of order $2p - 1$. In general, the equations of motion are obtained from internal *compatibility* conditions which guarantee that the continuum is deformed without superposing or splitting of matter, and from the *dynamic equilibrium* principle which guarantees that every portion of the system is in equilibrium with reference to the external and internal forces or moments applied to it, including fictitious terms which are known as inertia forces and moments (after d'Alembert). The internal compatibility and dynamic equilibrium conditions involve derivatives of the field variables of order p and $2p$, respectively. The boundary conditions result from displacement/rotation compatibility conditions at the boundaries, which are called 'geometric' or 'essential' boundary conditions, and involve spatial derivatives of the field variables of order $p - 1$; and from force/moment balance conditions at the boundaries, which are called 'natural' or 'additional' boundary conditions and involve spatial derivatives of the field variables of order $2p - 1$ (Meirovitch, 1967; Timoshenko *et al.*, 1992).

The set of differential equations for the equations of motion and boundary conditions states the problem in a 'strong form' since they must be met at every point of the continuum. Alternatively, the problem can be defined in a 'weak form' if the equations of motion and boundary conditions are met only in an average sense by considering an integral expression of a function that implicitly contains the equations of motion and boundary conditions. For vibrational problems of elastic systems, use can be made of the fundamental variational principle of dynamics known as Hamilton's principle, which can be formulated either in terms of displacement-like or force-like functions (Gladwell and Zimmermann, 1966). In the former case, Hamilton's principle states that among all displacements (linear and angular) which satisfy the prescribed kinematic boundary conditions and the prescribed conditions at times $t = t_1$ and $t = t_2$, the actual solution satisfies the equation

$$\int_{t_1}^{t_2} [\delta\,(T - U) + \delta W_{nc}]\,dt = 0 \qquad (8.1)$$

where T and U denote, respectively, the kinetic and potential energies of the system expressed in terms of displacement-like functions and δW_{nc} is the virtual work done by the non-conservative forces (e.g., external forces and moments and dissipation forces and moments): δ denotes the first variation of a quantity (Courant and Hilbert, 1962). In the latter case, Hamilton's principle states that among all forces and moments which satisfy the prescribed kinetic boundary conditions and the prescribed conditions at times $t = t_1$ and $t = t_2$, the actual solution satisfies Eq. (8.1), where T and U denote, respectively, the kinetic and potential energies of the system expressed in terms of force-like quantities and δW_{nc} is the virtual work done by the non-conservative forces/moments. In this chapter, structural and acoustic vibration problems are analysed using Hamilton's principle, formulated in terms of displacement-like and force-like quantities respectively.

In general, it is rather difficult to find the field variables that are both solutions to the equations of motion and satisfy the boundary conditions of a problem stated in strong form. This obstacle can be overcome by applying the Rayleigh–Ritz method to a problem stated in weak form, which results in a substitute problem that has a finite number of unknown functions of time and is described by ordinary differential equations rather than by partial differential equations. In the Rayleigh–Ritz method the vibrational field is approximated by a series of linearly independent 'prescribed functions' of space, each multiplied by an unknown function of time that is called a 'generalised coordinate'. In order to guarantee that the ordinary differential equations resulting from the Rayleigh–Ritz approximation have a solution, the prescribed functions should be 'comparison functions' that satisfy both the geometric and natural boundary conditions and be

$2p$-times differentiable. These are severe requirements that make the selection of the prescribed functions quite problematic. For this reason, some of the above conditions are relaxed and the prescribed functions are chosen to be 'admissible functions' which satisfy only the geometric-boundary conditions and are only p-times differentiable. This prevents the integrals for the energy and work functions in Hamilton's principle (which involve spatial derivatives up to order p), from becoming infinite, and the resulting algebraic equations from having no solution. However, in order to guarantee convergence to the exact solution, the set of prescribed functions must be 'complete'; that is they must arbitrarily closely match the field variables and their derivatives that appear in the energy and work functions to be used in Eq. (8.1) (Reddy, 1984b; Cook *et al.*, 1989).

Although finding prescribed functions to be used in the Rayleigh–Ritz method for the solution of the variational formulation given by Hamilton's principle is easier than solving the differential equations of motion derived from dynamic equilibrium and compatibility conditions, the number of problems to which the Rayleigh–Ritz method can be successfully applied is restricted to systems with rather regular geometries and uniformly distributed physical properties. This limitation has been overcome by the Finite Element Method which can be seen as an automatic procedure for obtaining a set of prescribed functions for use in the Rayleigh–Ritz formulation. In the Finite Element Method the physical domain of concern is divided into a finite set of contiguous elements defined by a gridwork of lines (mesh) which may be straight or curved. At discrete points (nodes) on this grid, which may lie solely at intersections or also at intermediate points, certain field functions, plus their spatial derivatives, which are pertinent to the problem under consideration, are selected as the generalised coordinates of the Rayleigh–Ritz series expansion. The distributions of the field functions within each element are then derived with reference to the nodal generalised coordinates using prescribed functions which are therefore defined within the element only. Thus, in this case, the problem of finding suitable prescribed functions is much easier since it is limited to the small portions of the structures with rather regular geometry and whose physical properties are uniformly distributed.

8.4   Finite Element Analysis of Vibrations in Solid Structures

Finite Element modelling is first discussed for the analysis of vibrations in solids. The general formulation is introduced by considering the flexural vibrations of uniform slender beams and thin flat plates. The approximate solution using the Rayleigh–Ritz approach is first introduced for the flexural vibration of a beam. Finite Element Analysis is then applied to the flexural vibration of slender beams and thin flat plates. Finally, a brief review of Finite Elements for other types of structure is presented. The equations of motion are derived using Hamilton's principle formulated using displacement-like functions.

8.4.1   Flexural Vibration of Slender Beams: Rayleigh–Ritz Method

Consider a one-dimensional structure of length L whose axial coordinate is x and transverse displacement $v(x, t)$.[1] The Rayleigh–Ritz method approximates the solution by a finite series expansion of the form

$$v(x, t) = \sum_{i=1}^{n} d_i(x) v_i(t) \tag{8.2}$$

where $d_i(x)$ represents the prescribed functions of x and $v_i(t)$ are the unknown coefficients of the expansion. A continuous deformable body consists of an infinite number of points, and therefore it has infinitely many degrees of freedom. By assuming that the displacement is given by expression (8.2), the continuous system has been reduced to an equivalent, stiffer system, with a finite number of degrees of freedom with $v_{n+1} = v_{n+2} = \cdots = 0$. Substituting (8.2) into the appropriate energy expressions T, U and δW_{nc} reduces the continuous structure to a multi-degree of freedom system with v_1, v_2, ..., v_n as degrees of freedom.

As discussed above, in order to guarantee convergence to the exact solution, the functions $d_i(x)$ must satisfy the following conditions (Meirovitch, 1967; Reddy, 1984a):

1. Be linearly independent;
2. Be internally compatible and therefore p times differentiable over the spatial domain;
3. Satisfy the geometric boundary conditions, and therefore $p - 1$ times differentiable over the boundaries;
4. Form a complete series, that is
$$\lim_{n \to \infty} \int_0^L (v(x, t) - \sum_{i=1}^{n} d_i(x) v_i(t))^2 \times dx = 0.$$

Polynomials in power of x, trigonometric functions, Legendre, Tchebycheff and Jacobi or hypergeometric polynomials are all complete series of functions.

---

[1] In accordance with the Finite Element convention for structural dynamics, in this chapter structural displacements and velocities are denoted by u, v, w and $\dot{u}$, $\dot{v}$, $\dot{w}$ in the x, y, z directions. Also, angular displacements and velocities around the x, y, z axes are denoted by θ_x, θ_y, θ_z and $\dot{\theta}_x$, $\dot{\theta}_y$, $\dot{\theta}_z$. Normalised coordinates for the x and y directions are denoted by ξ and η.

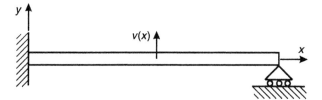

Fig. 8.1.   Clamped, simply supported beam.

It should be noted that in using the Rayleigh–Ritz method the equations of motion and boundary conditions will only be satisfied approximately.

As an example, the natural frequencies and associated natural modes of free flexural vibration of the uniform, undamped, clamped, simply supported beam structure in Fig. 8.1 are derived. In this case, $\delta W_{nc} = 0$ so that only the kinetic energy and potential energy expressions are required in Eq. (8.1). For structural vibrations, the potential energy is the strain energy. The kinetic and strain energy functions for pure flexural vibration of a slender beam (rotary inertia and shear strain effects neglected) are given by the following expressions (Petyt, 1998):

$$T = \frac{1}{2} \int_0^L \rho A \left(\frac{\partial v(x, t)}{\partial t} \right)^2 dx \tag{8.3}$$

$$U = \frac{1}{2} \int_0^L EI_z \left(\frac{\partial^2 v(x, t)}{\partial x^2} \right)^2 dx \tag{8.4}$$

where ρ and E are the density and Young's modulus of elasticity of the material respectively, A is the cross-sectional area of the beam, I_z is the second moment of area of the beam cross section about the z axis and L is the beam length. According to these two energy expressions, the integral in Eq. (8.1) involves derivatives in x up to the order $p = 2$. Therefore, the prescribed functions $d_i(x)$ to be used in the finite series expansion of Eq. (8.2) must be differentiable at least twice. Also these functions must be linearly independent, form a complete series and satisfy the geometric boundary conditions; that is, $v(0, t) = 0$, $\partial v(x, t)/\partial x|_{x=0} = 0$ and $v(L, t) = 0$, $\partial v(x, t)/\partial x|_{x=L} \neq 0$. A set of functions that satisfy all these conditions are, for example, power terms of order $i \geq 1$

$$d_i(x) = x^{i+1} (L - x) \tag{8.5}$$

so that the function in Eq. (8.2) corresponds to a polynomial of order n whose coefficients are indeed the $v_i(t)$ unknown functions of t.

When the displacement of the beam is approximated by a finite expansion, as in Eq. (8.2), the mathematical formulation of the kinetic energy and strain energy as given in Eqs. (8.3) and (8.4) is reduced to that of a discrete system with a finite number of degrees of freedom (the $v_i(t)$ unknown functions of t). If the finite series expansion of Eq. (8.2) is expressed in matrix form such that[2]

$$v(x, t) = \lfloor \mathbf{d}(x) \rfloor \{\mathbf{v}(t)\} = \lfloor d_1(x) \, d_2(x) \, \ldots \, d_n(x) \rfloor \begin{bmatrix} v_1(t) \\ v_2(t) \\ \vdots \\ v_n(t) \end{bmatrix} \tag{8.6}$$

the two energy formulae can be expressed as:

$$T(t) = \frac{1}{2} \{\dot{\mathbf{v}}(t)\}^T [\mathbf{M}] \{\dot{\mathbf{v}}(t)\} \tag{8.7}$$

and

$$U(t) = \frac{1}{2} \{\mathbf{v}(t)\}^T [\mathbf{K}] \{\mathbf{v}(t)\} \tag{8.8}$$

$[\mathbf{M}]$ and $[\mathbf{K}]$ are the inertia and stiffness matrices which, for flexural vibration of the clamped simply supported beam, are given by

$$[\mathbf{M}] = \rho A \int_0^L \lfloor \mathbf{d}(x) \rfloor^T \lfloor \mathbf{d}(x) \rfloor \, dx \tag{8.9}$$

with

$$M_{k,s} = \rho A \left\{ \frac{L^{k+s+5}}{k+s+3} - 2\frac{L^{k+s+5}}{k+s+4} + \frac{L^{k+s+5}}{k+s+5} \right\} \tag{8.10}$$

and

$$[\mathbf{K}] = EI_z \int_0^L \left[\frac{\partial^2 \mathbf{d}(x)}{\partial x^2} \right]^T \left[\frac{\partial^2 \mathbf{d}(x)}{\partial x^2} \right] \, dx \tag{8.11}$$

[2] In this Chapter $\lfloor \, \rfloor$ indicates row vectors while column vectors are represented either by [] or {}.

with

$$K_{k,s} = EI_z \left\{ \left[\frac{(k+1)k(s+1)sL^{k+s+1}}{k+s-1} \right] - \left[\frac{(k+2)(k+1)(s+1)sL^{k+s+1}}{k+s} \right] \right.$$

$$- \left[\frac{(k+1)k(s+2)(s+1)L^{k+s+1}}{k+s} \right]$$

$$\left. + \left[\frac{(k+2)(k+1)(s+2)(s+1)L^{k+s+1}}{k+s+1} \right] \right\} \tag{8.12}$$

Assuming the work done by external forces and dissipation forces to be zero, Hamilton's principle yields

$$\int_{t_1}^{t_2} \delta\,(T - U)dt = \int_{t_1}^{t_2} \delta\left(\tfrac{1}{2} \{\dot{v}(t)\}^T [M] \{\dot{v}(t)\} - \tfrac{1}{2} \{v(t)\}^T [K] \{v(t)\} \right)dt = 0 \tag{8.13}$$

which, assuming that the operators δ and $\partial(\ )/\partial t$ are commutative, after integrating the first term by parts, gives (Petyt, 1998)

$$- \int_{t_1}^{t_2} \{\delta v(t)\}^T ([M] \{\ddot{v}(t)\} + [K] \{v(t)\})dt = 0 \tag{8.14}$$

Since $\{\delta v(t)\}$ is arbitrary, Eq. (8.14) is satisfied only if

$$[M] \{\ddot{v}(t)\} + [K] \{v(t)\} = \{0\} \tag{8.15}$$

Considering harmonic motion in the form $\{v\} = \{a\} \sin \omega t$, where ω is the circular frequency and the complex amplitudes in $\{a\} = \lfloor A_1 \quad A_2 \quad \cdots \quad A_n \rfloor^T$ are independent of time, Eq. (8.15) gives

$$\left[K - \omega^2 M \right] \{a\} = \{0\} \tag{8.16}$$

which represent a set of n linear homogeneous equations in the unknowns $A_1, A_2, \ldots, A_n$. The condition for these equations to have a non-zero solution is that the determinant of coefficients should vanish: $|K - \omega^2 M| = 0$. This condition leads to a polynomial of degree n in ω^2. The n roots of this polynomial $\omega_1^2, \omega_2^2, \ldots, \omega_n^2$, which are all real and positive (Bishop *et al.*, 1965), are the squares of the approximate values of the first n natural frequencies of the system. These approximate natural frequencies will be greater than the true natural frequencies of the system because the expressions for the strain energy and kinetic energy have been calculated for an equivalent stiffer beam with only n degrees of freedom (in fact, the remaining degrees of freedoms have been set equal to zero).

Corresponding to each natural frequency ω_j, a unique solution (to within an arbitrary constant) exists to Eq. (8.16) for $\{\mathbf{a}\}_j$ which, when combined with the prescribed functions $d_i(x)$, gives the approximate shapes of the natural modes of vibration. The approximate shape of the j-th natural mode is therefore given by

$$\psi_j(x) = \sum_{i=1}^{n} d_i(x) A_{i,j} = \lfloor d_1(x)\, d_2(x) \cdots d_n(x) \rfloor \begin{bmatrix} A_1 \\ A_2 \\ \vdots \\ A_n \end{bmatrix}_j \tag{8.17}$$

Convergence to the true frequencies and mode shapes is obtained as the number of terms in the approximating expression in Eq. (8.2) is increased.

As an example, the first four natural frequencies and natural modes of a clamped simply supported aluminium beam of rectangular cross section have been derived assuming $\rho = 2700$ kg m^{-3}, $E = 7.1 \times 10^{10}$ N m^{-2}, length $L = 0.5$ m, width $b = 0.02$ m and thickness $h = 0.005$ m. Figure 8.2 compares the natural frequencies and natural modes derived with the exact analytical solution

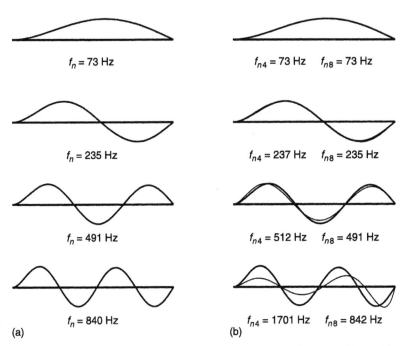

(a)   (b)

Fig. 8.2. Flexural natural frequencies and modes of a clamped simply supported beam: (a) exact solution (Bishop and Johnson, 1960); (b) Rayleigh–Ritz solution with the polynomial function in Eq. (8.6) containing 4 (faint line) and 8 (solid line) power terms as expressed in Eq. (8.5).

(Bishop and Johnson, 1960) and the approximate Rayleigh–Ritz formulation. As expected, the natural frequencies and natural modes converge to the exact ones as the order of the polynomial function in Eq. (8.6) is raised. Also the number of power terms to be used in (8.6) is directly linked to the highest order of the natural frequency/mode to be calculated.

8.4.2 Flexural Vibration of Slender Beams: Finite Element Analysis

The major drawback to the Rayleigh–Ritz method is the difficulty of constructing a set of prescribed functions, particularly for a built-up structure comprising various different forms of component (beams, plates, shells, solids) and possessing irregular geometries. This difficulty can be overcome by using the Finite Element Method that provides an automatic procedure for constructing such functions. In the Finite Element Method the prescribed functions are constructed as follows:

1. Divide the structure into contiguous one-, two- or three-dimensional elements defined by a gridwork of straight or curved lines (mesh);
2. Define a set of node points that are normally placed at the vertices of each element and, in some cases, also along the edges of each element;
3. Associate with each node point a given number of degrees of freedom (displacement, slope, etc.);
4. Construct a set of prescribed functions that satisfy the Rayleigh–Ritz convergence criteria such that each one gives a unit value for one degree of freedom and zero value for all the others.

The Finite Element procedure is illustrated for the flexural vibration of the slender beam in Fig. 8.1. As shown in Fig. 8.3(a), the beam has been subdivided into four elements of equal length that define five node points at their vertices. The highest derivative in x appearing in the energy expressions of Eqs. (8.3) and (8.4) is the second. Therefore, in order to satisfy the Rayleigh–Ritz convergence criteria, the prescribed functions themselves and their first derivatives in x need to be continuous so that the second derivative can be calculated at every point of the beam. Thus, as shown in Fig. 8.3(b), the transverse displacement v and the angular displacement θ_z are the degrees of freedom required at each node point so that the displacement field and its first derivative are guaranteed to be continuous across elements and thus the second derivative of the displacement function is defined at every point of the beam so that the Rayleigh–Ritz convergence criteria are satisfied. For the specific example at hand, a total of ten degrees of freedom is used in the model. As a result, the ten third-order polynomial prescribed functions shown in Fig. 8.4, which satisfy the Rayleigh–Ritz convergence criteria, have been constructed by giving each node point in turn a unit transverse or angular

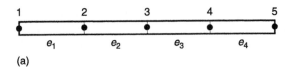

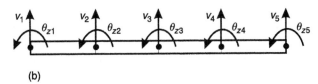

(a)

(b)

Fig. 8.3. (a) Beam divided into four elements; (b) transverse and angular degrees of freedom at the five nodal positions shown in (a).

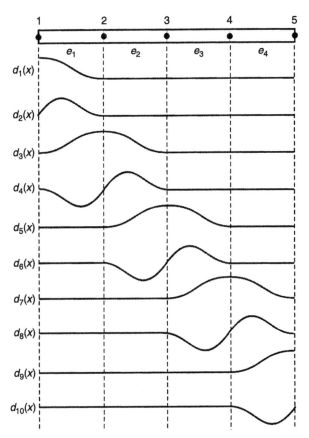

Fig. 8.4. Prescribed functions for bending vibration of a beam.

displacement, whilst maintaining zero displacements at all other nodes. In the Rayleigh–Ritz method, convergence is obtained as the number of prescribed functions is increased. To increase the number of prescribed functions in the Finite Element Method, the number of node points, and therefore the number of elements, is increased (Oden, 1972).

As shown in Fig. 8.4, the way the prescribed functions are constructed is such that each element deforms in only four deformation patterns and that these are the same for each element. The emphasis therefore is on determining deformation patterns for individual elements and not for the whole structure. These pre-scribed functions are referred to as 'element displacement functions' or 'element shape functions'. In order to satisfy the convergence criteria of the Rayleigh–Ritz method, the shape functions should comply with the following conditions:

1. Be linearly independent;
2. If the integral in Eq. (8.1) involves derivatives up to order p, be continuous and have continuous derivatives up to order $(p-1)$ both within the elements and across the elements boundaries (compatibility condition);
3. Satisfy the geometric boundary conditions;
4. Form a complete series (completeness condition).

As discussed in the introductory part of this section in relation to the Rayleigh–Ritz method, the compatibility condition is necessary to avoid shape functions which may result in any term in the integrals for the Hamilton's principle becom-ing infinite, and thus give a set of algebraic equations with no solution. Also, the completeness condition ensures that all the terms occurring in the integrals for the Hamilton's principle are, in the limit, capable of being approximated as nearly as possible and give single constant values over the infinitesimal parts of the elements' domains of integration. This is a necessary condition for con-vergence to the exact vibration field since, as seen above, the accuracy of the solution can be enhanced by dividing the structure into an increasing number of finite elements of decreasing size.

If the element shape functions are polynomial functions, then, in order to form a complete series, they must be complete polynomials of at least degree p. If any terms of degree greater than p are used, they need not be complete. However, the rate of convergence to the exact solution of the problem is governed by the order of completeness of the polynomial. When the prescribed functions satisfy the compatibility and completeness conditions the elements are said to be 'conforming'.

There are a number of ways of defining the shape function of an element. The most common of these are as follows (Petyt, 1998):

1. By inspection;
2. By assuming a polynomial function having the appropriate number of terms;

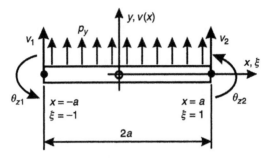

Fig. 8.5.   Flexural beam element.

3. By solving the equations of static equilibrium to determine the deformation of the element due to prescribed boundary displacement.

In practice, the most appropriate method is used for each type of element. As discussed above, the flexural beam element of Fig. 8.5 is characterised by transverse and angular displacements at the nodal positions. If, for example, a polynomial shape function is sought, then, although the Rayleigh–Ritz convergence criteria require that the shape function and its first derivative are continuous functions so that the shape function is twice differentiable in x, a polynomial of third order is used so that the vibratory field $v(x)$ within the element is uniquely defined by the linear, v_1 and v_2, and angular, θ_{z1} and θ_{z2}, nodal displacements.

As shown in Fig. 8.5, the polynomial shape function can be expressed as a function of the dimensionless coordinate $\xi = x/a$, where $2a$ is the element length, so that

$$v(\xi) = \alpha_1 + \alpha_2\xi + \alpha_3\xi^2 + \alpha_4\xi^3 \tag{8.18}$$

The transverse displacement given by Eq. (8.18) can be expressed in matrix form as

$$v(\xi) = \lfloor \mathbf{p}(\xi) \rfloor \{\mathbf{a}(\alpha)\} = \begin{bmatrix} 1 & \xi & \xi^2 & \xi^3 \end{bmatrix} \begin{bmatrix} \alpha_1 \\ \alpha_2 \\ \alpha_3 \\ \alpha_4 \end{bmatrix} \tag{8.19}$$

Thus, the transverse displacements v_1 and v_2, and angular displacements $\theta_{z1} = \left.\dfrac{\partial v}{\partial x}\right|_{x=-a} = \left.\dfrac{1}{a}\dfrac{\partial v}{\partial \xi}\right|_{\xi=-1}$ and $\theta_{z2} = \left.\dfrac{\partial v}{\partial x}\right|_{x=a} = \left.\dfrac{1}{a}\dfrac{\partial v}{\partial \xi}\right|_{\xi=1}$ at the two nodal positions,

($\xi = \pm 1$), are given by

$$
\begin{Bmatrix} v_1 \\ a\theta_{z1} \\ v_2 \\ a\theta_{z2} \end{Bmatrix} = \begin{bmatrix} 1 & -1 & 1 & -1 \\ 0 & 1 & -2 & 3 \\ 1 & 1 & 1 & 1 \\ 0 & 1 & 2 & 3 \end{bmatrix} \begin{Bmatrix} \alpha_1 \\ \alpha_2 \\ \alpha_3 \\ \alpha_4 \end{Bmatrix}
$$

(8.20a)

or

$$
\{\bar{\mathbf{v}}_e\} = [\mathbf{A}_e]\{\mathbf{a}(\alpha)\}
$$

(8.20b)

Solving for $\alpha_1, \ldots, \alpha_4$ gives

$$
\{\mathbf{a}(\alpha)\} = [\mathbf{A}_e]^{-1}\{\bar{\mathbf{v}}_e\}
$$

(8.21)

Equation (8.21) can be written in an alternative form as

$$
\{\mathbf{a}(\alpha)\} = [\mathbf{C}_e]\{\mathbf{v}_e\}
$$

(8.22)

where

$$
\{\mathbf{v}_e\}^T = \lfloor v_1 \quad \theta_{z1} \quad v_2 \quad \theta_{z2} \rfloor
$$

(8.23)

is the vector with the element displacements at the nodal points and

$$
[\mathbf{C}_e] = \begin{bmatrix} 2 & a & 2 & -a \\ -3 & -a & 3 & -a \\ 0 & -a & 0 & a \\ 1 & a & -1 & a \end{bmatrix}
$$

(8.24)

Substituting Eq. (8.22) into Eq. (8.19) gives

$$
v(\xi) = \lfloor \mathbf{p}(\xi) \rfloor [\mathbf{C}_e]\{\mathbf{v}_e\} = \lfloor \mathbf{n}(\xi) \rfloor \{\mathbf{v}_e\}
$$

(8.25)

with

$$
\lfloor \mathbf{n}(\xi) \rfloor = \lfloor N_1(\xi) \quad aN_2(\xi) \quad N_3(\xi) \quad aN_4(\xi) \rfloor
$$

(8.26)

where

$$
N_1(\xi) = \tfrac{1}{4}\left(2 - 3\xi + \xi^3\right), \qquad N_2(\xi) = \tfrac{1}{4}\left(1 - \xi - \xi^2 + \xi^3\right),
$$

$$
N_3(\xi) = \tfrac{1}{4}\left(2 + 3\xi - \xi^3\right) \quad \text{and} \quad N_4(\xi) = \tfrac{1}{4}\left(-1 - \xi + \xi^2 + \xi^3\right)
$$

(8.27a–d)

Once the transverse displacement for each element is found with Eq. (8.25) it is then straightforward to evaluate the energy expressions in Eqs. (8.3) and (8.4) by calculating the kinetic and strain energies for each element and then summing the contributions from the elements. When approached in this way, there is no reason why the elements should be identical. For example, in Fig. 8.3 the elements could have different cross-sectional areas and lengths as well as different physical properties. The energy expressions for a single element are

$$T_e = \tfrac{1}{2} \int_{-1}^{+1} \rho A \dot{v}^2(\xi, t) \, a \, d\xi$$

and   (8.28a,b)

$$U_e = \tfrac{1}{2} \int_{-1}^{+1} E I_z \frac{1}{a^4} \left(\frac{\partial^2 v(\xi, t)}{\partial \xi^2} \right)^2 a \, d\xi$$

Substituting Eq. (8.25) into Eqs. (8.28a,b) gives

$$T_e = \tfrac{1}{2} \{\dot{v}_e\}^T [M_e] \{\dot{v}_e\}$$

(8.29a,b)

$$U_e = \tfrac{1}{2} \{v_e\}^T [K_e] \{v_e\}$$

where

$$[M_e] = \rho A a \int_{-1}^{+1} \lfloor \mathbf{n}(\xi) \rfloor^T \lfloor \mathbf{n}(\xi) \rfloor \, d\xi = \frac{\rho A a}{105} \begin{bmatrix} 78 & 22a & 27 & -13a \\ 22a & 8a^2 & 13a & -6a^2 \\ 27 & 13a & 78 & -22a \\ -13a & -6a^2 & -22a & 8a^2 \end{bmatrix}$$

(8.30)

is the 'element inertia matrix', and

$$[K_e] = \frac{E I_z}{a^3} \int_{-1}^{+1} \left\lfloor \frac{\partial^2 \mathbf{n}(\xi)}{\partial \xi^2} \right\rfloor^T \left\lfloor \frac{\partial^2 \mathbf{n}(\xi)}{\partial \xi^2} \right\rfloor \, d\xi = \frac{E I_z}{2a^3} \begin{bmatrix} 3 & 3a & -3 & 3a \\ 3a & 4a^2 & -3a & 2a^2 \\ -3 & -3a & 3 & -3a \\ 3a & 2a^2 & -3a & 4a^2 \end{bmatrix}$$

(8.31)

is the 'element stiffness matrix'. The energies of the complete beam are obtained by adding together the energies of all the individual elements. Before carrying this out it is necessary to relate the displacements and rotations of a single element $\{v_e\}$ to the set of displacements and rotations for the complete beam $\{\bar{v}\}$, where

$$\{\bar{v}\}^T = \lfloor v_1 \quad \theta_{z1} \quad v_2 \quad \theta_{z2} \quad v_3 \quad \theta_{z3} \quad v_4 \quad \theta_{z4} \quad v_5 \quad \theta_{z5} \rfloor \qquad (8.32)$$

For element e the relation is

$$\{v_e\} = [a_e]\{\bar{v}\} \qquad (8.33)$$

where, for the first element, $[a_e]$ takes the form

$$[a_1] = \begin{bmatrix} 1 & 0 & 0 & 0 & 0 & 0 & 0 & 0 & 0 & 0 \\ 0 & 1 & 0 & 0 & 0 & 0 & 0 & 0 & 0 & 0 \\ 0 & 0 & 1 & 0 & 0 & 0 & 0 & 0 & 0 & 0 \\ 0 & 0 & 0 & 1 & 0 & 0 & 0 & 0 & 0 & 0 \end{bmatrix} \qquad (8.34)$$

Substituting Eq. (8.33) into Eqs. (8.29a,b) and summing over all the elements gives the total kinetic and strain energies for the beam:

$$T = \tfrac{1}{2}\{\dot{\bar{v}}\}^T [\bar{M}]\{\dot{\bar{v}}\}$$

and   (8.35a,b)

$$U = \frac{1}{2}\{\bar{v}\}^T [\bar{K}]\{\bar{v}\}$$

where

$$[\bar{M}] = \sum_{e=1}^{4} [a_e]^T [M_e][a_e]$$

$$\qquad\qquad (8.36a,b)$$

$$[\bar{K}] = \sum_{e=1}^{4} [a_e]^T [K_e][a_e]$$

The matrix products $[a_e]^T [M_e][a_e]$ and $[a_e]^T [K_e][a_e]$, in Eqs. (8.36a,b), effectively locate the positions in $[\bar{M}]$ and $[\bar{K}]$ to which the elements of $[M_e]$ and $[K_e]$, respectively, have to be added. However, in practice it is unnecessary to form the four matrices $[a_e]$ and carry out the matrix multiplication. The information required to form the matrices $[\bar{M}]$ and $[\bar{K}]$ can in fact be obtained from the indexes given to the nodal points. Nodes $4(e-1)+1$ to $4e$ correspond to the element e. Therefore, the four rows and columns of the element inertia or stiffness matrices [Eqs. (8.30) and (8.31)] are added into the rows and the columns $4(e - 1) + 1$ to $4e$ of the inertia and stiffness matrices respectively for the complete beam. This procedure is known as the 'assembly process'. Also, note that if the elements and nodes are numbered progressively as in Fig. 8.3, then the matrices $[\overline{\mathbf{M}}]$ and $[\overline{\mathbf{K}}]$ are symmetric and banded. These properties can be exploited when writing computer programs. All the information displayed in Eqs. (8.36a,b) can be stored in nine locations, whereas 25 locations would be required to store the complete matrix $[\overline{\mathbf{M}}]$ or $[\overline{\mathbf{K}}]$.

Once the inertia and stiffness matrices [Eqs. (8.36a,b)] are derived, so that the kinetic and strain energies are algebraically formulated, the geometric boundary conditions have to be satisfied. For example, if the Finite Element Model of Fig. 8.3 is used to analyse the dynamic response of the clamped simply supported beam as in Fig. 8.1, the nodal linear displacements v_1, v_5 and nodal angular displacement θ_{z1} are zero. This condition is simply introduced by omitting v_1, v_5 and θ_{z1} from the set of degrees of freedom for the complete beam [Eq. (8.32)]

$$\{\mathbf{v}\}^T = \lfloor v_2 \quad \theta_{z2} \quad v_3 \quad \theta_{z3} \quad v_4 \quad \theta_{z4} \quad \theta_{z5} \rfloor \tag{8.37}$$

and at the same time omitting the first two rows and columns and the penultimate row and column from the inertia and stiffness matrices in Eqs. (8.36a,b) for the complete beam:

$$[\mathbf{M}] = \begin{bmatrix} m_{13} & m_{14} & \cdots & m_{18} & m_{110} \\ m_{23} & m_{24} & \cdots & m_{28} & m_{210} \\ \vdots & \vdots & \ddots & \vdots & \vdots \\ m_{83} & m_{84} & \cdots & m_{88} & m_{810} \\ m_{103} & m_{104} & \cdots & m_{108} & m_{1010} \end{bmatrix} \quad [\mathbf{K}] = \begin{bmatrix} k_{13} & k_{14} & \cdots & k_{18} & k_{110} \\ k_{23} & k_{24} & \cdots & k_{28} & k_{210} \\ \vdots & \vdots & \ddots & \vdots & \vdots \\ k_{83} & k_{84} & \cdots & k_{88} & k_{810} \\ k_{103} & k_{104} & \cdots & k_{108} & k_{1010} \end{bmatrix}$$
$$\tag{8.38a,b}$$

As in the Rayleigh–Ritz method, the total energy expressions [Eqs. (8.35a,b)], derived using the displacements vector in Eq. (8.37) and the inertia and stiffness matrices in Eqs. (8.38a,b), are substituted into the integral in Eq. (8.1) for the Hamilton's principle so that, according to the formulation presented in Eqs. (8.13)–(8.15), the following set of second order ordinary differential equations is obtained:

$$[\mathbf{M}]\{\ddot{\mathbf{v}}(t)\} + [\mathbf{K}]\{\mathbf{v}(t)\} = \{\mathbf{0}\} \tag{8.39}$$

where $\{\mathbf{v}(t)\}$ represents a column matrix of nodal degrees of freedom as shown in Eq. (8.37).

In the Rayleigh–Ritz method the accuracy of the solution is increased by increasing the number of prescribed functions in the assumed series. To increase the number of prescribed functions in the Finite Element Method, the number of node points, and therefore the number of elements, is increased. The shape function for each element should satisfy the Rayleigh–Ritz convergence criteria. In particular, the functions and their derivatives up to order $(p - 1)$ should be continuous across element boundaries. This version of the Finite Element Method is often referred to as the h-method where h refers to the size of the element (Zienkiewicz and Taylor, 2000).

The natural frequencies and natural modes of the clamped simply supported beam in Fig. 8.1 considered in the example for the Rayleigh–Ritz formulation have been derived using the Finite Element formulation presented above, assuming that the beam is divided into either four or eight elements. Figure 8.6 compares the first four exact natural frequencies and natural modes with the approximate Finite Element solutions. This figure confirms that in order to increase the accuracy of the calculation it is necessary to increase the number of elements. As a 'rule of thumb', in order to correctly derive higher order natural frequencies,

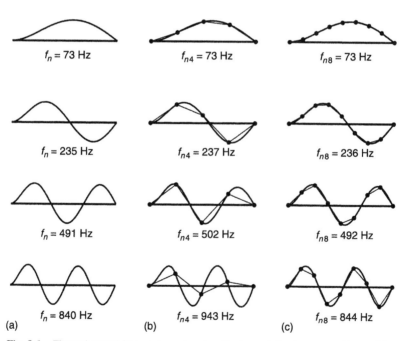

Fig. 8.6.   Flexural natural frequencies and modes of a clamped simply supported beam: (a) exact solution; Finite Element solution with (b) 4 and (c) 8 elements. Linear interpolation of nodal transverse displacements (thin line); cubic interpolation of transverse and angular displacements (thick line).

six elements per bending wavelength should be used: for beams and plates in flexure, the total number increases as the square root of frequency.

8.4.3   Flexural Vibration of Thin Flat Plates: Finite Element Analysis

Many vibroacoustic problems involve the vibration of closed shells and the acoustics of fluids in the enclosed volumes. In general, the structures take the form of folded flat plates or curved shell structures whose vibrations transverse to the mid-plane are coupled with the fluid in the cavity. The Finite Element Analysis for flexural vibration of a thin flat plate is presented here as an example since it will be used later to analyse the vibration of a thin rectangular plate that forms one wall of an otherwise rigid rectangular box coupled to a contained fluid. In this formulation, the effects of non-conservative forces due to external excitations and dissipative effects will also be taken into account. Therefore the virtual work done by external forces δW_f, and non-conservative dissipative effects δW_d, will be included in the application of Hamilton's principle.

Following the procedure introduced above, the flexural vibration of the rectangular plate shown in Fig. 8.7(a) can be analysed by dividing it up into an assemblage of two-dimensional finite elements called thin-plate flexural elements. As shown in Fig. 8.8, these elements may be either quadrilateral or triangular in shape with straight or curved sides and, four/eight or three/six nodes respectively. A comprehensive formulation of the Finite Element Models for these elements can be found, for example, in Cook et al. (1989), Bathe (1996), Petyt (1998), and Zienkiewicz and Taylor (2000). As an example, the simple case of a rectangular, thin-plate, non-conforming element with four nodes is introduced. Thus, as shown in Fig. 8.7b, a regular mesh of rectangular elements with four nodal points at the vertices has been constructed.

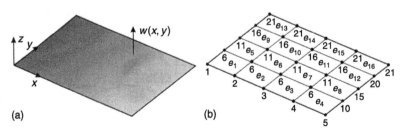

Fig. 8.7. (a) Thin flat rectangular plate; (b) 4×4 mesh of four-node rectangular thin-plate elements.

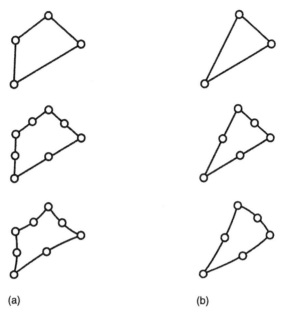

(a)   (b)

Fig. 8.8. (a) Quadrilateral and (b) triangular plate elements with straight edges and four/three nodes (top), eight/six nodes (centre) and curved edges and eight/six nodes (bottom).

The element kinetic and strain energies of flexural vibration of a thin flat plate are given by the following expressions (Petyt 1998):

$$T_e = \tfrac{1}{2} \int_A \rho h \dot{w}^2 \, dA = \tfrac{1}{2} \int_{-1}^{+1} \int_{-1}^{+1} \rho h \dot{w}(\xi, \eta, t)^2 ab \, d\xi \, d\eta \qquad (8.40a)$$

and

$$U_e = \tfrac{1}{2} \int_A I \, \{\chi\}^T [\mathbf{D}] \, \{\chi\} \, dA$$

$$= \tfrac{1}{2} \int_{-1}^{+1} \int_{-1}^{+1} I \, \{\chi(\xi, \eta, t)\}^T [\mathbf{D}] \, \{\chi(\xi, \eta, t)\} \, ab \, d\xi \, d\eta \qquad (8.40b)$$

while the virtual work done by dissipative effects and external forces on an element are given by

$$\delta W_{d,e} = \int_A -\mu \dot{w} \, \delta w \, dA = - \int_{-1}^{+1} \int_{-1}^{+1} \mu \dot{w}(\xi, \eta, t) \delta w(\xi, \eta, t) ab \, d\xi \, d\eta \qquad (8.40c)$$

and

$$\delta W_{f,e} = \int_A p_z \, \delta w \, dA = \int_{-1}^{+1} \int_{-1}^{+1} p_z(\xi, \eta, t) \delta w(\xi, \eta, t) \, ab \, d\xi \, d\eta \qquad (8.40d)$$

where $\xi = x/a$ and $\eta = y/b$ are dimensionless coordinates with reference to the dimensions $2a$ and $2b$ of the rectangular element, μ is the viscous damping coefficient per unit surface area, $p_z(\xi, \eta, t)$ is the distributed transverse force per unit area acting on the element surface, $I = h^3/12$ is the second moment of area per unit width of the cross section, h is the thickness of the plate, $\delta w(\xi, \eta, t)$ is the virtual transverse displacement (Meirovitch, 1967; Reddy, 1984a) and, assuming the material to be isotropic,

$$(\chi\xi, \eta, t) = \begin{bmatrix} \dfrac{\partial^2 w(x,y,t)}{\partial x^2} \\[2mm] \dfrac{\partial^2 w(x,y,t)}{\partial y^2} \\[2mm] 2\dfrac{\partial^2 w(x,y,t)}{\partial x \partial y} \end{bmatrix} = \begin{bmatrix} \dfrac{1}{a^2}\dfrac{\partial^2 w(\xi,\eta,t)}{\partial \xi^2} \\[2mm] \dfrac{1}{b^2}\dfrac{\partial^2 w(\xi,\eta,t)}{\partial \eta^2} \\[2mm] \dfrac{2}{ab}\dfrac{\partial^2 w(\xi,\eta,t)}{\partial \xi \partial \eta} \end{bmatrix}, \quad [\mathbf{D}] = \begin{bmatrix} E' & E'v & 0 \\ E'v & E' & 0 \\ 0 & 0 & G \end{bmatrix}$$

$$(8.41a,b)$$

with $E' = E/(1 - v^2)$ and $G = E/2(1 + v)$.

The highest spatial derivative appearing in the energy and virtual work expressions in Eqs. (8.40a–d) is the second. Thus, shape functions should be chosen to ensure that the function for the transverse displacement $w(x, y)$ can be differentiated twice with reference to x and y coordinates and w, $\partial w/\partial x$, $\partial w/\partial y$ are continuous functions within the element and at every point of a common boundary between two elements. Therefore, as shown in Fig. 8.9, the functions w, $\theta_x = \partial w/\partial y$ and $\theta_y = -\partial w/\partial x$ should be taken as degrees of freedom at each node. If a polynomial shape function is used, it should be a complete polynomial of at least degree two in order for it to be two times differentiable in x and y:

$$w = \alpha_1 + \alpha_2 x + \alpha_3 y + \alpha_4 x^2 + \alpha_5 xy + \alpha_6 y^2 + \text{higher order terms} \qquad (8.42)$$

The rectangular element of Fig. 8.9 has three degrees of freedom at each node point: the component of displacement normal to the plane of the plate w, and the two rotations $\theta_x = \partial w/\partial y$ and $\theta_y = -\partial w/\partial x$, which in terms of a-dimensional coordinates $\xi = x/a$ and $\eta = y/b$ become

$$\theta_x = \frac{1}{b}\frac{\partial w}{\partial \eta}, \quad \theta_y = -\frac{1}{a}\frac{\partial w}{\partial \xi} \qquad (8.43a,b)$$

Therefore, the element has twelve degrees of freedom so that, in order to uniquely derive the function in Eq. (8.42) with reference to the nodal degrees of freedom,

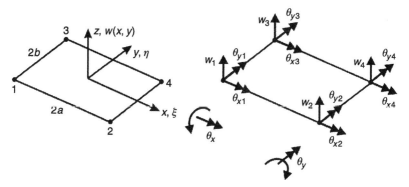

Fig. 8.9.   Four-node rectangular bending element with straight edges.

the shape function must be a polynomial having twelve terms. A two-dimensional complete cubic polynomial is chosen with the two additional cubic terms $x^3 y$ and xy^3 which have been selected in such a way as to ensure that the elements are geometrically invariant (Petyt, 1998). Thus, the polynomial in a-dimensional coordinates ξ and η is

$$w(\xi, \eta) = \alpha_1 + \alpha_2 \xi + \alpha_3 \eta + \alpha_4 \xi^2 + \alpha_5 \xi \eta + \alpha_6 \eta^2$$
$$+ \alpha_7 \xi^3 + \alpha_8 \xi^2 \eta + \alpha_9 \xi \eta^2 + \alpha_{10} \eta^3 + \alpha_{11} \xi^3 \eta + \alpha_{12} \xi \eta^3 \tag{8.44}$$

This expression and its derivatives with reference to ξ and η can be written in matrix form as follows:

$$w(\xi, \eta) = \lfloor 1 \quad \xi \quad \eta \quad \xi^2 \quad \xi\eta \quad \eta^2 \quad \xi^3 \quad \xi^2\eta \quad \xi\eta^2 \quad \eta^3 \quad \xi^3\eta \quad \xi\eta^3 \rfloor \{\alpha\}$$

$$= \lfloor \mathbf{p}(\xi, \eta) \rfloor \{\alpha\} \tag{8.45a}$$

$$\frac{\partial w}{\partial \xi} = \lfloor 0 \quad 1 \quad 0 \quad 2\xi \quad \eta \quad 0 \quad 3\xi^2 \quad 2\xi\eta \quad \eta^2 \quad 0 \quad 3\xi^2\eta \quad \eta^3 \rfloor \{\alpha\}$$

$$= \lfloor \partial\mathbf{p}(\xi, \eta)/\partial\xi \rfloor \{\alpha\} \tag{8.45b}$$

and

$$\frac{\partial w}{\partial \eta} = \lfloor 0 \quad 0 \quad 1 \quad 0 \quad \xi \quad 2\eta \quad 0 \quad \xi^2 \quad 2\xi\eta \quad 3\eta^2 \quad \xi^3 \quad 3\xi\eta^2 \rfloor \{\alpha\}$$

$$= \lfloor \partial\mathbf{p}(\xi, \eta)/\partial\eta \rfloor \{\alpha\} \tag{8.45c}$$

where

$$\{\boldsymbol{\alpha}\}^T = \lfloor \alpha_1 \quad \alpha_2 \quad \alpha_3 \quad \alpha_4 \quad \alpha_5 \quad \alpha_6 \quad \alpha_7 \quad \alpha_8 \quad \alpha_9 \quad \alpha_{10} \quad \alpha_{11} \quad \alpha_{12} \rfloor \quad (8.46)$$

Evaluating Eqs. (8.45a–c) at the four nodal points, i.e., at $\xi = \mp 1$ and $\eta = \mp 1$ gives

$$\{\mathbf{w}_e\} = [\mathbf{A}_e]\{\boldsymbol{\alpha}\} \qquad (8.47)$$

where

$$[\mathbf{A}_e] = \begin{bmatrix}
1 & -1 & -1 & 1 & 1 & 1 & -1 & -1 & -1 & -1 & 1 & 1 \\
0 & 0 & 1/b & 0 & -1/b & -2/b & 0 & 1/b & 2/b & 3/b & -1/b & -3/b \\
0 & -1/a & 0 & 2/a & 1/a & 0 & -3/a & -2/a & -1/a & 0 & 3/a & 1/a \\
1 & 1 & -1 & 1 & -1 & 1 & 1 & -1 & 1 & -1 & -1 & -1 \\
0 & 0 & 1/b & 0 & 1/b & -2/b & 0 & 1/b & -2/b & 3/b & 1/b & 3/b \\
0 & -1/a & 0 & -2/a & 1/a & 0 & -3/a & 2/a & -1/a & 0 & 3/a & 1/a \\
1 & 1 & 1 & 1 & 1 & 1 & 1 & 1 & 1 & 1 & 1 & 1 \\
0 & 0 & 1/b & 0 & 1/b & 2/b & 0 & 1/b & 2/b & 3/b & 1/b & 3/b \\
0 & -1/a & 0 & -2/a & -1/a & 0 & -3/a & -2/a & -1/a & 0 & -3/a & -1/a \\
1 & -1 & 1 & 1 & -1 & 1 & -1 & 1 & -1 & 1 & -1 & -1 \\
0 & 0 & 1/b & 0 & -1/b & 2/b & 0 & 1/b & -2/b & 3/b & -1/b & -3/b \\
0 & -1/a & 0 & 2/a & -1/a & 0 & -3/a & 2/a & -1/a & 0 & -3/a & -1/a
\end{bmatrix}$$

$$(8.48)$$

and $\{\mathbf{w}_e\}$ is a column vector with the nodal displacements and rotations

$$\{\mathbf{w}_e\}^T = \lfloor w_1 \quad \theta_{x1} \quad \theta_{y1} \quad \cdots \quad w_4 \quad \theta_{x4} \quad \theta_{y4} \rfloor \qquad (8.49)$$

where, according to Eqs. (8.43a,b) $\theta_x = \frac{1}{b}\frac{\partial w}{\partial \eta}$ and $\theta_y = -\frac{1}{a}\frac{\partial w}{\partial \xi}$. Substituting Eq. (8.45a–c) into Eqs. (8.40a,b) and (8.40c,d) the kinetic and strain energy terms for one element and the virtual work done by dissipative and external forces on the element are found to be given by

$$T_e = \tfrac{1}{2}\{\dot{\mathbf{w}}_e\}^T [\mathbf{M}_e]\{\dot{\mathbf{w}}_e\}$$

$$U_e = \tfrac{1}{2}\{\mathbf{w}_e\}^T [\mathbf{K}_e]\{\mathbf{w}_e\}$$

$$(8.50a,b)$$

and

$$\delta W_{d,e} = -\{\delta\mathbf{w}_e\}^T [\mathbf{C}_e]\{\dot{\mathbf{w}}_e\}$$

$$\delta W_{f,e} = \{\delta\mathbf{w}_e\}^T \{\mathbf{f}_e\}$$

$$(8.50c,d)$$

where, knowing from Eq. (8.47) that $\{\alpha\} = [A_e]^{-1}\{w_e\}$,

$$[M_e] = [A_e]^{-T} \int_{-1}^{+1} \int_{-1}^{+1} \rho h \lfloor \mathbf{p}(\xi,\eta) \rfloor^T \lfloor \mathbf{p}(\xi,\eta) \rfloor \, ab \, d\xi \, d\eta \, [A_e]^{-1}$$

$$[K_e] = [A_e]^{-T} \int_{-1}^{+1} \int_{-1}^{+1} I_z [\mathbf{P}(\xi,\eta)]^T [D] [\mathbf{P}(\xi,\eta)] \, ab \, d\xi \, d\eta \, [A_e]^{-1}$$

$$(8.51a,b)$$

$$[C_e] = [A_e]^{-T} \int_{-1}^{+1} \int_{-1}^{+1} \mu \lfloor \mathbf{p}(\xi,\eta) \rfloor^T \lfloor \mathbf{p}(\xi,\eta) \rfloor \, ab \, d\xi \, d\eta \, [A_e]^{-1}$$

$$(8.51c,d)$$

$$[f_e] = [A_e]^{-T} \int_{-1}^{+1} \int_{-1}^{+1} \lfloor \mathbf{p}(\xi,\eta) \rfloor^T p_z(\xi,\eta) ab \, d\xi \, d\eta$$

where $[\ ]^{-T} = \left([\ ]^{-1}\right)^T$ and $[\mathbf{P}(\xi,\eta)]$ is the following 3×12 matrix:

$$[\mathbf{P}(\xi,\eta)] = \begin{bmatrix} \frac{1}{a^2} \left\lfloor \dfrac{\partial^2 \mathbf{p}(\xi,\eta)}{\partial\xi^2} \right\rfloor \\ \frac{1}{b^2} \left\lfloor \dfrac{\partial^2 \mathbf{p}(\xi,\eta)}{\partial\eta^2} \right\rfloor \\ \frac{2}{ab} \left\lfloor \dfrac{\partial^2 \mathbf{p}(\xi,\eta)}{\partial\xi\partial\eta} \right\rfloor \end{bmatrix} \qquad (8.52)$$

The integrals of the elements of the matrices in Eqs. (8.51a–d) are relatively simple to derive since they have the following forms $\int_{-1}^{+1}\xi^m d\xi$, $\int_{-1}^{+1}\eta^n d\eta$, $\int_{-1}^{+1}\int_{-1}^{+1}\xi^m\eta^n d\xi d\eta$ which for m or n odd give zeros and for m and n even give respectively $2/(m+1)$, $2/(n+1)$ and $4/(m+1)(n+1)$. Assuming p_z to be constant over the element surface, the integral in Eq. (8.51d) gives

$$\{f_e\}^T = p_z \frac{ab}{3} [3 \quad b \quad -a \quad 3 \quad b \quad a \quad 3 \quad -b \quad a \quad 3 \quad -b \quad -a] \qquad (8.53)$$

As in the previous example for the bending vibrations of a beam, the kinetic and strain energy terms and, in this case, the virtual work done by the dissipative and external transverse forces, on each element can be combined together to give the total energies and virtual works. These expressions can then be introduced into the integral in Eq. (8.1) for the Hamilton principle:

$$\int_{t_1}^{t_2} \delta(T-U) + \delta W_d + \delta W_f \, dt = \int_{t_1}^{t_2} \delta\left(\tfrac{1}{2}\{\dot{\mathbf{w}}\}^T [M]\{\dot{\mathbf{w}}\} - \tfrac{1}{2}\{\mathbf{w}\}^T [K]\{\mathbf{w}\}\right)$$

$$-\{\delta\mathbf{w}\}^T [C]\{\dot{\mathbf{w}}\} + \{\delta\mathbf{w}\}^T \{\mathbf{f}\} \, dt = 0 \qquad (8.54)$$

After integrating the first term by parts and, because $\{\delta\mathbf{w}\}$ is arbitrary, following the procedure used to derive Eqs. (8.13)–(8.15), the following system of n ordinary differential equations is obtained:

$$[\mathbf{M}]\{\ddot{\mathbf{w}}(t)\} + [\mathbf{C}]\{\dot{\mathbf{w}}(t)\} + [\mathbf{K}]\{\mathbf{w}(t)\} = \{\mathbf{f}(t)\} \tag{8.55}$$

where $[\mathbf{M}]$, $[\mathbf{C}]$ and $[\mathbf{K}]$ are the global inertia, damping and stiffness matrices when the rows and columns associated with constrained degrees of freedom have been eliminated following the procedure discussed in the previous example. Similarly, $\{\mathbf{w}\}$ and $\{\mathbf{f}\}$ represent, respectively, the column vectors with the nodal degrees of freedom and nodal excitations generated by the transverse forces acting on the elements surfaces when the terms associated with constrained degrees of freedom have been eliminated.

In general, the damping matrix is not derived from the integral in Eq. (8.51c). For mathematical convenience, but not necessarily physical validity it is expressed using a spectral damping scheme, known as Rayleigh or proportional damping, so that

$$[\mathbf{C}] = a_1 [\mathbf{M}] + a_2 [\mathbf{K}] \tag{8.56}$$

where the a_1 and a_2 coefficients are derived with reference to the damping ratios ζ_1 and ζ_2 relative to two resonance frequencies ω_1 and ω_2 measured on a prototype structure. Following the procedure presented by Petyt (1998), the two coefficients a_1 and a_2 are given by:

$$a_1 = 2\omega_1\omega_2 \frac{\zeta_1\omega_2 - \zeta_2\omega_1}{\omega_2^2 - \omega_1^2} \tag{8.57}$$

and

$$a_2 = 2\frac{\zeta_2\omega_2 - \zeta_1\omega_1}{\omega_2^2 - \omega_1^2} \tag{8.58}$$

In general, ω_1 and ω_2 correspond to the lowest and highest resonance frequencies considered in the model.

As an example, the first nine natural frequencies and natural modes are derived for an aluminium plate ($\rho = 2700$ kg m^{-3} and $E = 7.1 \times 10^{10}$ N m^{-2}), which has four clamped edges, dimensions $L_x \times L_y = 0.414$ m $\times$ 0.314 m and thickness $h = 1$ mm. Figure 8.10 presents the first nine natural frequencies and natural modes of the clamped plate derived using the Rayleigh–Ritz formulation, where the modes are taken to be the product of the hyperbolic trigonometric beam functions described by Warburton (1951). Figure 8.11 shows the same

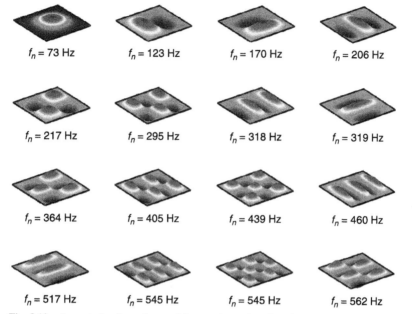

Fig. 8.10. Lowest nine flexural natural frequencies and modes of a clamped plate (Warburton solution).

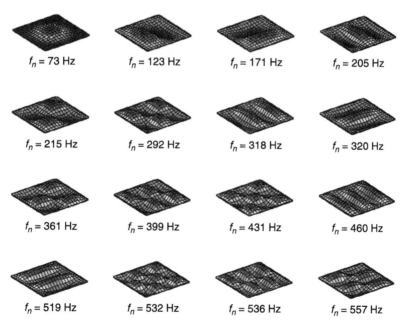

Fig. 8.11. Lowest nine flexural natural frequencies and modes of a clamped rectangular plate (Finite Element solution).

natural frequencies and natural modes when derived with the plate Finite Element Model made of a 16×16 mesh of elements with inertia and stiffness matrices given in Eqs. (8.51a,b). It is interesting to note that, in this case, the Finite Element Model tends to give lower natural frequencies than that derived with the Warburton formulation. This is because the formulation presented above gives a non-conforming element which is more compliant than thin plate structures (Petyt, 1998).

The set of Eqs. (8.55) can be used to derive the response to various types of excitation such as steady state harmonic excitations, steady state periodic excitations, transient 'shock' excitations and random excitations. In general, the so-called 'modal analysis' solution approach is used where the set of Eqs. (8.55) are transformed in terms of modal coordinates so that a reduced set of equations is derived with size equal to the number of natural frequencies/modes taken into account in the analysis. This approach is based on the derivation of the natural frequencies and natural modes. Thus, the first step in any type of study is free vibration analysis which, as shown above, implies an eigenvalue–eigenvector problem. In order to model complex practical structures, Finite Element Models with many elements, and therefore degrees of freedom, must be used. This implies the solution of large eigenvalue–eigenvector problems including many thousand degrees of freedom which can be accelerated by the use of iterative algorithms.

The number of degrees of freedom is often reduced by eliminating 'unwanted' or 'unnecessary' degrees of freedom with automatic procedures where master and slave degrees of freedom are selected to be retained and neglected respectively. In many cases, the time histories of periodic or transient excitations cannot be described by analytical expressions. It is then necessary to derive the response by means of numerical models where the excitation time history is sampled at a number of intervals and then either finite Fourier series or finite difference methods are implemented for the two types of problems. Good introductions to the analysis of free and forced vibration problems with Finite Element Models are given by Cook *et al.* (1989), Bathe (1996) and Petyt (1998).

8.4.4   Finite Element Models for Other Types of Structure

The vibration of one-dimensional structures is normally modelled with one-dimensional elements with six degrees of freedom per node so that the bending vibration effects in two cross-sectional planes can be modelled, as well as axial and torsional vibration effects. This type of element is often referred to as a 'beam element', although it could be equally used for the analysis of axial vibration of rods or torsional vibrations of shafts. The uncoupled out-of-plane bending vibrations and in-plane axial and shear vibrations of two-dimensional flat structures are modelled by two-dimensional 'plate elements' and 'membrane elements'. In some cases the two elements are merged into one element which is normally referred as plate element although it accounts for both the bending and axial/shear vibrations. When curved two-dimensional structures are considered, 'shell elements' are used where out-of-plane and in-plane vibrational motions are assumed to be coupled. Finally, the vibrations of systems with complex geometries are modelled with three-dimensional 'solid elements', which normally have three degrees of freedom per node. These types of element replicate the axial and shear vibratory field of an unbounded body only. Thus, the mesh must be constructed in such a way to have at least four elements on each surface of the body, even the smallest one. In principle it is possible to use solid elements to model beams, plates and shells. However, because of their intrinsic inability to reproduce the effect of boundary surfaces a certain number of elements must be used to model the cross-sectional areas of the beam, plate and shell structures.

In practice, there are many types of other elements that have been specifically developed to model particular types of structure. For example, axisymmetric elements have been constructed to model circular shells such as fluid containment tanks or aircraft fuselage structures. Multi-layer plate and shell elements have been developed to model composite panels. Comprehensive presentations of different Finite Element Models for structural vibrations are available in the following books: (Reddy 1984a,b; Cook *et al.*, 1989; Bathe, 1996; Reddy, 1997; Petyt, 1998; Zienkiewicz and Taylor, 2000).

8.5   Finite Element Analysis of Acoustic Vibrations of Fluids in Cavities

The Finite Element approach for the analysis of acoustic vibration of fluids in cavities is now presented. The general formulation is introduced for two examples of interest in coupled vibroacoustic problems which involve the acoustic vibrations either in one- or three-dimensional cavities. The approximate solution using the Rayleigh–Ritz approach is first presented for the acoustic vibration of a fluid in a pipe. The Finite Element Analysis is then presented for the acoustic vibrations in one- and three-dimensional cavities. The equations of motion are derived using Hamilton's principle formulated using force-like functions. (Gladwell and Zimmermann, 1966).

The kinetic and potential energy densities for small-amplitude oscillations in fluids are related to the particle velocity vector,[3] $\mathbf{\dot{u}}^T = \{\dot{u}_x \quad \dot{u}_y \quad \dot{u}_z\}$ and sound pressure p, respectively, by

$$T = \frac{1}{2}\rho_0 |\mathbf{\dot{u}}|^2 \quad U = \frac{1}{2}\kappa p^2 \qquad (8.59\text{a,b})$$

where ρ_0 is the mean density of the fluid and κ is the fluid compressibility (Morse and Ingard, 1968). For non-harmonic time dependence, the particle velocity components cannot be expressed as functions of sound pressure derivatives. Therefore, the velocity potential ϕ is used to derive the integral in Eq. (8.1) for the Hamilton's principle since, using the relations

$$p = -\rho_0 \frac{\partial \phi}{\partial t} \quad \text{and} \quad \mathbf{\dot{u}} = \text{grad } \phi \qquad (8.60,61)$$

where $\text{grad}\phi = \left[\frac{\partial \phi}{\partial x} \quad \frac{\partial \phi}{\partial y} \quad \frac{\partial \phi}{\partial z} \right]^T$, a homogeneous formulation is obtained. The equations of motion are then expressed in terms of sound pressure by differentiating the integral in Eq. (8.1) with reference to time.

8.5.1 One-Dimensional Acoustic Vibration of a Fluid in a Uniform Straight Pipe: Rayleigh–Ritz Method

Consider a uniform straight pipe whose axial coordinate is x and in which a one-dimensional (plane) acoustic field is defined in terms of sound pressure $p(x, t)$. The Rayleigh–Ritz method approximates the solution by a finite series expansion which is expressed in terms of the velocity potential function $\phi(x, t)$ so that the integrals for the kinetic and potential energies are defined with reference to the same set of prescribed functions of x, $d_i(x)$ and unknown functions $\phi_i(t)$ in the form

$$\phi(x, t) = \sum_{i=1}^{n} d_i(x)\phi_i(t) \qquad (8.62)$$

In this case, the prescribed functions are chosen in such a way as to satisfy the four conditions listed in the earlier section on the beam problem except that force-like (sound pressure) rather than displacement-like (particle velocity) boundary conditions are imposed.

---

[3] According to Finite Element convention, in this chapter fluid particle velocities in the x, y, z directions are denoted by $\dot{u}_x$, $\dot{u}_y$, $\dot{u}_z$. Normalised coordinates denoted by ξ_1, ξ_2, ξ_3 for the x, y, z directions are used in the fluid domain.

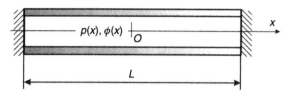

Fig. 8.12.   Pipe with closed ends.

As an example, the natural frequencies and associated natural modes of a fluid in a pipe with closed ends are determined (Fig. 8.12): one-dimensional (plane) motion is assumed. The kinetic and potential energies of one-dimensional acoustic vibration of a fluid are given by the following expressions (Craggs, 1971):

$$T = \frac{1}{2} \int_0^L \rho_0 A \dot{u}_x^2(x, t)\, dx = \frac{1}{2} \int_0^L \rho_0 A \left(\frac{\partial \phi(x, t)}{\partial x}\right)^2 dx$$

$$U = \frac{1}{2} \int_0^L \kappa A p^2(x, t)\, dx = \frac{1}{2} \int_0^L \frac{\rho_0}{c^2} A \left(\frac{\partial \phi(x, t)}{\partial t}\right)^2 dx$$

$$(8.63\text{a,b})$$

where c is the speed of sound in the fluid and A is the cross-sectional area of the pipe. According to the energy expressions in Eq. (8.63), the integral in Eq. (8.1) involves derivatives in x of $\phi(x, t)$ up to order $p = 1$. Therefore, the prescribed functions $d_i(x)$ to be used in the finite series expansion of Eq. (8.62) must be at least once differentiable. Also these functions must be linearly independent, form a complete series and satisfy the force-like boundary conditions for ϕ; which, assuming the origin of the system of reference in the middle of the pipe (Fig. 8.10), are given by $p\left(-\frac{L}{2}\right) \neq 0$ and $p\left(+\frac{L}{2}\right) \neq 0$, that is, $\phi\left(-\frac{L}{2}\right) \neq 0$ and $\phi\left(+\frac{L}{2}\right) \neq 0$ for a pipe with closed ends. One example of a set of functions that satisfy these conditions is,

$$d_i(x) = x^i \tag{8.64}$$

where $i \geq 1$. Thus the velocity potential in Eq. (8.62) corresponds to a polynomial of order n whose coefficients are given by unknown functions of t, $\phi_i(t)$. When the velocity potential of the one-dimensional sound field is approximated by a finite expansion so that

$$\phi(x, t) = \lfloor \mathbf{d}(x) \rfloor \{\boldsymbol{\phi}(t)\} = \lfloor d_1(x)\, d_2(x) \; \cdots \; d_n(x) \rfloor \begin{bmatrix} \phi_1(t) \\ \phi_2(t) \\ \vdots \\ \phi_n(t) \end{bmatrix} \tag{8.65}$$

the mathematical formulation of the kinetic and potential acoustic energies, as given in Eqs. (8.63a,b), is reduced to that of a discrete system with a finite number of degrees of freedom:

$$T = \frac{1}{2}\rho_0 \{\boldsymbol{\phi}(t)\}^T [\mathbf{H}] \{\boldsymbol{\phi}(t)\},$$

$$U = \frac{1}{2}\rho_0 \{\dot{\boldsymbol{\phi}}(t)\}^T [\mathbf{Q}] \{\dot{\boldsymbol{\phi}}(t)\} \tag{8.66a,b}$$

where

$$[\mathbf{H}] = A \int_{-\frac{L}{2}}^{+\frac{L}{2}} \left[\frac{\partial \mathbf{d}(x)}{\partial x} \right]^T \left[\frac{\partial \mathbf{d}(x)}{\partial x} \right] dx \tag{8.67a}$$

and

$$[\mathbf{Q}] = \frac{A}{c^2} \int_{-\frac{L}{2}}^{+\frac{L}{2}} \lfloor \mathbf{d}(x) \rfloor^T \lfloor \mathbf{d}(x) \rfloor\, dx \tag{8.67b}$$

with

$$H_{k,s} = A \int_{-\frac{L}{2}}^{+\frac{L}{2}} \frac{\partial d_k(x)}{\partial x} \frac{\partial d_s(x)}{\partial x} dx = \frac{Aks}{(k+s-1)} \left[\left(\frac{L}{2}\right)^{k+s-1} - \left(-\frac{L}{2}\right)^{k+s-1} \right] \tag{8.68a}$$

and

$$Q_{k,s} = \frac{A}{c^2} \int_{-\frac{L}{2}}^{+\frac{L}{2}} d_k(x)d_s(x)dx = \frac{A}{c^2(k+s+1)} \left[\left(\frac{L}{2}\right)^{k+s+1} - \left(-\frac{L}{2}\right)^{k+s+1} \right] \tag{8.68b}$$

In this case, assuming the dissipation function and generalised forces to be zero, Hamilton's principle gives

$$\int_{t_1}^{t_2} \delta (T - U) dt = \int_{t_1}^{t_2} \delta \left(\frac{1}{2}\rho_0 \{\boldsymbol{\phi}(t)\}^T [\mathbf{H}] \{\boldsymbol{\phi}(t)\} - \frac{1}{2}\rho_0 \{\dot{\boldsymbol{\phi}}(t)\}^T [\mathbf{Q}] \{\dot{\boldsymbol{\phi}}(t)\} \right) dt = 0 \tag{8.69}$$

which, after integrating the second term by parts, becomes

$$-\int_{t_1}^{t_2} \{\delta\boldsymbol{\phi}(t)\}^T \left(\rho_0\,[\mathbf{H}]\,\{\boldsymbol{\phi}(t)\} + \rho_0\,[\mathbf{Q}]\,\{\ddot{\boldsymbol{\phi}}(t)\}\right)dt = 0. \qquad (8.70)$$

Since $\{\delta\boldsymbol{\phi}(t)\}$ is arbitrary, the following system of n ordinary differential equations must be satisfied:

$$-\rho_0\,[\mathbf{H}]\,\{\boldsymbol{\phi}(t)\} - \rho_0\,[\mathbf{Q}]\,\{\ddot{\boldsymbol{\phi}}(t)\} = \{\mathbf{0}\} \qquad (8.71)$$

This equation can be expressed in terms of the unknown sound pressures by differentiating it with respect to time and using the relation $p_i = -\rho_0\partial\phi_i/\partial t$ (Craggs, 1971), so that

$$[\mathbf{Q}]\,\{\ddot{\mathbf{p}}(t)\} + [\mathbf{H}]\,\{\mathbf{p}(t)\} = \{\mathbf{0}\} \qquad (8.72)$$

where $\{\mathbf{p}\} = -\rho_0\left\{\frac{\partial\phi}{\partial t}\right\}$ are the unknown field variables which can be used to derive the sound pressure in the pipe by combining Eqs. (8.65) and (8.60). Because of the similarity between Eqs. (8.72) and (8.15), the matrices $[\mathbf{Q}]$ and $[\mathbf{H}]$ are by convention referred to as the acoustic inertia and stiffness matrices, respectively. However, the reader should be aware that the formulation presented above clearly indicates that the matrix $[\mathbf{Q}]$ is related to the acoustic potential energy, and thus should be compared to the stiffness matrix $[\mathbf{K}]$, while the $[\mathbf{H}]$ matrix is related to the kinetic energy, and therefore should be compared to the inertia matrix $[\mathbf{M}]$.

Considering harmonic vibration of the fluid in the pipe, such that $\{\mathbf{p}\} = \{\mathbf{a}\}\sin\omega t$ and the complex amplitudes $\{\mathbf{a}\}$ are independent of time, Eq. (8.72) results in the same eigenvalue–eigenvector problem as Eq. (8.16) so that, for each natural frequency ω_j, a solution $\{\mathbf{a}\}_j$, to within an arbitrary constant, can be found. The solution $\{\mathbf{a}\}_j$ can then be combined with the prescribed functions in Eq. (8.65) and Eq. (8.60) to give the j-th natural mode $\psi_j(x)$.

As an example, the first four acoustic natural frequencies and natural modes of a 0.5 m long pipe with closed ends filled with air ($\rho_0 = 1.21$ kg m^{-3}, $c = 343$ m s^{-1}) are derived. Figure 8.13 compares the derived natural frequencies and natural modes with the exact analytical solution. It is interesting to note that in this case the third natural frequency and natural mode derived with the 4-th order polynomial function do not converge to the exact solution, while the fourth one does. This is because the fourth natural frequency has been derived as if it were the first one of a quarter length cavity and thus converges better than the third one which is actually the second odd mode of the full length cavity.

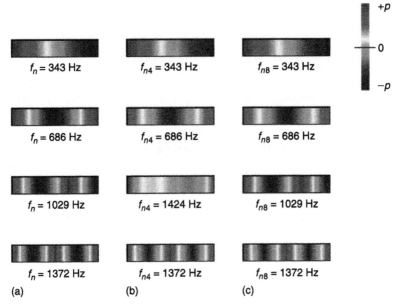

Fig. 8.13. Acoustic natural frequencies and modes of a fluid in a pipe with closed ends: (a) exact solution: Rayleigh–Ritz solution with the polynomial function of Eq. (8.65) and with (b) four and (c) eight power terms.

8.5.2   One-Dimensional Acoustic Vibration of Fluid in a Uniform Straight Pipe: Finite Element Analysis

The same procedure as that described in Section 8.4.2 is implemented to derive the Finite Element Model of the acoustic oscillations in a pipe with closed ends. Five node points have been selected at equal intervals and thus, as shown in Fig. 8.14, the fluid is divided into four equal elements.

The highest order spatial derivative in the energy expressions in Eqs. (8.63a,b) for the one-dimensional sound field is the first. Therefore, only the prescribed functions themselves need to be continuous, and so the velocity potential $\phi(x, t)$, and thus sound pressure $p(x, t)$, is the only degree of freedom required at each node point. As illustrated in Fig. 8.15, five prescribed functions that satisfy the Rayleigh–Ritz criteria have been constructed by giving each node point in turn a unit velocity potential whilst maintaining zero velocity potential at all other nodes.

As already seen for the structural case, Fig. 8.15 shows that the velocity potential exhibits two characteristic distributions within the elements. Thus emphasis is on determining velocity potential patterns for individual elements via element

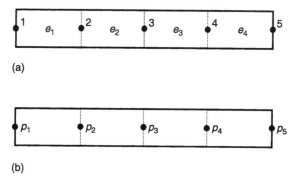

Fig. 8.14.   Pipe volume modelled with a mesh of four one-dimensional acoustic finite elements.

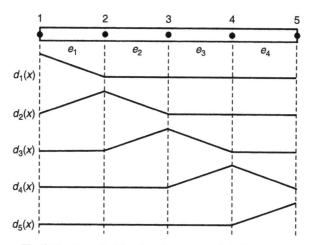

Fig. 8.15.   Prescribed functions for the acoustic field in a pipe.

shape functions. These element shape functions here are determined by assuming a polynomial function having the appropriate number of terms.

In order to satisfy the Rayleigh–Ritz convergence criteria the element shape function should be a first order polynomial so that if, as shown in Fig. 8.16, the dimensionless coordinate $\xi_1 = x/a_1$ is used, where $2a_1$ is the length of the element, the velocity potential function within the element can be expressed

$$\phi(\xi_1) = \alpha_1 + \alpha_2\xi_1 \qquad (8.73)$$

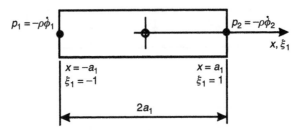

Fig. 8.16.   Fluid element showing non-dimensional coordinates.

Evaluating this function at $\xi_1 = \mp 1$ gives $\phi_1 = \alpha_1 - \alpha_2$ and $\phi_2 = \alpha_1 + \alpha_2$, where ϕ_1, ϕ_2 are the velocity potential of nodes 1 and 2. Solving for α_1 and α_2 gives

$$\alpha_1 = \frac{1}{2}(\phi_1 + \phi_2), \quad \alpha_2 = \frac{1}{2}(\phi_2 - \phi_1) \qquad (8.74\text{a,b})$$

Therefore the velocity potential for the one-dimensional acoustic element is given by

$$\phi(\xi_1, t) = \frac{1}{2}(1 - \xi_1)\,\phi_1(t) + \frac{1}{2}(1 + \xi_1)\,\phi_2(t) \qquad (8.75\text{a})$$

This expression can be rewritten in matrix form as

$$\phi(\xi_1, t) = \lfloor G_1(\xi_1)\ G_2(\xi_1) \rfloor \begin{bmatrix} \phi_1(t) \\ \phi_2(t) \end{bmatrix} = \lfloor g(\xi_1) \rfloor \{\phi_e(t)\} \qquad (8.75\text{b})$$

where

$$G_1(\xi_1) = \frac{1}{2}(1 - \xi_1), \quad G_2(\xi_1) = \frac{1}{2}(1 + \xi_1) \qquad (8.76\text{a,b})$$

and

$$\{\phi_e(t)\} = \begin{bmatrix} \phi_1(t) \\ \phi_2(t) \end{bmatrix} \qquad (8.77)$$

Once the matrix relation in Eq. (8.75b) has been found for each element it is straightforward to evaluate for each element the kinetic and potential energies given in Eqs. (8.63a,b). The energy expressions for a single element are

$$T_e = \frac{1}{2}\rho_0 \int_{-1}^{+1} \frac{A}{a_1^2} \left(\frac{\partial \phi(\xi_1, t)}{\partial \xi_1}\right)^2 a_1\, d\xi_1 \qquad (8.78\text{a})$$

and

$$U_e = \frac{1}{2}\rho_0 \int_{-1}^{+1} \frac{A}{c^2} \left(\frac{\partial\phi(\xi_1, t)}{\partial t} \right)^2 a_1 d\xi_1 \tag{8.78b}$$

where A is the cross-sectional area. Substituting Eq. (8.75b) into Eqs. (8.78a) and (8.78b) gives

$$T_e = \frac{1}{2}\rho_0 \{\phi_e\}^T [H_e] \{\phi_e\},$$

$$U_e = \frac{1}{2}\rho_0 \{\dot{\phi}_e\}^T [Q_e] \{\dot{\phi}_e\} \tag{8.79a,b}$$

where

$$[H_e] = \frac{A}{a_1} \int_{-1}^{+1} \left[\frac{\partial g(\xi_1)}{\partial \xi_1} \right]^T \left[\frac{\partial g(\xi_1)}{\partial \xi_1} \right] d\xi_1 = \frac{A}{a_1} \begin{bmatrix} 1/2 & -1/2 \\ -1/2 & 1/2 \end{bmatrix} \tag{8.80a}$$

is the element 'acoustic stiffness matrix', and

$$[Q_e] = \frac{Aa_1}{c^2} \int_{-1}^{+1} \lfloor g(\xi_1) \rfloor^T \lfloor g(\xi_1) \rfloor d\xi_1 = \frac{Aa_1}{c^2} \begin{bmatrix} 2/3 & 1/3 \\ 1/3 & 2/3 \end{bmatrix} \tag{8.80b}$$

is the element 'acoustic inertia matrix'. Following the procedure described for the beam problem, the energy expressions for the complete fluid volume are then obtained by adding together the energies for all the individual elements. If the velocity potentials at the nodal points are grouped into the following vector:

$$\{\overline{\phi}\}^T = \lfloor \phi_1 \quad \phi_2 \quad \phi_3 \quad \phi_4 \quad \phi_5 \rfloor \tag{8.81}$$

then the velocity potential for element e is given by

$$\{\phi_e\} = [a_e] \{\overline{\phi}\} \tag{8.82}$$

where, for the four elements, $[a_e]$ takes the form

$$[a_1] = \begin{bmatrix} 1 & 0 & 0 & 0 & 0 \\ 0 & 1 & 0 & 0 & 0 \end{bmatrix} \qquad [a_2] = \begin{bmatrix} 0 & 1 & 0 & 0 & 0 \\ 0 & 0 & 1 & 0 & 0 \end{bmatrix}$$

$$[a_3] = \begin{bmatrix} 0 & 0 & 1 & 0 & 0 \\ 0 & 0 & 0 & 1 & 0 \end{bmatrix} \qquad [a_4] = \begin{bmatrix} 0 & 0 & 0 & 1 & 0 \\ 0 & 0 & 0 & 0 & 1 \end{bmatrix} \tag{8.83a–d}$$

Substituting Eq. (8.82) into Eqs. (8.79a,b) and summing over all the elements gives the total kinetic and potential energies of the fluid:

$$T = \tfrac{1}{2}\rho_0 \{\overline{\phi}\}^T [\overline{\mathbf{H}}] \{\overline{\phi}\}$$

$$U = \tfrac{1}{2}\rho_0 \{\dot{\overline{\phi}}\}^T [\overline{\mathbf{Q}}] \{\dot{\overline{\phi}}\}$$

(8.84a,b)

where

$$[\overline{\mathbf{H}}] = \sum_{e=1}^{4} [\mathbf{a}_e]^T [\mathbf{H}_e] [\mathbf{a}_e] \quad \text{and} \quad [\overline{\mathbf{Q}}] = \sum_{e=1}^{4} [\mathbf{a}_e]^T [\mathbf{Q}_e] [\mathbf{a}_e] \qquad (8.85a,b)$$

As in the case of structural vibration analysis, the matrix products $[\mathbf{a}_e]^T [\mathbf{H}_e] [\mathbf{a}_e]$ and $[\mathbf{a}_e]^T [\mathbf{Q}_e] [\mathbf{a}_e]$ in Eqs. (8.85a,b) effectively locate the positions in $[\overline{\mathbf{H}}]$ and $[\overline{\mathbf{Q}}]$ to which the elements of $[\mathbf{H}_e]$ and $[\mathbf{Q}_e]$, respectively, have to be added. However this is usually carried out following the 'assembly process' described above. The last step in the formulation concerns the boundary conditions which are set by cancelling the rows and columns in the $[\overline{\mathbf{H}}]$ and $[\overline{\mathbf{Q}}]$ matrices and $\{\overline{\phi}\}$ vector related to the nodal sound pressure which is zero.

For example, if the Finite Element Model is used to analyse the acoustic response of the fluid when the left end of the pipe is closed and the right end is open, then the nodal sound pressure p_5 should be set equal to zero [Note: this is an approximation since the external fluid loads the internal fluid and the pressure is small but not actually zero]. This condition is simply introduced by omitting ϕ_5 from the set of degrees of freedom for the complete one-dimensional cavity so that,

$$\{\phi\}^T = \lfloor \phi_1 \quad \phi_2 \quad \phi_3 \quad \phi_4 \rfloor \tag{8.86a}$$

and at the same time omitting the last row and column from the acoustic inertia and stiffness matrices so that the matrices $[\mathbf{H}]$ and $[\mathbf{Q}]$ are obtained. If, instead, both ends are closed, the sound pressure at both ends must be retained and the whole set of velocity potential degrees of freedom should also be retained:

$$\{\phi\}^T = \lfloor \phi_1 \quad \phi_2 \quad \phi_3 \quad \phi_4 \quad \phi_5 \rfloor \tag{8.86b}$$

The total energy expressions in terms of the velocity potential vector in Eq. (8.86) are substituted into the integral in Eq. (8.1) that, implementing the mathematical formulation given in Eqs. (8.69)–(8.72), gives the following set of second order ordinary differential equations

$$[\mathbf{Q}] \{\ddot{\mathbf{p}}(t)\} + [\mathbf{H}] \{\mathbf{p}(t)\} = \{\mathbf{0}\} \tag{8.87}$$

where $\{\mathbf{p}\} = -\rho_0 \left\{ \frac{\partial \phi}{\partial t} \right\}$ are the nodal sound pressures.

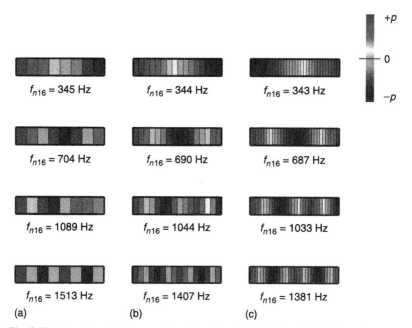

Fig. 8.17. Acoustic natural frequencies and modes of a closed pipe: FE solutions with (a) eight, (b) sixteen and (c) thirty two elements.

Figure 8.17 presents the first four modes of the fluid in the pipe with both ends closed, calculated using meshes of 8, 14 and 32 elements. Comparing these results with those derived with the exact formula and given in the left column of Fig. 8.13 shows that although only 8 elements are sufficient to accurately predict the first natural frequency and mode, at least 32 elements are necessary to predict the fourth natural frequency and mode with accuracy.

8.5.3   Acoustic Vibration of a Fluid in a Three-Dimensional Cavity: Finite Element Analysis

As pointed out above, many vibroacoustic problems involve structural vibration of closed shells or folded plates and acoustic vibration of the enclosed fluid. Three-dimensional acoustic elements can be used to model any form of three-dimensional cavity. The general formulation is presented first which takes into account the non-conservative effects of boundary surface dissipation and volume-distributed sources of sound. The virtual work done by dissipative mechanisms (δW_d) and by non-conservative volumetric sources (δW_q) will be included in the integral of Eq. (8.1). The Finite Element Analysis of a rectangular cavity with

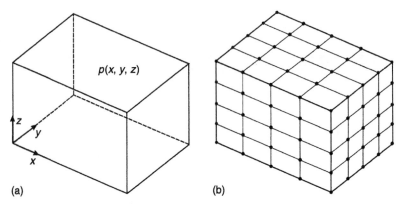

Fig. 8.18. (a) Rectangular cavity; (b) Finite Element mesh of hexahedral elements with rectangular faces and eight nodes at the vertices of the edges.

rigid walls, which is shown in Fig. 8.18(a), is presented as an example that will be used later to derive the response of fluid in such a cavity when closed on one face by a flexible plate.

Following the Finite Element procedure introduced above for a one-dimensional acoustic field, the Finite Element Model is built up by dividing the rectangular cavity into an assemblage of elements called 'acoustic three-dimensional elements'. As shown in Fig. 8.19, these elements could be hexahedra

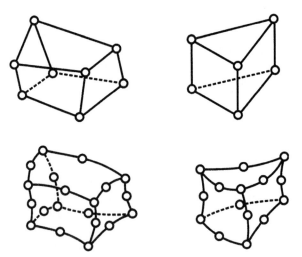

Fig. 8.19. Three-dimensional acoustic hexahedral (left) and pentahedral (right) elements with either straight or curved edges and either eight or twenty and six or fifteen nodes.

or pentahedra with either straight or curved edges and with a total of either eight or six nodes at the vertices and in some cases another twelve and eight nodes (respectively for the hexahedra or pentahedra elements) placed in the middle of the edges.

In this example, as shown in Fig. 8.18(b), the cavity is divided into a regular, three-dimensional mesh with hexahedral elements which, as shown in Fig. 8.20, have eight node points at the vertices. The acoustic kinetic and potential energy for each element are given by the following expressions (Craggs, 1971; Shuku and Ishihara, 1973):

$$T_e = \frac{1}{2} \int_V \rho_0 |\dot{u}|^2 \, dV = \frac{1}{2} \int_V \rho_0 |grad\phi|^2 \, dV$$

$$= \frac{1}{2} \int_{-1}^{+1} \int_{-1}^{+1} \int_{-1}^{+1} \rho_0 \left[\frac{1}{a_1^2} \left(\frac{\partial\phi(\xi_{1,2,3}, t)}{\partial\xi_1} \right)^2 + \frac{1}{a_2^2} \left(\frac{\partial\phi(\xi_{1,2,3}, t)}{\partial\xi_2} \right)^2 \right.$$

$$\left. + \frac{1}{a_3^2} \left(\frac{\partial\phi(\xi_{1,2,3}, t)}{\partial\xi_3} \right)^2 \right] a_1 a_2 a_3 \, d\xi_1 d\xi_2 d\xi_3$$

$$U_e = \frac{1}{2} \int_V \kappa p^2 dV = \frac{1}{2} \int_V \frac{\rho_0}{c^2} \dot{\phi}^2 dV$$

$$\tag{8.88a,b}$$

$$= \frac{1}{2} \int_{-1}^{+1} \int_{-1}^{+1} \int_{-1}^{+1} \frac{\rho_0}{c^2} \dot{\phi}^2(\xi_{1,2,3}, t) a_1 a_2 a_3 \, d\xi_1 d\xi_2 d\xi_3$$

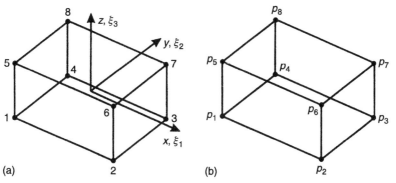

Fig. 8.20. Hexahedral acoustic element with eight nodes and degrees of freedom.

where V is the volume of the element, $\xi_1 = x/a_1$, $\xi_2 = y/a_2$ and $\xi_3 = z/a_3$ are dimensionless coordinates with reference to the dimensions $2a_1$, $2a_2$ and $2a_3$ of the element and $\xi_{1,2,3}$ stands for ξ_1, ξ_2, ξ_3.

Dissipative mechanisms in gases are small and can generally be neglected. Most acoustic energy dissipation occurs at boundary surfaces of cavities that are either flexible or porous. For each element located with one or more sides coincident with the boundary surface of the cavity, the virtual work done over the r-th dissipative face by the sound pressure p and virtual displacement δu_r, in direction orthogonal to this face, is given by

$$\delta W_{d,er} = \int_{S_r} p\,\delta u_r\,dS_r = \int_{-1}^{+1}\int_{-1}^{+1} p(\xi_{r1,2,3}, t)\delta u_r(\xi_{r1,2,3}, t)a_{r1}a_{r2}\,d\xi_{r1}d\xi_{r2}$$

(8.88c)

where $\xi_{r1,2,3}$ stands for ξ_{r1}, ξ_{r2}, ξ_{r3} and ξ_{r1} and ξ_{r2} are the a-dimensional coordinates of the r-th dissipative surface of the element with area $S_r = a_{r1} \times a_{r2}$. The a-dimensional coordinate ξ_{r3} locates the position of the face; thus, $\xi_{r3} = \pm 1$. For example, the a-dimensional coordinates of the lateral face denoted by the nodal points 3,4,7,8 of the element in Fig. 8.20 are given by: $\xi_{r1} = \xi_1$, $\xi_{r2} = \xi_3$ and $\xi_{r3} = \xi_2 = +1$. In the case of harmonic vibration, such that $p = \mathrm{Re}\,\{\tilde{p}\exp(j\omega t)\}$ and $u_r = \mathrm{Re}\,\{\tilde{u}_r\exp(j\omega t)\}$, the work can be expressed in terms of the complex specific acoustic impedance, $\tilde{z}_r(\omega) = \tilde{p}/j\omega\tilde{u}_r$ of the $r = 1, ..., R$ dissipating elemental faces (Gladwell, 1966):

$$\delta W_{d,er}(\omega) = \int_{S_r} p\,\delta u_r\,dS_r = -\frac{j\omega\rho_0^2}{\tilde{z}_r}\int_{-1}^{+1}\int_{-1}^{+1}\tilde{\phi}(\xi_{r1,2,3})\delta\tilde{\phi}(\xi_{r1,2,3})a_{r1}a_{r2}\,d\xi_{r1}d\xi_{r2}$$

(8.88d)

where $\tilde{\phi}(\xi_{r1,2,3})$ is the complex of the velocity potential. The virtual work done on a fluid element by volumetric acoustic sources is given by,

$$\delta W_{q,e} = \int_V p\,\delta q\,dV = -\int_{-1}^{+1}\int_{-1}^{+1}\int_{-1}^{+1}\rho_0\dot{\phi}(\xi_{1,2,3})\delta q(\xi_{1,2,3})a_1a_2a_3\,d\xi_1d\xi_2d\xi_3$$

(8.88e)

where $q(\xi_1, \xi_2, \xi_3)$ is the distributed rate of volume displacement per unit volume of the acoustic source. This expression can also be used to describe acoustic sources acting at a boundary surface, in which case the volumetric integral will become a surface integral for the element faces in common with a vibrating boundary surface (Gladwell, 1966; Craggs, 1971).

According to these energy expressions, the integral in Eq. (8.1) involves derivatives in ξ_1, ξ_2 and ξ_3 of $\phi(\xi_1, \xi_2, \xi_3, t)$ of order $p = 1$. Therefore, in order to satisfy the Rayleigh–Ritz convergence criteria, the element shape function should be a first order polynomial with four terms: $\phi(\xi_1, \xi_2, \xi_3) = \beta_1 + \beta_2\xi_1 + \beta_3\xi_2 + \beta_4\xi_3$. Also, the velocity potential must be a continuous function across the element boundaries so that the velocity potential, and thus sound pressure, is chosen as the unknown degree of freedom at each nodal point.

In order unambiguously to define the velocity potential within the element and at the element boundaries from the eight nodal velocity potentials or sound pressures, four additional terms should be added to the first order polynomial. These additional terms are chosen to be symmetric polynomial functions and thus guarantee that the element will be geometric invariant:

$$\phi(\xi_1, \xi_2, \xi_3) = \beta_1 + \beta_2\xi_1 + \beta_3\xi_2 + \beta_4\xi_3 + \beta_5\xi_1\xi_2 + \beta_6\xi_1\xi_3 + \beta_7\xi_2\xi_3 + \beta_8\xi_1\xi_2\xi_3$$

$$(8.89a)$$

This expression and its derivatives with reference to ξ_1, ξ_2 and ξ_3 can be written in matrix form as

$$\phi(\xi_1, \xi_2, \xi_3) = \lfloor 1 \quad \xi_1 \quad \xi_2 \quad \xi_3 \quad \xi_1\xi_2 \quad \xi_1\xi_3 \quad \xi_2\xi_3 \quad \xi_1\xi_2\xi_3 \rfloor \{\boldsymbol{\beta}\}$$

$$= \lfloor \mathbf{q}(\xi_1, \xi_2, \xi_3) \rfloor \{\boldsymbol{\beta}\} \qquad (8.89b)$$

and

$$\frac{\partial\phi(\xi_{1,2,3})}{\partial\xi_1} = \lfloor 0 \quad 1 \quad 0 \quad 0 \quad \xi_2 \quad \xi_3 \quad 0 \quad \xi_2\xi_3 \rfloor \{\boldsymbol{\beta}\} = \lfloor \mathbf{q}_1(\xi_1, \xi_2, \xi_3) \rfloor \{\boldsymbol{\beta}\}$$

$$(8.89c)$$

$$\frac{\partial\phi(\xi_{1,2,3})}{\partial\xi_2} = \lfloor 0 \quad 0 \quad 1 \quad 0 \quad \xi_1 \quad 0 \quad \xi_3 \quad \xi_1\xi_3 \rfloor \{\boldsymbol{\beta}\} = \lfloor \mathbf{q}_2(\xi_1, \xi_2, \xi_3) \rfloor \{\boldsymbol{\beta}\}$$

$$(8.89d)$$

$$\frac{\partial\phi(\xi_{1,2,3})}{\partial\xi_3} = \lfloor 0 \quad 0 \quad 0 \quad 1 \quad 0 \quad \xi_1 \quad \xi_2 \quad \xi_1\xi_2 \rfloor \{\boldsymbol{\beta}\} = \lfloor \mathbf{q}_3(\xi_1, \xi_2, \xi_3) \rfloor \{\boldsymbol{\beta}\}$$

$$(8.89e)$$

where

$$\{\boldsymbol{\beta}\}^T = \lfloor \beta_1 \quad \beta_2 \quad \beta_3 \quad \beta_4 \quad \beta_5 \quad \beta_6 \quad \beta_7 \quad \beta_8 \rfloor \qquad (8.90)$$

Evaluating the function $\phi(\xi_{1,2,3})$ in (8.89b) at $\xi_1 = \mp 1$, $\xi_2 = \mp 1$ and $\xi_3 = \mp 1$ gives

$$\{\phi_e\} = [\mathbf{B}_e]\{\beta\} \tag{8.91}$$

where

$$[\mathbf{B}_e] = \begin{bmatrix} 1 & -1 & -1 & -1 & 1 & 1 & 1 & -1 \\ 1 & 1 & -1 & -1 & -1 & -1 & 1 & 1 \\ 1 & 1 & 1 & -1 & 1 & -1 & -1 & -1 \\ 1 & -1 & 1 & -1 & -1 & 1 & -1 & 1 \\ 1 & -1 & -1 & 1 & 1 & -1 & -1 & 1 \\ 1 & 1 & -1 & 1 & -1 & 1 & -1 & -1 \\ 1 & 1 & 1 & 1 & 1 & 1 & 1 & 1 \\ 1 & -1 & 1 & 1 & -1 & -1 & 1 & -1 \end{bmatrix} \tag{8.92}$$

and $\{\phi_e\}$ is a column vector with the nodal velocity potentials

$$\{\phi_e\}^T = \lfloor \phi_1 \quad \phi_2 \quad \phi_3 \quad \phi_4 \quad \phi_5 \quad \phi_6 \quad \phi_7 \quad \phi_8 \rfloor \tag{8.93}$$

The velocity potential within the element can then be expressed in terms of the velocity potentials at the eight nodal potentials by substituting the vector $\{\beta\}$ derived from Eq. (8.91) into Eq. (8.89). Substituting Eq. (8.89) into Eqs. (8.88a,b), and (8.88d,e), respectively, the kinetic and potential energy terms for one element, the virtual work done by dissipative effects on the boundary surfaces and the virtual work done by volume acoustic sources are found to be given by

$$T_e = \tfrac{1}{2}\{\phi_e\}^T [\mathbf{H}_e]\{\phi_e\}$$
$$U_e = \tfrac{1}{2}\{\dot{\phi}_e\}^T [\mathbf{Q}_e]\{\dot{\phi}_e\} \tag{8.94a,b}$$

and

$$\delta W_{d,er} = -\rho_0 \{\delta\dot{\phi}_e\}^T [\mathbf{D}_{er}]\{\dot{\phi}_e\}$$
$$\delta W_{q,e} = -\rho_0 \{\delta\dot{\phi}_e\}^T \{\mathbf{q}_e\} \tag{8.94c,d}$$

In this case the element 'acoustic stiffness matrix' and 'acoustic inertia matrix' are given by

$$[\mathbf{H}_e] = [\mathbf{B}_e]^{-T} \int_{-1}^{+1} \int_{-1}^{+1} \int_{-1}^{+1} \frac{1}{a_1^2} \lfloor \mathbf{q}_1 \rfloor^T \lfloor \mathbf{q}_1 \rfloor$$

$$+ \frac{1}{a_2^2} \lfloor \mathbf{q}_2 \rfloor^T \lfloor \mathbf{q}_2 \rfloor + \frac{1}{a_3^2} \lfloor \mathbf{q}_3 \rfloor^T \lfloor \mathbf{q}_3 \rfloor a_1 a_2 a_3 \, d\xi_{1,2,3} [\mathbf{B}_e]^{-1} \qquad (8.95a,b)$$

$$[\mathbf{Q}_e] = \frac{1}{c^2} [\mathbf{B}_e]^{-T} \int_{-1}^{+1} \int_{-1}^{+1} \int_{-1}^{+1} \lfloor \mathbf{q} \rfloor^T \lfloor \mathbf{q} \rfloor a_1 a_2 a_3 \, d\xi_{1,2,3} [\mathbf{B}_e]^{-1}$$

The 'acoustic dissipation matrix' for the r-th dissipative face $[\mathbf{D}_{er}]$ is given by

$$[\mathbf{D}_{er}] = [\mathbf{B}_e]^{-T} \frac{j\omega\rho_0}{\tilde{z}_r} \int_{-1}^{+1} \int_{-1}^{+1} \lfloor \overline{\mathbf{q}}_r \rfloor^T \lfloor \overline{\mathbf{q}}_r \rfloor a_{r1} a_{r2} \, d\xi_{r1} d\xi_{r2} [\mathbf{B}_e]^{-1} \quad (8.95c)$$

where $\lfloor \overline{\mathbf{q}}_r \rfloor$ is the vector in Eq. (8.89b) with ξ_{r3} set to ± 1 according to the location of the r-th absorbing face of the element. Finally, the 'acoustic excitation vector', $\{\mathbf{q}_e\}$, for s-volume acoustic sources, is given by

$$\{\mathbf{q}_e\} = [\mathbf{B}_e]^{-T} \int_{-1}^{+1} \int_{-1}^{+1} \int_{-1}^{+1} \lfloor \mathbf{q} \rfloor^T q(\xi_1, \xi_2, \xi_3) a_1 a_2 a_3 \, d\xi_{1,2,3} \qquad (8.95d)$$

Typical integrals in Eqs. (8.95a)–(8.95d) have the following forms $\int_{-1}^{+1} \xi_1^m d\xi_1$, $\int_{-1}^{+1} \int_{-1}^{+1} \xi_1^m \xi_2^n d\xi_1 d\xi_2$ and $\int_{-1}^{+1} \int_{-1}^{+1} \int_{-1}^{+1} \xi_1^m \xi_2^n \xi_3^l d\xi_1 d\xi_2 d\xi_3$ which for m or n or l odd give zeros and for m, n and l even give $2/(m+1)$, $4/(m+1)(n+1)$ and $8/(m+1)(n+1)(l+1)$ respectively.

As in the previous example of acoustic vibration of fluid in a pipe, the kinetic and potential energy terms and, in this case, the virtual work done by the dissipative effects and volume displacement acoustic source on each element, can be combined to give the total energies and virtual work. These expressions can then be introduced into Eq. (8.1) to give

$$\int_{t_1}^{t_2} \left(\delta \left(T - U \right) + \delta W_d + \delta W_q \right) dt$$

$$= \int_{t_1}^{t_2} \left[\delta \left(\frac{1}{2} \rho_0 \{ \boldsymbol{\phi}(t) \}^T [\mathbf{H}] \{ \boldsymbol{\phi}(t) \} - \frac{1}{2} \{ \dot{\boldsymbol{\phi}}(t) \}^T [\mathbf{Q}] \{ \dot{\boldsymbol{\phi}}(t) \} \right) \right.$$

$$\left. - \rho_0 \{ \delta \dot{\boldsymbol{\phi}}(t) \} [\mathbf{D}] \{ \dot{\boldsymbol{\phi}}(t) \} - \rho_0 \{ \delta \dot{\boldsymbol{\phi}}(t) \} \{ \mathbf{q}(t) \} \right] dt = 0 \qquad (8.96a)$$

After integrating the second, third and fourth terms by parts, and differentiating the expression in the resulting integral, which is valid for any arbitrary $\{\delta\phi(t)\}$, as shown by Eqs. (8.69)–(8.72), the following set of equations in terms of the nodal pressures $\{\mathbf{p}\} = -\rho \left\{ \frac{\partial \phi}{\partial t} \right\}$ is obtained:

$$[\mathbf{Q}]\{\ddot{\mathbf{p}}(t)\} + [\mathbf{D}]\{\dot{\mathbf{p}}(t)\} + [\mathbf{H}]\{\mathbf{p}(t)\} = \{\mathbf{q}(t)\} \qquad (8.96b)$$

where $[\mathbf{Q}]$, $[\mathbf{D}]$ and $[\mathbf{H}]$ are the global acoustic inertia, damping and stiffness matrices when the rows and columns associated with the constrained degrees of freedom have been eliminated and $\{\mathbf{q}\}$ represents the column vector with the nodal excitations generated by the volumetric sources in the cavity when the terms associated with the constrained degrees of freedom have been eliminated.

As an example, the first nine natural frequencies and natural modes of a rigid-walled cavity with dimensions $L_x \times L_y \times L_z = 0.414 \text{ m} \times 0.314 \text{ m} \times 0.360 \text{ m}$ filled with air, for which $\rho_0 = 1.21 \text{ kg m}^{-3}$ and $c = 343 \text{ m s}^{-1}$ have been derived. Figure 8.21 presents the exact natural frequencies and natural modes (Beranek and Vér, 1992) while Fig. 8.22 shows the results obtained with a Finite Element Model constructed with a mesh of $16 \times 16 \times 16$ hexahedral acoustic elements. Good agreement between the analytical and numerical results is found for all modes except the eighth one which has an exact natural frequency of 829 Hz that is overestimated as 834 Hz by the Finite Element Model. This is due to the fact that this is the first mode with one mode index greater than one. The $16 \times 16 \times 16$ mesh produces a very good estimate of the modes with indices of either 0 or 1 but a less accurate estimate for higher order mode indices.

8.6   Coupled Fluid–Structure Analysis

If the top face of the rectangular cavity shown in Fig. 8.18 is not a rigid wall but is instead a flexible plate, as shown in Fig. 8.23(a), the acoustic field in the cavity is influenced by the motion of the plate and the vibration of the plate is perturbed by the fluid pressure loading. The vibrational behaviour of this coupled system can be derived using Hamilton's principle. However, in this case the integral in Eq. (8.1) for the acoustic system must include the work done by the plate on the fluid. Likewise, the integral in Eq. (8.1) for the structural system must include the complementary work done by the fluid on the plate. The Finite Element Analysis involves three-dimensional acoustic elements and two-dimensional structural plate elements. As shown in Fig. 8.23(b), the mesh of

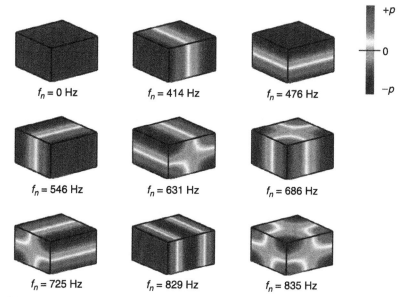

Fig. 8.21. Lowest nine acoustic natural frequencies and modes of a 0.414 m × 0.314 m × 0.360 m air-filled cavity with rigid walls (exact solution).

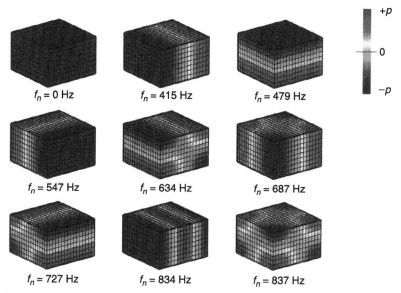

Fig. 8.22. Lowest nine acoustic natural frequencies and modes of a 0.414 m × 0.314 m × 0.360 m air-filled cavity with rigid walls (FE solution with a mesh of 16×16×16 hexahedral acoustic elements).

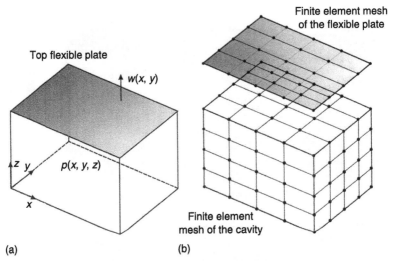

Fig. 8.23.   (a) Rectangular cavity with five rigid walls and a flexible top plate; (b) FE mesh of hexahedral acoustic elements with rectangular faces and eight nodes at the vertices of the edges and four-node rectangular bending element with straight edges.

acoustic elements is constructed in such a way that the upper faces of the elements facing the flexible panel match the plate elements. Therefore, the transverse displacements of the plate elements produce volumetric acoustic excitation acting on the corresponding top acoustic elements; likewise the sound pressure on the upper faces of the top acoustic elements represents the distributed force excitation acting on the corresponding plate elements.

The Finite Element Model presented in this section is based on the two formulations described at the ends of Sections 8.5 and 8.4, respectively, for acoustic vibration in cavities and structural vibration of plates. For the cavity system, the elemental kinetic and potential energies and the elemental virtual work done by boundary dissipative effects of the non-flexible faces and by volume acoustic sources are given by Eqs. (8.94a–d); for the plate system the elemental kinetic and strain energies and the elemental virtual work done by dissipative and external forces are given by Eqs. (8.50a–d). However, in order to model the coupling between the fluid and plate, two additional terms must be derived. First, the virtual work done by the volume displacement of the plate on the upper layer of acoustic elements; second, the virtual work done by the acoustic pressure on the plate elements.

Considering the general problem of a cavity with a number of walls modelled as flexible plates, for each acoustic coupling element located with, either, one (middle surface elements), two (edge elements) or three (corner elements) side(s) coincident with the plate elements, the element virtual work done over the c-th coupling face by the pressures $-p_{z,c}$ and virtual displacements δw_c of the matching plate element is given by (Craggs, 1971)

$$\delta W_{ac,e} = -\int_{S_c} p_{z,c}\,\delta w_c\,dS_c$$

$$= \rho_0 \int_{-1}^{+1}\int_{-1}^{+1} \dot{\phi}(\xi_{c1,2,3}, t)\delta w(\xi_{c1,2,3}, t)a_{c1}a_{c2}\,d\xi_{c1}d\xi_{c2} \quad (8.97a)$$

where $\xi_{c1,2,3}$ stands for $\xi_{c1}, \xi_{c2}, \xi_{c3}$ and ξ_{c1} and ξ_{c2} are the a-dimensional coordinates of the c-th coupling surface of the acoustic element with area $S_c = a_{c1} \times a_{c2}$ matched to a plate element such that $\xi_{c1} = \xi$ and $\xi_{c2} = \eta$. The a-dimensional coordinate ξ_{c3} locates the position of the face and thus, $\xi_{c3} = \pm 1$. For example, the a-dimensional coordinates of the top face denoted by the nodal points 5,6,7,8 of the element in Fig. 8.20 are given by: $\xi_{c1} = \xi_1, \xi_{c2} = \xi_2$ and $\xi_{c3} = \xi_3 = +1$. The virtual work done by the acoustic pressure p_c and virtual displacement δu_c through the c-th face of the coupling element on the matching plate element is given by (Craggs, 1971)

$$\delta W_{sc,e} = \int_A p_c\delta w_c\,dA = -\rho_0 \int_{-1}^{+1}\int_{-1}^{+1} \dot{\phi}(\xi, \eta)\delta w(\xi, \eta)\,ab\,d\xi\,d\eta \quad (8.97b)$$

where $\xi_{r1} = \xi$ and $\xi_{r2} = \eta$. Substituting Eqs. (8.89b), with $\{\beta\}$ derived from Eq. (8.91) and (8.45a), and with $\{\alpha\}$ derived from Eq. (8.47), respectively into Eqs. (8.97a) and (8.97b), yields the following matrix expressions for the virtual works done on the fluid and the plate, respectively:

$$\delta W_{ac,e} = \rho_0 \left\{\delta\dot{\phi}_e\right\}^T [R_e]\{w_e\},$$

$$\delta W_{sc,e} = \rho_0 \{\delta w_e\}^T [S_e]\{\dot{\phi}_e\} \qquad (8.98a,b)$$

where $\{\phi_e\}$ is the vector of nodal acoustic velocity potentials of the acoustic element and $\{w_e\}$ is the vector of nodal linear and angular displacements of the structural element. The coupling matrices $[R_e]$ and $[S_e]$ are given by

$$[R_e] = [B_e]^{-T} \int_{-1}^{+1}\int_{-1}^{+1} \lfloor \overline{q}_c \rfloor^T \lfloor p \rfloor\,a_{c1}a_{c2}\,d\xi_{c1}d\xi_{c2}[A_e]^{-1},$$

$$[S_e] = [A_e]^{-T} \int_{-1}^{+1}\int_{-1}^{+1} \lfloor p \rfloor^T \lfloor \overline{q}_c \rfloor\,ab\,d\xi d\eta[B_e]^{-1} \qquad (8.99a,b)$$

where $\lfloor \overline{\mathbf{q}}_c \rfloor$ is the vector in Eq. (8.89b) with ξ_{c3} set to ± 1 according to the location of the c-th coupling face of the acoustic element. As seen for the uncoupled plate and cavity models, the integrals in (8.99a,b) are straightforward to derive since they have the following forms: $\int_{-1}^{+1} \xi^m d\xi$, $\int_{-1}^{+1} \eta^n d\eta$, $\int_{-1}^{+1} \int_{-1}^{+1} \xi^m \eta^n d\xi d\eta$ which, for m or n odd, give zeros, and for m and n even give $2/(m+1)$, $2/(n+1)$ and $4/(m+1)(n+1)$, respectively.

Following the procedure described in the previous section, the kinetic and strain energy terms in Eqs. (8.94a,b), the virtual work done by the dissipative effects of the non flexible faces and volume displacement acoustic sources in Eqs. (8.94c,d) and, in this case, the virtual work done by the plate on the fluid [Eq. (8.98a)] can be combined to give the total energies and virtual works for the acoustic cavity. These expressions can then be introduced into Eq. (8.1) so that

$$\int_{t_1}^{t_2} \left(\delta\left(T-U\right) + \delta W_d + \delta W_q + \delta W_{ac} \right) dt$$

$$= \int_{t_1}^{t_2} \left(\delta\left(\tfrac{1}{2}\rho_0 \{\phi(t)\}^T [\mathbf{H}]\{\phi(t)\} - \tfrac{1}{2} \{\dot{\phi}(t)\}^T [\mathbf{Q}]\{\dot{\phi}(t)\} \right) \right.$$

$$\left. -\rho_0 \{\delta\dot{\phi}(t)\}[\mathbf{D}]\{\dot{\phi}(t)\} - \rho_0 \{\delta\dot{\phi}(t)\}\{q(t)\} + \rho_0 \{\delta\dot{\phi}(t)\}[\mathbf{R}]\{\mathbf{w}(t)\} \right) dt = 0$$
(8.100)

After integrating the second, third, fourth and fifth terms by parts, and differentiating the expression to be integrated, which is valid for any arbitrary $\{\delta\phi(t)\}$, as shown with Eqs. (8.69)–(8.72), the following set of equations in terms of the nodal pressures $\{\mathbf{p}\} = -\rho \left\{\frac{\partial\phi}{\partial t}\right\}$ and displacements $\{\mathbf{w}\}$ is obtained:

$$[\mathbf{Q}]\{\ddot{\mathbf{p}}(t)\} + [\mathbf{D}]\{\dot{\mathbf{p}}(t)\} + [\mathbf{H}]\{\mathbf{p}(t)\} - [\mathbf{R}]\{\ddot{\mathbf{w}}(t)\} = \{q(t)\} \qquad (8.101)$$

where $[\mathbf{R}]$ is the global acoustic–structural coupling matrix when the rows and columns associated with constrained degrees of freedom have been eliminated.

Similarly, following the procedure described in Section 8.4 for the plate system, the kinetic and strain energy terms in Eqs. (8.50a,b), the virtual work done by the dissipative and external transverse forces in Eqs. (8.50c,d) and, in this case, the virtual work done by the fluid on the plate [Eq. (8.98b)] can be combined to give the total energies and virtual works. These expressions can then be introduced into the Eq. (8.1) to give

$$\int_{t_1}^{t_2} \left(\delta \left(T - U \right) + \delta W_d + \delta W_f + \delta W_{sc} \right) dt$$

$$= \int_{t_1}^{t_2} \left[\delta \left(\tfrac{1}{2} \{\dot{\mathbf{w}}\}^T [\mathbf{M}] \{\dot{\mathbf{w}}\} - \tfrac{1}{2} \{\mathbf{w}\}^T [\mathbf{K}] \{\mathbf{w}\} \right) \right. \tag{8.102}$$

$$\left. + \{\delta \mathbf{w}\}^T [\mathbf{C}] \{\dot{\mathbf{w}}\} + \{\delta \mathbf{w}\}^T \{\mathbf{f}\} + \rho_0 \{\delta \mathbf{w}\}^T [\mathbf{S}] \{\dot{\boldsymbol{\phi}}\} \right] dt = 0$$

In this case, after integrating the first term by parts and, because $\{\delta \mathbf{w}\}$ is arbitrary, following the same procedure used to derive Eqs. (8.13)–(8.15), the following matrix expression is obtained:

$$[\mathbf{M}] \{\ddot{\mathbf{w}}(t)\} + [\mathbf{C}] \{\dot{\mathbf{w}}(t)\} + [\mathbf{K}] \{\mathbf{w}(t)\} + [\mathbf{S}] \{\mathbf{p}(t)\} = \{\mathbf{f}(t)\} \tag{8.103}$$

where $[\mathbf{S}]$ is the global structural–acoustic coupling matrix when the rows and columns associated with constrained degrees of freedom have been eliminated.

Because of reciprocity, the two coupling matrices are related such that $[\mathbf{S}] = [\mathbf{R}]^T$. Thus, Eqs. (8.101) and (8.103) can be combined into one matrix to derive the free and forced vibrational behaviour of a coupled structural–fluid system:

$$\begin{bmatrix} \mathbf{M} & \mathbf{0} \\ -\mathbf{S}^T & \mathbf{Q} \end{bmatrix} \begin{Bmatrix} \ddot{\mathbf{w}}(t) \\ \ddot{\mathbf{p}}(t) \end{Bmatrix} + \begin{bmatrix} \mathbf{C} & \mathbf{0} \\ \mathbf{0} & \mathbf{D} \end{bmatrix} \begin{Bmatrix} \dot{\mathbf{w}}(t) \\ \dot{\mathbf{p}}(t) \end{Bmatrix} + \begin{bmatrix} \mathbf{K} & \mathbf{S} \\ \mathbf{0} & \mathbf{H} \end{bmatrix} \begin{Bmatrix} \mathbf{w}(t) \\ \mathbf{p}(t) \end{Bmatrix} = \begin{Bmatrix} \mathbf{f}(t) \\ \mathbf{q}(t) \end{Bmatrix} \tag{8.104}$$

As an example, the first eighteen natural frequencies and natural modes of the coupled cavity and plate systems considered in the previous examples have been derived. The cavity has been modelled by a mesh of $16 \times 16 \times 8$ hexahedral acoustic elements as described in Section 8.5, while the plate has been modelled by a mesh of 16×16 plate elements as described in Section 8.4. The first 18 plate and cavity coupled mode shapes are plotted in Figs. 8.24 and 8.25. Also the natural frequencies of the plate–cavity coupled system and plate and cavity uncoupled systems, presented in Sections 8.4.3 and 8.5.3, are summarised in Table 8.1.

Comparison of the results shown in Figs. 8.24 and 8.25 with those presented in Figs. 8.21 and 8.10 for the uncoupled cavity and plate systems indicate that the first ten lowest natural frequencies and modes of the coupled cavity–plate system are 'controlled' by the vibration of the plate coupled with the volumetric vibration of the fluid in the cavity. As a result, the shapes of these natural modes are similar to the *in-vacuo* plate modes and to the (0,0,0) rigid-wall cavity mode

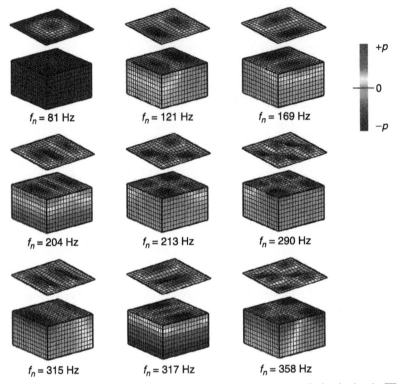

Fig. 8.24.   First nine natural modes of a coupled plate-cavity system calculated using the FEM. (All maximum modal amplitudes scaled to the same value).

slightly perturbed in the upper region to conform with the displacement field of the plate mode. This type of effect can also been seen in other higher order modes where *in-vacuo* plate type mode shapes are coupled with higher order acoustic modes slightly perturbed in the upper region. However, the modes with natural frequencies 418 Hz, 487 Hz and 552 Hz appear to be 'controlled' by the acoustic vibration of the fluid in the cavity. In these cases the plate mode is distorted to conform more closely to the pressure distribution of the cavity mode. Comparing the natural frequencies of the uncoupled plate and cavity systems with those of the coupled plate–cavity system summarised in Table 8.1, it is seen that the modes 'controlled' by the vibration of the plate have natural frequencies slightly lower than those of the corresponding uncoupled plate modes. The natural frequencies of the modes 'controlled' by the acoustic vibrations in the cavity are instead slightly higher than those of the corresponding uncoupled cavity modes.

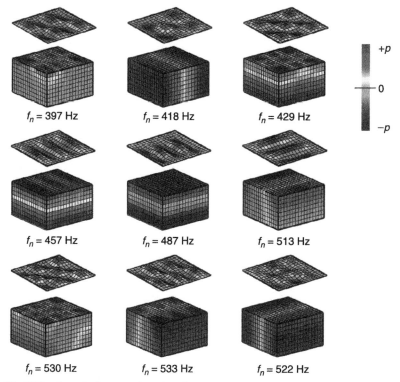

$f_n = 397$ Hz $f_n = 418$ Hz $f_n = 429$ Hz

$+p$

0

$-p$

$f_n = 457$ Hz $f_n = 487$ Hz $f_n = 513$ Hz

$f_n = 530$ Hz $f_n = 533$ Hz $f_n = 522$ Hz

Fig. 8.25. Second nine natural modes of a coupled plate-cavity system calculated using the FEM. (All maximum modal amplitudes scaled to the same value.)

8.7   Boundary Element Analysis for Vibroacoustic Problems

It has already been mentioned that FE analysis, in which an acoustic field is described by means of the values of various field quantities at a finite number of nodal points distributed throughout the volume, together with assumed internodal distributions, is most practicable when the extent of the region of fluid concerned is limited by a containing boundary surface. It is, however, often necessary to evaluate sound pressures in acoustic fields that are created in virtually infinite expanses of fluid by the action of sound sources located in the fluid. In this book we are concerned, in particular, with sound fields generated by the mechanically induced vibration of structures in contact with fluids and with the interaction of incident sound fields and structures immersed in fluids. As shown in Section 3.13,

TABLE 8.1

Natural Modes and Frequencies of the Uncoupled Panel and Cavity and Coupled Panel and Cavity

In-vacuo plate 0.414 × 0.314		Rigid-walled cavity 0.414 × 0.314 × 0.360		Coupled plate–cavity 0.414 × 0.314 × 0.360	
Mode	Frequency (Hz)	Mode	Frequency (Hz)	Modes involved	Frequency (Hz)
		(0,0,0)	0		
(1,1)	73			(1,1)-(0,0,0)	81
(2,1)	123			(2,1)-(0,0,0)	121
(1,2)	171			(1,2)-(0,0,0)	169
(3,1)	205			(3,1)-(0,0,0)	204
(2,2)	215			(2,2)-(0,0,0)	213
(3,2)	292			(3,2)-(0,0,0)	290
(4,1)	318			(4,1)-(0,0,0)	316
(1,3)	320			(1,3)-(0,0,0)	317
(2,3)	361			(2,3)-(1,0,0)	358
(4,2)	399			(4,2)-(0,0,0)	397
		(1,0,0)	415	(4,2)-(1,0,0)	418
(3,3)	431			(3,3)-(0,0,1)	429
(5,1)	460			(5,1)-(0,0,1)	457
		(0,0,1)	479	(5,1)-(0,0,1)	487
(1,4)	519			(1,4)-(0,1,0)	513
(4,3)	532			(4,3)-(1,0,1)	530
(5,2)	536	(0,1,0)	547	(5,2)-(0,1,0)	533
(2,4)	557			(2,4)-(0,1,0)	552

the Kirchhoff–Helmholtz integral equation can be used to evaluate the radiation and scattering of sound by objects in terms of the distributions of sound pressure and normal velocity on their surfaces. In Section 7.3 we have seen that it also applies to fully bounded acoustic problems. As indicated in Section 3.13, the Kirchhoff–Helmholtz integral equation is a general formulation for any type of vibroacoustic problem irrespective of the geometrical complexity of the surface of the object. However, the integration process is difficult to implement when the surface is irregular and cannot be represented by a simple analytical formula. For this reason a numerical approach, known as the 'Boundary Element Method' (BEM), has been developed which approximates the Kirchhoff–Helmholtz integral equation by a summation over a finite number of small surface elements with regular shapes. BE analysis can be applied to both exterior and interior problems. However, BE models of interior problems are not as efficient as FE models but are all suited to the calculation of sound radiation into free field.

8.7.1 Direct Boundary Element Method

Both direct and indirect boundary integral equations derived in Section 3.13 [Eqs. (3.183) and (3.184)] represent general formulations that enable analysis of the scattering/radiation of sound by surfaces of any geometric form. However, apart from a few cases with simple bounding surfaces such as parallelepipeds, circular cylinders and spheres, the integration of either the sound pressure and particle velocity or single and double layer potentials over the scattering/radiating surface is not trivial. For this reason, the 'Boundary Element Method' (BEM) has been formulated whereby the sound field is related to a discretised distribution of either the sound pressure and particle velocity or single and double layer potentials on the surface of the scattering/radiating object. Two separate approaches have been developed which are based on the direct and indirect boundary integral formulations. Both approaches follow a two-step procedure where the 'boundary variables' are first evaluated at a discrete set of points and then the 'field variable', that is, the sound pressure in the fluid domain, is obtained from the direct integral formulation, given in Eq. (7.21) for fully bounded acoustic domains and Eq. (3.183) for unbounded acoustic domains, and indirect integral formulation given by Eq. (3.184). In this section we focus our attention on the sound radiation and scattering problem in an unbounded fluid, although the formulation that is presented below can equally be used for fully bounded acoustic problems provided the positive orientation for the unit normal vector **n** to the boundary surface S_a is correctly specified.

As in FEM, in the first step of the BE method the boundary variables of the scattering/radiating object are approximated by a set of prescribed functions which are locally defined over small elements into which the surface has been discretised. In this way, the problem is reworked into a discrete set of equations with the sound pressure p_i and normal velocities v_{ni} as variables at the nodal positions of the mesh of small elements. The boundary surface S_a is divided into a grid of small subsurfaces S_e, also known as 'boundary elements', which are connected at a finite number of nodal points that are normally placed at the vertices of the elements, but in some cases at middle points of the edges or elemental area. In general, the elements have triangular or quadrilateral shapes with nodal points at the vertices. The two variables necessary to implement the direct boundary integral formulation are then approximated by a set of prescribed functions. The functions are chosen in such a way that either the sound pressure or normal velocity is equal to unity at a specific nodal position and equal to zero at all other nodes of the grid of boundary elements. Since the integral expressions in Eq. (3.183) do not involve spatial derivatives of the particle velocity or sound pressure, uniform prescribed functions could in principle be used. In which case, in order to have a determined problem, the elements should have only one node which is normally located at the centre of the element itself.

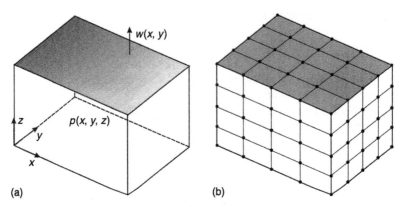

(a)   (b)

Fig. 8.26. Rectangular box with rigid lateral and bottom walls: (a) vibrating top face; (b) boundary element mesh of rectangular-linear elements with four nodes at the vertices of the edges.

However, uniform prescribed functions are rarely used since a rather large number of elements is necessary in order to produce rapid convergence to the exact solution.

For example, the radiation properties of the box in Fig. 8.26, which has rigid lateral and bottom walls and a vibrating top face, can be modelled by a regular array of rectangular boundary elements which has four nodal points at the vertices. In this case, the variation of sound pressure and normal particle velocity within each element can be uniquely expressed in terms of the sound pressures and normal particle velocities at the nodal positions using linear prescribed functions which are chosen to give a unit value at a specific nodal position and to be equal to zero at all other nodal points.

As shown by Fig. 8.27, the prescribed functions are defined to be equal to zero over the whole boundary surface except within the four elements surrounding the node point where the function is set to be non-zero. Also, as highlighted by the darker-grey element, the same four types of prescribed functions are used over each element of the boundary element mesh. Therefore, as in the FEM, the problem is reduced to determining shape functions for individual elements and not for the whole surface. These prescribed functions are referred to as 'element field functions' or 'element shape functions'.

Consider a general rectangular element of dimensions $a \times b$ (Fig. 8.28). Positions in the plane of the element are defined in terms of non-dimensional coordinates $\xi = f(x, y, z)/a$ and $\eta = g(x, y, z)/b$ such that the nodal positions correspond to $\xi = \pm 1$ and $\eta = \pm 1$. Then the sound pressure and normal velocity at a generic position (ξ, η) can be expressed in terms of linear polynomial

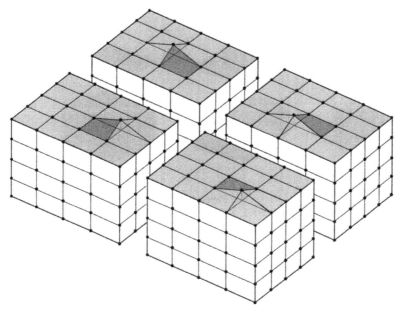

Fig. 8.27. Four prescribed functions on the top face of the rectangular box which show the four characteristic shape functions that are obtained for each element.

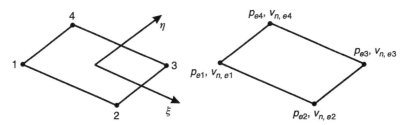

Fig. 8.28. Plane rectangular boundary element.

functions in ξ, η as follows:

$$p(\xi, \eta) = \alpha_1 + \alpha_2\xi + \alpha_3\eta + \alpha_4\xi\eta \tag{8.105a}$$

$$v_n(\xi, \eta) = \beta_1 + \beta_2\xi + \beta_3\eta + \beta_4\xi\eta \tag{8.105b}$$

These two expressions can also be written in matrix form as follows:

$$p(\xi, \eta) = \lfloor \mathbf{r}(\xi, \eta) \rfloor \{\boldsymbol{\alpha}\}, \quad v_n(\xi, \eta) = \lfloor \mathbf{r}(\xi, \eta) \rfloor \{\boldsymbol{\beta}\} \tag{8.106a,b}$$

where

$$\lfloor \mathbf{r}(\xi, \eta) \rfloor = \lfloor 1 \quad \xi \quad \eta \quad \xi\eta \rfloor \tag{8.107}$$

and

$$\{\boldsymbol{\alpha}\} = \begin{bmatrix} \alpha_1 \\ \alpha_2 \\ \alpha_3 \\ \alpha_4 \end{bmatrix}, \quad \{\boldsymbol{\beta}\} = \begin{bmatrix} \beta_1 \\ \beta_2 \\ \beta_3 \\ \beta_4 \end{bmatrix} \tag{8.108a,b}$$

The coefficients $\alpha_1, \dots, \alpha_4$ and $\beta_1, \dots, \beta_4$ can then be derived by evaluating the two equations at the four nodal points, for $\xi = \pm 1$ and $\eta = \pm 1$, so that

$$\{\mathbf{p}_e\} = [\mathbf{D}_e]\{\boldsymbol{\alpha}\} \quad \text{and} \quad \{\mathbf{v}_{n,e}\} = [\mathbf{D}_e]\{\boldsymbol{\beta}\} \tag{8.109a,b}$$

where

$$[\mathbf{D}_e] = \begin{bmatrix} 1 & -1 & -1 & 1 \\ 1 & 1 & -1 & -1 \\ 1 & 1 & 1 & 1 \\ 1 & -1 & 1 & -1 \end{bmatrix} \tag{8.110}$$

and

$$\{\mathbf{p}_e\} = \begin{Bmatrix} p_{e1} \\ p_{e2} \\ p_{e3} \\ p_{e4} \end{Bmatrix}, \quad \{\mathbf{v}_{n,e}\} = \begin{Bmatrix} v_{n,e1} \\ v_{n,e2} \\ v_{n,e3} \\ v_{n,e4} \end{Bmatrix} \tag{8.111a,b}$$

are the element sound pressure and normal velocity vectors. Using Eqs. (8.109a,b) it is then straightforward to find the values for $\alpha_1, \dots, \alpha_4$ and $\beta_1, \dots, \beta_4$, since

$$\{\boldsymbol{\alpha}\} = [\mathbf{D}_e]^{-1}\{\mathbf{p}_e\} \quad \text{and} \quad \{\boldsymbol{\beta}\} = [\mathbf{D}_e]^{-1}\{\mathbf{v}_{n,e}\} \tag{8.112a,b}$$

Substituting these values for $\alpha_1, \dots, \alpha_4$ and $\beta_1, \dots, \beta_4$ respectively into Eqs. (8.106a,b), the distributions of sound pressure and normal velocity within the element are then derived in terms of the sound pressure and particle velocity functions at the nodal degrees of freedom via the following vector equations:

$$p(\xi, \eta) = \lfloor B_1(\xi, \eta) \quad B_2(\xi, \eta) \quad B_3(\xi, \eta) \quad B_4(\xi, \eta) \rfloor \begin{bmatrix} p_{e1} \\ p_{e2} \\ p_{e3} \\ p_{e4} \end{bmatrix}$$

$$= \lfloor \mathbf{b}(\xi, \eta) \rfloor \{\mathbf{p}_e\} \tag{8.113a}$$

$$v_n(\xi, \eta) = \lfloor B_1(\xi, \eta) \quad B_2(\xi, \eta) \quad B_3(\xi, \eta) \quad B_4(\xi, \eta) \rfloor \begin{bmatrix} v_{n,e1} \\ v_{n,e2} \\ v_{n,e3} \\ v_{n,e4} \end{bmatrix}$$

$$= \lfloor \mathbf{b}(\xi, \eta) \rfloor \{\mathbf{v}_{n,e}\} \tag{8.113b}$$

where

$$B_1(\xi, \eta) = \frac{1}{4}(1 - \xi)(1 - \eta), \quad B_2(\xi, \eta) = \frac{1}{4}(1 + \xi)(1 - \eta)$$

$$\tag{8.114a–d}$$

$$B_3(\xi, \eta) = \frac{1}{4}(1 + \xi)(1 + \eta), \quad B_4(\xi, \eta) = \frac{1}{4}(1 - \xi)(1 + \eta)$$

The sound pressures and particle velocities at all nodal positions are then determined by the so-called collocation scheme where the direct boundary integral formulation in Eq. (3.183) is evaluated at the nodal positions using the boundary variable expansions given by Eqs. (8.113a,b). The integration over the boundary surface S_a is carried out on an element by element basis so that for each collocation node c the following expression is obtained:

$$c_c p_c = \sum_{e=1}^{N_e} \lfloor \mathbf{h}_{ce} \rfloor \{\mathbf{p}_e\} + j\omega\rho_0 \lfloor \mathbf{g}_{ce} \rfloor \{\mathbf{v}_{n,e}\} + p_{i,c} \tag{8.115}$$

where p_c is the sound pressure at the collocation point, $p_{i,c}$ is the sound pressure at the collocation point due to the incident acoustic waves generated by other distributed sources $\tilde{Q}(\mathbf{r}_s)$ in the fluid domain as specified in Eq. (3.165) and

$$c_c = 1 + \frac{1}{4\pi} \sum_{e=1}^{N_e} \int_{-1}^{+1} \int_{-1}^{+1} \frac{\partial}{\partial n}\left(\frac{1}{|\mathbf{r}_c - \mathbf{r}_e|}\right) d\xi d\eta \tag{8.116a}$$

where the vectors $\mathbf{r}_c = (x_c, y_c, z_c)$ and $\mathbf{r}_e = (x_e, y_e, z_e)$ give, respectively, the positions at the collocation node c and within the e-th element, $\mathbf{n}$ is the unit normal to the boundary surface at the collocation node and N_e is the number of elements. Also, the four coefficients in the row vectors $\lfloor h_{ce} \rfloor = \lfloor h_{ce1} \quad \cdots \quad h_{ce4} \rfloor$ and $\lfloor g_{ce} \rfloor = \lfloor g_{ce1} \quad \cdots \quad g_{ce4} \rfloor$ for the e-th element are

$$h_{cej} = \int_{-1}^{+1} \int_{-1}^{+1} \frac{\partial G(\mathbf{r}_c, \mathbf{r}_{ej})}{\partial n} B_j(\xi, \eta) \, d\xi \, d\eta \qquad (8.116b)$$

$$g_{cej} = \int_{-1}^{+1} \int_{-1}^{+1} G(\mathbf{r}_c, \mathbf{r}_{ej}) B_j(\xi, \eta) \, d\xi \, d\eta \qquad (8.116c)$$

where, in this case, $\mathbf{r}_{ej} = (x_{ej}, y_{ej}, z_{ej})$ gives the position of the j-th node of the e-th element. When the collocation point c is not on the element e the integrals in Eqs. (8.116) can be evaluated numerically by standard Gauss–Legendre quadrature (Wu, 2000). However, when the collocation point is located on a node j of the e-th element the integrals for h_{cej} and g_{cej} become singular; that is they have an infinite value as $|\mathbf{r}_c - \mathbf{r}_{ej}| \to 0$. In this case the Gauss–Legendre quadrature is carried out by considering a polar system of coordinates with origin at the nodal point j. In this way it is possible to integrate the two functions in Eqs. (8.116) that behave as $1/|\mathbf{r}_c - \mathbf{r}_{ej}|$.

When the collocation scheme is repeated for all nodal points N of the boundary element mesh, a set of N expressions in the nodal field variables is obtained which can be assembled into the following matrix equation:

$$[\mathbf{A}]\{\mathbf{p}\} = j\omega\rho_0 [\mathbf{B}]\{\mathbf{v}_n\} + \{\mathbf{p}_i\} \qquad (8.117)$$

where $\{\mathbf{p}\}$ and $\{\mathbf{v}_n\}$ are the vectors with the sound pressure and velocity in the normal direction to the boundary surface at the nodal position of the boundary element mesh while $\{\mathbf{p}_i\}$ is the vector with the incident sound pressure generated by other acoustic sources $\tilde{Q}(\mathbf{r}_s)$ in the fluid domain at the nodal positions of the boundary element mesh:

$$\{\mathbf{p}\} = \begin{Bmatrix} p_1 \\ p_2 \\ \vdots \\ p_N \end{Bmatrix}, \qquad \{\mathbf{v}_n\} = \begin{Bmatrix} v_{n,1} \\ v_{n,2} \\ \vdots \\ v_{n,N} \end{Bmatrix}, \qquad \{\mathbf{p}_i\} = \begin{Bmatrix} p_{i,1} \\ p_{i,2} \\ \vdots \\ p_{i,N} \end{Bmatrix} \qquad (8.118a,b,c)$$

and the elements in the two matrices in Eq. (8.117) are given by

$$A_{cj} = c_c - \sum_e h_{cej} \quad \text{and} \quad B_{cj} = \sum_e g_{cej} \qquad (8.119a,b)$$

where $c = 1,..., N$, $j = 1,..., N$ and the two summations are for the four elements around the nodal point j. The matrix expression in (8.117) represents a set of N algebraic expressions in the N pressure and N normal velocity unknown nodal degrees of freedom. However, the boundary conditions impose either a prescribed sound pressure or a prescribed normal velocity or a ratio between the two. Therefore, taking into account the prescribed sound pressure or normal velocity at the nodal positions of the boundary element mesh, the matrix expression in Eq. (8.117) can be formulated in such a way as to give a set of N independent equations in N unknown nodal variables which could be either the sound pressure or normal velocity depending on the type of boundary condition associated with each nodal position.

Once the sound pressure and normal velocity at the nodal positions of the boundary element mesh have been derived the second post-processing step is implemented. The sound pressure at any position $\mathbf{r}$ in the acoustic domain V, which is not located on the boundary surface S_a, is derived as the superposition of the direct pressure field, which is due to other distributed acoustic sources $\widetilde{Q}(\mathbf{r}_s)$ in the fluid domain radiating into free space without taking into account the effects of the scattering/radiating body with surface S_a, and the scattered/radiated pressure field, which is due to the scattering and radiation of sound by the body of surface S_a. The first term is derived directly from the integral $\int_{V_s} \widetilde{Q}(\mathbf{r}_s)\widetilde{G}(\mathbf{r}, \mathbf{r}_s)dV_s$ in Eq. (3.183) while the second term is derived from Kirchhoff–Helmholtz integral in Eq. (3.183) using the sound pressure on the scattering/radiating surface S_a obtained in the first step of the boundary element analysis. Therefore, following the same procedure used to derive Eq. (8.117), the approximated scattered/radiated sound pressure generated by the surface S_a can be expressed by the following matrix relation

$$p(\mathbf{r}) = \lfloor \mathbf{a} \rfloor \{\mathbf{p}\} + j\omega\rho_0 \lfloor \mathbf{b} \rfloor \{\mathbf{v}_n\} + \lfloor \mathbf{c} \rfloor \{\mathbf{f}\} + j\omega\rho_0 \lfloor \mathbf{d} \rfloor \{\mathbf{q}\} : \qquad (8.120)$$

where $\{\mathbf{p}\}$ and $\{\mathbf{v}_n\}$ are the vectors with the sound pressure and particle velocity at the nodal positions of the Boundary Element mesh derived from the first step with Eq. (8.117). The coefficients a_i and b_i on the row vectors $\lfloor \mathbf{a} \rfloor$ and $\lfloor \mathbf{b} \rfloor$ are

$$a_i = \sum_e \int_{-1}^{+1} \int_{-1}^{+1} \frac{\partial G(\mathbf{r}, \mathbf{r}_{ei})}{\partial n} B_i(\xi, \eta)\, d\xi d\eta$$

and   (8.121a,b)

$$b_i = \sum_e \int_{-1}^{+1} \int_{-1}^{+1} G(\mathbf{r}, \mathbf{r}_{ei}) B_i(\xi, \eta)\, d\xi d\eta$$

and the elements in the row vectors $\lfloor \mathbf{c} \rfloor$ and $\lfloor \mathbf{d} \rfloor$ can be derived from the integral $\int_{V_s} \tilde{Q}(\mathbf{r}_s)\tilde{G}(\mathbf{r}, \mathbf{r}_s)dV_s$, with $\tilde{Q}(\mathbf{r}_s) = j\rho_0\omega\tilde{q}\,(\mathbf{r}_s) - \nabla\tilde{\mathbf{f}}(\mathbf{r}_s)$ (see Eq. (3.164)), in Eq. (3.183):

$$c_i = \int_{V_q} \nabla\tilde{G}(\mathbf{r}, \mathbf{r}_{fi})dV, \quad d_i = \int_{V_q} \tilde{G}(\mathbf{r}, \mathbf{r}_{qi})\,dV \qquad (8.121\text{c,d})$$

where $\mathbf{r}_{fi}$ and $\mathbf{r}_{qi}$ define the positions of the i-th distributed force or volume velocity sources which are defined by the elements in the two vectors $\{\mathbf{f}\} = [f_1 \cdots f_F]^T$ and $\{\mathbf{q}\} = [q_1 \cdots q_Q]^T$.

The direct boundary integral formulation in Eq. (3.183) does not involve spatial derivatives of the boundary pressure and boundary normal velocity. Therefore, the convergence of a boundary element discretisation is ensured even when the sound pressure and normal velocity expansions in Eqs. (8.109a,b) produce discontinuities between adjacent elements. Brebbia *et al.* (1984) have demonstrated that the convergence of a boundary element model with uniform shape functions, which incorporates discontinuities at the boundaries of the elements, is guaranteed. However, the convergence rate is found to be relatively slow with uniform elements and 3-noded linear triangular or 4-noded linear quadrilateral elements are normally used. As shown in Fig. 8.27, the nodes of these types of elements are located at the vertices of the triangular or rectangular element. When linear shape functions are used the sound pressure and normal velocity field variables are approximated by continuous functions that guarantee a better convergence rate than the uniform functions. In general, a critical problem for convergence rate is associated with geometrical discontinuities of the scattering/radiating surface such as sharp edges or corners as, for example, the edges of the rectangular box considered in this section and shown in Fig. 8.26. As Fig. 8.29 shows, this problem is often solved by duplicating the nodes along edges or at corner points where the normal to the scattering/radiating surface is not uniquely defined. In this case, the unit variation of the field variable at the two collocated nodes occurs with reference to the normals to the top and front surfaces of the box. As a result, two shape functions are obtained, each of which involves variations of the field variables only over two elements instead of four elements as for the 'internal nodes' in Fig. 8.27. This technique is also used to improve the modelling of scattering/radiating surfaces characterised by two or more types of boundary condition. A typical case is represented by the sound radiation of a flat plate in a rigid baffle which is normally modelled with double nodes at the edges of the plate where the boundary condition is not uniquely defined.

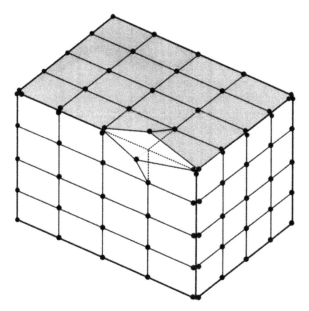

Fig. 8.29. Shape functions for unit variation of two collocated nodes on an edge of the rectangular box with five rigid walls and flexible top plate.

When multiple-nodes meshing is used, then, for a collocation point c with M multiple nodes, M equations of the type in (8.115) are obtained:

$$
\begin{cases}
c_{sc} p_c = \sum_{e=1}^{N_e} \lfloor \mathbf{h}_{ce,1} \rfloor \{ \mathbf{p}_e \} + j\omega\rho_0 \lfloor \mathbf{g}_{ce} \rfloor \{ \mathbf{v}_{n,e} \} \\
\qquad\qquad\qquad \vdots \\
c_{sc} p_c = \sum_{e=1}^{N_e} \lfloor \mathbf{h}_{ce,M} \rfloor \{ \mathbf{p}_e \} + j\omega\rho_0 \lfloor \mathbf{g}_{ce} \rfloor \{ \mathbf{v}_{n,e} \}
\end{cases}
\tag{8.122}
$$

where the c_c coefficients and the elements in the row vectors $\lfloor \mathbf{g}_{ce} \rfloor = \lfloor g_{ce1} \cdots g_{ce4} \rfloor$ are all given by Eqs. (8.116a) and (8.116b), respectively. The elements in the row vectors $\lfloor \mathbf{h}_{ce,m} \rfloor = \lfloor h_{ce1,m} \cdots h_{ce4,m} \rfloor$ depend instead on the multiple node and are given by

$$
h_{cej,m} = \int_{-1}^{+1} \int_{-1}^{+1} \frac{\partial G(\mathbf{r}_c, \mathbf{r}_{ej,m})}{\partial n} B_{j,m}(\xi, \eta) \, d\xi d\eta
\tag{8.123}
$$

where, in this case, $\mathbf{r}_{ej,m} = (x_{ej,m}, y_{ej,m}, z_{ej,m})$ gives the position of the j-th node with multiplicity M. When the collocation scheme is repeated for all nodal points N of the boundary element mesh, the matrix expression in Eq. (8.117) is derived where the matrices $[\mathbf{A}]$ and $[\mathbf{B}]$ are in this case characterised by groups of rows which correspond to collocation points where multiple nodes are defined.

The boundary element formulation presented above can be used either for exterior or interior problems provided the direction of the normal is appropriately defined. Thus, for any closed boundary surface S_a, an interior problem can be defined which, for example, is characterised by a Dirichlet boundary condition such that the sound pressure is zero over the whole boundary surface. This type of system is obviously characterised by a set of undamped natural modes so that, at the corresponding natural frequencies, the Kirchhoff–Helmholtz integral Eq. (7.20) does not have a unique solution. Since the solution for the interior problem applies to both interior and exterior domains (except for a sign inversion) a modal behaviour is also found for the sound radiation problem which obviously does not have any physical meaning. Indeed, this is merely a mathematical problem where the global matrices in Eq. (8.117) become poorly conditioned at frequencies close to the natural frequencies of the corresponding interior problem. The acoustic modal density of acoustic enclosures rises asymptotically with the square of frequency so that at higher frequencies it becomes impossible to predict the sound radiation by closed surfaces using the Kirchhoff–Helmholtz integral formulation. Several methods have been developed to overcome this mathematical problem. For example, Burton and Miller (1971) have reformulated the standard Kirchhoff–Helmholtz integral equation by taking its derivative with reference to the normal $\mathbf{n}$. Such a normal-derivative integral equation will still give the solution of both interior and exterior problems for a given closed surface S_a. Also, the solution will be still characterised by a set of undamped natural modes which however have different natural frequencies than those derived with the standard integral equation. As a result, the sound radiation properties of a closed surface can be correctly derived at all frequencies by mutually excluding the two sets of natural frequencies derived with the standard and normal-derivative integral equations. This can be done in a simple way by combining the results obtained with the standard and normal-derivative integrals using a factor j/ω which stabilises the solution. An alternative method, which is widely used by software writers, was proposed by Schenck (1968) in which a number I of points $\mathbf{r}_{int}$ at which the pressure is set to zero are defined within the closed space delimited by the radiating surface S_a. These points are treated as collocation points so that for each of them a matrix expression as in Eq. (8.115) is obtained, but with the sound pressure at the collocation point equal to zero:

$$0 = \sum_{e=1}^{N_e} \lfloor \mathbf{h}_{int,e} \rfloor \{\mathbf{p}_e\} + j\omega\rho_0 \lfloor \mathbf{g}_{int,e} \rfloor \{\mathbf{v}_{n,e}\} \tag{8.124}$$

with $int = 1,\ldots, I$. These expressions can then be assembled together with those relative to the surface collocation points so that the resulting equation

$$[\overline{\mathbf{A}}]\{\mathbf{p}\} = j\omega\rho_0\,[\overline{\mathbf{B}}]\{\mathbf{v}_n\} \tag{8.125}$$

is characterised by two matrices $[\overline{\mathbf{A}}]$ and $[\overline{\mathbf{B}}]$ which have $(N + I)$ rows and (N) columns. Thus this is an over-determined problem which, for example, can be solved by a least square method. This technique is commonly known as the 'CHIEF' method which stands for Combined Helmholtz Integral Equation Formulation. The selection of the number and positions of the internal points is very important. Although there are no rules, one can intuitively see that the points should not be regularly spaced; more importantly, they should not be located in positions that are likely to coincide with nodal planes of the interior normal modes where the sound pressure would be zero in any case. Also, the number of internal points should be a small fraction of the nodal points used for the mesh of the boundary surface.

As an illustration, the sound radiation generated by the box shown in Fig. 8.30 has been modelled assuming that the vibrations of the top face correspond to the first four *in-vacuo* modes of a simply-supported panel.

8.8   Coupled Fluid–Structure Analysis

When, as shown in Fig. 8.31, the top face of the box considered in the previous section consists of a flexible plate then the scattering/radiation into free field of the box is determined by the coupled response of the panel with the fluid surrounding the box. The sound field in the external fluid is generated by the vibration of the plate and the vibrations of the plate are perturbed by the pressure exerted by the fluid. A coupled FE–BE model is therefore necessary in which the coupling between the plate and the fluid is expressed through the equality of transverse plate velocity and the normal component of fluid particle velocity.

Assuming that the finite element mesh for the top flexible plate is constructed with the four-nodes rectangular thin plate elements considered in Section 8.4 and assuming that, as shown in Fig. 8.31(b), such a finite element mesh matches the boundary element mesh for the top face of the rectangular box, then a coupled FE–BE model can be derived by taking into account the compatibility condition over the surface of concurrent structural FE and acoustic BE elements. Using the formulation presented in section 8.4 for the plate elements, the work done by the sound pressure loading exerted over the surface of an element of the plate can be determined with Eq. (8.40d) which, considering the expansion of the sound

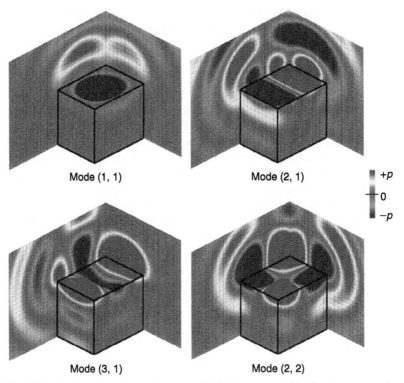

Fig. 8.30. Sections through sound pressure distributions generated by four *in-vacuo* modes of the top plate of a rectangular box at 1000 Hz.

pressure on the face of the element given by the boundary element formulation in Eq. (8.109a) and the expansion of the transverse displacement over the element surface given by the finite element formulation in Eq. (8.45a) with $\{\alpha\}$ derived from Eq. (8.47), becomes

$$\delta W_{sc,e} = \rho_0 \{\delta\mathbf{w}_e\}^T [\mathbf{L}_e] \{\mathbf{p}_e\} \tag{8.126}$$

where $\{\mathbf{p}_e\}$ is the vector with the nodal acoustic pressures of the acoustic BE element and $\{\mathbf{w}_e\}$ is the vector with the nodal linear and angular displacements of the structural FE element. The coupling matrix $[\mathbf{L}_e]$ is given by

$$[\mathbf{L}_e] = [\mathbf{A}_e]^{-T} \int_{-1}^{+1} \int_{-1}^{+1} \lfloor\mathbf{p}\rfloor^T \lfloor\mathbf{b}\rfloor \, ab \, d\xi \, d\eta \tag{8.127}$$

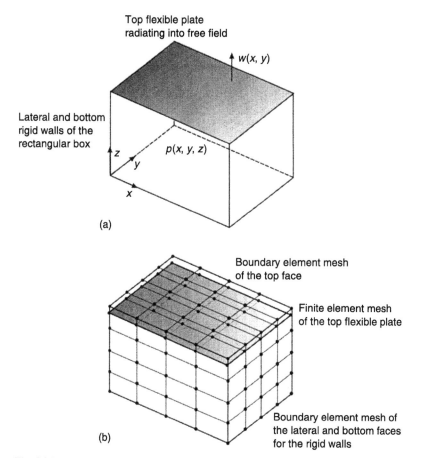

Fig. 8.31.   (a) Rectangular box with five rigid walls and a flexible plate closure and (b) its mesh of rectangular-linear boundary elements and four-node rectangular-linear bending elements.

Following the procedure described in Section 8.4.3 for the plate system, the kinetic and strain energy terms in Eqs. (8.50a,b), the virtual work done by the dissipative and external transverse forces in Eqs. (8.50c,d) and, in this case, the virtual work done by the acoustic loading exerted on the plate [Eq. (8.126)], can be combined to give the total energies and virtual works. These expressions can then be introduced into the integral in Eq. (8.1):

$$\int_{t_1}^{t_2} \left[\delta \left(\tfrac{1}{2} \{\dot{\mathbf{w}}\}^T [\mathbf{M}] \{\dot{\mathbf{w}}\} - \tfrac{1}{2} \{\mathbf{w}\}^T [\mathbf{K}] \{\mathbf{w}\} \right) \right.$$

$$\left. + \{\delta\mathbf{w}\}^T [\mathbf{C}] \{\dot{\mathbf{w}}\} + \{\delta\mathbf{w}\}^T \{\mathbf{f}\} + \{\delta\mathbf{w}\}^T [\mathbf{L}] \{\mathbf{p}_c\} \right] dt = 0 \qquad (8.128)$$

where $\{p_c\}$ is the vector with the sound pressure at the nodes of the BE acoustic elements concurrent to the FE structural elements. Equation (8.128) can be integrated by parts so that, following the same procedure expressed by Eqs. (8.13)–(8.15), the following system of N ordinary differential equations is derived:

$$[\mathbf{M}]\{\ddot{\mathbf{w}}(t)\} + [\mathbf{C}]\{\dot{\mathbf{w}}(t)\} + [\mathbf{K}]\{\mathbf{w}(t)\} + [\mathbf{L}]\{\mathbf{p}_c(t)\} = \{\mathbf{f}(t)\} \qquad (8.129)$$

where $[\mathbf{L}]$ is the global structural–acoustic coupling matrix which is derived by assembling the element coupling matrices given by Eq. (8.127).

The compatibility condition for the normal velocity over concurrent BE acoustic and FE structural elements is expressed by the following equation:

$$\{\mathbf{v}_{n,e}\} = j\omega\,[\mathbf{T}_e]\{\mathbf{w}_e\} \qquad (8.130)$$

where, considering the definition for the structural transverse displacement given by Eq. (8.45a) and using the expression in Eq. (8.47) for the vector $\{\boldsymbol{\alpha}\}$, the matrix $[\mathbf{T}_e]$ is found to be given by

$$[\mathbf{T}_e] = \begin{bmatrix} \mathbf{p}(\xi_1, \eta_1) \\ \vdots \\ \mathbf{p}(\xi_4, \eta_4) \end{bmatrix} [\mathbf{A}_e]^{-1} \qquad (8.131)$$

The compatibility condition between all concurrent BE acoustic and FEM structural elements is then derived by assembling the element compatibility matrices $[\mathbf{T}_e]$ into a global matrix $[\mathbf{T}]$ so that

$$\{\mathbf{v}_n\} = j\omega\,[\mathbf{T}]\{\mathbf{w}\} \qquad (8.132)$$

Substituting this expression into the acoustic BE matrix expression in Eq. (8.117), with $\{p_i\} = \{0\}$, yields

$$\begin{bmatrix} \mathbf{A}_{11} & \mathbf{A}_{12} \\ \mathbf{A}_{21} & \mathbf{A}_{22} \end{bmatrix}\begin{Bmatrix} \mathbf{p}_c \\ \mathbf{p}_b \end{Bmatrix} = \begin{bmatrix} \mathbf{B}_{11} & \mathbf{B}_{12} \\ \mathbf{B}_{21} & \mathbf{B}_{22} \end{bmatrix}\begin{Bmatrix} -\rho_0\omega^2 \mathbf{T}\mathbf{w} \\ j\rho_0\omega\mathbf{v}_b \end{Bmatrix} \qquad (8.133)$$

where the vectors $\{p_b\}$ and $\{v_b\}$ give the nodal sound pressures and normal velocities of the acoustic BE elements not facing the structural mesh of elements. The equations governing the FE structural and BE acoustic coupled models can then be derived by combining the modified FE model in Eq. (8.129) and the modified BE acoustic model in Eq. (8.133) which yields the following matrix formulation:

$$\begin{bmatrix} (\mathbf{K} + j\omega\mathbf{C} - \omega^2\mathbf{M}) & \mathbf{L} & \mathbf{0} \\ \rho_0\omega^2\mathbf{B}_{11}\mathbf{T} & \mathbf{A}_{11} & \mathbf{A}_{12} \\ \rho_0\omega^2\mathbf{B}_{21}\mathbf{T} & \mathbf{A}_{21} & \mathbf{A}_{22} \end{bmatrix} \begin{Bmatrix} \mathbf{w} \\ \mathbf{p}_c \\ \mathbf{p}_b \end{Bmatrix} = \begin{Bmatrix} \mathbf{f} \\ j\rho_0\omega\mathbf{B}_{12}\mathbf{v}_b \\ j\rho_0\omega\mathbf{B}_{22}\mathbf{v}_b \end{Bmatrix} \qquad (8.134)$$

Problems

8.1 Knowing that the kinetic and strain energy functions for axial vibrations in rods are given by $T = \frac{1}{2}\int_0^l \rho A(\frac{\partial u}{\partial t})^2\, dx$ and $U = \frac{1}{2}\int_0^l EA(\frac{\partial u}{\partial x})^2\, dx$, where ρ and E are the density and Young's modulus of elasticity of the material A and l and are the cross-sectional area and length of the beam, derive the elements in the mass and stiffness matrices for the Rayleigh–Ritz energy method in the case where the beam is blocked at one end and free at the other end (use a polynomial prescribed function $d_j(x) = x^i$).

8.2 Using the elemental kinetic and strain energy expressions for axial vibrations of rods given in Question 8.1, derive expressions for the elemental kinetic and strain energies of axial vibrations of rods in terms of the non-dimensional coordinate $\xi = x/a$ where a is half the element length.

8.3 Using the expressions for the elemental kinetic and strain energies of axial vibrations of rods given in Question 8.1, derive the element mass and stiffness matrices for axial vibrations of rods assuming that the element shape function is given by the following polynomial function: $u = \alpha_1 + \alpha_2\xi$.

9 | Introduction to Active Control of Sound Radiation and Transmission

9.1 Introduction to Active Control

We have seen in previous chapters that passive treatments can be successfully used to reduce sound radiation and transmission by vibrating structures in the so-called mid- and high audio frequency ranges. Active vibroacoustic control systems, in which externally activated sources are driven in such a manner so as to minimise sound and vibration, are most effective at low audio frequencies. The intention of this chapter is to introduce the basic concepts of control theory and to describe and explain the physics and technology of active structural acoustic control (ASAC) and active vibration control (AVC) for the reduction of sound transmission and radiation. The principles, physical mechanisms and limitations of active structural acoustic control and active vibration control will be illustrated with reference to a simple model problem of sound transmission through a baffled rectangular panel.

9.2   Fundamentals of Active Control Theory

The principle of active sound control was first introduced by Lueg (1936) in a patent for the single channel feed-forward control of tonal disturbances propagating in a one-dimensional acoustic waveguide or in a three-dimensional free field as is shown in Fig. 9.1. However, it took about two decades before the first single-channel analogue control systems were developed by Olson and May (1953) and Conover (1956). Practical applications became available much later in the 1980s when digital signal processing (DSP) enabled the implementation of adaptive digital controllers (Chaplin, 1983; Roure, 1985).

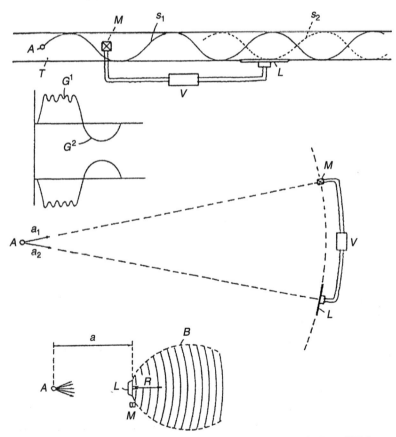

Fig. 9.1.   Ilustrations page from the active sound control patent by Lueg (1936).

During the past two decades much of progress has been made in the design and development of both digital and analogue control systems. Several control schemes have been proposed which can be classified into two main families: 'feed-forward' and 'feedback' control architectures. Also, the initial studies on 'single-input single-output' control systems (SISO) have been extended to more complex 'multi-input multi-output' (MIMO) controllers which enable the implementation of control systems that can operate on complex multi-resonant systems. The feed-forward and feedback control architectures of SISO control systems are reviewed in the following sections. Although the extension to MIMO control systems is not trivial, the main features of MIMO controllers are then briefly introduced without entering into the mathematical details of the formulations. Comprehensive presentations of feed-forward and feedback SISO/MIMO control approaches can be found in specialised text books such as those by Nelson and Elliott (1992), Kuo and Morgan (1996), Fuller *et al.* (1996), Hansen and Snyder (1997), Clark *et al.* (1998), Elliott (2001) and Preumont (2002).

9.2.1 Feed-Forward Control

The feed-forward control architecture can be introduced by considering the example illustrated by Lueg where the sound propagation along a duct is controlled with a SISO feed-forward control system. As schematically shown in the top part of Fig. 9.1, the waveform of the primary wave described by the solid line S_1 is detected by a microphone M and used to drive the control loudspeaker L via the electronic controller V. The loudspeaker generates a secondary acoustic wave, whose waveform is defined by the dotted line S_2. The control system V is set to manipulate the detected signal from the microphone in such a way that the secondary waveform destructively interferes with the primary wave. Thus, the secondary wave is generated to have the same frequency and amplitude but opposite phase to the primary wave. Although this example suggests that active control is a relatively simple concept, the practical implementation is not that simple for a number of reasons which are discussed below.

Since the focus of this chapter is on the vibration control of flexible structures in order to reduce sound radiation and transmission, the system shown in Fig. 9.2(a) will be considered as a model example. The system is composed of a flexible structure (specifically we will consider a plate) excited by a primary source to which a secondary source is applied and driven in such a way as to control the flexural vibration at the error sensor position.

Let us consider first the case where, as illustrated in Fig. 9.2(a), there is access to a reference harmonic signal $\tilde{r}(\omega_0)$ of frequency ω_0 directly derived from the primary source. Although, in theory, many types of disturbance can be treated as being deterministic *within the timescales of active control*, only harmonic or

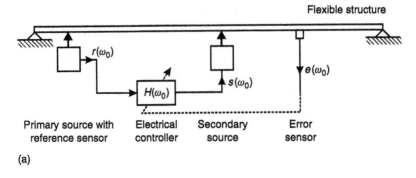

Primary source with · Electrical · Secondary · Error
reference sensor · controller · source · sensor

(a)

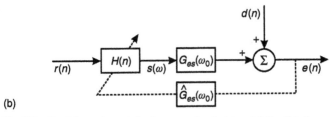

(b)

Fig. 9.2. Feed-forward control of narrow band deterministic disturbances: (a) schematic; (b) equivalent block diagram.

other periodic disturbances are truly deterministic in practice. These forms of disturbance are most commonly generated by rotating or reciprocating machines. In these cases the deterministic reference signal can be measured directly on the machines, for example by using a tachometer that measures the angular speed of a rotating shaft.

Assuming the system is linear and considering the disturbance to be harmonic then, as illustrated by the block diagram in Fig. 9.2(b), the complex velocity $\tilde{e}(\omega_0)$ signal, measured by the error sensor located where the control action is desired is given by the superposition of the velocity signals generated by the primary disturbance $\tilde{d}(\omega_0)$ and by the control source $\tilde{G}_{es}(\omega_0)\tilde{s}(\omega_0)$ so that

$$\tilde{e}(\omega_0) = \tilde{d}(\omega_0) + \tilde{G}_{es}(\omega_0)\tilde{s}(\omega_0) \tag{9.1}$$

where $\tilde{G}_{es}(\omega_0)$ is the frequency response function between the error sensor signal $\tilde{e}(\omega_0)$ and the secondary signal $\tilde{s}(\omega_0)$, which is normally referred as the 'plant response'. As shown in Chapter 2, if the control actuator is an ideal point force and the error sensor is an ideal transverse velocity sensor at position (x_s, y_s), the plant response can be expressed in terms of a modal summation

$$\tilde{G}_{es}(\omega) = j\omega \sum_{p=1}^{P} \sum_{q=1}^{Q} \frac{\phi_{pq}(x_e, y_e)\phi_{pq}(x_s, y_s)}{M_{pq}\left[\omega_{pq}^2\,(1+j\eta) - \omega^2\right]} \tag{9.2}$$

where $\phi_{pq}(x, y)$ is the p,q-th mode shape, ω_{pq} is the p,q-th natural frequency, M_{pq} is the modal mass and η is the modal-average structural loss factor. If we now assume the electrical controller has a frequency response $\tilde{H}(\omega_0)$, so that

$$\tilde{s}(\omega_0) = \tilde{H}(\omega_0)\tilde{r}(\omega_0) \tag{9.3}$$

the error sensor output is

$$\tilde{e}(\omega_0) = \tilde{d}(\omega_0) + \tilde{G}_{es}(\omega_0)\tilde{H}(\omega_0)\tilde{r}(\omega_0) \tag{9.4}$$

The reference signal $\tilde{r}(\omega_0)$ can be generated in many ways. For example, if it has unit amplitude then, assuming the control system is set to cancel the error sensor signal, the feed-forward controller must implement the complex control function

$$\tilde{H}_{opt}(\omega_0) = -\frac{\tilde{d}(\omega_0)}{\tilde{G}_{es}(\omega_0)} \tag{9.5}$$

In principle, the control function $\tilde{H}_{opt}(\omega_0)$ can be easily derived by measuring the spectrum of the primary disturbance $\tilde{d}(\omega_0)$ at the error sensor and the frequency response function $\tilde{G}_{es}(\omega_0)$ between the error sensor and the secondary source signal. However, in practice, the measures of the primary disturbance $\tilde{d}(\omega_0)$ and plant response $\tilde{G}_{es}(\omega_0)$ are never accurate enough to guarantee convergence. Thus, an iterative method must be implemented which, as schematically shown in Fig. 9.2(b), can also track non-stationary effects in the primary disturbance and adapt the control filter accordingly (adaptive controller) (Elliott, 2001). This gives an extra advantage, provided these effects occur on a longer time scale than the period of the disturbance. If the reference signal is periodic but non-harmonic then, provided the system is linear, separate controllers can be implemented for the fundamental and higher order harmonics of the reference signal. K independent control functions are therefore implemented which do not interact in the steady state.

When the primary disturbance is random (stochastic) and distributed in space, there is generally no direct access to a reference electrical signal that uniquely represents the source action and the signal from a reference vibration sensor must be used, such as that shown in Fig. 9.3(a). The block diagram for this

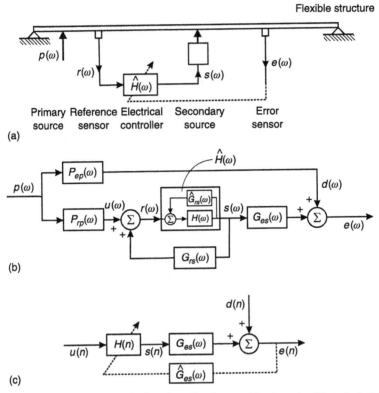

Fig. 9.3. Feed-forward control of random disturbances: (a) schematic; (b) equivalent block diagram and (c) equivalent block diagram for the design of a dummy optimal controller $H(\omega)$.

control arrangement shown in Fig. 9.3(b) illustrates how the reference signal $\tilde{r}(\omega)$ is influenced by a back effect of the control source as indicated by the feedback loop via the transfer function $\tilde{G}_{rs}(\omega)$. Therefore, the reference signal is given by $\tilde{r}(\omega) = \tilde{u}(\omega) + \tilde{G}_{rs}(\omega)\tilde{s}(\omega)$ where $\tilde{u}(\omega) = \tilde{P}_{rp}(\omega)\tilde{p}(\omega)$ is the reference signal generated by the primary disturbance and $\tilde{P}_{rp}$ and $\tilde{G}_{rs}(\omega)$ are the frequency response functions between the reference sensor and the primary and secondary sources, respectively . If, as shown in Fig. 9.3(b), the controller is arranged in such a way as to implement an internal feedback cancellation architecture in which an internal model of the feedback path $\overset{\approx}{G}_{rs}(\omega)$ is used to cancel the back effect of the control actuator at the reference sensor then, as shown in the block diagram of Fig. 9.3(c), the design of a dummy optimal controller $\tilde{H}(\omega)$ can be treated in the same way as for the feed-forward control of deterministic disturbances. The dummy optimal controller is therefore set to cancel the error signal $\tilde{e}(\omega)$ generated by the superposition of the primary disturbance $\tilde{d}(\omega) = \tilde{P}_{ep}(\omega)\tilde{p}(\omega)$ and control source $\tilde{G}_{es}(\omega)\tilde{s}(\omega)$, where $\tilde{P}_{ep}$ and $\tilde{G}_{es}(\omega)$ are the frequency response functions between the error sensor and respectively the primary and secondary sources, so that $\tilde{e}(\omega) = \tilde{d}(\omega) + \tilde{G}_{es}(\omega)\tilde{s}(\omega)$ as in Eq. (9.1). Thus, as found in Eq. (9.5), the dummy optimal controller is given by $\tilde{H}_{opt}(\omega) = -\tilde{d}(\omega)/\tilde{G}_{es}(\omega)$. Once the dummy optimal controller has been designed, the response of the physical controller $\hat{H}(\omega)$ can be derived from the frequency response function of the estimated feedback path $\hat{G}_{es}(\omega)$ as $\hat{H}(\omega) = \tilde{H}_{opt}(\omega)/1 + \hat{G}_{es}(\omega)\tilde{H}_{opt}(\omega)$. The main features of feed-forward control of stochastic disturbances can be found in specialised books such as those by Nelson and Elliott (1992); Fuller *et al.* (1996); Kuo and Morgan (1996); Hansen and Snyder (1997) and Elliott (2001).

The formulations presented above can be extended to systems with multiple control sensors and actuators. The MIMO feed-forward control of harmonic disturbance at frequency ω_0 can be represented by the standard feed-forward block diagram of Fig. 9.2(b), which in this case accounts for the presence of multiple error sensors and control sources as denoted by the broad arrows in the block diagram in Fig. 9.4(a).

Assuming there are n error sensors, whose outputs are grouped in the vector $\tilde{\mathbf{e}}(\omega_0)$, and m control actuators, whose inputs are grouped in the vector $\tilde{\mathbf{s}}(\omega_0)$, then the error sensor outputs generated by the superposition of the primary disturbance and control actuators can be expressed as

$$\tilde{\mathbf{e}}(\omega_0) = \tilde{\mathbf{d}}(\omega_0) + \tilde{\mathbf{G}}_{es}(\omega_0)\tilde{\mathbf{s}}(\omega_0) \qquad (9.6)$$

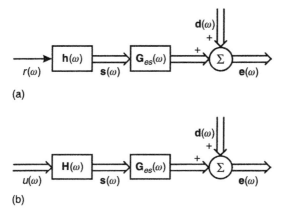

(a)

(b)

Fig. 9.4. Block diagrams for feed-forward control of: (a) deterministic; (b) random disturbances.

where $\tilde{\mathbf{d}}(\omega_0)$ and $\tilde{\mathbf{G}}_{es}(\omega_0)\tilde{\mathbf{s}}(\omega_0)$ are the vectors containing the error signals due to the action of the primary disturbance and control sources, respectively. The matrix $\tilde{\mathbf{G}}_{es}(\omega_0)$ contains the $n \times m$ plant responses between the n sensors and m actuators. In this case, the electric controller is characterised by m control filters which generate the signals driving the m control actuators

$$\tilde{\mathbf{s}}(\omega_0) = \tilde{\mathbf{h}}(\omega_0)\tilde{r}(\omega_0) \qquad (9.7)$$

If we consider the control objective to be the minimisation of the sum of the mean-square error signals, then the following quadratic cost function must be minimised:

$$J = \sum_{r=1}^{n} |\tilde{e}_r|^2 = \tilde{\mathbf{e}}^H \tilde{\mathbf{e}} = \tilde{\mathbf{h}}^H \tilde{\mathbf{G}}_{es}^H \tilde{\mathbf{G}}_{es} \tilde{\mathbf{h}} + \tilde{\mathbf{h}}^H \tilde{\mathbf{G}}_{es}^H \tilde{\mathbf{d}} + \tilde{\mathbf{d}}^H \tilde{\mathbf{G}}_{es} \tilde{\mathbf{h}} + \tilde{\mathbf{d}}^H \tilde{\mathbf{d}} \qquad (9.8)$$

This is a Hermitian quadratic form whose minimum is given by the following vectors with the optimal control filters (Nelson and Elliott, 1992):

$$\tilde{\mathbf{h}}_{opt} = -\left[\tilde{\mathbf{G}}_{es}^H \tilde{\mathbf{G}}_{es}\right]^{-1} \tilde{\mathbf{G}}_{es}^H \tilde{\mathbf{d}} \qquad n > m \qquad (9.9a)$$

$$\tilde{\mathbf{h}}_{opt} = -\tilde{\mathbf{G}}_{es}^{-1} \tilde{\mathbf{d}} \qquad n = m \qquad (9.9b)$$

$$\tilde{\mathbf{h}}_{opt} = -\tilde{\mathbf{G}}_{es}^H \left[\tilde{\mathbf{G}}_{es} \tilde{\mathbf{G}}_{es}^H\right]^{-1} \tilde{\mathbf{d}} \qquad n < m \qquad (9.9c)$$

for respectively (a) the over-determined case with more sensors than actuators; (b) the fully-determined case with equal numbers of sensors and actuators; (c) the under-determined case with fewer sensors than actuators. In reality, for the under-determined case, there is no unique solution to the minimisation problem since there is an infinite set of vectors $\tilde{\mathbf{h}}_{opt}$ that set the cost function in Eq. (9.8) to zero. The solution given in Eq. (9.9c) yields a particular set of control filters that also minimises the sum of the modulus-squared control signals $J = \sum_{k=1}^{m} |\tilde{s}_k|^2$. As seen for SISO feed-forward control systems, the practical implementation of the control system requires some form of adaptation of the control filters so that the non-stationary effects or other changes of the plant responses can be tracked down. This is done with a multiple channel steepest descent algorithm which requires a model of the matrix $\tilde{\mathbf{G}}_{es}(\omega_0)$ comprising the $n \times m$ plant responses between the n sensors and m actuators (Nelson and Elliott, 1992).

Stochastic disturbances can also be controlled with MIMO feed-forward control systems. In this case k-reference signals are used to detect the primary random disturbance. The structure of the control system is again similar to that of the SISO feed-forward control system, which, as shown in Fig. 9.3(b), is primarily characterised by the back effect of the control sources on the reference signals so that, in order to implement an ordinary feed-forward control system, multiple feedback cancellation loops are assumed to be working in the controller. In this way, the control problem is reduced to the standard design of a MIMO feed-forward control system as shown in Fig. 9.4(b). The reader is referred to the specialised books on active control of noise and vibration listed above, where the control algorithms for stochastic disturbances are presented in great detail.

The formulation presented above focuses on the main control features of feed-forward systems with no reference to the selection of the numbers, types and positions of the control transducers. This part of the study is often referred as the 'physics of active control' since it sets the best control performance that can be achieved independently of the limitations introduced by the electric controller. In order to better appreciate the importance of this aspect of active control, a model problem is now considered where, as shown in Fig. 9.5, the flexural vibration of a rectangular flat plate generated by a transverse point force F_p is controlled with a feed-forward control system using one or two transverse control forces F_{s1} and F_{s2}.

The primary source is assumed to be harmonic so that a reference signal is available for the implementation of feed-forward SISO or MIMO control schemes as shown in Fig. 9.2(b) and Fig. 9.4(a). The simply supported aluminium panel is 1 mm thick and has dimensions 414 mm × 314 mm. A loss factor $\eta = 0.02$ is assumed. The primary force is positioned at $x_p = 0.15l_x$ and $y_p = 0.44l_y$. A SISO control system is first considered where a control force f_{s1} positioned at $x_s = 0.33l_x$ and $y_s = 0.71l_y$ is driven to cancel the vibration at the control point itself, so that $x_e = 0.33l_x$ and $y_e = 0.71l_y$. As a measure of the effectiveness of the system, the time-averaged total kinetic energy of flexural vibration of the panel is evaluated in a frequency range up to 1 kHz. The solid lines in Figs. 9.6(a,b) show the spectrum of the time-averaged total kinetic energy of the

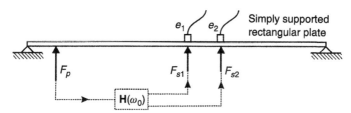

Fig. 9.5. Simply supported rectangular plate excited by a primary harmonic point force and controlled by a feed-forward system with either one or two secondary point forces minimising the response of the plate measured either at the error sensor or over the whole surface in terms of its kinetic energy.

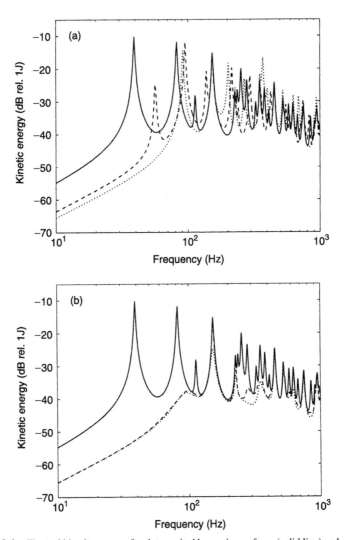

Fig. 9.6.   Flexural kinetic energy of a plate excited by a primary force (solid line) and controlled by: (a) one (dashed line) or two (dotted line) secondary forces set to cancel the vibration at the control position; (b) one or two secondary forces set to minimise the kinetic energy of the plate.

panel which is characterised by well separated resonance peaks approximately below 300 Hz where the modal overlap is smaller than unity. The dashed line in Fig. 9.6(a) indicates that, despite the control system cancelling the vibration at the control position, the overall vibration of the panel is not always reduced by the ideal feed-forward control system, but instead is rearranged into a new set of resonance frequencies. This is due to the fact that the control system generates an extra boundary condition at the error position where the panel is 'actively pinned' so that the overall response of the panel is characterised by a new set of modes. As a result, the control system produces overall vibration reductions of the panel at certain frequency bands, such as for example below 40 Hz or between 60 and 80 Hz, but also it increases the overall vibration in other frequency ranges embracing the new resonances, such as for example between 40 and 60 Hz. Therefore, regardless of the limitations that would arise from the implementation of the electrical controller, the performance of the control system could be compromised by an inappropriate selection of the error sensor and control source positions. If, for example, the error position is moved to $x_e = 0.63l_x$ and $y_e = 0.17l_y$, then as shown by the dotted line in Fig. 9.6(a) the vibration field of the panel is rearranged in such a way as that the first resonance occurs at about 92 Hz in place of 48 Hz. As a result, good control reductions are achieved below about 80 Hz.

Harmonic excitations are commonly generated by rotating machines which may operate over a range of different speeds. Thus, in order to guarantee controllability at all frequencies, the control system should be set in such a way to reduce vibration without the spillover effect produced by the pinning effect at the error sensor. This can be achieved by using an error sensor that measures the global flexural vibration of the panel rather than that at a single point. For instance, a distributed sensor or an array of sensors that samples the transverse velocity over the whole surface of the panel can be used to estimate the total flexural kinetic energy of the panel.

The dashed line in Fig. 9.6(b) shows the reduction of the flexural kinetic energy when the SISO system drives the control force F_{s1} so as to minimise the flexural kinetic energy of the panel measured by an ideal distributed sensor. In this case, as expected, there is no spillover effect and good reductions are achieved at low frequencies. The control performance can be improved by increasing the numbers of secondary forces. For example, when a MIMO control system with two secondary forces F_{s1} and F_{s2} positioned at $x_{s1} = 0.33l_x$, $y_{s1} = 0.71l_y$ and $x_{s2} = 0.63l_x$, $y_{s2} = 0.17l_y$ is used to minimise the total flexural kinetic energy of the panel then, as shown by the dotted line in Fig. 9.6(b), the reduction of vibration is extended to higher frequencies. In general, for structures exhibiting multi-modal response, a feed-forward control system with n control forces and error sensors suppress the vibration contribution of the first n-modes of the structure. Thus, an estimate of the modal density of the structure to be controlled gives a direct estimate of the number of control sources and error sensors necessary to produce the desired control for a certain frequency band.

This simple example has brought to light the importance of considering the control effectiveness of configurations of the sensors–actuators regardless of the limitations imposed by the controller. This phase of the exposition sets out the best strategy in terms of types, number and positions of the sensor actuator transducers. However, it must be emphasised that the implementation of the control filter(s) is also strongly linked to the types, numbers and positions of the transducers. In fact, as discussed above, a key feature for the implementation of adaptive feed-forward control is the plant response or matrix of plant responses which must be 'minimum phase' or 'positive definite' in order to guarantee a stable and fast converging adaptive feed-forward algorithm. Therefore, the designer must find the best balance between the requirements for the implementation of the control filters and the requirements for obtaining the desired control effects. This is not an easy problem, principally because the number of unknowns is rather large and therefore several solutions can be conceived which give similar results. For instance, SISO controllers can be developed with distributed sensors and actuators specifically designed to produce the objective control effect and a minimum phase plant response. Alternatively, large arrays of point sensors and point actuators specifically arranged to give the desired control effect and a set of minimum phase plant responses can be used in MIMO control systems.

The physics of the control system is of great importance in the sound transmission and radiation control problems which are considered in this chapter. The aim of designers concerned with vibroacoustic problems is to develop control systems incorporating actuators and sensors applied to structures in such a way as to reduce sound radiation or transmission. A major complication derives from the fact that a reduction of vibration is no guarantee of acoustic effectiveness, since a reduction of overall vibration can, in some cases, lead to increased sound radiation/transmission. A specific analysis of this problem is presented in the second part of this chapter.

9.2.2   Feedback Control

When a reference signal of the primary disturbance is not available or cannot be measured, a feedback control approach must be used in which the error sensor signal is also used as the reference signal for the control filter. In this case, as shown in Fig. 9.7(a) for the SISO control model problem considered above, the vibration of a flexible structure generated by a random primary source is controlled by a secondary source which is driven to minimise the vibration measured by the error sensor via a classic 'disturbance rejection' feedback loop, as illustrated by the block diagram in Fig. 9.7(b) (Meirovitch, 1990; Clark et al., 1998; Preumont, 2002).

Assuming the system is linear, the complex amplitude $\tilde{e}(\omega)$ of the velocity measured by the error sensor where the control action is desired is given by the superposition of the effects generated by the primary disturbance,

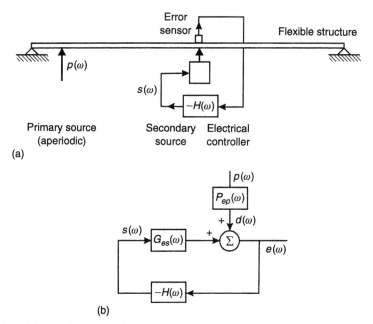

Fig. 9.7.   Single-input single-output (SISO) feedback control of random disturbances.

$\tilde{d}(\omega) = \widetilde{P}_{ep}(\omega)\tilde{p}(\omega)$, and by the control source $\widetilde{G}_{es}(\omega)\tilde{s}(\omega)$. Hence,

$$\tilde{e}(\omega) = \tilde{d}(\omega) + \widetilde{G}_{es}(\omega)\tilde{s}(\omega) \tag{9.10}$$

where $\widetilde{G}_{es}(\omega)$ is the plant response which, as seen in the previous section, is given by the frequency response function between the error sensor signal and the secondary signal $\tilde{s}(\omega)$. $\widetilde{P}_{ep}(\omega)$ is the frequency response function between the error sensor signal and primary disturbance $\tilde{p}(\omega)$. If we now assume the electric controller implements a control function $-\widetilde{H}(\omega)$, where the phase inversion indicates that a negative feedback loop is implemented, then the ratio between the error sensor signal $\tilde{e}(\omega)$ and the primary disturbance $\tilde{d}(\omega)$ becomes

$$\frac{\tilde{e}(\omega)}{\tilde{d}(\omega)} = \widetilde{S}(\omega) = \frac{1}{1 + \widetilde{G}_{es}(\omega)\widetilde{H}(\omega)} \tag{9.11}$$

This response function is normally known as the 'sensitivity function' $S(\omega)$, and indicates the effectiveness of the control loop. In this specific case, the objective of the control loop is to reduce the amplitude of the response of the structure at the error sensor. If, for example, the plant response of the structure under control is rather uniform and free from phase lag at frequencies below ω_c, then the electrical controller should amplify the error signal in such a way as to generate a relatively high 'loop gain' $\widetilde{G}_{es}(\omega)\widetilde{H}(\omega)$ so that

$$\left|1 + \widetilde{G}_{es}(\omega)\widetilde{H}(\omega)\right| \gg 1 \qquad \omega < \omega_c \qquad (9.12)$$

and thus

$$\frac{\tilde{e}(\omega)}{\tilde{d}(\omega)} \ll 1 \qquad \omega < \omega_c \qquad (9.13)$$

In practice, when a small primary perturbation is detected by the error sensor, then a large loop gain is generated with opposite time history to the primary disturbance, which is therefore balanced. As shown in Chapters 1 and 2, the plant responses $\widetilde{G}_{es}(\omega)$ of lightly damped flexible structures are characterised by a sequence of resonance peaks and antiresonance troughs which, for non-collocated sensor–actuator pairs, produces relatively large phase lags. This has a major effect on the control performance of the feedback loop as we shall see in the latter part of this section.

    In general, there are two approaches to the design of feedback control functions $\widetilde{H}(\omega)$ which lead to the 'classic control' and 'modern control' theories. In the first case, the control function is selected *a priori* with reference to the control objective and normally it refers to simple laws, normally called 'compensators' since they are chosen in such a way as to compensate for instabilities. Although the control functions are chosen in advance, the amplitude of the control signal is determined by the control gain which can be tuned in such a way as to ensure the stability of the control loop. Thus, the control gain is a key feature of any feedback loop function. Indeed, the simplest control function is given by a pure gain in which case the so called 'direct feedback control' is implemented. The stability of the control loop is normally assessed on the basis of the root-locus or Nyquist criteria. In contrast, modern feedback control sets a more ambitious objective; that is to find the optimum control function $\widetilde{H}_{opt}(\omega)$ for a given control objective. This design approach is normally carried out using a state-space formulation which therefore requires a model for the many states that characterise the system to be controlled. The states of a system are defined as the minimum number of functions necessary to describe the response of a system with first order linear differential equations, thus their number is equal to the order of the system and involves both displacement and velocity functions (Meirovitch, 1990; Preumont, 2002). For example, a single-degree-of-freedom system with mass, damper and spring lumped elements connected in parallel is characterised by two states, namely the displacement and velocity of the mass. In practical control systems, the functions are synthesised from the smaller number of observed (measured) error signals.

Let us first consider the classic feedback control theory which is often referred as the 'frequency method' since the stability of the control loop is assessed with respect to the open-loop frequency response function of the control system. The sensitivity function $\widetilde{S}(\omega) = 1/\big(1 + \widetilde{G}_{es}(\omega)\widetilde{H}(\omega)\big)$ can be expressed as the ratio of two polynomial functions which can be factored in terms of first order and quadratic factors so that

$$\widetilde{S}(\omega) = K \frac{\displaystyle\prod_{n=1}^{Z} (j\omega - z_n)}{\displaystyle\prod_{m=1}^{P} (j\omega - p_m)} \tag{9.14}$$

where p_n and z_m are the complex poles and zeros of the system. According to Eq. (9.11) the poles are given by the roots of the characteristic equation

$$1 + \widetilde{G}_{es}(\omega)\widetilde{H}(\omega) = 0 \tag{9.15}$$

Thus the stability of the closed-loop control system is guaranteed provided the real part of the zeros of the characteristic function, i.e., the poles of the sensitivity function, are negative or zero. This definition is based on the hypothesis that both the plant of the controlled system and the response of the controller are inherently stable. Normally, the stability is assessed graphically with the so called 'root-locus method' in which the roots of the characteristic equation are plotted in the complex plane as the compensator gains varies from 0 to $+\infty$. The roots determine smooth curves which are known as 'loci' that all together form the 'root-locus plot'. The loci obtained by varying the compensator gains from 0 to $-\infty$ are called 'complementary loci'. The system is then established to be unconditionally stable provided the loci occupy the left-hand side (real negative side) of the plot. In many practical applications, it is rather difficult to derive an explicit pole/zero expression for the sensitivity function of the system under study since the model can be of very high order or can be characterised by delays and large variability effects. Therefore, an alternative approach is normally used that, as shown in Fig. 9.8(a), considers the polar plot as ω varies between $-\infty$ and $+\infty$ of the open-loop sensor–actuator frequency response $\widetilde{G}_{es}(\omega)\widetilde{H}(\omega)$, which is referred to as the Nyquist plot. This method is based on the 'Nyquist criterion' which states that, if the open loop response function $\widetilde{G}_{es}(\omega)\widetilde{H}(\omega)$ has poles on the right-hand side (real positive side) of the complex plane, i.e., it is intrinsically unstable, then the close-loop feedback system is stable provided that the Nyquist plot of $\widetilde{G}_{es}(\omega)\widetilde{H}(\omega)$ encircles the point $-1 + j0$ in a counterclockwise sense as

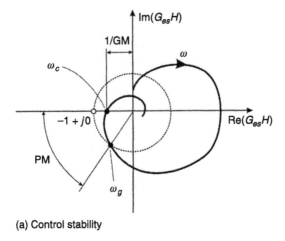

(a) Control stability

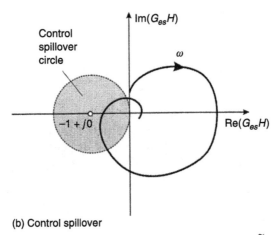

(b) Control spillover

Fig. 9.8.   Nyquist plots of the open-loop frequency response function $\widetilde{G}_{es}(\omega)\widetilde{H}(\omega)$: (a) the Nyquist critical point $-1 + j0$, the phase and gain crossover points, ω_g and ω_c respectively, and the phase PM and gain GM margins; (b) the control spillover circle.

many times as there are poles of $\widetilde{G}_{es}(\omega)\widetilde{H}(\omega)$ in the right half of the complex plane. When the open-loop response function $\widetilde{G}_{es}(\omega)\widetilde{H}(\omega)$ has no poles in the right side of the complex plane, i.e., it is intrinsically stable, then the closed-loop feedback system is stable provided that the Nyquist plot of $\widetilde{G}_{es}(\omega)\widetilde{H}(\omega)$ does not encircle the point $(-1) + j0$, and then it is termed 'minimum phase' (see Chapter 2). However, the Nyquist criterion applies to both minimum phase and non-minimum phase control systems. The point $(-1) + j0$ is normally referred as the Nyquist critical point.

It is important to emphasise that, at those frequencies where the polar plot of $\widetilde{G}_{es}(\omega)\widetilde{H}(\omega)$ enters the circle of unit radius centred in the Nyquist critical point $(-1) + j0$ [dotted circle in Fig. 9.8(b)], $\left|1/1 + \widetilde{G}_{es}(\omega)\widetilde{H}(\omega)\right| > 1$. As a result, even if the polar plot does not encircle the Nyquist point, and thus the system is stable, the control loop enhances the response rather than reducing it, which leads to the so called 'control spillover effect'. It is important to stress that 'control spillover' does not implies instability, although the proximity of the polar plot of the open-loop response function $\widetilde{G}_{es}(\omega)\widetilde{H}(\omega)$ to the Nyquist critical point clearly indicates a danger of instability. The Nyquist criterion provides a clear means of assessing whether or not a feedback control system is stable. However, it does not provide indications about the degree of stability: in other words, it does not reveal how close the feedback loop is to instability. This is normally assessed in terms of gain and phase margins which, as shown in Fig. 9.8(a), are defined by the closest amplitude and phase of the open-loop response function $\widetilde{G}_{es}(\omega)\widetilde{H}(\omega)$ to the amplitude and phase of the Nyquist critical point, respectively -1 and $180°$. In general, the open-loop response of a control system is subject to variations which can be due to changes of operation of the system, such as loading effects on the structure, changes of physical conditions, such as temperature variation, or changes in the controller due to non-linear effects. These changes produce uncertainties that can produce large enough variations of the open-loop response function to cause instabilities. The main task of control engineers is to design sensor–actuator pairs that give rise to a positive real open-loop response function. This is a rather difficult task, particularly for distributed flexible systems which always introduce phase lag effects at higher frequencies. Normally, engineers aim to design systems with an open-loop response function $\widetilde{G}_{es}(\omega)\widetilde{H}(\omega)$ which is positive real in the frequency range where control is desired and whose magnitude gradually decreases at higher frequencies. This type of design is normally done by considering the magnitude and phase plot of the $\widetilde{G}_{es}(\omega)\widetilde{H}(\omega)$ response function which, in the case where logarithmic scales are used, is known as a 'Bode plot'.

As mentioned above, in the classic feedback control approach the control functions are chosen in advance. These control functions are normally called compensators since they are specifically designed to compensate for excessive phase lags in the feedback loop that may lead to instabilities. Thus, the choice of the control function is specific to the structure under control and sensor–actuator transducers used in the loop. This is a logical design approach for those systems where the objective of the controller is to cancel the vibration at the control position. However, as discussed in the previous example of the feed-forward control of the vibration of a panel, the control of the overall response of a flexible structure in order to minimise its sound radiation cannot be achieved with control systems that perfectly cancel the vibration at the control position.

As we shall see in the second part of this chapter, there are two principal control approaches to this problem. The first aims to cancel the vibration components of the panel that constructively contribute to the sound radiation, i.e., the radiation modes. This requires specially designed distributed sensors and actuators transducers that detect and excite the radiation modes of the structure under control (Fuller *et al.*, 1996).

The second aims to specify and design arrays of localised control systems which produce a reduction of the sound radiation through specific dynamic effects on the whole structure, rather than by directly cancelling its radiation modes. In this second case, the control function or compensator is not designed exclusively to achieve greater stability of the feedback loop. On the contrary, it is designed to produce a specific effect on the structure. For this reason, the introduction of compensator functions is presented in a rather unusual form in which the dynamic effects of the compensator, that will be more appropriately referred to as the control function, are discussed in terms of the physical effects that it produces on the structure, rather than emphasising its amplitude and phase properties that are normally used to judge the effects on stability. It is important to emphasise that stability remains a fundamental issue even when the feedback loop is not designed to produce perfect cancellation at the error sensor, although it is reasonable to anticipate that relatively smaller control gains will be implemented than in the cancellation control problem, and thus a trade-off between stability and control requirements will be the key issue for the designer.

The three fundamental control functions that can be implemented in feedback loops are the so-called 'proportional', 'integral' and 'derivative' controls. If we assume a collocated velocity sensor and force actuator pair then, as shown in Fig. 9.9, these three control functions synthesise an active damper, an active stiffness and an active mass. In fact the three control forces generated by the three feedback control loops are given by

Proportional Control $\qquad \widetilde{F}_s(\omega) = -h\tilde{v}_e(\omega)$ $\qquad$ (9.16a)

Integral Control $\qquad \widetilde{F}_s(\omega) = -\dfrac{h}{j\omega}\tilde{v}_e(\omega)$ $\qquad$ (9.16b)

Derivative Control $\qquad \widetilde{F}_s(\omega) = -j\omega h\tilde{v}_e(\omega)$ $\qquad$ (9.16c)

where $\tilde{v}_e(\omega)$ is the sensor velocity, $\widetilde{F}_s(\omega)$ is the control force and h is the control gain.

When, in place of one sensor–actuator pair, L sensors and M actuators are used in such a way as to implement a multi-input multi-output (MIMO) feedback control as illustrated by the block diagram shown in Fig. 9.10, then the control

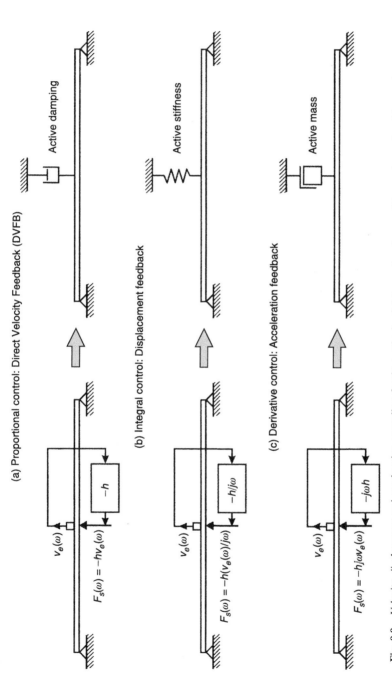

(a) Proportional control: Direct Velocity Feedback (DVFB)

(b) Integral control: Displacement feedback

(c) Derivative control: Acceleration feedback

Fig. 9.9. Velocity, displacement and acceleration controllers for the synthesis of: (a) an active damper; (b) an active spring; (c) an active mass.

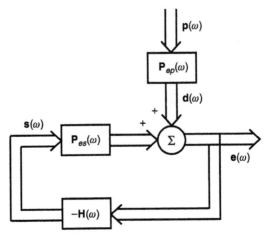

Fig. 9.10. Multiple-input multiple-output (MIMO) feedback control of random disturbances.

functions between all sensor–actuator pairs have to be selected in order to generate the $L \times M$ matrix of control gains $\mathbf{H}$. In this case, the design process is rather more complicated than for the SISO case. Considering the block diagram shown in Fig. 9.10, and following the same algebraic formulation as for the SISO problem, the closed loop error signals can be expressed by the matrix relation

$$\tilde{\mathbf{e}}(\omega) = \left[\mathbf{I} + \tilde{\mathbf{G}}_{es}(\omega)\tilde{\mathbf{H}}(\omega)\right]^{-1}\tilde{\mathbf{d}}(\omega) \qquad (9.17)$$

where the matrix $\left[\mathbf{I} + \tilde{\mathbf{G}}_{es}(\omega)\tilde{\mathbf{H}}(\omega)\right]$ is called the 'return difference matrix'. Equation (9.17) is valid provided the matrix is non-singular. By analogy with the SISO formulation, the matrix $\left[\mathbf{I} + \tilde{\mathbf{G}}_{es}(\omega)\tilde{\mathbf{H}}(\omega)\right]^{-1}$ is called the 'matrix sensitivity function'. Assuming that both the plant and control systems are stable, the closed-loop stability can be estimated with the generalised form of the Nyquist criterion, which for a MIMO system states that the closed-loop system is stable provided the locus of the function

$$\det\left[\mathbf{I} + \tilde{\mathbf{G}}_{es}(\omega)\tilde{\mathbf{H}}(\omega)\right] = 0 \qquad (9.18)$$

does not encircle the origin as ω varies between $-\infty$ and $+\infty$. In general, this locus tends to be rather complex in shape and the identification of the conditions for stability is rather complicated, since it is not easy to recognise how the locus changes with reference to the various control gains. This problem can be overcome by noting that

$$\det\left[\mathbf{I} + \widetilde{\mathbf{G}}_{es}(\omega)\widetilde{\mathbf{H}}(\omega)\right] = [1 + \lambda_1][1 + \lambda_2]\ldots[1 + \lambda_L] \tag{9.19}$$

where λ_i are the eigenvalues of the matrix $\widetilde{\mathbf{G}}_{es}(\omega)\widetilde{\mathbf{H}}(\omega)$. This relation suggests that the locus of $\det\left[\mathbf{I} + \widetilde{\mathbf{G}}_{es}(\omega)\widetilde{\mathbf{H}}(\omega)\right] = 0$ does not encircle the origin provided none of the eigenvalues loci, called 'characteristic loci' encircles the point $(-1, 0)$. However, one must be cautious in using the characteristic loci since the concepts of phase and gain margin still applies to the MIMO problem in the case where there is a simultaneous change in the gain and phase of all elements of the plant response.

An interesting MIMO control application is that in which an array of SISO feedback loops is implemented independently; this can be treated as a standard MIMO control problem where, nevertheless, the control matrix $\widetilde{\mathbf{H}}(\omega)$ is diagonal. This greatly simplifies the design of the system since the number of control gains corresponds exactly to the number of sensor–actuator pairs and not to the product of the sets. Also, the practical implementation of decentralised control loops is much simpler than that of fully coupled MIMO controllers in particular because there is no need for a multi-channel, centralised, fast-acting controller with an intricate set of cables connected to the sensor and actuator transducers, which tends to be an expensive and fragile solution. Moreover, the decentralised solution is robust against changes in operation, particularly when one of the sensor–actuator transducers fails.

An alternative approach to the design of SISO feedback control loops is the so called 'modern method' which aims to derive the optimal control function for a given problem. We have seen that the classic frequency analysis of SISO feedback control loops is based on the frequency response functions which, as described in Chapter 2, are normally expressed in closed form or with a modal summation approach. In contrast, the state-space formulation refers to a set of first order linear differential equations in terms of the states (displacements and velocities) of the system which are normally collected in a vector $\{\mathbf{x}\}$, so that

$$\dot{\mathbf{x}} = \mathbf{A}\mathbf{x} + \mathbf{B}s \tag{9.20a}$$

$$e = \mathbf{C}\mathbf{x} + Ds \tag{9.20b}$$

where $\mathbf{A}$ is the system matrix, $\mathbf{B}$ the input vector, $\mathbf{C}$ the output line vector and D the feed-through term. The input control function is defined by s and the output error function is defined by e.

The basic principle of state-space feedback control is to synthesise a full state feedback control loop which, as shown by the block diagram in Fig. 9.11(a)

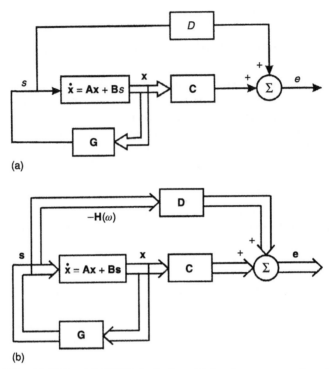

Fig. 9.11. (a) SISO and (b) MIMO feedback models based on state-space formulations.

for the SISO feedback control problem, involves a vector of control gains $\mathbf{G} = \lfloor G_1 \quad G_2 \quad \cdots \quad G_n \rfloor$ such that

$$s = -\mathbf{Gx} \tag{9.21}$$

Therefore, in contrast to the classic control method, which is illustrated by the block diagram in Fig. 9.7(b), the modern control approach aims to design a set of control functions, normally called control gains, which refers to the states used to model the system under control.

Therefore, the implementation of the control scheme shown in Fig. 9.11(a) is based on the assumption that there is access to the states of the system. If that is the case then substituting Eq. (9.21) in Eq. (9.20a) yields

$$\dot{x} = (\mathbf{A} - \mathbf{BG})\,\mathbf{x} \tag{9.22}$$

As for any type of system which is modelled with a state-space formulation, the system is asymptotically stable if the closed-loop eigenvalues of the system matrix $(\mathbf{A} - \mathbf{BG})$ are real and negative or complex with negative real part. Also, the system is said to be merely stable if all the closed-loop eigenvalues of $(\mathbf{A} - \mathbf{BG})$ are pure imaginary. Alternatively, the system is unstable if at least one eigenvalue of $(\mathbf{A} - \mathbf{BG})$ is real and positive, or complex with positive real part (Meirovitch, 1990). The eigenvalues of the closed-loop system correspond to its closed-loop poles which are given as solutions of the characteristic equation (Preumont, 2002)

$$\det[j\omega\mathbf{I} - (\mathbf{A} - \mathbf{BG})] = 0 \qquad (9.23)$$

Therefore the state-space design of feedback control loops can be reduced to the problem of 'pole allocation' (also called 'pole assignment' or 'pole placement') where the control gains are chosen in such a way as to locate the poles in the stability region. [There are several approaches for the pole placement which are extensively discussed in specialised books of control theory such as, for example, that by Meirovitch (1990)] It is important to emphasise that the pole allocation process is constrained by the controllability and the required control effort. Controllability is a measure of the ability of the control actuator to control all the states of the system. In those cases where the control system configuration would enable the control of all states of the system then the implementation of the control loop could still be impossible in practice because of excessive control effort requirements.

An alternative approach to the design of the full state feedback control regulator is called 'optimal control' where a single scalar parameter is used to assess the effectiveness of the controller. Optimal control is specifically suited for MIMO control problems which are normally analysed in terms of the block diagram shown in Fig. 9.11(b). Modern control theory is rather involved and will not be introduced in this chapter. For those readers interested in it, a comprehensive presentation of modern control methods can be found in many reference books such as for example those by Meirovitch (1990); Clark *et al.* (1998) and Preumont (2002) which specifically deals with vibroacoustic problems.

To conclude this section, we now consider the plate problem used in the previous section to describe the main features of the physics of feed-forward control. As shown in Fig. 9.12, in place of the feed-forward control system, a SISO feedback control loop is now implemented using a collocated ideal transverse velocity sensor and ideal point force actuator where the control function is designed according to the classic feedback control theory. The three fundamental compensators illustrated in Fig. 9.9 for the implementation of proportional, integral and derivative velocity feedback are studied in order to review the fundamental physical features of classic feedback control that also apply to optimal feedback

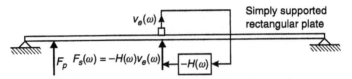

Fig. 9.12. Simply supported rectangular plate excited by a primary harmonic point force and controlled by a SISO feedback system implementing proportional, integral and derivative control.

implementation with modern control theory. The stability of these three control approaches is assessed with the Bode and Nyquist plots shown in Fig. 9.13 for the open-loop frequency response functions $\widetilde{G}_{es}(\omega)\widetilde{H}(\omega)$ assuming $\widetilde{H}(\omega) = 1$. The three Bode plots show the classic response functions of mobility $\widetilde{Y} = \tilde{v}_e/\widetilde{F}_s$, receptance $\tilde{\alpha} = (\tilde{v}_e/j\omega)/\widetilde{F}_s$ and inertance $\widetilde{A} = j\omega\tilde{v}_e/\widetilde{F}_s$, for the proportional, integral and derivative velocity feedback loops, respectively. Therefore, the open-loop frequency response function of proportional velocity feedback shown in Fig. 9.13(a) is characterised by a sequence of resonances and antiresonances (poles and zeros) such that the response is minimum-phase (Preumont, 2002). The ideal transverse velocity sensor and ideal transverse point force excitation forms a collocated[1] and dual[2] sensor–actuator pair so that the phase is confined between $+90°$ and $-90°$ (Jayachandran and Sun, 1997; Sun, 1996). Therefore, the response function is constrained to be real positive so that the Nyquist plot in Fig. 9.13(a) occupies the right-hand side and the control loop is unconditionally stable since, for any control gain, the Nyquist critical point cannot be encircled (Balas, 1979).

In Chapter 2, we have seen that the frequency-average value of the driving point mobility function for uniform flat plates is equal to that of an infinite plate which, according to Eq. (2.68a), is independent of frequency. The sequence of resonances tends to produce velocity peaks whose amplitudes decrease at higher frequencies because of combined higher damping and modal overlap effects. As a result a 'handy' open-loop response function is obtained whose polar plot is characterised by relatively large circles in the real part of the Nyquist plot for the low frequency resonances where the control action is desired. At higher frequencies, the circles become smaller and smaller so that that the inevitable higher frequency phase lag effects of the sensor–actuator transducers and compensator

[1] Collocation is a geometrical condition where point sensors and actuators are placed in the same position on the structure and distributed sensors and actuators with equal spatial sensitivity are placed over the same area of the structure.

[2] Duality is a physical property where the actuator and sensor excite and detect the vibrations of a structure in the same manner so that the product of the excitation and sensed response is proportional to the power supplied to the system. Dual sensor–actuator pairs are also said to be 'matched', 'compatible' or 'reciprocal'.

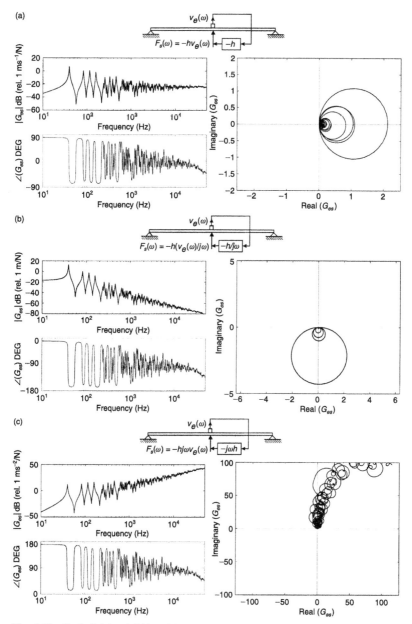

Fig. 9.13. Bode (left-hand side) and Nyquist (right-hand side) plots for the open-loop frequency response function of: (a) proportional; (b) integral; (c) derivative velocity feedback.

circuit that move the circles to the left-hand side quadrants can be tolerated since there is no danger of instability or large spillover effects, even when large control gains are implemented.

The open-loop frequency response functions of the integral and derivative velocity feedback, depicted respectively in Figs. 9.13(b,c), also show an alternating sequence of resonances and antiresonances, but the implementation of either integral or derivative control introduces respectively a phase lag or phase lead of 90° so that the Nyquist plots also occupies the real negative quadrants. As a result, even if both loci do not cross over the real negative axis, and thus the systems are bound to be unconditionally stable, the gain/phase margins are relatively tight so that even small variations in the response functions due to changes of operation will lead to instabilities. Also, the dynamic response of practical sensor and actuator transducers and control circuits are likely to introduce additional phase lags which are likely to impair the stability of integral control. Moreover, the amplitude of the open-loop response function for the integral control tends to decrease at a rate of $1/\omega$ so that the control effect is limited only to the first few resonances. In contrast, the amplitude of the open-loop response function for the derivative control tends to rise at a rate of ω, which is normally an undesired effect since the higher frequency phase lags introduced by the sensor–actuator and control circuit systems are likely to drift the loci at higher frequency close to the Nyquist critical point.

Turning to the performance of the three control loops, Fig. 9.14(a) shows that, as the control gain of the proportional velocity feedback control loop is increased from zero, the response of the panel tends to be reduced in the range of the first few resonance frequencies. This is indeed the effect that one would expect from the damping control force [see Eq. (9.16a)] generated by the control loop. The active damper synthesised by controller enhances the damping of some of the first few modes so that the response peaks at some of the lower resonance frequencies are reduced and broadened. However, when the control gain exceeds an optimal value, then a pinning effect is produced which constrains the vibration at the control position. As a result the vibration field of the panel is rearranged into a new set of modes which are as lightly damped as those of the uncontrolled panel, since the pinning effect at the control position suppress the damping action (Gardonio and Elliott, 2005a). This is a very important feature of proportional velocity feedback control which indicates that the best control is achieved with an *ad hoc* control gain. It must be emphasised that, in principle, the proportional velocity feedback control is unconditionally stable and thus any value of control gain can be implemented in order to get the optimal control effect.

The implementation of the integral and derivative velocity control loops respectively synthesises an active spring and active mass. Thus, as shown in Fig. 9.14(b), when the control gain of the integral velocity feedback loop is augmented the resonance frequencies increase. However, when the control gain

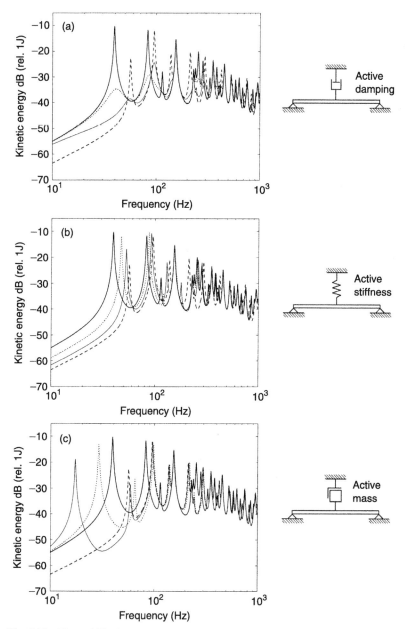

Fig. 9.14. Flexural kinetic energy of a panel without control (thick solid line) and with increasing levels of control gain (dotted, solid faint and dashed lines) implemented in: (a) proportional; (b) derivative; (c) velocity feedback control loops.

is very large, a pinning effect is produced so that the response of the panel is char-acterised by the same new, lightly damped, modes of the panel obtained with the proportional velocity feedback control loop. Vibration reduction is only obtained at frequencies below the first resonance. Figure 9.14(c) shows that when the derivative velocity feedback control gain is increased, the resonance frequencies are reduced. For very large control gains a blocking mass is synthesised which is suspended via the flexible plate so that a very low frequency first resonance is obtained. It is important to emphasise that although integral and derivative velocity feedback are unconditionally stable, they are characterised by relatively small phase margins so that in practice the necessary feedback gains that would lead to the active-spring pinning effect or active-mass blocking effect are not achievable in practice. With integral and proportional feedback control loops, the resonance frequencies are moved up or down even for small control gains well before the pinning and blocking mass conditions are met. This phenomenon has been physically interpreted as the result of the stiffening and mass effects. However, it can also be interpreted from the control point of view as a spillover effect which is confirmed by the fact that part of the polar plots of the open-loop sensor–actuator response functions enter the spillover circle.

9.3   Sensor–Actuator Transducers

In the previous section, the principal issues of feed-forward and feedback control approaches have been discussed with reference to ideal systems where the dynamics of the actuator and sensor transducers are not taken into account. In this section, the response functions of the most common transducers used for the construction of 'smart panels'[3] are briefly reviewed. For the purpose of active control, mechanical vibrations can be both generated and detected by strain transducers, such as piezoelectric laminae or films, or generated by electrody-namic inertial actuators and detected by inertial sensors (accelerometers). The main features of strain and inertial actuators are first reviewed. In the follow-ing two subsections the principal characteristics of strain and inertial sensors are presented. In particular the principal effects of these transducers in SISO direct velocity feedback control loops is analysed with reference to the open-loop sensor actuator response function $\widetilde{G}_{es}(\omega)\widetilde{H}(\omega)$ assuming $\widetilde{H}(\omega) = 1$.

---

[3] Smart panels are adaptive structures with sensor and actuator transducers directed by a controller capable of modifying the dynamic response of the structure in presence of time-varying environmental and operational conditions (Clark et al., 1998).

9.3.1 Strain Actuators

Flexural vibrations can be generated in a panel by activating strain transducers bonded to the surface. As shown in Fig. 9.15, these transducers are
formed by a piezoelectric lamina/film polarised along its thickness (defined as
direction 3), sandwiched between two thin sheets of metal acting as electrodes.
When a voltage is applied across the two electrodes, an electric field is generated across the piezoelectric material which, because of the inverse piezoelectric
effect (Fuller *et al.*, 1996), produces in-plane deformations in directions 1 and
2 and a comparatively smaller deformation in direction 3.[4] Since the lamina is
bonded to the panel, an in-plane stress field is then produced on the surface of the
panel, which causes it to bend, twist and stretch (Lee, 1990). The actuation effect
is generated only on the 'active' portion of the piezoelectric transducer where the
electric field is produced. The actuation strength per unit input voltage is determined by the polarisation of the piezoelectric material. Complex distributions of
excitation fields can be generated by shaping the electrodes or by shading the
poling of the piezoelectric lamina according to a specific spatial function.

The forced equations of uncoupled out-of-plane [Eq. (9.24c)] and in-plane
[Eqs. (9.24a,b)] vibrations of a panel generated by a layer of piezoelectric
material bonded to one surface of the panel can be written as (Lee, 1990;

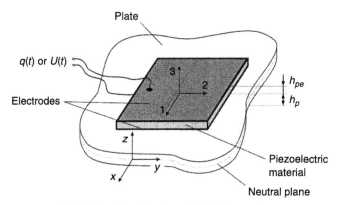

Fig. 9.15. Piezoelectric patch bonded on a plate.

---

[4] Note that in order to use the IEEE notation, the transverse displacement due to flexural vibration
in a flat plate is taken in z-direction and therefore identified by ζ while the in-plane displacements
due to longitudinal and shear waves are defined by ξ and η in x- and y-directions, respectively.

Gardonio and Elliott, 2004):

$$
\frac{E_s h_s}{1 - v_s^2} \left(\frac{\partial^2 \xi}{\partial x^2} + \frac{1 - v_s}{2} \frac{\partial^2 \xi}{\partial y^2} \right) + \frac{E_s h_s}{2(1 - v_s)} \frac{\partial^2 \eta}{\partial x \partial y} - \rho_s h_s \frac{\partial^2 \xi}{\partial t^2}
$$

$$
= \delta_{ep} \left(e_{31}^0 \frac{\partial S(x, y)}{\partial x} + e_{36}^0 \frac{\partial S(x, y)}{\partial y} \right) U_s(t) \tag{9.24a}
$$

$$
\frac{E_s h_s}{2(1 - v_s)} \frac{\partial^2 \xi}{\partial y \partial x} + \frac{E_s h_s}{1 - v_s^2} \left(\frac{1 - v_s}{2} \frac{\partial^2 \eta}{\partial x^2} + \frac{\partial^2 \eta}{\partial y^2} \right) - \rho_s h_s \frac{\partial^2 \eta}{\partial t^2}
$$

$$
= \delta_{ep} \left(e_{36}^0 \frac{\partial S(x, y)}{\partial x} + e_{32}^0 \frac{\partial S(x, y)}{\partial y} \right) U_s(t) \tag{9.24b}
$$

$$
\frac{E_s h_s^3}{12(1 - v_s^2)} \left(\frac{\partial^4 \zeta}{\partial x^4} + 2 \frac{\partial^4 \zeta}{\partial x^2 \partial y^2} + \frac{\partial^4 \zeta}{\partial y^4} \right) + \rho_s h_s \frac{\partial^2 \zeta}{\partial t^2}
$$

$$
= -\frac{h_s}{2} \delta_{ep} \left(e_{31}^0 \frac{\partial^2 S(x, y)}{\partial x^2} + 2 e_{36}^0 \frac{\partial^2 S(x, y)}{\partial x \partial y} + e_{32}^0 \frac{\partial^2 S(x, y)}{\partial y^2} \right) U_s(t) \tag{9.24c}
$$

where ξ, η, ζ are the displacements in x, y, z directions, $U_s(t)$ is the voltage applied across the piezoelectric transducer, δ_{ep} is equal to ± 1 depending whether the electric field and poling direction point in the same or opposite directions, $h_s = h_p + h_{pe}$ is the thickness of the panel h_p and piezoelectric lamina h_{pe} and ρ_s, E_s, v_s are the smeared density, Young's modulus and Poisson's ratio of the smart panel, respectively. $S(x, y) = F(x, y) P_0(x, y)$ is the spatial sensitivity function which takes into account the effective surface electrode function $F(x, y)$ and the poling shedding function $P_0(x, y)$ which normally is assumed to be equal for the stretching directions 1 and 2, and the twisting direction 6 (rotation about the axis 3) (Lee, 1990).

The three parameters e_{31}^0, e_{32}^0 and e_{36}^0 [5] are the piezoelectric stress/charge constants which for a flat lamina can be derived from the following

---

[5] 'The superscript '0' means that the quantity is a constant part of the piezoelectric constant' (Lee, 1990).

relation (Lee, 1990):

$$
\begin{Bmatrix} e^0_{31} \\ e^0_{32} \\ e^0_{36} \end{Bmatrix} =
\begin{bmatrix}
\dfrac{E_{pe}}{(1-v^2_{pe})} & \dfrac{v_{pe}E_{pe}}{(1-v^2_{pe})} & 0 \\[2mm]
\dfrac{v_{pe}E_{pe}}{(1-v^2_{pe})} & \dfrac{E_{pe}}{(1-v^2_{pe})} & 0 \\[2mm]
0 & 0 & \dfrac{E_{pe}}{2(1+v_{pe})}
\end{bmatrix}
\begin{Bmatrix} d^0_{31} \\ d^0_{32} \\ d^0_{36} \end{Bmatrix}
\tag{9.25}
$$

where the density ρ_{pe}, Young's modulus E_{pe}, Poisson's ratio v_{pe} and the piezoelectric strain/charge parameters d^0_{31}, d^0_{32} and d^0_{36}, for typical piezo-electric materials such as PZT (lead zirconate, titanate) and PVDF or PVF$_2$ (polyvinylidene fluoride polymer) are given in Table 9.1.

As an example, let us consider the bending excitation generated by a small square PZT patch bonded on the surface of a panel as shown in Fig. 9.16. Since d^0_{36} is normally zero for PZT materials, the bending excitation can been derived by introducing the two terms $\partial^2 S(x, y)/\partial x^2$ and $\partial^2 S(x, y)/\partial y^2$ with the sensitivity function $S(x, y)$ for the square patch, into the differential expression on the left-hand side of Eq. (9.24c). Sullivan et al. (1996), have proposed a mathematical approach where the spatial sensitivity function $S(x, y)$ is represented as the product of two generalised functions in orthogonal coordinates (Hoskins, 1979) using the Macauly notation (Pilkey, 1964). In this way, the derivative terms on the right-hand side of Eq. (9.24c) can be described in terms of step, doublet and delta functions from which it is possible to reconstruct the equivalent bending forces and moments generated on the panel by a specific sensitivity function of the piezoelectric transducer bonded onto it.

For instance, as shown by Gardonio and Elliott (2004), if the principal axes $1, 2, 3$ of the piezoelectric material are aligned along the x, y, z axes of the panel,

TABLE 9.1

Physical Constants for Typical PVDF and PZT Piezoelectric Materials

Parameter	PVDF	PZT
Density	$\rho_{pe} = 1780\,\mathrm{kgm^{-3}}$	$\rho_{pe} = 7600\,\mathrm{kgm^{-3}}$
Young's modulus	$E_{pe} = 2 \times 10^9\,\mathrm{Nm^{-2}}$	$E_{pe} = 6.3 \times 10^{10}\,\mathrm{Nm^{-2}}$
Poisson's ratio	$v_{pe} = 0.29$	$v_{pe} = 0.31$
Strain/charge constants	$d^0_{31} = 23 \times 10^{-12}\,\mathrm{mV^{-1}}$	$d^0_{31} = 166 \times 10^{-12}\,\mathrm{mV^{-1}}$
	$d^0_{32} = 3 \times 10^{-12}\,\mathrm{mV^{-1}}$	$d^0_{32} = 166 \times 10^{-12}\,\mathrm{mV^{-1}}$
	$d^0_{36} = 0$	$d^0_{36} = 0$

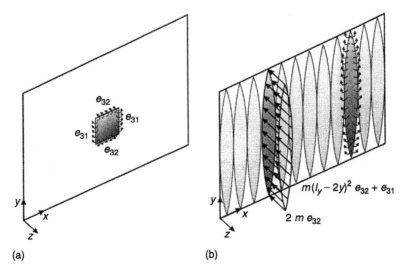

Fig. 9.16.   (a) Square piezoelectric patch for the generation of localised bending excitation;
(b) quadratically shaped array of piezoelectric strips for the generation of forces and moments that
produce an approximation to a uniform force distribution over the panel surface.

as shown in Fig. 9.16(a), the square piezoelectric patch actuator produces bending
moment excitations

$$m_x(x_{e1,3}, y_{e1,3}, t) = \pm \frac{h_s}{2} e_{32}^0 U_s(t), \quad m_y(x_{e2,4}, y_{e2,4}, t) = \pm \frac{h_s}{2} e_{31}^0 U_s(t)$$

$$(9.26a,b)$$

respectively along the horizontal edges and along the vertical edges as shown in
Fig. 9.16(a). If the piezoelectric material has a skew excitation effect ($d_{36}^0 \neq 0$)
then the term with mixed derivatives of the sensitivity function in Eq. (9.24c),
$-h_s \delta_{ep} e_{36}^0 (\partial^2 S/\partial x \partial y)$ gives two pairs of forces with equal amplitudes and
opposite directions acting at opposing vertices of the panel:

$$f_y(x_{v1,3}, y_{v1,3}, t) = -f_y(x_{v2,4}, y_{v2,4}, t) = h_s e_{36}^0 U_s(t) \qquad (9.26c)$$

In this way, as one would expect for strain actuators, there is no net transverse
force excitation on the panel.

Having found the bending moment excitations generated by the square piezoelectric patch bonded to the panel, the response of the panel can be derived
using the modal expansion formulation described in Chapter 2. For example,
if as shown in Fig. 9.17 the square patch actuator is used in conjunction with a

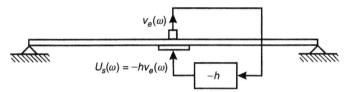

Fig. 9.17. Simply supported rectangular plate with a velocity feedback control loop using a piezoelectric patch actuator and a velocity sensor at its centre.

velocity sensor located at its centre in order to implement a disturbance rejection velocity feedback control loop, then the open-loop response function of this control system is given by

$$\tilde{G}_{es}(\omega) = \frac{\tilde{v}_e(x_e, y_e)}{\tilde{U}_s} = j\omega \sum_{p=1}^{\infty} \sum_{q=1}^{\infty} \frac{\phi_{pq}(x_e, y_e)F_{pq}}{M_{pq}\left[\omega_{pq}^2(1+j\eta) - \omega^2\right]} \tag{9.27}$$

According to the formulation presented in Sections 2.3.3 and 2.4.2, the modal excitation terms $F_{pq}(\omega)$ (force/volt) are given by the sum of the integrals along the four edges of the piezoelectric patch of the first derivatives of the panel natural modes in directions orthogonal to the edges (outward from the patch) multiplied by the appropriate piezoelectric excitation coefficients given in Eqs. (9.26a,b); that is

$$F_{pq}(\omega) = \frac{h_s}{2}e_{32}^0\left\{+\int_{x_{v1}}^{x_{v2}}\frac{\partial\phi_{pq}(x, y_{e1})}{\partial y}dx - \int_{x_{v3}}^{x_{v4}}\frac{\partial\phi_{pq}(x, y_{e3})}{\partial y}dx\right\}$$

$$+ \frac{h_s}{2}e_{31}^0\left\{-\int_{y_{v1}}^{y_{v4}}\frac{\partial\phi_{pq}(x_{e4}, y)}{\partial x}dy + \int_{y_{v2}}^{y_{v3}}\frac{\partial\phi_{pq}(x_{e2}, y)}{\partial x}dy\right\} \tag{9.28}$$

The Bode and Nyquist plots of the open-loop frequency response functions $\tilde{G}_{es}(\omega)$ of this feedback control loop are shown in Fig. 9.18. The increase with frequency of magnitude of the Bode plot illustrates a characteristic feature of strain actuators which produce increased flexural excitation as the frequency rises. The phase of the Bode plot indicates that the frequency response function generated by the transducer system is positive definite up to about 10 kHz beyond which the phase lag exceeds $-90°$. This is due to the time $t_d = a/2c_b$ that it takes the bending waves generated around the edges of the patch to reach the velocity sensor located at the centre of the patch, where a is the dimension of the patch.

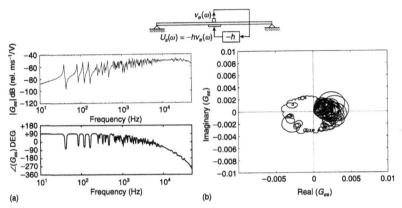

Fig. 9.18. (a) Bode plots; (b) Nyquist plots of the open-loop frequency response function of the feedback control system using a piezoelectric patch actuator and a velocity sensor at its centre.

Recalling the formula for the phase speed of bending waves given in Chapter 1, the frequency-dependent phase lag is given by $\phi = \omega t_d = (a/2)\omega^{1/2}(m/D)^{1/4}$. Therefore the non-perfect collocation between actuation and detection makes this sensor–actuator pair only conditionally stable. In fact, the Nyquist plot in Fig. 9.18(b) suggests that this control system would be unstable even with small feedback control gains since the high frequency range of the frequency response functions, with the larger magnitude, would encircle the stability point $-1 + j0$. Therefore, the combination of the rising magnitude and rising phase lag of the open-loop response function makes this sensor–actuator pair inappropriate for feedback control loops unless the control function is properly designed in such a way as to suppress the high frequency response, for example with a low-pass filter.

Another type of strain actuator of interest for ASAC smart panels is the distributed transducer made of a series of quadratically shaped strips which is shown in Fig. 9.16(b). As shown by Gardonio and Elliott (2004), assuming the quadratic shape of each strip is specified by the two functions $S_k(x, y) = \Gamma_k = x_k \pm m \left(l_y y - y^2\right)$, where $m = 2\delta/l_y^2$, δ is the width of the strips and x_k is the coordinate of the y-symmetry axis of the k-th strip, each strip produces line forces along the two edges given by

$$f_{z,k}(\Gamma, t) = m e_{32}^0 h_s U_s(t) \tag{9.29a}$$

and two line moments along the two edges given by

$$m_{\mathbf{n},k}(\Gamma, t) = \frac{1}{2}\left\{m^2\left(l_y - 2y\right)^2 e_{32}^0 + e_{31}^0\right\} h_s U_s(t) \tag{9.29b}$$

where $\mathbf{n}$ is the normal to the tangent along the two quadratic curves in the plane of the plate of the k-th strip. If the panel is freely suspended, the two lines of forces would be balanced by two point forces of amplitude $1/2 \oint_{\Gamma_k} f_{z,k}(x, y)d\Gamma$ acting at the tips of the strip so that, as one would expect for strain actuators, there is no excitation of rigid body motion. If, instead, the panel is constrained along the perimeter so that the forces at the tips of the strips are reacted by the constraint, it is possible to generate parallel lines of force over the panel surface. At frequencies such that the bending wavelength is larger than the width of a quadratic strip, the lines of forces effectively act as a uniformly distributed transverse force over the panel surface which can be used to excite the first radiation mode of the panel. As for the square patch, the modal excitation terms F_{pq} to be used in the modal summation of Eq. (9.27) are obtained by integrating along the boundaries of the quadratic strips the products of the force and moment excitations with, respectively, the mode shape $\phi_{pq}(x, y)$ and its derivative with respect to $\mathbf{n}(x, y)$:

$$\tilde{F}_{pq}(\omega) = -h_s \sum_{k=1}^{K} \oint_{\Gamma_k} me_{32}^0 \phi_{pq}(x, y)d\Gamma_k$$

$$- \frac{h_s}{2} \oint_{\Gamma_k} \left(m^2(l_y - 2y)^2 e_{32}^0 + e_{31}^0 \right) \frac{\partial \phi_{pq}(x, y)}{\partial n(x, y)} d\Gamma_k \qquad (9.30)$$

9.3.2 Inertial Electrodynamic Actuators

One way to generate a 'sky-hook' force excitation on structures is to use linear actuators which react off a mass. If, for example, the form of electrodynamic linear motor as shown in Fig. 9.19 is used, the inertial actuation mechanism can be obtained by fixing the coil assembly to the case of the actuator on which the permanent magnet is mounted via soft springs and acts as mass. In the first part of the following analysis, it is assumed that the actuator is mounted on a rigid base.

The current I_s in the coil of this classic electrodynamic motor is given by

$$R_e I_s + L_e \frac{dI_s}{dt} = U_s - \psi \frac{d\eta}{dt} \qquad (9.31)$$

where ψ is the frequency-independent voice-coil coefficient, U_s is the applied voltage or electromotive force (e.m.f.), R_e and L_e are the resistance and inductance of the coil and $-\psi(d\eta/dt)$ is the back e.m.f. generated by the axial velocity $d\eta/dt$ of the mass with respect to the case of the actuator which is assumed

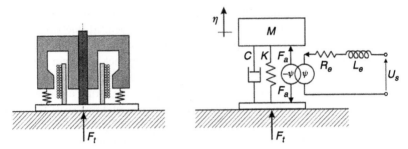

Fig. 9.19.   Electrodynamic inertial actuator mounted on a rigid base.

to be fixed in this analysis. The equation of motion for the mass with respect to the base is given by

$$M\frac{d^2\eta}{dt^2} + C\frac{d\eta}{dt} + K\eta = F_a \qquad (9.32)$$

where M, C, K are the mass, damping and stiffness of the mount and F_a is the electromagnetic force, which is proportional to the current in the coil:

$$F_a = \psi I_s \qquad (9.33)$$

Assuming harmonic functions, substitution of Eq. (9.31) into Eq. (9.33) gives

$$\tilde{F}_a = \psi\frac{(\tilde{U}_s - j\omega\psi\tilde{\eta})}{R_e + j\omega L_e} \qquad (9.34)$$

Combining Eqs. (9.32) and (9.34) gives the relationship between the displacement $\tilde{\eta}$ of the moving mass and the driving voltage U_s:

$$\left\{-\omega^2 M + j\omega\left(C + \frac{\psi^2}{R_e + j\omega L_e}\right) + K\right\}\tilde{\eta} = \frac{\psi}{R_e + j\omega L_e}\tilde{U}_s \qquad (9.35)$$

Therefore the back e.m.f. generates an additional electrodynamic force $j\omega\psi^2/(R_e + j\omega L_e)$ whose real part acts as a frequency-dependent damping force. The vibration of the mass produces an inertia force given by $\omega^2 M\tilde{\eta}$, which is transmitted to the case structure via the spring-dashpot mounting system. Therefore, deriving $\tilde{\eta}$ from Eq. (9.35), the force transmitted to the base per unit voltage input to the actuator is given by

$$\frac{\widetilde{F}_t}{\widetilde{U}_s} = \frac{\omega^2 M \psi / (R_e + j\omega L_e)}{-\omega^2 M + j\omega \left[C + \psi^2 / (R_e + j\omega L_e)\right] + K} \qquad (9.36)$$

The plot in Fig. 9.20 shows the modulus and phase of the ratio of the force F_t transmitted to the base per unit driving voltage U_s when $M = 0.01\,\text{kg}$, $K = 88.8\,\text{Nm}^{-1}$, $C = 0.14\,\text{Nsm}^{-1}$, $R_e = 80\,\Omega$, $L_e = 0.001\,\text{H}$ and $\psi = 2.6\,\text{Vsm}^{-1}$. This plot indicates that for frequencies above the fundamental mechanical resonance frequency of the actuator, $f_{res} \approx 1/2\pi \sqrt{K/M} = 15\,\text{Hz}$, the magnitude of the transmitted force F_t is frequency independent and in phase with the driving voltage $\widetilde{U}_s$. Therefore, this type of actuator can be used as a sky-hook force actuator provided its fundamental resonance frequency is well below that of the structure under control. This requirement cannot always be satisfied since, in order to achieve a sufficiently low natural frequency, a small value of K/M must be selected and undesirably large static displacements of the mass are possible. If the structure to be controlled possesses very low resonance frequencies, the actuator must be designed with a very soft mounting system so that the mass can undergo very large displacements in order to generate sufficiently

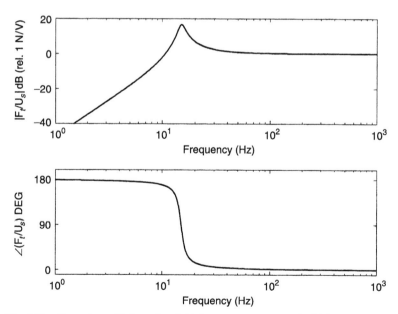

Fig. 9.20. Force transmitted to the rigid base structure per unit driving voltage by an electrodynamic actuator.

Fig. 9.21. Inertial actuator with circular springs: (a) photograph; (b) schematic.

large forces to suppress low frequency disturbances of vehicles such as aircraft, helicopters, trains and automobiles that can be generated during manoeuvres.

Figure 9.21(a) shows a prototype actuator specifically built for feedback control purposes by Paulitsch (2006). As shown schematically in Fig. 9.21(b), the actuator comprises an annular magnet which is guided by a vertical pin and seats on three circular springs arranged in such a way as to produce a very low stiffness effect in the axial direction and a much higher stiffness in lateral directions, so that the mass is constrained to vibrate in the axial direction with a low fundamental natural frequency. The driving coil is fixed to the base case and only a small fraction of the winding is visible below the magnet in Fig. 9.21(a).

If, as shown in Fig. 9.22, an inertial actuator is used on a plate to implement a direct velocity feedback control loop, the response of the structure affects the force generated by the actuator. The sensor–actuator open-loop frequency response function can be derived with a mobility formulation which includes the fully coupled response of the inertial actuator and the plate.

The vector of complex amplitudes of the velocity $\tilde{v}_e$ at the base of the actuator where the error velocity sensor is positioned and the velocity of the mass $\tilde{v}_a$ can

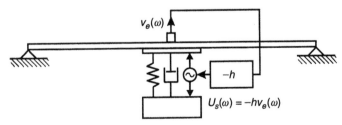

Fig. 9.22. Simply supported rectangular plate with a velocity feedback control loop using an electrodynamic inertial actuator and a velocity sensor at its centre.

be related to the forces $\tilde{f}_e$ on the base and $\tilde{f}_a$ on the mass by the following mobility matrix expression:

$$\left\{ \begin{array}{c} \tilde{v}_e \\ \tilde{v}_a \end{array} \right\} = \left[\begin{array}{cc} \tilde{Y}_e & 0 \\ 0 & \tilde{Y}_a \end{array} \right] \left\{ \begin{array}{c} \tilde{f}_e \\ \tilde{f}_a \end{array} \right\} \quad \Rightarrow \quad \tilde{\mathbf{v}} = \tilde{\mathbf{Y}}\tilde{\mathbf{f}} \tag{9.37}$$

where $\tilde{Y}_e$ is the driving point mobility of the plate where the actuator is mounted, which can be calculated with Eq. (9.2), and $\tilde{Y}_a = 1/j\omega M_a$ the mobility of the mass $\tilde{Y}_a$. These velocity and force vectors can also be related with reference to the impedance functions of the mounting system and the reactive forces generated by the coil-magnet actuator $\pm \tilde{F}_a$:

$$\left\{ \begin{array}{c} \tilde{f}_e \\ \tilde{f}_a \end{array} \right\} = \left[\begin{array}{cc} \tilde{Z}_{11} & \tilde{Z}_{12} \\ \tilde{Z}_{21} & \tilde{Z}_{22} \end{array} \right] \left\{ \begin{array}{c} \tilde{v}_e \\ \tilde{v}_a \end{array} \right\} + \left[\begin{array}{c} -1 \\ +1 \end{array} \right] \tilde{F}_a \quad \Rightarrow \quad \tilde{\mathbf{f}} = \mathbf{Z}\tilde{\mathbf{v}} + \mathbf{u}\tilde{F}_a \tag{9.38}$$

where $\tilde{Z}_{12} = \tilde{Z}_{21} = -\tilde{Z}_{11} = -\tilde{Z}_{22} = K/j\omega + C$. Since in this case the base of the actuator is not fixed, then the back e.m.f. term on the right-hand side of Eq. (9.31) must be written with reference to the difference between the velocities of the mass and base so that

$$\tilde{Z}_e \tilde{I}_s = \tilde{U}_s - \psi \lfloor -1 \quad 1 \rfloor \left\{ \begin{array}{c} \tilde{v}_e \\ \tilde{v}_a \end{array} \right\} \quad \Rightarrow \quad \tilde{Z}_e \tilde{I}_s = \tilde{U}_s - \psi \mathbf{u}^T \tilde{\mathbf{v}} \tag{9.39}$$

where $\tilde{Z}_e = (R_e + j\omega L_e)$. Thus, remembering that the complex amplitude of the actuation force is given by $\tilde{F}_a = \psi \tilde{I}_s$, the ratio between the complex velocity at the control position $\tilde{v}_e$ and the driving voltage to the actuator $\tilde{U}_s$, which represents the open-loop response function for this velocity feedback control loop, can be derived from Eqs. (9.37–9.39):

$$\tilde{G}_{es} = \frac{\tilde{v}_e}{\tilde{U}_s} = \mathbf{h} \left(\mathbf{I} - \mathbf{Y} \left[\mathbf{Z} - \frac{\psi^2}{\tilde{Z}_e} \mathbf{u}\mathbf{u}^T \right] \right)^{-1} \frac{\psi}{\tilde{Z}_e} \mathbf{Y}\mathbf{u} \tag{9.40}$$

where $\mathbf{h} = \lfloor 0 \quad 1 \rfloor$. Figure 9.23 shows the Bode and Nyquist plots of the open-loop frequency response function $\tilde{G}_{es}(\omega)$ of this feedback control loop when the inertial actuator considered above is attached to the same aluminium plate and at the same position as considered in examples in the previous sections. The axial natural frequency of the actuator is at 15 Hz so that the first peak in the amplitude Bode plot is due to the resonance of the actuator and those following are due to the panel.

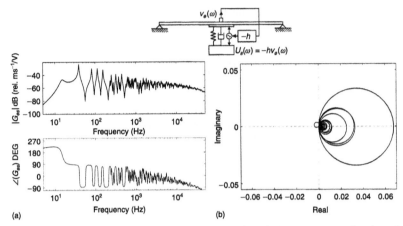

Fig. 9.23. (a) Bode; (b) Nyquist plots for the open-loop frequency response function of the feedback control system using an electrodynamic inertial actuator and a velocity sensor at its centre.

Indeed, apart from the first peak, the magnitude of the Bode plot is similar to that in Fig. 9.13(a) where the resonances are separated by antiresonance frequencies indicating the collocation of actuation and detection. The only difference is produced by the low frequency dynamics of the inertial actuator which produces an excitation in phase opposition to the driving voltage so that the phase Bode plot starts from about $+270°$ and undergoes a $-180°$ phase lag in correspondence to the resonance of the actuator. As a result, the open-loop sensor–actuator response function is not strictly positive definite. This time the stability problem occurs at low frequency where, because of the dynamics of the inertial actuator, the phase is confined to the range $+270°$ to $+90°$. This effect is better seen in the Nyquist plot of Fig. 9.13(b), which is now characterised by a low frequency circle that occupies the real negative part. Therefore, the active damping control will be effective only for those resonances which are characterised by the larger diameter circles than those of the actuator resonance. The only solution to this problem is to have an actuator that produces a resonant response of small amplitude. Indeed this can be achieved by lowering the fundamental natural frequency of the actuator and by augmenting the internal damping effect, although this has to be tuned to an optimal value rather than just being maximised (Paulitsch *et al.*, 2006). The large frequency range of the plot in Fig. 9.23(a) shows a particular feature of voltage-driven inertial actuators, which produce an amplitude roll off and phase lag of the open-loop response function at higher frequencies. This effect, which is due to the coil inductance, can cause control spillover effects, or even instability problems, at higher frequencies since, as shown in Fig. 9.23(b), the locus enters the real negative quadrant. This problem can be avoided by implementing a feedback loop with a current control amplifier, in which case the open-loop sensor–actuator response function is not affected by the inductive effect of the coil-magnet actuator.

9.3.3 Strain Sensors

Flexural vibrations of panels can be measured by bonding strain sensors onto the surface. As with strain actuators, the simplest type of these sensors comprises a piezoelectric lamina/film sandwiched between two thin layers of metal working as electrodes. When, as shown in Fig. 9.15, the transducer is bonded to the surface of a panel, in-plane (longitudinal and shear) and out-of-plane (bending) vibrations of the panel deform the piezoelectric material so that it generates a distribution of charges on the surfaces of the two electrodes by means of the piezoelectric effect (Fuller $et\ al.$, 1996). A charge output $q_e(t)$ proportional to the plate vibration can therefore be measured by connecting a charge amplifier to the two electrodes. If the charge output signal is differentiated with respect to time, then the current, $i_e(t) = dq_e(t)/dt$, is measured, which is proportional to the velocity of panel vibration.

According to Lee's formulation (1990), and assuming that the piezoelectric transducer is polarised along the thickness, that is direction 3 in Fig. 9.15, the total current output can be expressed as the sum of two components: the current generated by the bending vibration of the panel $i_b(t)$ and the current due to the in-plane vibration of the panel $i_i(t)$ such that

$$i_e(t) = i_b(t) + i_i(t) \tag{9.41a}$$

where

$$i_b(t) = -\frac{h_s}{2} \int_0^{l_x} \int_0^{l_y} S(x,y) \left[e_{31}^0 \frac{\partial^2 w(x,y,t)}{\partial x^2} + e_{32}^0 \frac{\partial^2 w(x,y,t)}{\partial y^2} \right.$$

$$\left. + 2e_{36}^0 \frac{\partial^2 w(x,y,t)}{\partial x \partial y} \right] dxdy \tag{9.41b}$$

$$i_i(t) = \int_0^{l_x} \int_0^{l_z} S(x,z) \left[e_{31}^0 \frac{\partial u(x,z,t)}{\partial x} + e_{32}^0 \frac{\partial v(x,y,t)}{\partial y} \right.$$

$$\left. + e_{36}^0 \left(\frac{\partial u(x,y,t)}{\partial y} + \frac{\partial v(x,z,t)}{\partial x} \right) \right] dxdy \tag{9.41c}$$

and h_s is the total thickness of the smart panel and $u(x, y, t)$, $v(x, y, t)$ and $w(x, y, t)$ are the velocities of the plate in the x, y and z directions.[6]

The simplicity and small mass per unit surface area of these transducers makes them particularly well-suited for the construction of distributed sensors. For instance, the transducer with quadratically shaped strips described in Section 9.3.1 can also be used as a sensor to measure the volumetric vibration of the panel which, at low frequencies, is a good approximation to the first radiation mode of the panel. Figure 9.24 shows a schematic and picture of a prototype panel with such a matched pair of quadratically shaped sensor–actuator transducers. Assuming the panel is vibrating only in bending, then substituting in Eq. (9.41b) the sensitivity function of the strips represented by the same pair of functions $S_k(x, y)$ specified for the uniform force actuator, shown in Fig. 9.16(b), the sensor output for the k-th strip is found to be proportional to the integrals along the two quadratically shaped edges of the transverse, $w(x, y, t)$, and angular, $\theta_\mathbf{n}(x, y, t)$, velocities:

$$i_{b,k}(t) = -h_s \oint_{\Gamma_k} me_{32}^0 w(x, y, t)\, d\Gamma_k$$

$$-\frac{h_s}{2} \oint_{\Gamma_k} \left(m^2(l_y - 2y)^2 e_{32}^0 + e_{31}^0\right) \theta_\mathbf{n}(x, y, t)\, d\Gamma_k \qquad (9.42)$$

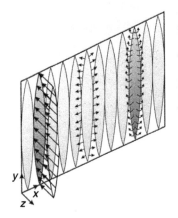

Fig. 9.24. Smart panel with two distributed piezoelectric transducers whose electrodes are shaded in quadratic strips so that they form matched volume velocity sensor and uniform force actuator.

---

[6] Also in this case, in order to use the IEEE notation, the transverse velocity due to flexural vibration in a flat plate is taken in z-direction and therefore identified by w while the in-plane velocities due to longitudinal and shear waves are defined by u and v in x- and y-directions, respectively.

where $\theta_{\mathbf{n}}(x, y) = \partial w(x, y)/\partial n(x, y)$ is the angular velocity with direction normal to the tangent along the two quadratic curves in the plane of the plate of the k-th strip. The total output signal is then given by the summation of the output contributions from all quadratic strips, that is

$$i_b(t) = \sum_{k=1}^{K} i_{b,k}(t) \tag{9.43}$$

If this sensor is used in conjunction with the uniform force actuator to form a matched sensor–actuator pair, a direct velocity feedback control loop can be implemented to drive to zero the volume velocity of the panel that we have seen in Section 3.7 to be the most efficient radiating component at low frequencies. The open-loop frequency response function of this sensor–actuator pair can be derived from Eq. (9.42) by expressing the transverse velocity $w(x, y, t)$ and angular velocity $\theta_{\mathbf{n}}(x, y, t)$ generated by the lines of transverse forces and moments given in Eqs. (9.29a,b) along the borders of the quadratically shaped strips in terms of a modal summation as described in Section 2.4.2, so that the following expression is obtained

$$\tilde{G}_{es}(\omega) = \frac{\tilde{i}_e(x_e, z_e)}{\tilde{U}_s} = j\omega \sum_{p=1}^{\infty} \sum_{q=1}^{\infty} \frac{\sigma_{pq} F_{pq}}{M_{pq}\left[\omega_{pq}^2\,(1 + j\eta) - \omega^2\right]} \tag{9.44}$$

where, since the sensor and actuator transducers are matched, the integrals in Eq. (9.30) give modal sensing terms which are the same as the modal excitations, that is $\sigma_{pq} = F_{pq}$. In summary, the formulation presented above indicates that this sensor actuator pair can be considered to be collocated and dual so that the feedback control loop is unconditionally stable. Indeed, the faint line in the Bode plot of Fig. 9.25(a), shows that the phase of the open-loop sensor–actuator response function $\tilde{G}_{es}(\omega)$ remains in the range $\pm 90°$ which indicates that $\tilde{G}_{es}(\omega)$ is strictly positive real. The magnitude is characterised by a large resonance peak at low frequency which is generated by the first natural mode of the panel whose shape is close to that of the first radiation mode. At higher frequencies, the magnitude tends to decrease as the frequency rises. No peaks are visible in correspondence to the resonances of the even natural modes. Also, the peaks produced by the resonances of the higher order odd modes are smaller than that of the fundamental. This is due to the fact that the volumetric vibration component of these modes tends to decrease as the mode-order increases. Since, in principle, this sensor–actuator pair is unconditionally stable, it should be possible to implement large control gains so that the volumetric vibration of the panel, and thus its sound radiation, is greatly reduced at low frequencies.

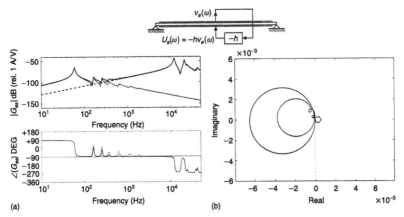

Fig. 9.25. (a) Bode plot; (b) Nyquist plot of the open-loop frequency response function of the system shown in Fig. 9.24.

The analysis presented above assumes that the volume velocity sensor and uniform force actuator transducers are coupled only via the flexural waves. In practice, as shown in Fig. 9.24(a), the actuator exerts an in-plane excitation on the surface of the panel which also generates longitudinal and shear waves. These in-plane vibrations of the panel are then picked up by the sensor. Therefore, the open-loop sensor–actuator response function should be formulated taking into account both the coupling via out-of-plane flexural waves and via in-plane longitudinal and shear waves. The dashed line in Fig. 9.25(a) shows the in-plane coupling effect (Gardonio *et al.*, 2001b). The fundamental resonances associated with in-plane vibrations of a panel occur at much higher frequencies than fundamental flexural resonances so that the amplitude plot is characterised by a rising trend up to the first resonance at about 11 kHz. This coupling is still in minimum phase but it is out-of-phase with the flexural coupling. In fact, when the actuator transducer stretches on the surface of the panel, the sensor transducer is also stretched under the effect of the in-plane coupling while it is compressed under the effect of the out-of-plane coupling. When the two coupling effects are combined then, as shown by the thick solid line in Fig. 9.25(a), at low frequencies the in-plane coupling is negligible compared with the out-of-plane coupling. However, at higher frequencies the situation is reversed and the out-of-plane coupling effect is negligible. In principle, this is not a problem since the smart panel is meant to control low frequency sound radiation. However, this effect has an important impact on the stability of the control loop because it adds a 180° phase lag combined with a rising amplitude of the open-loop response function which is therefore no longer minimum phase. At higher frequencies the phase exceeds $-90°$ so that $\widetilde{G}_{es}(\omega)$ becomes real negative. This is a particularly undesiderable effect since it is associated with the rising in-plane coupling effect, so that the real negative part of $\widetilde{G}_{es}(\omega)$ is characterised by relatively larger amplitudes than the low frequencies where the flexural waves control the coupling between the sensor and actuator transducers. Thus, as clearly indicated by Fig. 9.25(b), the Nyquist plot of the open-loop sensor–actuator response function is characterised by a relatively large loop in the left-hand quadrants which indicates how critical the stability issue would be for this control arrangement. This is a very unfortunate complication, which, however, represents a characteristic feature of strain sensor–actuator pairs that must be taken into account in the design and development of these types of smart panel.

9.3.4 Inertial Sensors (Accelerometers)

One practical sensor that can be used to measure the vibration of structures at a specific point is the accelerometer. As shown in Fig. 9.26, this type of device comprises a seismic mass M_a connected to a vibrating base via a spring of stiffness K_a whose damping effect is modelled with a dashpot with a damping coefficient C_a. The spring normally consists of a piezoelectric element which generates a voltage signal U_a proportional to the relative displacement between the mass η_a and the case η_e, respectively,

$$U_a = c_\sigma (\eta_a - \eta_e) \tag{9.45}$$

where c_σ is the detection constant of the piezoelectric elastic element.

Assuming harmonic motion of the base, the complex relative displacement $\tilde{\eta} = \tilde{\eta}_a - \tilde{\eta}_e$ with reference to the complex base acceleration $-\omega^2 \tilde{\eta}_e(\omega)$ is

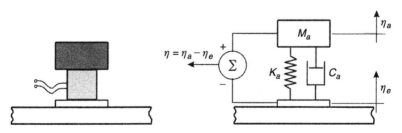

Fig. 9.26. Accelerometer.

given by (Rao, 1995):

$$\frac{\tilde{\eta}(\omega)}{-\omega^2 \tilde{\eta}_e(\omega)} = \frac{-M_a}{(K_a - \omega^2 M_a) + j\omega C_a} = \frac{-1/\omega_a^2}{\left(1 - (\omega^2/\omega_a^2)\right) + j2\zeta_a \left(\omega/\omega_a\right)}$$

(9.46)

where $\omega_a = \sqrt{K_a/M_a}$ and $\zeta_a = C_a/2\sqrt{K_a M_a}$ are respectively the natural frequency and damping ratio of the mass–spring accelerometer transducer.

As shown in Fig. 9.27, for frequencies well below the fundamental resonance frequency $\omega \ll \omega_a$, $\tilde{\eta}(\omega) \approx \left(\omega^2/\omega_a^2\right) \tilde{\eta}_e(\omega)$ so that the accelerometer voltage output is proportional and opposite to the acceleration of the base: $\tilde{U}_a(\omega) \approx -\left(c_\sigma/\omega_a^2\right)\left(-\omega^2 \tilde{\eta}_e(\omega)\right)$. In summary, with this transducer it is possible to measure the acceleration at a point of a structure provided its natural frequency is well above the measurement range. The effective range of measurement also depends on the damping of the transducer since larger damping broadens the magnitude peak and spreads phase transition around the resonance frequency. Thus, it is preferable to have a very lightly damped transducer. It is important to emphasise that the sensitivity of the transducer is inversely proportional to the square of its natural frequency and thus, in order to maximise

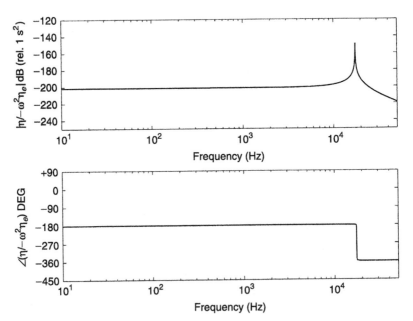

Fig. 9.27. Frequency response function of an accelerometer.

the output signal-to-noise ratio, it is vital to keep the resonance frequency as close as possible to the range of measurement.

If this type of sensor is used to implement a feedback control loop with the piezoelectric patch discussed in Section 9.3.1 and shown in Fig. 9.17, the open-loop sensor–actuator response function, that considers the coupled response of the panel with the piezoelectric patch actuator and accelerometer sensor can be derived with a similar mobility model to that developed for the coupled response of the inertial actuator and panel. The vector containing the complex velocity $\tilde{v}_e$ at the error point where the accelerometer is positioned and the complex velocity $\tilde{v}_a$ of the accelerometer mass can be related to the complex force $\tilde{f}_e$ on the base of the accelerometer and $\tilde{f}_a$ on the mass by the following mobility and impedance matrix expression:

$$
\left\{ \begin{array}{c} \tilde{v}_e \\ \tilde{v}_a \end{array} \right\} = \left[\begin{array}{cc} \tilde{Y}_e & 0 \\ 0 & \tilde{Y}_a \end{array} \right] \left\{ \begin{array}{c} \tilde{f}_e \\ \tilde{f}_a \end{array} \right\} + \left[\begin{array}{c} \tilde{Y}_{es} \\ 0 \end{array} \right] \tilde{U}_s \quad \Rightarrow \quad \tilde{\mathbf{v}} = \tilde{\mathbf{Y}}\tilde{\mathbf{f}} + \tilde{\mathbf{Y}}_s\tilde{U}_s \quad (9.47a)
$$

$$
\left\{ \begin{array}{c} \tilde{f}_e \\ \tilde{f}_a \end{array} \right\} = \left[\begin{array}{cc} \tilde{Z}_{11} & \tilde{Z}_{12} \\ \tilde{Z}_{21} & \tilde{Z}_{22} \end{array} \right] \left\{ \begin{array}{c} \tilde{v}_e \\ \tilde{v}_a \end{array} \right\} \quad \Rightarrow \quad \tilde{\mathbf{f}} = \mathbf{Z}\tilde{\mathbf{v}} \quad (9.47b)
$$

where $\tilde{Y}_e$ is the driving point mobility function of the plate where the accelerometer is mounted, which can be calculated from Eq. (9.2), $\tilde{Y}_a = 1/j\omega M_a$ is the mobility function of the mass, $\tilde{Y}_{es} = \tilde{v}_e(x_e, y_e)/\tilde{U}_s$ is the mobility function that gives the velocity at the error position generated by a unit input voltage to the piezoelectric patch actuator, which can be calculated from Eq. (9.27). $\tilde{Z}_{12} = \tilde{Z}_{21} = -\tilde{Z}_{11} = -\tilde{Z}_{22} = C_a - jK_a/\omega$ are the point and transfer impedance terms of the piezoelectric element in the accelerometer. The coupled response of the panel and accelerometer sensor is therefore given by

$$
\tilde{\mathbf{v}} = \left(\mathbf{I} - \tilde{\mathbf{Y}}\tilde{\mathbf{Z}} \right)^{-1} \tilde{\mathbf{Y}}_s \tilde{U}_s \quad (9.48)
$$

The output voltage signal from the accelerometer can be expressed as,

$$
\tilde{U}_e = \frac{c_\sigma}{j\omega} \lfloor -1 \quad 1 \rfloor \left\{ \begin{array}{c} \tilde{v}_e \\ \tilde{v}_a \end{array} \right\} \quad \Rightarrow \quad \tilde{U}_e = \frac{c_\sigma}{j\omega} \mathbf{h}\tilde{\mathbf{v}} \quad (9.49)
$$

Assuming that the output signal of the accelerometer is passed through an ideal integrator with transfer function $1/j\omega$, the open-loop sensor–actuator frequency response function is given by

$$
\tilde{G}_{es}(\omega) = \frac{\tilde{U}_e/j\omega}{\tilde{U}_s} = -\frac{c_\sigma}{\omega^2} \mathbf{h} \left(\mathbf{I} - \tilde{\mathbf{Y}}\tilde{\mathbf{Z}} \right)^{-1} \tilde{\mathbf{Y}}_s \quad (9.50)
$$

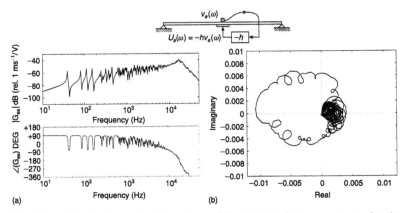

Fig. 9.28. (a) Bode plot and (b) Nyquist plot of the open-loop frequency response function of the feedback control system using a piezoelectric patch actuator and an accelerometer at its centre with an ideal time integrator.

At first sight, it might seem unnecessary to take into account the dynamic influence of the presence of the accelerometer whose dimensions (diameter and height of about 4 mm) and mass (about 3–4 g) are normally much smaller that those of the panel structure. However, comparing the Bode and Nyquist plots in Fig. 9.28 for the open-loop sensor–actuator function that takes into account the dynamics effects of the accelerometer and those in Fig. 9.18 for the equivalent open-loop response function when an ideal and massless velocity sensor is considered, it is clear that the dynamics of the accelerometer play a fundamental role at higher frequencies. In fact, the amplitude and phase Bode plots shows that the resonance of the accelerometer sensor at about 20 kHz produces a large peak in $\tilde{G}_{es}(\omega)$ which is associated with a $-180°$ phase lag. Above the resonance frequency, the accelerometer acts as a low-pass filter so that the amplitude of the response function decreases monotonically. These two effects are of great importance since, as can be seen in in Fig. 9.28(b), a large loop is generated in the negative real quadrants of the Nyquist plot. This loop is much greater than the low frequency loops due to the resonances of the panel. Therefore, the accelerometer sensor makes the control loop unstable even for very small feedback control gains. The simulation result plotted in Fig. 9.28 has been derived assuming a very lightly damped accelerometer.

The problem is less critical if a heavily damped accelerometer is used, in which case the useful frequency band is reduced since the increasing phase lag of the accelerometer resonance is spread out over a wider frequency band. In principle, this is not a stability problem but it affects the control action since the sensor output gives the desired feedback control signal and thus desired control

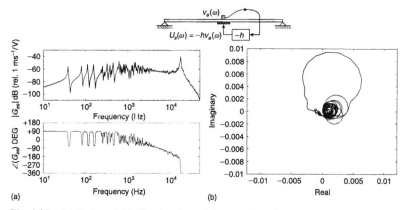

Fig. 9.29.   (a) Bode and (b) Nyquist plots of the open-loop frequency response of the system shown in Fig. 9.28 when the mass loading effects of the piezoelectric patch and accelerometer are taken into account.

effect only over a narrow low frequency band. The analysis presented above does not take into account another important effect of the accelerometer sensor and piezoelectric patch actuator, namely the mass that they add to the panel. This effect can be straightforwardly accounted for by adding to the mobility function $\widetilde{Y}_e$ in Eq. (9.47a) the mobility function for the mass of the piezoelectric patch and part of the accelerometer: $\widetilde{Y}_{PZT} = 1/j\omega(M_{PZT} + M_{case})$. The Bode plot in Fig. 9.29(a) shows that this mass effect is beneficial for the stability of the control loop. In fact, as described in Section 2.2, the mass effect tends to reduce the magnitude of $\widetilde{G}_{es}(\omega)$. Also, it produces an additional phase lag so that the big loop generated by the accelerometer sensor is rotated in a clockwise direction. The three analyses which demonstrate the actuation effect of the piezoelectric patch actuator (Fig. 9.18), the resonance effect of the accelerometer sensor (Fig. 9.28), and the mass effect of the sensor–actuator transducers (Fig. 9.29) indicate how the stability of the feedback control loop with these two transducers could be enhanced by properly selecting the resonance frequency of the accelerometer and the size and mass of the piezoelectric patch.

9.4   From Active Noise Control to Active Structural Acoustic Control and Active Vibration Control

Before entering into the details of smart panels for the control of sound radiation and transmission, a brief review of the various steps in the development of this technology is presented. It is rather important to consider the technical background behind the various control approaches that have been developed during the past two decades. In particular, the approaches to the development of active structural acoustic control (ASAC) and active vibration control (AVC) control are contrasted. These two strategies achieve sound reduction in totally different ways. They are complementary in the sense that the ASAC control is most favourable for feed-forward control of tonal disturbances while AVC is more suited for broadband random disturbances. Most of this review has been taken from a paper (Gardonio *et al.*, 2004) which includes a comprehensive list of references related to these approaches.

9.4.1   Feed-Forward Active Noise Control and Active Noise-Vibration Control

Active noise control (ANC) and active noise-vibration control (ANVC) systems have been successfully applied to the control of tonal noise in the passenger and crew compartments of a range of vehicles (Gardonio, 2002). As shown schematically in Fig. 9.30, these control systems operate with large numbers of sound error sensors and control actuators distributed within the enclosure via a MIMO adaptive feed-forward controller. In the ANC system, standard voice-coil loudspeakers are used as acoustic control sources whereas ANVC systems use inertial or strain actuators mounted on the flexible walls of the enclosure (for example trim panels) which are excited in such a way as to produce the desired acoustic control source.

In general, ANC and ANVC systems are bulky, heavy, invasive (they require lots of cables and sensor–actuator transducers) and costly (both to install and to operate) and can control only tonal disturbances for which a causal reference signal unaffected by any control input can be fed through to the controller. As a result, they have been successfully implemented in only few applications, such as the control of tonal disturbances in propeller-driven aircraft and high speed boats and engine noise in cars.

The ANC systems using adaptive feed-forward controllers have also been developed for the control of stationary random disturbances. The success of these control systems depends on three features: first, the possibility of modelling within the controller the feedback effects of the secondary sources on the reference sensors so that the control filters can be derived from the design of an optimal dummy controller; second, the possibility of reconstructing the randomly distributed disturbance which, as shown in Fig. 9.31, requires very large numbers of reference sensors; third, the possibility of measuring the primary disturbance well in advance so that the optimal controller has a causal impulse response. These three problems have made the development of ANC or ANVC systems

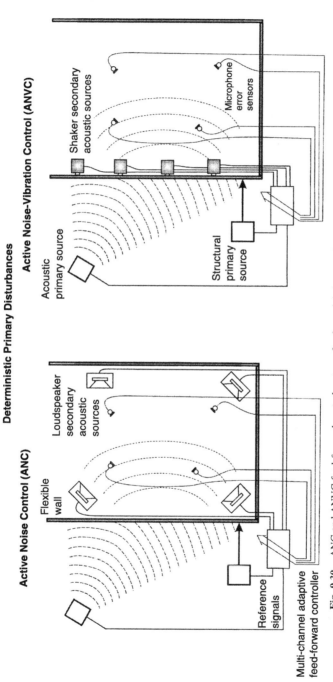

Deterministic Primary Disturbances

Active Noise Control (ANC)

Active Noise-Vibration Control (ANVC)

Flexible wall

Loudspeaker secondary acoustic sources

Shaker secondary acoustic sources

Acoustic primary source

Microphone error sensors

Structural primary source

Reference signals

Multi-channel adaptive feed-forward controller

Fig. 9.30. ANC and ANVC feed-forward control systems for deterministic acoustic and structural primary disturbances.

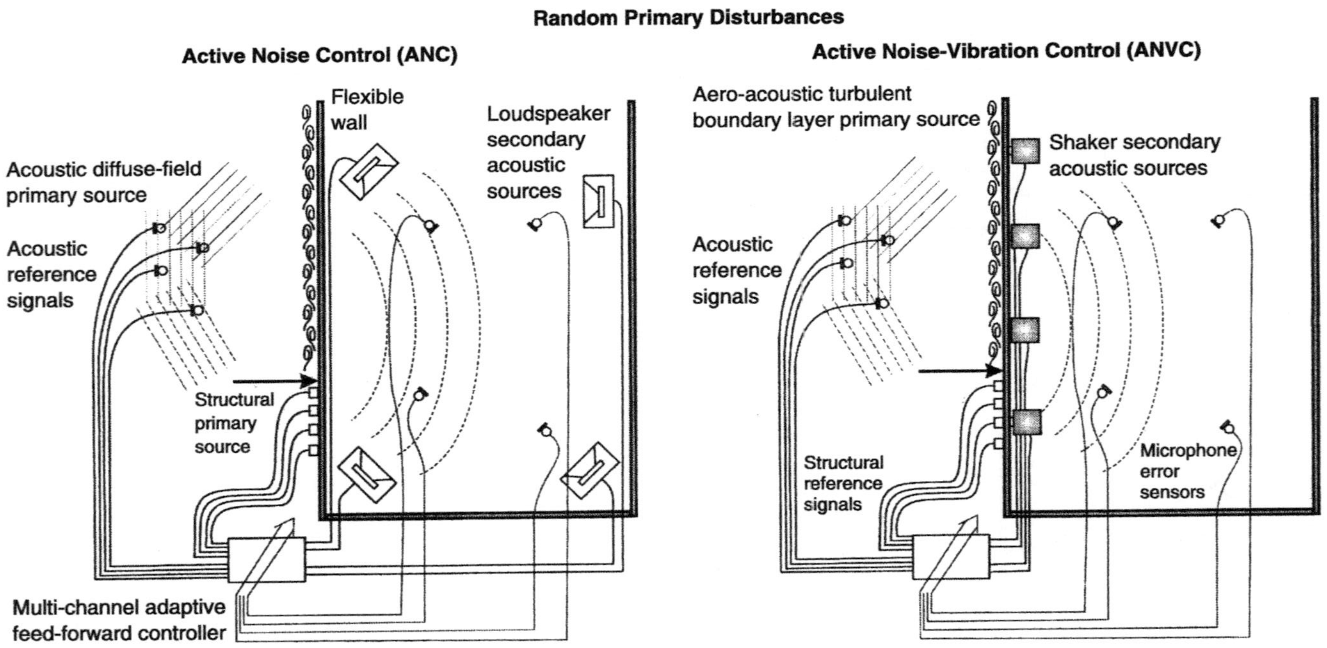

Random Primary Disturbances

Active Noise Control (ANC)

Active Noise-Vibration Control (ANVC)

Fig. 9.31. ANC and ANVC feed-forward control systems for random acoustic, aerodynamic and structural primary disturbances.

very difficult and challenging for the control of stationary random noise such as jet and boundary layer noise in aircraft and tyre and aerodynamic noise in road vehicles.

9.4.2   Feed-Forward Active Structural Acoustic Control

In active structural acoustic control (ASAC), structural actuators are integrated into the walls in such a way as to modify or reconstruct the vibration of the partitions and reduce the sound radiation or transmission. This control approach was conceived and initially developed within the same scientific community that developed ANC systems. As a result, the first ASAC systems were built using adaptive feed-forward controllers which required a set of error sensors distributed in the receiver space for the detection of the total sound power radiation as an error quantity to be minimised, and enabled only the control of tonal disturbances for which causal reference signals are available. Thus, although the actuators were integrated into the walls, these systems still had all the practical drawbacks noted for ANC and ANVC systems. Moreover, it was found that, in some frequency bands, the minimisation of the sound radiation was achieved by reconstructing the modal response of the partition so that the vibration field was slightly enhanced rather than reduced.

An important step forward for this technology was achieved by integrating error sensors within the enclosure boundaries which estimate the signal forms that can be related to the radiated sound field (Fig. 9.32). This allowed the construction of compact and light control systems with a rather small number of input–output channels that were therefore more suitable for practical applications. The principal application target was the aerospace sector where there was a clear requirement for reducing the sound transmission and radiation of fuselage walls which are made of thin and lightly damped panels. It was found that at low frequencies, where the acoustic wavelength is larger than the dimensions of the panels that make up the fuselage walls, the vibration of each panel can be considered to be the superposition of a number of frequency-dependent radiation modes introduced in Section 3.7, of which by far the most efficient closely corresponds to the net volume velocity of the panel over a relatively large frequency band (see Fig. 3.16). Much work has therefore been carried out to develop smart panels with integrated distributed strain sensors or with arrays of sensors that measure the vibration components of a panel that predominantly contribute to the far-field sound radiation: in particular, the first radiation mode and the volumetric velocity of the panel. The ASAC tends to be most effective at relatively low frequencies where the first radiation mode, or its volume velocity equivalent, produces most of the sound radiation. Thus, the output of just one error sensor can provide a good estimate of the total sound radiation by a panel.

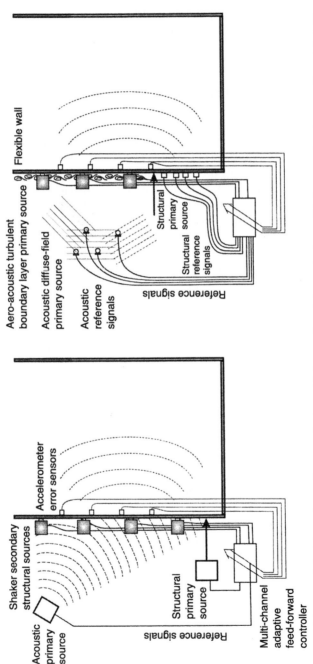

Fig. 9.32. ASAC feed-forward control systems for deterministic acoustic and structural primary disturbances and random acoustic, aerodynamic and structural primary disturbances.

Efforts have been made to devise single input strain actuators, made either with arrays of small piezoceramic patches or distributed piezoelectric films, that efficiently couple with the sensor transducer. If the sensor–actuator transducers are collocated and dual then, as discussed in Section 9.3.3, the real part of their response function is constrained to be positive real, so that a SISO adaptive feed-forward controller can be implemented with a very fast-acting controller. Technological progress in the construction of smart panels with collocated and dual sensor–actuator pairs has promoted the development of compact, light and non-invasive control systems that are very effective for the control of tonal disturbances.

Some attempts have also been made to develop adaptive feed-forward controllers for broadband random disturbances by placing detection sensors in the vicinity of the panel if not on the panel itself. In this case, the main problem is that having closely placed detection sensor and control actuator transducers prevents the possibility of having a causal controller. This type of control system has been developed with some success in double panel partitions, such as aircraft double wall constructions, by placing the detection transducers on the excited panel and the control and error transducers on the radiating panel.

9.4.3 Feedback Active Structural Acoustic Control

In parallel with the work on ASAC systems with feed-forward controllers, research has been carried out to develop ASAC systems using feedback controllers. The implementation of feedback control systems does not require a reference signal so that, when used for disturbance rejection applications, they can be used to control both tonal and random, wide-band, disturbances. Baumann *et al.* (1991) and Baumann *et al.* (1992) first proposed a methodology to design a feedback control system for the reduction of broadband sound radiation by a panel which uses structural error sensor and control actuator transducers acting on the panel.

Although the work carried out to develop distributed strain sensors and actuators for smart panels was initially aimed at the implementation of feed-forward control systems, it was soon recognised that these type of transducers would also have great advantages in the design and implementation of feedback control systems. Indeed, the possibility of devising strain sensors which measure the radiation modes of panels has provided the possibility of implementing output-feedback control systems where the state variables to be minimised can be directly derived from the measured radiation mode outputs without the need for a state estimator. In particular, at low frequency, sound radiation is controlled by the first radiation mode, which can be approximated by the volume velocity vibration of the panel component, so that a SISO feedback control system can be implemented.

Classic feedback control theory has been used to design the sensor–actuator transducers for the implementation of simple SISO feedback control systems (Fig. 9.33). As discussed in Section 9.2.2 for the specific case of disturbance rejection feedback control, the most suitable strategy is active damping, which reduces the response of the structure at resonance frequencies and, as a result, the steady-state response to wide-band disturbances. Active damping does not produce control effects at off-resonance frequency bands but, on the contrary, tends to fill in the response spectrum at antiresonance frequencies. Therefore, active damping is not particularly suitable for the control of tonal disturbances except in the particular case where the excitation frequency corresponds to a resonance frequency of the smart panel which does not vary as the physical or operational conditions change (for example, variation of temperature or change of static loading).

The simplest and most robust way of achieving active damping is by implementing direct velocity feedback. In this case the output of the velocity sensor is directly fed back to the actuator via a fixed control gain. This control scheme is constrained to be stable for any value of the control gain if the sensor–actuator frequency response is strictly positive real which, as discussed in Section 9.2.2, happens when the sensor–actuator transducers are collocated and dual. Most of the research and development work is therefore focused on the design of collocated and dual sensor–actuator pairs so that a relatively simple feedback controller which is unconditionally stable can be implemented. If the sensor–actuator transducers are not collocated and dual, then only a limited range of control gains can be implemented in order to guarantee stability. This results in a reduced performance capability of the system. In this case, the control performance can be enhanced by shaping the control function with standard compensator circuits (Preumont, 2002).

9.4.4 Decentralised Feedback Active Vibration Control

The problems encountered in the development of smart panels with SISO feedback ASAC control using large-area distributed transducers have motivated the development of alternative control solutions. Active damping feedback control tends to be effective only in the vicinity of resonance frequencies, where the sound radiation is controlled by the self-radiation of the resonant mode, while it tends to enhance the vibration at antiresonance frequencies, where the sound radiation is controlled by the self- and mutual-radiation effects of a set of modes. Therefore, in order to implement active damping for the control of sound radiation, it is sufficient to have sensor–actuator transducers that are well coupled to

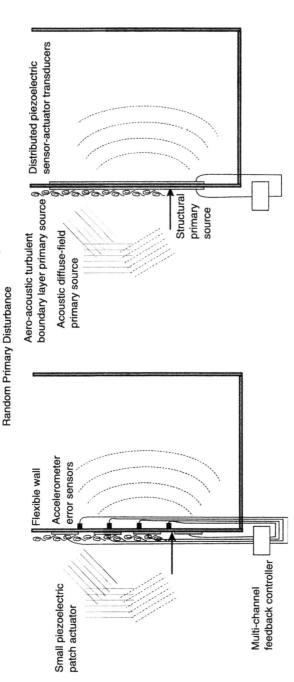

Fig. 9.33. ASAC feedback control systems for random acoustic, aerodynamic and structural primary disturbances.

the low order structural modes, which determined the sound radiation of the low order resonances. Moving the sensor–actuator design problem from the control of radiation modes to that of structural modes is not a real improvement since distributed sensor and actuator transducers that cover the entire surface of the panel are still required. However, Elliott *et al.* (2002) have shown that active damping of structural modes can be achieved with arrays of small localised control units that implement SISO velocity feedback control. They initially considered the ideal case of having a collocated and dual point-velocity and point-force sensor–actuator system which is positioned away from the nodal lines of the modes to controlled. These two transducers enabled the implementation of unconditionally stable direct velocity feedback control. It was noticed that, as the control gain was gradually increased from zero, the active damping effect of the structural modes increased and consequently both the total kinetic energy and total sound radiation of the panel in a certain frequency band decreased. However, it was also found that this behaviour was only sustained up to an optimal feedback control gain, above which the damping effect faded away and so the kinetic energy or sound radiation by the panel increased again and could even exceed the uncontrolled value. The analysis of this phenomenon showed that the velocity feedback control unit worked as a sky-hook damper which absorbed energy from the structural modes. However, for large control gains the action of the feedback controller was to pin the panel at the error sensor position, so that the vibration of the panel was that of a lightly damped structure with extra pinning points. Therefore a set of new lightly damped structural modes were created which could be excited at new resonance frequencies and radiated sound even more effectively than the original modes [see Gardonio and Elliott (2005b)].

As shown in Fig. 9.34, and recalling the discussion presented in Sections 9.3.1 and 9.3.2, practical systems can be built using grids of either electrodynamic inertial actuators or piezoelectric patch actuators and closely located accelerometer inertial sensors. Although these arrangements are not properly collocated and dual, a proper design of the two types of decentralised control units enables relatively large feedback control gains so that the equivalent of a grid of sky-hook dampers can be implemented, provided the control gains do not exceed the optimal values. The large number of control units which are evenly scattered over the panel surface allows the damping of a number of low frequency resonant modes and consequently facilitates low frequency, wide-band, sound radiation control. This type of vibration and sound radiation control effect was first observed by Petitjean and Legrain (1996) in an experiment where a grid of 5×3 collocated piezoceramic patches mounted on either side of a panel was used either to implement a fully coupled MIMO LQG feedback controller or fifteen SISO rate feedback controllers. For this particular type of panel they found that the vibration reductions in the two control cases were very similar. This type of experiment was then repeated on a different smart panel and it was again

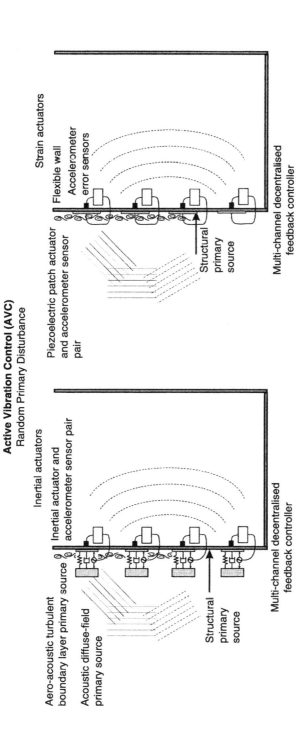

Fig. 9.34. AVC feedback control systems for random acoustic, aerodynamic and structural primary disturbances.

found that, despite its simplicity, decentralised feedback control was giving similar control effects to those of a MIMO optimal feedback controller. The apparent contradiction arises from the fact that radiation mode theory is important at off-resonance frequencies where the mutual radiation effects cannot be neglected. If broadband steady-state sound radiation is controlled by the resonances of the structure, the structural control effect produced by an array of active dampers is comparable to that produced by an optimal sensor–actuator transducers pair designed to control the most effective radiation mode of the structure. It is important to emphasise that in case of tonal disturbances it is instead necessary to use sensor–actuator transducers that act on radiation modes so that control is achieved independently of the excitation frequency with a feed-forward system.

9.5   Smart Panels for ASAC and AVC Systems

The Section 9.3 has introduced the most important control features of inertial and strain actuators and sensors. In this section, the reduction of noise transmission and radiation is considered for a set of smart panels that exemplifies the most interesting and most promising combinations of sensor–actuator pairs and control architectures. We shall consider a baffled flat rectangular panel with strain or inertial sensor–actuator transducers mounted on both sides of the panel. Both SISO or MIMO arrangements are considered with reference to feed-forward and feedback control architectures.

9.5.1   Models of Smart Panels

The flexural response and sound radiation of a smart panel which, as shown in Fig. 9.35, is excited by an acoustic plane wave, will be used as a model problem to illustrate the main features of a set of control arrangements for the ASAC and AVC approaches. The panel is assumed to be baffled and simply supported along the perimeter with dimensions $l_x \times l_y = 414$ mm $\times 314$ mm and thickness $h_p = 1$ mm. It is made of aluminium whose density, Young's modulus and Poisson's ratio are $\rho_p = 2720 \, \text{kgm}^{-3}$, $E_p = 7 \times 10^{10} \, \text{Nm}^{-2}$ and $v_p = 0.33$, respectively. The loss factor η has been assumed to be 0.02 for all modes.

Although the total sound power radiated by the panel usefully quantifies the far-field pressure that it generates, high vibration levels in weakly radiating modes can give rise to significant pressure levels in the near field of the panel. The kinetic energy of a panel provides a better measure of near-field pressure than radiated sound power, and so if there is any possibility that some listeners may be in

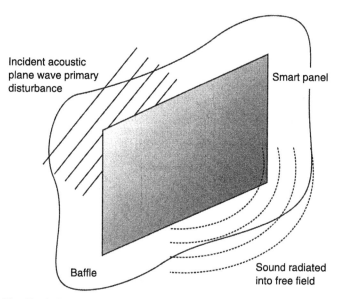

Fig. 9.35. Physical arrangement considered in the simulation study in which a simply supported panel is excited by an incident acoustic plane wave and radiates into an anechoic half-space.

close proximity to the panel, and others further away, both of these quantities are important for active structural acoustic control.

The response of the panel has been derived by assuming the incident acoustic plane wave to be harmonic with azimuthal and elevation angles $\phi = 45°$ and $\theta = 45°$ relative to the normal so that all the structural modes of the panel are excited. The incident sound pressure field $p_i(x, y, t)$ in the plane of the plate is therefore given by

$$p_i(x, y, t) = \mathrm{Re}\left\{ p_i(\omega)e^{j(\omega t - k_x x - k_y y)} \right\} \tag{9.51}$$

where $p_i(\omega)$ is the complex amplitude of the incident wave which has a wavenumber vector component in x direction given by $k_x = k\sin(\theta)\cos(\phi)$ and a wavenumber vector component in y direction given by $k_y = k\sin(\theta)\sin(\phi)$, where $k = \omega/c$ and $c = 343\,\mathrm{ms}^{-1}$.

As introduced in Section 3.7.4, both the time-averaged kinetic energy and time-averaged total sound power radiated have been derived by dividing the panel into a grid of rectangular tiles whose dimensions have been taken to be $l_{xe} = l_x/(4P)$ and $l_{ye} = l_y/(4Q)$, where P and Q are the higher modal orders used in the calculations. The complex amplitudes of the transverse velocities $\tilde{v}_{tj}(\omega)$

at the centres of these elements have been grouped into the column vector $\tilde{\mathbf{v}}_t(\omega) \equiv \{\tilde{v}_{t1} \quad \tilde{v}_{t1} \quad \ldots \quad \tilde{v}_{tT}\}$. The flexural vibration of the smart panel is synthesised by the superposition of the response to primary acoustic excitation generated by the incident plane wave and the secondary structural excitation(s) generated by the control transducer(s). Thus, assuming that the system is linear, and that the radiated pressure has negligible effect on the panel response, which is a reasonable assumption in air for this thickness and damping of panel, the vector $\mathbf{v}_t(\omega)$ of complex amplitudes of velocities at the centres of the elements is given by

$$\tilde{\mathbf{v}}_t(\omega) = \tilde{\mathbf{Y}}_{tp}(\omega)\tilde{\mathbf{f}}_p(\omega) + \tilde{\mathbf{Y}}_{ts}(\omega)\tilde{\mathbf{u}}_s(\omega) \tag{9.52}$$

The vector $\tilde{\mathbf{f}}_p(\omega) = \{\tilde{F}_{t1} \quad \tilde{F}_{t2} \quad \ldots \quad \tilde{F}_{tT}\}$ represents the resultant forces generated on each element by the primary acoustic wave which have been assumed to be the blocked sound pressure $2p_i(\omega)$ at the centre of the element multiplied by the area of the element and $\tilde{\mathbf{u}}_s(\omega) = \{\tilde{U}_{s1} \quad \tilde{U}_{s2} \quad \ldots \quad \tilde{U}_{sS}\}$ is the vector of complex input voltage signals $\tilde{U}_{sj}(\omega)$ to the S control transducers. As shown in Section 2.4.2, the elements in the two matrices of Eq. (9.52) are given by

$$\tilde{Y}_{ij} = j\omega \sum_{p=1}^{P} \sum_{q=1}^{Q} \frac{\phi_{pq}(x_i, y_i)\phi_{pq}(x_j, y_j)}{M_{pq}\left[\omega_{pq}^2(1 + j\eta) - \omega^2\right]} \tag{9.53}$$

According to formulation presented in Section 3.7.4, the time-averaged kinetic energy of the panel is given by

$$\overline{E}(\omega) = \frac{M_e}{4}\tilde{\mathbf{v}}_t^H(\omega)\tilde{\mathbf{v}}_t(\omega) \tag{9.54}$$

where M_e is the mass of each element and H denotes the Hermitian. Also, using the formulation presented in Section 3.6.2, the time-average total sound power radiated by the baffled panel is given by

$$\overline{P}_r(\omega) = \tilde{\mathbf{v}}_t^H(\omega)\mathbf{R}(\omega)\tilde{\mathbf{v}}_t(\omega) \tag{9.55}$$

where the radiation matrix $\mathbf{R}(\omega)$ is given in Eq. (3.57). The ratio of the time-averaged total sound power radiated to the time-averaged incident sound power, which is termed the sound power transmission coefficient

$$\tau(\omega) = \overline{P}_r(\omega)/\overline{P}_i(\omega) \tag{9.56}$$

is used as a measure of performance. The time-average incident sound power is given by:

$$\overline{P}_i(\omega) = |\tilde{p}_i(\omega)|^2 \, l_x l_y \cos(\theta)/2\rho_0 c \tag{9.57}$$

The complex amplitudes of the error signals, $\tilde{U}_e(\omega)$, from the E control sensors, can also be grouped in a vector $\tilde{\mathbf{u}}_e(\omega) = \{\tilde{U}_{e1} \quad \tilde{U}_{e2} \quad \ldots \quad \tilde{U}_{eF}\}$ so that

$$\tilde{\mathbf{u}}_e(\omega) = \tilde{\mathbf{Y}}_{ep}(\omega)\tilde{\mathbf{f}}_p(\omega) + \tilde{\mathbf{Y}}_{es}(\omega)\tilde{\mathbf{u}}_s(\omega) \tag{9.58}$$

where the elements of the two matrices $\tilde{\mathbf{Y}}_{ep}(\omega)$ and $\tilde{\mathbf{Y}}_{es}(\omega)$ are respectively given by

$$\tilde{Y}_{ij} = j\omega \sum_{p=1}^{P} \sum_{q=1}^{Q} \frac{\sigma_{pq} \psi_{pq}(x_j, y_j)}{M_{pq}\left[\omega_{pq}^2 (1 + j\eta) - \omega^2\right]} \tag{9.59a}$$

and

$$\tilde{Y}_{ij} = j\omega \sum_{p=1}^{P} \sum_{q=1}^{Q} \frac{\sigma_{pq} F_{pq}}{M_{pq}\left[\omega_{pq}^2 (1 + j\eta) - \omega^2\right]} \tag{9.59b}$$

where σ_{pq} and F_{pq} are respectively the modal sensing and excitation terms generated by the sensor(s) and actuator(s) on the panel which are derived for each type of transducer according to the formulations presented in Section 9.3.

Considering the MIMO block diagram for feed-forward control architectures shown in Fig. 9.4, and assuming that the number of sensor and actuator transducers is equal, then the vector of the optimal control sources that cancels the error sensors is given by

$$\tilde{\mathbf{u}}_s(\omega) = -\tilde{\mathbf{Y}}_{es}^{-1}(\omega)\tilde{\mathbf{Y}}_{ep}(\omega)\tilde{\mathbf{f}}_p(\omega) \tag{9.60}$$

If the control system is set to minimise the time-average total sound power radiated by the baffled panel, then, substitution of Eq. (9.52) in Eq. (9.55) gives a quadratic matrix expression whose minimum is given by the following vector of optimal control sources:

$$\tilde{\mathbf{u}}_s(\omega) = -\left(\tilde{\mathbf{Y}}_{ts}^{H}(\omega)\mathbf{R}(\omega)\tilde{\mathbf{Y}}_{ts}(\omega)\right)^{-1} \tilde{\mathbf{Y}}_{ts}^{H}(\omega)\mathbf{R}(\omega)\tilde{\mathbf{Y}}_{tp}(\omega)\tilde{\mathbf{f}}_p(\omega) \tag{9.61}$$

Thus, for these two control cases the vector of complex amplitudes of the elemental velocities is given by

$$\tilde{\mathbf{v}}_t(\omega) = \left[\tilde{\mathbf{Y}}_{tp}(\omega) - \tilde{\mathbf{Y}}_{ts}(\omega)\tilde{\mathbf{Y}}_{es}^{-1}(\omega)\tilde{\mathbf{Y}}_{ep}(\omega)\right]\tilde{\mathbf{f}}_p(\omega) \qquad (9.62a)$$

or

$$\tilde{\mathbf{v}}_t(\omega) = \left[\tilde{\mathbf{Y}}_{tp}(\omega) - \tilde{\mathbf{Y}}_{ts}(\omega)\left(\tilde{\mathbf{Y}}_{ts}^H(\omega)\mathbf{R}(\omega)\tilde{\mathbf{Y}}_{ts}(\omega)\right)^{-1}\tilde{\mathbf{Y}}_{ts}^H(\omega)\mathbf{R}(\omega)\tilde{\mathbf{Y}}_{tp}(\omega)\right]\tilde{\mathbf{f}}_p(\omega)$$
$$(9.62b)$$

Considering now the MIMO block diagram for feedback control architectures shown in Fig. 9.10, the vector of the complex control sources is given by

$$\tilde{\mathbf{u}}_s(\omega) = -\mathbf{H}(\omega)\tilde{\mathbf{u}}_e(\omega) \qquad (9.63)$$

where, assuming an equal number of sensors and actuators, $\mathbf{H}(\omega)$ is a square matrix of feedback control filters. Provided the control system is stable, the vector with the elemental velocities in terms of the primary excitation vector $\tilde{\mathbf{f}}_p(\omega)$ and matrix $\mathbf{H}(\omega)$ of control functions is then obtained by substituting into Eq. (9.52) the vector with the control signals derived by substituting Eq. (9.63) into Eq. (9.58):

$$\tilde{\mathbf{v}}_t(\omega) = \left[\tilde{\mathbf{Y}}_{tp}(\omega) + \tilde{\mathbf{Y}}_{ts}(\omega)\mathbf{H}(\omega)\left(\mathbf{I} + \tilde{\mathbf{Y}}_{es}(\omega)\mathbf{H}(\omega)\right)^{-1}\tilde{\mathbf{Y}}_{ep}(\omega)\right]\tilde{\mathbf{f}}_p(\omega) \quad (9.64)$$

For a SISO, the vectors and matrices reduce to scalars, but in general $\tilde{\mathbf{Y}}_{es}(\omega)$ is a fully populated matrix of point and transfer responses between the actuators and sensors on the panel and $\mathbf{H}(\omega)$ is either fully populated for MIMO control or is a diagonal matrix for decentralised local control. When the SISO or MIMO feed-forward or feedback control systems are implemented, the kinetic energy and sound power transmission coefficient can still be derived from Eqs. (9.54) to (9.57) with the vector of velocities at the centre of the elements given by Eqs. (9.62a,b) or Eq. (9.64).

9.5.2   Smart Panels with Feed-Forward MIMO and SISO Control Systems

The principal features of smart panels with feed-forward control systems are illustrated in this section by considering four types of smart panels which are schematically illustrated in Fig. 9.36. Two actuation arrangements

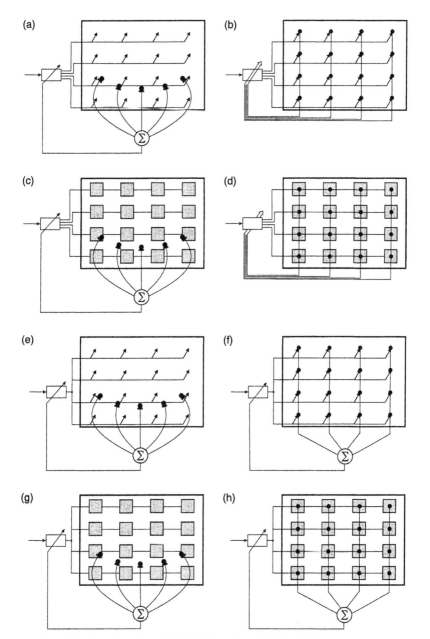

Fig. 9.36. Smart panels with MIMO (a,b,c,d) and SISO (e,f,g,h) control systems.

are studied: first, 4×4 grids of idealised sky-hook force actuators (panels a,b,e,f); second, a 4×4 grids of piezoelectric patch actuators (panels c,d,g,h). Their control effectiveness is assessed by considering the ideal case where the time-averaged total sound power radiation by the panel is minimised (panels a,c,e,g) and considering the most practical case where the vibration of the panel is measured with a 4×4 grid of ideal velocity sensors (panels b,d,f,h). Both MIMO (panels a,b,c,d) and SISO (panels e,f,g,h) approaches are considered where, in the latter case, the outputs from the 4×4 array of velocity sensors are summed and the same control signal is fed to the 4×4 array of control actuators. All analysis considered in this and the following sections are presented in terms of the kinetic energy and total sound power transmission coefficient (in logarithmic form) of the panel when it is excited by an oblique plane wave of unit pressure amplitude.

The solid line in Figs. 9.37(a,c,e,g) of the kinetic energy exhibits the typical response of a lightly damped panel which is characterised by a series of resonances whose amplitudes gradually decrease as the frequency rises. The peaks of the first few resonances are relatively high and sharp because of the low damping effects at lower frequencies associated with the assumption of uniform modal loss factor. The solid line in Figs. 9.37(b,d,f,h) of the sound power transmission coefficient shows a similar behaviour, although there are fewer peaks because of the relatively low sound radiation efficiency of even–even or even–odd mode shapes of the panel. Above about 300 Hz, mass controlled behaviour is observed.

Let us consider first the control effects of the MIMO control systems which are shown in Figs. 9.36(a–d). When these control systems are driven to minimise the total sound radiation, as shown by the dashed lines in Figs. 9.37(b,d), large reductions of the total sound power transmission coefficient itself are generated with both force and piezoelectric actuators. However, as shown by the dashed lines in Figs. 9.37(a,c), this is obtained at the expense of a significant increment of the kinetic energy of the panel, and thus its near-field sound radiation, in some frequency bands. This is due to the fact that control actuators are driven to reconstruct the modal response in such a way as to reduce the vibration component of the modes with large self-radiation efficiencies (low order odd–odd modes as shown in Fig. 3.12) and to enhance the vibrations of those modes with small self-radiation efficiencies (as shown in Fig. 3.12, even modes) but contribute to the negative mutual radiation efficiency terms (see Fig. 3.13) (Fuller et al., 1996).

The dotted lines in Figs. 9.37(a–d) show poor control effects that are obtained when the MIMO control systems are driven to cancel the vibration at the 4×4 grid of velocity sensors. As discussed in Section 9.2.1, this is due to the fact that the control system produces a pinning effect at the error sensor positions so that the modal response of the panel is changed to that of a simply supported panel with the grid of 16 constraints. These new natural modes are characterised by higher natural frequencies and, more importantly, by higher radiation efficiencies.

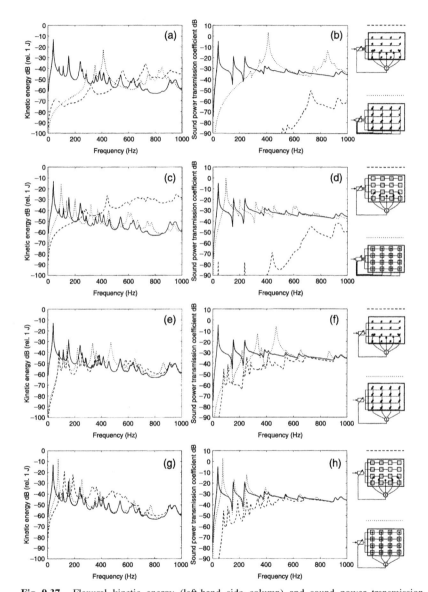

Fig. 9.37. Flexural kinetic energy (left-hand side column) and sound power transmission coefficient (right-hand side column) of smart panels excited by a plane sound wave with no control (solid line) and with the feed-forward control systems shown in Fig. 9.36 which are either set to minimise the total radiated sound power (dashed lines) or panel velocities (dotted lines).

In conclusion, the minimisation of the velocities at the error sensor positions produces a stiffening effect which raises the resonance frequencies so that both the kinetic energy and total radiated sound power, i.e., near-field and far-field sound radiation, are reduced at low frequency below the first new resonance frequency only. At higher frequency there is no clear reduction of kinetic energy and sound power transmission coefficient; in effect there is just a different distribution of the resonances which may lead to some reduction of the vibration and sound radiation in some narrow bands of frequency.

We consider now the SISO control systems shown in Figs. 9.36(e–h) where the 16 sky-hook force actuators or 16 piezoelectric patch actuators are driven by a single control signal obtained from a feed-forward control system set to minimise either the total sound power radiation or to cancel the sum of the 16 velocity signals. If we consider first the case where the total radiated sound power is minimised then, as shown by the dashed lines in Figs. 9.37(f,h), compared with MIMO control systems, smaller reductions of the far-field sound radiation are obtained up to about 500 Hz. However, the dashed lines in Figs. 9.37(e,g), show that the vibration of the panel is not modified as much as for the MIMO control cases, particularly when the point force actuators are used. This is because, by driving all the actuators in phase, there is no possibility of exciting the odd modes, which produce the beneficial effect of enhancing the negative mutual radiation efficiencies, and thus reduces the sound radiation, but at the cost of increasing the vibration of the panel itself.

When the SISO control systems are driven to cancel the error signal from the 16 velocity sensors then, as shown by the dotted lines in Figs. 9.37(e–h), control is very weak. As seen above, this is due to the pinning effect at the error sensor positions which generates new natural modes which have higher natural frequencies and higher radiation efficiencies. Thus, apart from a reduction of the kinetic energy and total radiated sound power below the first resonance of the actively constrained panel, there is no effective control at higher frequencies.

9.5.3   Smart Panel with Feed-Forward SISO Control Systems Using a Volume Velocity Distributed Sensor and Uniform Force Distributed Actuator

In Section 3.7, we have seen that the vibration of thin structures, in this case of the smart panel, can be formulated in terms of radiation modes which independently radiate sound power. Therefore, it is likely that, using a feed-forward control system with sensor–actuator pairs that independently measure and excite these vibration components substantial reductions of the sound radiation can be obtained without the vibration-spillover effect revealed in the previous section (Elliott and Johnson, 1993; Johnson and Elliott, 1995).

For example, let us consider a SISO feed-forward system that controls the first radiation mode, which, according to Fig. 3.15, is the most efficient radiator at low frequencies below the critical frequency. According to Fig. 3.16, at low frequencies, this radiation mode is defined by the volumetric velocity of the panel. We have seen in Sections 9.3.1 and 9.3.3 that a distributed piezoelectric transducer with quadratically shaped electrode strips can be used as a uniform force actuator or volume velocity sensor. Thus, as schematically shown in Figs. 9.38(a,b), we now consider the control effects generated by a feed-forward control system, which is (a) set to minimise the total sound power radiated using a uniform force actuator and (b) set to cancel the volume velocity with a matched uniform force actuator.

The dashed line in Fig. 9.39b shows that when a uniform force actuator is driven to minimise the total sound power radiated, then large reductions of the total sound power radiation coefficient are generated at low frequency up to about 600 Hz, which gradually decrease as the frequency rises. The dashed line

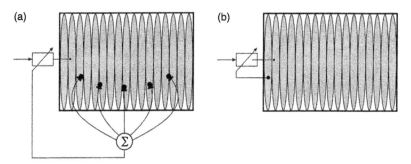

Fig. 9.38. Smart panels with SISO feed-forward control systems using a uniform force actuator for: (a) the minimisation of the total radiated sound power measured with an array of microphones (type *i*); (b) the cancellation of the first radiation mode using a volume velocity sensor (type *l*).

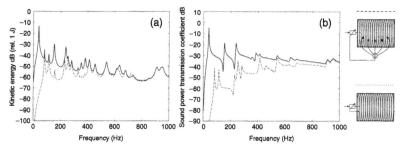

Fig. 9.39. Flexural kinetic energy and sound power transmission coefficient of the panel excited by a plane acoustic wave with no control (solid line) and with the feed-forward control systems type *i* (dashed line) and type *l* (dotted line) shown in Fig. 9.38. Note: latter two lines coincide.

in Fig. 9.39a shows that this control system also reduces the kinetic energy of the panel at low frequencies up to about 400 Hz with little spillover at higher frequencies. Spillover is due to the non-perfect uniform force excitation generated by the transducer with quadratically shaped electrodes and also to the fact that, as the frequency raises, the shape of the first radiation mode changes toward a smoothened volumetric distribution and thus the uniform actuator becomes increasingly mismatched with the first vibration component that radiates sound independently.

The dotted lines in Figs. 9.39(a,b) show the vibration and sound radiation control effects produced by the SISO control system with the matched volume velocity sensor and uniform force actuator. These two lines overlap quite closely the equivalent dashed lines for the optimal control of the total sound power radiation. Therefore this sensor–actuator pair constitutes an ideal solution to construct light, compact and non-invasive smart panels for the control of low frequency near-field and far-field sound radiation and transmission. The advantage of this solution is given by the simplicity of the SISO controller which uses a control actuator and error sensor directly bonded on the panel. The control effects shown in Fig. 9.39, can be further improved if multiple sensor–actuator pairs are embedded on a smart panel in order to control the most efficient radiation modes. Since these transducer pairs are matched to vibration components that radiate sound independently, there is no need to implement fully coupled MIMO feed-forward control systems. On the contrary it is sufficient to implement independent, decentralised, feed-forward SISO control systems for each sensor–actuator pair.

9.5.4 Smart Panels with Feedback MIMO and SISO Control Systems

The close location of sensor and actuator transducers in smart panels offers the possibility that these control systems may also be successfully used for the implementation of feedback control architecture, which, as discussed in Section 9.2.2, does not require reference sensors and thus is particularly suited for the control of random disturbances. In particular, as shown in Figs. 9.40(a,b), decentralised MIMO control systems can be built with either a 4 × 4 grid of idealised sky-hook force actuators and collocated velocity sensors or a 4 × 4 grid of piezoelectric patch actuators with velocity sensors at their centres. Both control systems can be arranged to implement independent direct velocity feedback loops that act as an array of active dampers. Also, as shown in Figs. 9.40(c,d), the outputs from the sixteen velocity sensors can be summed in order to implement a SISO velocity feedback loop which drives the sixteen force or piezoelectric actuators with the same signal (Gardonio and Elliott, 2005b).

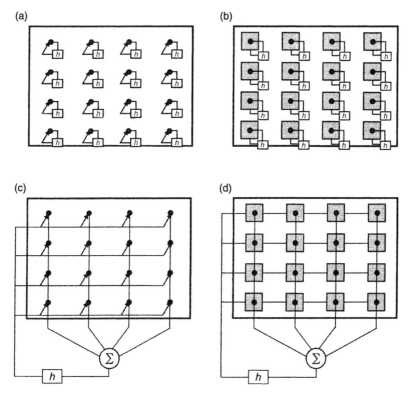

Fig. 9.40. Smart panels with decentralised-MIMO (types m,n) and SISO (types o,p) feedback control systems using 4×4 grids of idealised inertial force actuators and collocated velocity sensors (types m,o) and 4×4 grids of piezoelectric patch actuators and closely located sensors (types n,p).

Figures 9.41(a,c) and Figs. 9.41(b,d) show, respectively, the kinetic energy and the sound power transmission coefficient of the smart panel with MIMO decentralised feedback control systems using the 4×4 array of point force actuators or square piezoelectric patch actuators with at their centres velocity sensors. The dashed and faint lines in the kinetic energy and sound power transmission coefficient plots for the two control arrangements show that when the gains of the sixteen control units are raised, the resonance peaks are lowered. This is due to the active damping effect generated by the sixteen direct velocity feedback control loops that indeed increase the overall damping of the lower frequency resonant modes of the smart panel (Elliott *et al.*, 2002). However, when higher control gains are implemented this trend is reversed and, as shown by the dotted lines in the two sets of plots, the kinetic energy and the sound power transmission coefficient are once more characterised by a new set of low frequency resonances

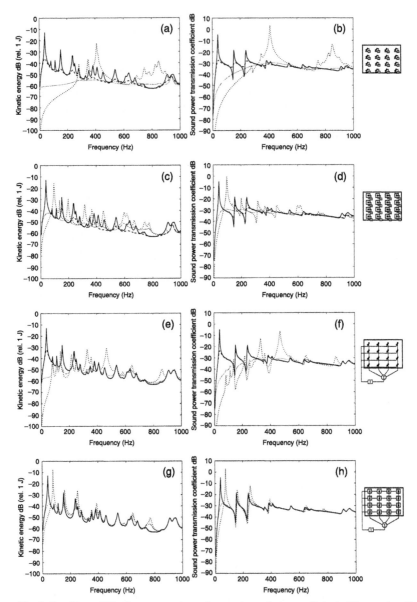

Fig. 9.41. Flexural kinetic energy and sound power transmission coefficient of the panel excited by a plane acoustic wave with no control (solid line) and with the velocity feedback control systems type *m,n,o,p* in Fig. 9.40 when increasing levels of control gains (respectively, dotted, solid–faint and dashed lines) are implemented.

whose amplitudes are similar, if not higher, than those of the panel with no control. Again this is due to the pinning effect at the control positions seen for the feed-forward control systems, so that the new modes with higher natural frequencies are generated. The two sets of plots show that the new natural modes generated by the force actuators resonate at much higher frequencies than those generated by the piezoelectric patch actuators (Gardonio and Elliott, 2005b).

This type of behaviour is illustrated by the solid and faint lines in the two plots of Fig. 9.42 which show how the normalised[7] kinetic energy and normalised[8] sound power transmission coefficient, integrated from 0 to 1 kHz, vary with the feedback gain. Both plots indicate that, when the control gains are raised from zero, the frequency-average vibrational response and sound radiation of the smart panel decrease producing reductions of the normalised kinetic energy and normalised sound transmission coefficient respectively of 22 and 7 dB for the panel with force actuators and 14 and 5 dB for the panel with piezo actuators. The faint lines in the plots of Fig. 9.41 show the kinetic energy and sound power transmission coefficient when the optimal control gains are implemented in the control systems.

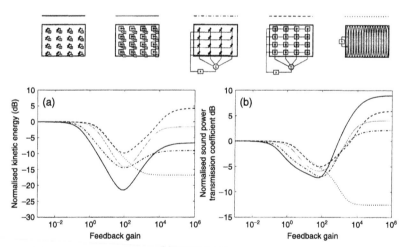

Fig. 9.42. (a) Normalised kinetic energy and (b) normalised sound power transmission coefficient, integrated from 0 to 1 kHz, plotted against the control gain for the control systems of type m (solid lines), n (faint lines), o (dash–dotted lines) and p (dashed line) in Fig. 9.40 and the control system q in Fig. 9.42 (dotted line).

---

[7] Normalised to the total kinetic energy when there is no control.

[8] Normalised to the sound power transmission coefficient when there is no control.

If the control gains are further increased then, because of the pinning effect described above, the response of the smart panel is brought back to the levels with no control while its sound radiation is even increased by about 8 and 4 dB compared with no control. This is due to the fact that the new resonant modes of the smart panel generated by the pinning effects at the 16 control positions have relatively higher sound radiation efficiency than the lower order modes of the unconstrained simply supported panel (Gardonio and Elliott, 2005b).

Figures 9.41(e,g) and Figs. 9.41(f,h) show respectively the kinetic energy and the sound power transmission coefficient of the smart panel with the 4×4 array of point force actuators or square piezoelectric patch actuators with at the centres velocity sensors SISO feedback control systems. As found with the MIMO control arrangement, the dashed and faint lines in the two sets of plots show that as the SISO control gains are raised up to an optimal control gain as the active damping generated by the feedback loops level down most of the resonance peaks. However, in this case, the two SISO control systems do not damp down the resonances of the natural modes of the panel which have no volumetric vibration component, since the sum of the 16 sensors outputs goes to zero and thus the SISO control loop becomes ineffective. For example, no damping is applied to the second, third and fourth resonances which are related to the (2,1), (1,2) and (2,2) natural modes of the panel. Beyond the optimal control gain, this trend is inverted and, as shown by the dotted lines, a new set of low frequency resonances emerge whose amplitudes are similar, if not higher, than those of the panel with no control. Also in this case, this phenomenon results from the control spillover effect where large control gains tend to pin the smart panel at the control positions so that the response of the smart panel is rearranged into that of a lightly damped panel which is pinned at the 16 control positions.

Figure 9.42 indicates that, as the control gain is raised, the normalised kinetic energy and normalised sound power transmission coefficient, integrated from 0 to 1 kHz, decrease and reach maximum reductions respectively of 15 and 7 dB for the panel with force actuators and 8 and 4 dB for the panel with piezoelectric actuators. Therefore, the SISO control arrangement is not able to match the 22 or 14 dB reductions of the kinetic energy produced by the equivalent MIMO control systems. In contrast they are capable of replicating the reductions of the sound power transmission coefficient produced by the equivalent MIMO control systems. This is due to the fact that the error signals used in the SISO feedback control loops are proportional to the volumetric vibration of the smart panel which generates most of the sound radiation at low frequency. When relatively high control gains are implemented, the pinning effect generated at the control positions brings the response of the smart panel back to the levels with no control while its sound radiation is increased relative to the case of no control. This is because of the increased sound radiation efficiency of the new resonant modes compared with that of the lower order modes of the unconstrained simply supported panel.

In summary, the decentralised MIMO control arrangements give better results than the equivalent SISO configurations. Also, the system with point force actuators produces better results than the equivalent systems with piezoelectric patch actuators.

9.5.5 Smart Panel with a Feedback SISO Control System Using a Volume Velocity Sensor and Uniform Force Actuator

As seen for the feed-forward control systems, an interesting application to consider is a SISO feedback control loop using a volume velocity sensor and uniform force actuator pair which, as shown in Fig. 9.43, is made with two distributed piezoelectric transducers with quadratically shaped electrode strips bonded on either side of the panel (Johnson and Elliott, 1995; Gardonio and Elliott, 2004).

The two plots in Fig. 9.44 show that the greater is the feedback gain the greater is the reduction in the kinetic energy and sound power transmission coefficient. Thus, there is a consistent reduction of both the near-field and far-field sound radiation by the panel. For high feedback gains, the results are similar to those predicted for feed-forward control. The dotted lines in Fig. 9.42 show that over this frequency range the normalised kinetic energy and sound power transmission coefficient, fall with feedback gain, giving a maximum reduction respectively of 17 and 12.5 dB for large control gains.

This control arrangement offers an interesting solution which is simple and compact. However, as discussed in Section 9.3.3, there is a major problem in terms of stability due to the coupling between the two strain transducers via in-plane longitudinal and shear vibrations.

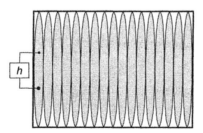

Fig. 9.43. Smart panel with a SISO velocity feedback control system using a matched uniform force actuator and volume velocity pair for the control of the first radiation mode at low frequencies.

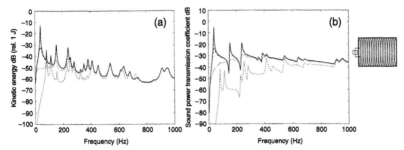

Fig. 9.44. Flexural kinetic energy and sound power transmission coefficient of the panel excited by a plane acoustic wave with no control (solid line) and with the velocity feedback control system shown in Fig. 9.42 when increasing levels of control gains (respectively, dotted, solid–faint and dashed lines) are implemented.

Problems

9.1 A feed-forward control system is to be implemented on a beam to control harmonic flexural wave propagation. Assuming the distance between the reference sensor and control actuator is L, derive the expression for the minimum speed of the controller necessary in order to implement a causal control system.

9.2 Consider a collocated point velocity-sensor and force-actuator system acting on a flexible structure for the implementation of a direct velocity feedback control. Demonstrate the fact that the system is unconditionally stable.

9.3 What is the difference between active vibration control (AVC) and active structural acoustic control (ASAC)?

9.4 Which are the best active sound radiation control systems for (a) deterministic tonal disturbances and (b) random disturbances?

Answers

Chapter 1

1.1 $c^2 = \dfrac{E_1}{\rho_1}\left[\dfrac{1 + [(d_2/d_1)^2 - 1](E_2/E_1)}{1 + [(d_2/d_1)^2 - 1](\rho_2/\rho_1)}\right]$

1.2 Central force per unit axial length $= 2\pi[Eh/(1 - v^2)](w/a)$, $\omega_r = c_l'/a$

1.3 Strip thickness $= b$: strip width $= h$

Torsional stiffness $= GKb^3h$ (Table 1.2)

$k_t = \omega[\rho_s bh^3/12Kb^3hG]^{1/2} = (\omega h/b)[\rho_s/12GK]^{1/2}$

$k_b = (\omega/b)^{1/2}[12\rho_s(1 - v^2)/E]^{1/4}$

At the critical frequency $k_t = k_b$

$\omega_c = 12\sqrt{3}bKc_l''(1 - v^2)^{1/2}/h^2(1 + v)$

1.4 Tip mass M: beam mass per unit length m

$\tilde{B} = -\tilde{A}\left[\dfrac{1 + 2k_bM/m + j}{1 + 2k_bM/m - j}\right]$

$$\tilde{D} = -\tilde{A}\left[\frac{2j}{1 + 2k_b M/m - j}\right]$$

$M = 0$: free end

$\tilde{B} = -\tilde{A}j$: $\left|\tilde{B}/\tilde{A}\right|^2 = 1$

$\tilde{D} = \tilde{A}(1 - j)$

$M = \infty$: simple support

$\tilde{B} = -\tilde{A}$: $\left|\tilde{B}/\tilde{A}\right|^2 = 1$

$\tilde{D} = 0$

Spring stiffness K

$$\tilde{B} = -\tilde{A}\left[\frac{2K/EIk_b^3 - 1 - j}{2K/EIk_b^3 - 1 + j}\right]$$

$$\tilde{D} = \tilde{A}\left[\frac{2j}{(2K/k_b^3 - 1) + j}\right]$$

$K = 0$: free end

$\tilde{B} = -\tilde{A}j$: $\left|\tilde{B}/\tilde{A}\right|^2 = 1$

$\tilde{D} = \tilde{A}(1 - j)$

$K = \infty$: simple support

$\tilde{B} = -\tilde{A}$: $\left|\tilde{B}/\tilde{A}\right|^2 = 1$

$\tilde{D} = 0$

Chapter 2

2.1 Impedance of mounted system 'seen' by floor $= \tilde{Z}_I$

$$\tilde{Z}_I = \frac{j\omega m(s + j\omega r)}{-\omega^2 m + s + j\omega r}$$

Velocity of floor in presence of system $= \tilde{v}_1$

Velocity of floor in absence of system $= \tilde{v}_0$

$\tilde{v}_1/\tilde{v}_0 = [1 + \tilde{Z}_I/\tilde{Z}_F]^{-1}$, where $\tilde{Z}_F$ is the point impedance of floor

Velocity of mass $= \tilde{v}_m$

$$\tilde{v}_m/\tilde{v}_0 = (\tilde{Z}_I/j\omega m)(1 + \tilde{Z}_I/\tilde{Z}_F)^{-1}$$

$\tilde{Z}_F$ purely real: real part of $\tilde{Z}_I$ positive

Maximum response of mass when $\mathrm{Im}\{\tilde{Z}_I\} = 0$: $\omega^2 = s/(m - r^2/s)$,

Unless $|\tilde{Z}_F| \ll sm/r$

2.2 $\overline{P} = \dfrac{1}{2}|\tilde{F}|^2 \mathrm{Re}\{1/Z^*\}$

$Z = A\omega^{1/2}(1 + j) - jK/\omega$

$$\mathrm{Re}\{1/Z^*\} = \frac{A\omega^{5/2}}{A^2\omega^3 + (A\omega^{3/2} - K)^2}$$

$\mathrm{Re}\{1/Z^*\}$ is maximum when $\omega = 0.91(K/A)^{2/3}$

2.3 Show that the rotational (moment) mobility at the simply supported end of a semi-infinite beam is given by $Y_b = \omega(1{+}j)/2EIk_b$.

Show that the rotational velocity at the simply supported end of a semi-infinite beam carrying an incident bending wave of transverse displacement amplitude A is given by $2\omega k_b A$.

Assume that e in Eq. (2.68b) equals half the beam depth.

At 100 Hz: $Y_b = 0.01(1 + j)\,\mathrm{s}^{-1}\mathrm{N}^{-1}\mathrm{m}^{-1}$; $Y_p = 0.079 + 0.26j\,\mathrm{s}^{-1}\mathrm{N}^{-1}\mathrm{m}^{-1}$; $P_{12}/W = 0.040$.

At 1000 Hz: $Y_b = 0.03(1 + j)\,\mathrm{s}^{-1}\,\mathrm{N}^{-1}\mathrm{m}^{-1}$; $Y_p = 0.794 + 1.42j\,\mathrm{s}^{-1}\mathrm{N}^{-1}\mathrm{m}^{-1}$; $P_{12}/W = 0.036$.

2.4 $Y_p = 8.2\mathrm{e}{-}4(1 - j)\,\mathrm{m\,s}^{-1}\mathrm{N}^{-1}$; $Y_w = 9.4\mathrm{e}{-}6\,\mathrm{m\,s}^{-1}\mathrm{N}^{-1}$; $Y_I = 0.003 + 0.031j\,\mathrm{m\,s}^{-1}\mathrm{N}^{-1}$; $[P_{12}]_I/[P_{12}]_R = 1.4\mathrm{e}{-}3$.

Chapter 3

3.1 $\overline{p^2} = \omega^2\rho_0^2|\tilde{Q}|^2/32\pi^2r^2 = 4 \times 10^{-4}\,\mathrm{Pa}^2$

$|\tilde{Q}| = 3.7 \times 10^{-3}\mathrm{m}^3\mathrm{s}^{-1}$

3.2 Piston radius a: axial distance R_0

$\tilde{p} = -(\omega\rho_0\tilde{v}_n/k)[\exp(-jkR') - \exp(-jkR_0)]$,

where R' is the distance from the piston edge to the observer position

$$\frac{\tilde{p}_{approx}}{\tilde{p}_{exact}} = -(ja^2k/2R_0)\{\exp\left[-jk(R'-R_0)\right]-1\}^{-1}$$

$R_0 = 1$ m, $f = 1$ kHz,   magnitude error $= 3\%$

3.5 Water: $\eta_{rad}/\sigma = 20.4$

Air: $\eta_{rad}/\sigma = 5.6 \times 10^{-3}$

3.6 $\overline{P} = \omega^2\rho_0|\tilde{Q}|^2/4\pi c$ (one side)

$\tilde{Q} = 4(a/\pi)^2\tilde{v}_0$

$\eta_{rad} = \overline{P}/\omega M \langle\overline{v^2}\rangle = 8\overline{P}/\omega M|\tilde{v}_0|^2$

$\qquad = 32\omega\rho_0 a^2/\pi^5 c\rho_s h$

Water: $\eta_{rad} = 0.28$

Air: $\eta_{rad} = 1.4 \times 10^{-3}$

3.11 $\tilde{v}_0/\tilde{v}_1 = \left[1 + \tilde{Z}_c/\tilde{Z}_p\right]^{-1}$

$\left|\tilde{v}_i\right|^2 = 0.097 \text{ m}^2\text{s}^{-2}$

$\tilde{Z}_c = j\omega m = 314.6j$

$\tilde{Z}_p = 860.6 \text{ Nm}^{-1}\text{s}$

$\left|\tilde{v}_0/\tilde{v}_i\right|^2 = 0.881$

$\left|\tilde{v}_0\right|^2 = 0.17 \text{ m}^2\text{s}^{-2}$

$\overline{P} = 16\rho_0|\tilde{v}_0|^2c^3/\pi\omega_c^2$

$\qquad = 1.8 \times 10^{-4}$

3.12 $I_z/\text{unit length} = \frac{1}{\lambda}\int_0^\lambda A^2 h \sin^2\left(\pi x/\lambda\right)ds$

where $ds^2 = dx^2 + dy^2$

Can approximate ds by dx because y^2 is greatest where dy/dx is least

$I_z/\text{unit length} \approx hA^2/2$, $I_x/\text{unit length} = h^3/12$

$D_x^{1/2}k_x^2 + D_z^{1/2}k_z^2 = \omega m^{1/2}$ and $k_x^2 + k_z^2 = k^2$ at intersection

Hence $k_x^2 = \dfrac{\omega m^{1/2} - D_z^{1/2}k^2}{D_x^{1/2} - D_z^{1/2}}$

$D_x = 3500$ Nm, $\quad D_z = 64.1$ Nm

Hence $k_x = 17.4$ m^{-1}, $\quad k_z = 34.5$ m^{-1}, $\quad \theta = 28.4°$ from z axis

Chapter 4

4.1 $m = 8\rho a_0^3/3$

$\omega_0' = \omega_0(M/(M+m))^{1/2}$

$f_0' = 31.06$ Hz

4.2 100 Hz, $k_b = 9.05$ m^{-1}, $m = 39$ kg m^{-2}, $c_b = 69.4$ m s^{-1}

$m_{air} \approx 0.27$ kg m^{-2}, $m_{water} \approx 221$ kg m^{-2}

$c_{b(water)} \approx 43.2$ m s^{-1}

1 kHz: $\quad k_b = 28.63$ m^{-1}, $c_b = 219.4$ m s^{-1}

$c_{b(water)} \approx 169.8$ m s^{-1}

2,2 mode

$\quad f_0 = 289.8$ Hz: $\quad f_{0(air)} = 288.7$ Hz, $\quad f_{0(water)} = 139.4$ Hz

10,10 mode

$\quad f_0 = 7243$ Hz: $\quad f_{0(air)} = 7243$ Hz, $\quad f_{0(water)} = 5614$ Hz

4.3 Equation (4.26) gives

$$\tilde{p}_1(a)/\tilde{v}_0 = \frac{-j\rho_0 c H_0(ka)}{H_0'(ka)}$$

For $ka \ll 1$, this may be written approximately as

$\tilde{p}_1(a)/\tilde{v}_0 \approx -j\rho_0 cka \, \ln(ka)$.

Hence the effective inertial loading per unit area m' is $-\rho_0 a \, \ln(ka)$, which is added to the cylinder mass per unit area $\rho_s h$. The ring frequency is reduced in the ratio $(1 + m'/\rho_s h)^{-1/2}$.

Generally, in water, $-\rho_0 a \, \ln(ka)/\rho_s h \gg 1$

Then

$$k'a = \left(\frac{c_l'}{c}\right)\left(\frac{\rho_s h}{-\rho_0 a \, \ln(k'a)}\right)^{1/2}:$$ no solution for typical values of parameters.

Chapter 5

5.1 $\tilde{v} = \dfrac{\tilde{p}_{bl} S}{\tilde{Z}_m + \tilde{Z}_{rad}}$, $\tilde{p}_{bl} = 2\tilde{p}_i$ (Eq. 4.10)

$\tilde{P}_{rad} = \frac{1}{2}|\tilde{v}|^2 \mathrm{Re}\{\tilde{Z}_{rad}\}$

$\tilde{P}_{inc} = S|\tilde{p}_i|^2 \cos\phi / 2\rho_0 c$

$\tau = 4\rho_0 c S \mathrm{Re}\{\tilde{Z}_{rad}\} / |\tilde{Z}_m + \tilde{Z}_{rad}|^2 \cos\phi$

$\tilde{Z}_m = j(\omega m - s/\omega)$

$\tilde{Z}_{rad} = 2\rho_0 c S\left[(ka)^2/2 + j8ka/3\pi\right]$

$\omega_0 = 314 \text{ rads}^{-1}$, $s = \omega_0^2 m = 1972 \text{ Nm}^{-1}$

At fluid-loaded resonance $\tilde{Z}_m + \tilde{Z}_{rad} = \mathrm{Re}\{\tilde{Z}_{rad}\}$

$\tau_{res} = 4\rho_0 c S / \mathrm{Re}\{\tilde{Z}_{rad}\} \cos\phi = 2/(ka)^2$

$\tau_{res} = 954!$

At frequencies much greater than ω_0, the inertial impedance of the piston controls transmission:

$\tau \approx 2(\rho_0 c S)^2 (ka)^2 / \omega^2 m^2 \cos\phi.$

τ is frequency independent and varies as sec ϕ as long as $ka \ll 1$.

5.2 To reduce the transmission coefficient by 10, the mechanical loss factor must be given by $\eta_m = 2.16\rho_0 c S(ka)^2 / \omega_0 m$.

5.3 Normal particle velocity $\tilde{v}$

$\tilde{v} = 2\tilde{p}_i\left[j\rho_2 c_2\left[(k_1 \sin\phi_i/k_2 - 1)^2\right]^{-1/2} + \rho_1 c_1 \sec\phi_i\right]^{-1}$

$\tilde{p}_t = 0.312\tilde{p}_i : (-9.8 \text{ dB})$

Incidence angle beyond critical: no energy transmission

5.4 Refraction causes effective incidence angle to increase

$f < f_c$: transmission increased

$f > f_c$: stiffness-controlled transmission decreased

: mass-controlled transmission increased

: overall effect is to increased diffuse field transmission

5.5 Eq. (5.56): $\tau_\infty/\tau_r \approx 200\left(A/P\lambda_c\right)^2 \eta_{tot} = 1$

$\lambda_c = 2.63 \times 10^{-2}, \quad \eta_{tot} = 4.6 \times 10^{-5}$

Eq. (5.58): $3/2 + \ln(2\omega/\Delta\omega) = 2.52, \quad \omega/\omega_c = 3.83 \times 10^{-2}$

Field incidence $R = R(0) - 5\,\text{dB}$

Therefore, $2.52 \mid 2.42 \times 10^{-3}/\eta = 10^{0.8}$, or $\eta = 6.7 \times 10^{-4}$

The difference lies in approximation for σ in Eq. (5.55)

5.6 $\omega_1 = 219\,\text{rad s}^{-1}, \quad \delta_1 = 9.4 \times 10^{-2}$

$\omega_0 = 1015.9\,\text{rad s}^{-1}, \quad f_0 = 161.7\,\text{Hz}$

5.7 $\tilde{\xi}_1/\tilde{\xi}_2 = -m_2/m_1$

5.8 Power radiated by enclosure per unit area $= \frac{1}{2}\rho_0 c|\tilde{v}|^2$

Power dissipated by enclosure per unit area $= \frac{1}{2}r|\tilde{v}|^2$

Power radiated by source/power radiated by enclosure $= 1 + r/\rho_0 c$

At frequencies far above ω_1, the enclosure mass controls transmission.

Maxima occur when $\sin kl = 1$.

Therefore, $\overline{P_e}/\overline{P} \approx (\rho_0 c/\omega m^2), IL \approx R(0) + 6\,\text{dB}$

Chapter 6

6.1 Incident field varies slightly in phase over the surface of cylinder

6.2 $ka \ll \pi$: a is larger panel dimension

All other modes have a direction of maximum response because the peak in the wavenumber spectrum is not at $k_x = 0$

6.4 $\overline{P_{rad}} = \rho_0 c S\sigma \langle v^2\rangle = \langle p^2\rangle a/4\rho_0 c$

where $a = 0.16 V/T$

$\sigma = 4.73 \times 10^{-4}$

$\eta_{rad} = 2.6 \times 10^{-6}, \eta_{rad} \ll \eta_{int}$

$n_s(\omega) = 3.67 \times 10^{-2}\,\text{rad}^{-1}\text{s}$

$\langle v^2\rangle = 9.2 \times 10^{-10}\,\text{m}^2\text{s}^{-2}$

6.5 $\eta_{int} = 2.2/f T_{mech}$

$\overline{P_{diss}} = \eta_{int}\omega M \langle v^2 \rangle$

$\overline{P_{inc}} = S \langle \overline{p^2} \rangle /4\rho_0 c$

$\alpha = 4\eta_{int}\omega m\rho_0 c \langle \overline{v^2} \rangle /\langle \overline{p^2} \rangle$

6.6 $\dfrac{\langle v^2 \rangle}{\langle p^2 \rangle} = \dfrac{2\pi^2 c n_s(\omega)}{M\rho_0\omega^2}\left[\dfrac{(\rho_0 c/\omega m)\sigma}{(\rho_0 c/\omega m)\sigma + \eta_{int}}\right]$

Assume $\eta_{int} \gg \eta_{rad}$ in test $\rightarrow n_s(\omega)\sigma/\eta_{int}$

Assume $\langle \varepsilon^2 \rangle \propto \langle v^2 \rangle$

Assume dependence of σ on f/f_c and $P\lambda c/A$ [see Eq. (5.55)]: extrapolate to operating conditions.

Major uncertainty in the form of σ.

Chapter 7

7.1 For $\psi_n = \cos(p\pi x/a)\cos(q\pi z/b)\cos(r\pi y/c)$,

$\phi_p = \sin(m\pi x/a)\sin(n\pi z/b)$ and

$\phi_q = \sin(m'\pi x/a)\sin(n'\pi z/b)$

$$\Lambda_n(k_n^2 - k^2)\alpha_{ppn} = \left[\frac{(-1)^{m+p} - 1}{1 - (m/p)^2}\right]^2\left[\frac{(-1)^{n+q} - 1}{1 - (n/q)^2}\right]^2\left[\frac{mnab}{p^2q^2\pi}\right]^2$$

$$\Lambda_n(k_n^2 - k^2)\alpha_{pqn} = \left[\frac{(-1)^{m+p} - 1}{1 - (m/p)^2}\right]\left[\frac{(-1)^{n+q} - 1}{1 - (n/q)^2}\right]\left[\frac{(-1)^{m'+p} - 1}{1 - (m'/p)^2}\right]$$

$$\times\left[\frac{(-1)^{n'+q} - 1}{1 - (n'/q)^2}\right]\left[\frac{mnm'n'a^2b^2}{p^4q^4\pi^2}\right]$$

7.5 $\omega_p = 146.6$ rad s^{-1}, $f_1 = 23.3$ Hz

$\omega_n = 0$ (bulk compression) because C_{np} is greatest for this mode

$C_{np} = (2/\pi)^2$, $K_{np} = 16\rho_0 ab/M\pi^2$, $G_{np} = 4c^2 S/\pi^2 V$

$\Lambda_n = V$, $\Lambda_p = M/4$

$\omega_p = 401.9$ rad s^{-1}, $f = 64$ Hz

No change of frequency will occur because $\rho_0 c^2$ is unchanged.

7.6 $(M_p/4)(-\omega^2 + \omega_0^2) + 4j\omega b^2 \tilde{Z}_0 S/\pi^2 = 0$

$$\tilde{Z}_0 = \frac{\rho_0 c}{S}\left[\frac{\tilde{Z}_l + j(\rho_0 c/S)\tan kl}{(\rho_0 c/S) + j\tilde{Z}_l \tan kl}\right] = \frac{\tilde{p}}{\tilde{Q}}$$

$\tilde{Q} = 4b^2 \tilde{v}_0/\pi^2$: b is the panel side length

$a = b/\sqrt{\pi} = 0.17$ m

$\omega_0 = 668$ rad s^{-1}

For a coupled resonance below ω_0, the fluid must present an inertial load to the panel. A good approximation to impedance of an open-ended tube is $\tilde{Z}_0 \simeq j(\rho_0 c)\tan kl$. Assume $kl = \pi/2$, $\eta = 0.5$

7.7 $\tilde{Z}_0 = R_0 + jX_0$, $\quad \tilde{Z}_l = R + jX$

$$\overline{P} = \tfrac{1}{2}S^2|\tilde{u}|^2 \text{Re}\{\tilde{Z}_0\} = \tfrac{1}{2}S^2|\tilde{u}|^2 R_0$$

$$R_0 = \frac{R(\rho_0 c/S)^2(1 + \tan^2 kl)}{(\rho_0 c/S - X\tan kl)^2 + R^2\tan^2 kl}$$

Putting $\partial R_0/\partial(kl) = 0$ yields stationary values given by

$$\tan 2kl = \frac{2(\rho_0 c/S)X}{|\tilde{Z}_l|^2 - (\rho_0 c/S)^2}$$

or $X(1 - \tan^2 kl)(\rho_0 c/S) = (|\tilde{Z}_l|^2 - (\rho_0 c/S)^2)\tan kl$.

Therefore $\tan kl = \dfrac{-b \pm \sqrt{b^2 - 4ac}}{2a}$

where $a = -c = X/(\rho_0 c/S)$ and $b = |\tilde{Z}_l|^2/(\rho_0 c/S)^2$

$R/(\rho_0 c/S) = 5$, $\quad X/(\rho_0 c/S) = -10$, $\quad |\tilde{Z}_l|^2/(\rho_0 c/S)^2 = 125$

$$\tan kl = \frac{-125 \pm \sqrt{1.56 \times 10^4 + 400}}{-20}$$

$\quad = 12.57 \quad$ or $\quad -7.45 \times 10^{-2}$

$kl = 1.49 \quad$ or $\quad 7.44 \times 10^{-2}$

$f_1 = 8.12$ Hz, $\quad f_2 = 162.7$ Hz

Maximum power at f_1, minimum power at f_2.

7.8 Let the pressure difference across hole equal $R\tilde{u}$, where $\tilde{u}$ is the particle velocity through the hole. Continuity of particle velocity at the piston and volume velocity at the termination yield the following expression for the total impedance:

$$\tilde{z}_t = \tilde{z}_m + \rho_0 c \left[\frac{1 - \beta^2 - 2j\beta \sin 2kl}{1 + \beta^2 - 2\beta \cos 2kl} \right]$$

where $\tilde{z}_m$ is the *in-vacuo* piston impedance, $\beta = (R + \alpha\rho_0 c)/(R - \alpha\rho_0 c)$, and α is the hole/tube cross-sectional area ratio. The real part of the expression in the square brackets represents dissipation(damping); the imaginary part represents stiffness or inertia, depending upon the sign. The damping is zero if $R = \infty$ (closed end).

Chapter 8

8.1 $M_{ij} = \int_0^l \rho A x^{j+k} dx = \dfrac{1}{(j+k-1)} \rho A l^{j+k+1}$ and

$\quad\; K_{ij} = \int_0^l EAjkx^{j+k-2} dx = \dfrac{jk}{(j+k-1)} EA l^{j+k-1}$

8.2 $T = \frac{1}{2}\int_0^l \rho A\left(\frac{\partial u}{\partial t}\right)^2 dx = \frac{1}{2}\int_{-1}^{+1} \rho A\left(\frac{\partial u}{\partial t}\right)^2 a d\xi$ and

$\quad\; U = \frac{1}{2}\int_0^l EA\left(\frac{\partial u}{\partial x}\right)^2 dx = \frac{1}{2}\int_{-1}^{+1} EA\frac{1}{a}\left(\frac{\partial u}{\partial \xi}\right)^2 a d\xi.$

8.3 $[M] = \rho Aa \begin{bmatrix} \frac{2}{3} & \frac{1}{3} \\ \frac{1}{3} & \frac{2}{3} \end{bmatrix}$ and $[K] = \dfrac{EA}{a} \begin{bmatrix} \frac{1}{2} & -\frac{1}{2} \\ -\frac{1}{2} & \frac{1}{2} \end{bmatrix}.$

Chapter 9

9.1 $t = \dfrac{L}{c_b} = L\left(\dfrac{\rho A}{EI}\right)^{1/4} \omega^{-1/2}.$

9.2 For the system to be stable, $P > 0$. Since the power input is given by $P = \frac{1}{2}\text{Re}\{\tilde{F}^*\tilde{v}\} = \frac{1}{2}|\tilde{F}|\text{Re}\{\tilde{Y}_{vF}\}$, then $\text{Re}\{\tilde{Y}_{vF}\} > 0$: thus the real part of the sensor–actuator plant response must be positive real which is the case for collocated sensor–actuator configurations.

9.3 Active Vibration Control systems are set to minimise a cost function linked to the vibration of a structure while Active Structural Acoustic Control systems are set to minimise a cost function linked to the sound radiation of a structure.

9.4 (a) Feed-forward Active Structural Acoustic Control systems and (b) Feed-back Active Vibration Control systems.

References

Allard, J.F., 1993, *Propagation of Sound in Porous Media: Modelling Sound Absorbing Materials*, Elsevier Applied Science, New York.

Anderson, J.S. and M. Bratos-Anderson, 2005, Radiation efficiency of rectangular orthotropic plates, *Acta Acustica with Acustica* 91, 61.

ANSI/IEEE Standard 176, 1987, Piezoelectricity (IEEE, New York).

Argyris, J.H., 1954, A study of thin-walled structures such as interspar wing, cut-outs and open-section stringers, *Aircraft Engineering* 36, 102.

Ashwell, D.J. and R.H. Gallagher (Eds.), 1976, *Finite Elements for Thin Shells and Curved Members*, John Wiley & Sons, Inc., London.

Au, A.C.K. and K.P. Byrne, 1987, On the insertion losses produced by plane acoustic lagging structures, *J. Acoust. Soc. Am.* 82, 1325.

Au, A.C.K. and K.P. Byrne, 1990, On the insertion losses produced by acoustic lagging structures which incorporate orthotropic impervious barriers, *Acustica* 70, 284.

Balas, M.J., 1979, Direct velocity feedback of large space structures, *Journal of Guidance and Control* 2, 252.

Bank, G. and G.T. Hathaway, 1981, A three-dimensional interferometric vibrational mode display, *J. Audio. Eng. Soc.* 29, 314.

Bathe, K.J., 1996, *Finite Element Procedures*, Prentice-Hall International, Upper Saddle River, New Jersey.

Baumann, W.T., W.R Saunders and H.H. Robertshaw, 1991, Active suppression of acoustic radiation from impulsively excited structures, *J. Acoust. Soc. Am.* 90, 3202.

607

Baumann, W.T., F.S. Ho and H.H. Robertshaw, 1992, Active structural acoustic control of broadband disturbance, *J. Acoust. Soc. Am.* 92, 1998.

Belousov, Yu.I. and A.V. Rimskii-Korsakov, 1975, The reciprocity principle in acoustics and its application to the calculation of the sound fields of bodies, *Sov. Phys. Acoust.* 21, 103.

Beranek, L.L. and I. Vér (Eds.), 1992, *Noise and Vibration Control Engineering*, John Wiley & Sons, Inc., New York.

Berry, A., J-L. Guyader and J. Nicolas, 1990, A general formulation for the sound radiation from rectangular, baffled plates with arbitrary boundary conditions, *J. Acoust. Soc. Am.* 88, 2792.

Bies, D.L., 1971, Acoustical properties of porous materials, In *Noise and Vibration Control*, L.L. Beranek, Ed., McGraw-Hill, New York.

Bishop, R.E.D. and D.C. Johnson, 1960, *The Mechanics of Vibration*, Cambridge University Press, Cambridge.

Bishop, R.E.D. and S. Mahalingham, 1981, An elementary investigation of local vibration, *J. Sound Vib.* 77, 149.

Bishop, R.E.D., G.M.L. Gladwell and S. Michaelson, 1965, *The Matrix Analysis of Vibration*, Cambridge University Press, Cambridge, England.

Blevins, R.D., 1995, *Formulas for Natural Frequency and Mode Shape*, Krieger Publishing Company, Melbourne, Florida.

Bolotin, V.V., 1984, *Random Vibrations of Elastic Systems*, Martinus Nijhoff Publishers, The Hague, The Netherlands.

Bolton, J.S., 2005, Porous materials for sound absorption and transmission control, Proceedings Inter-noise 2005, Rio de Janeiro, Brasil (CD ROM).

Bolton, J.S., N.-M. Shiau and Y.J. Kang, 1996, Sound transmission through multi-panel structures lined with elastic porous materials, *J. Sound Vib.* 191, 317.

Bonilha, M. and F.J. Fahy, 1998, On the vibration field correlation of randomly excited flat plate structures, II: experimental verification, *J. Sound Vib.* 214, 469.

Bonilha, M. and F.J. Fahy, 1999, An approximation to the frequency-average radiation efficiency of flat plates, Acoustical Society of America, 138th meeting, Columbus, Ohio.

Borgiotti, G.V., 1990, The power radiated by a vibrating body in an acoustic fluid and its determination from boundary measurements, *J. Acoust. Soc. Am.* 88, 1884.

Brebbia, C.A., J.C.F. Telles and L.C. Wrobel, 1984, *Boundary Element Techniques, Theory and Applications in Engineering*, Springer Verlag, Berlin.

Brekke, A., 1981, Calculation methods for the transmission loss of single, double and triple partitions, *Applied Acoustics* 14, 225.

Brennan, M.J. and W. Variyart, 2003, Simplified mobility expressions for infinite in-vacuo pipes, *J. Sound Vib.* 260, 329.

Bull, M. and M.P. Norton, 1982, On coincidence in relation to prediction of pipe wall vibration and noise radiation due to turbulent pipe flow disturbed by fittings, Paper H2 in Proceedings of the International Conference on Flow Induced Vibrations in Fluid Engineering, BHRA Fluid Engineering, Cranfield, Bedfordshire, England.

Burton, A.J. and G. Miller, 1971, The application of integral equation methods to the numerical solution of some exterior boundary value problems, Proceedings Roy. Soc. Lond. Series A 323, 201.

Byrne, K.P., H.V. Fuchs and H.M. Fischer, 1988, Sealed, close fitting, machine-mounted acoustic enclosures with predictable performance, *Noise Contr. Eng. Jour.* 31, 7.

Cederfeldt, L., 1974, Sound Insulation of Corrugated Plates. A Summary of Laboratory Measurements, Rep. 55, Division of Building Technology, Lund Institute of Technology, Sweden.

Chaplin, G.B.B., 1983, Anti-sound – The Essex breakthrough, *Chartered Mechanical Engineer* 30, 41.

Clark, R.L., W.R. Saunders and G.P. Gibbs, 1998, *Adaptive Structures*, John Wiley & Sons, Inc., New York.

Conover, W.B., 1956, Fighting noise with noise, *Noise Control* 2, 78.

Cook, R.D., D.S. Malkus and M.E. Plesha, 1989, *Concepts and Applications of Finite Element Analysis*, John Wiley & Sons, Inc., New York.

Courant, R. and D. Hilbert, 1962, *Methods of Mathematical Physics*, Vol. 2, Wiley (Interscience), New York.

Craggs, A., 1971, The transient response of a coupled plate-acoustic system using plate and acoustic finite elements, *J. Sound Vib.* 15, 509.

Craggs, A. and G. Stead, 1976, Sound transmission between enclosures – a study using plate and acoustic finite elements, *Acustica*, 35, 89.

Craik, R.M., 1996, *Sound Transmission through Buildings using Statistical Energy Analysis*, Gower Publishers, Aldershot, England.

Cremer, L., 1942, Theorie der Schalldämmung dünner Wände bei Schrägen einfall, *Akust. Ztg.* 7, 81.

Cremer, L., 1968, Schallschutz in Gebäuden, Berichte aus der Bauforschung No.56, Von Wilhelm Erst und Sohn, Berlin.

Cremer, L., M. Heckl and E.E. Ungar, 1988, *Structure-borne Sound (2nd Ed.)*, Springer Verlag, Berlin.

Cremer, L., M. Heckl and B.A.T. Petersson, 2005, *Structure-borne Sound: Structural Vibration and Sound Radiation (3rd Ed.)*, Springer Verlag, Berlin.

Crighton, D., 1972, Force and moment admittance of plates under arbitrary fluid loading, *J. Sound Vib.* 20, 209.

Crighton, D., 1977, Point admittance of an infinite thin elastic plate under fluid loading, *J. Sound Vib.* 54, 389.

Crighton, D., 1988, The Rayleigh Medal lecture, Fluid Loading: the interaction between sound and vibration, *J. Sound Vib.* 133, 1.

Cummings, A., 2001, Sound radiation from a plate into a porous medium, *J. Sound Vib.* 247, 389.

Cunefare, K.A., 1991, The minimum multimodal radiation efficiency of baffled finite beams, *J. Acoust. Soc. Am.* 90, 2521.

Cunefare, K.A. and G.H. Koopmann, 1991, Global optimum active noise control: surface and far-field effects, *J. Acoust. Soc. Am.* 90, 365.

Davies, H.G., 1971, Low frequency random excitation of water-loaded rectangular plates, *J. Sound Vib.* 15, 107.

Davis, E., 1999, Designing honeycomb panels for noise control, *J. Amer. Inst. of Aeronautics and Astronautics* 38, 1917.

Desmet, W., 1998, A wave based prediction technique for coupled vibro-acoustic analysis, PhD Thesis, Katholieke Universiteit, Leuven, Belgium.

Dowling, A.P., and J.E. Ffowcs-Williams, 1983, *Sound and Sources of Sound*, Ellis Horwood, Chicester, England.

Drotleff, H., 1996, Sound Radiation and Transmission by Simply-Stiffened and Corrugated Plates Treated with Multi-layer Insulation, MSc Dissertation, University of Southampton, England.

Duncan, W.J., A.S. Thom and A.D. Young, 1970, *Mechanics of Fluids (2nd Ed.)*, Edward Arnold, London.

Egle, D.M. and J.L. Sewall, 1968, An analysis of free vibration of orthogonally stiffened cylindrical shells with stiffeners treated as discrete elements, *AIAA Journal* 6, 518.

Egle, D.M. and K.E. Soder, Jnr., 1969, A theoretical analysis of the free vibration of discretely stiffened cylindrical shells with arbitrary end conditions, NASA CR-1316, US Government Printing Office, Washington DC.

Eichelberger, E.C., 1981, Point admittance of cylindrical shells with and without ring-stiffening, *ASME Journal of Engineering for Industry* 103, 293.

Einarsson, S. and J. Söderquist, 1982, A sound insulating element featuring high stiffness and low weight, Proceedings of Inter-Noise 82, 467.

Elliott, S.J., 2001, *Signal Processing for Active Control*, Academic Press, London.

Elliott, S.J. and M.E. Johnson, 1993, Radiation modes and the active control of sound power, *J. Acoust. Soc. Am.* 94, 2194.

Elliott, S.J., P. Gardonio, T.C. Sors and M.J. Brennan, 2002, Active vibro-acoustic control with multiple feedback loops, *J. Acoust. Soc. Am.* 111, 908.

Erickson, L.L., 1969, Modal densities of sandwich panels: theory and experiment, *Shock and Vibration Bulletin*, 39, Part 3, 1.

ESDU Data Item 80040, 1983, Free vibration of thin-walled, orthogonally stiffened, circular cylindrical shells, Engineering Sciences Data Unit, London.

ESDU, 2004, Data Item 04010, Mobilities and Impedances of Structures. Part II: compendium of point mobilities of infinite structures, IHS Engineering Sciences Data Unit, London.

Everstine, G.C., 1981, A symmetric potential formulation for fluid-structure interaction, *J. Sound Vib.* 79, 157.

Fagerlund, A.C.A., 1980, Sound transmission through a cylindrical pipe, ASME Paper 80-WA/NC-3, In Proceedings of the Winter Annual Meeting of the Amer. Soc. Mech. Eng., Chicago, Illinois, USA.

Fahy, F.J., 1969a, Acoustic Excitation of Containing Structures, PhD Thesis, University of Southampton, England.

Fahy, F.J., 1969b, Vibration of containing structures by sound in the contained fluid, *J. Sound Vib.* 10, 490.

Fahy, F.J., 1970, Response of a cylinder to random sound in the contained fluid, *J. Sound Vib.* 13, 171.

Fahy F.J., 1987, *Sound and Structural Vibration*, Academic Press, London.

Fahy, F.J., 1994, Statistical energy analysis: A Critical Review, In *Statistical Energy Analysis: An Overview with Applications in Structural Dynamics*, Eds. A.J. Keane and W.G. Price, Cambridge University Press, England.

Fahy, F.J., 1995a, The vibro-acoustic reciprocity principle and applications to noise control, *Acustica* 81, 544.

Fahy, F.J., 1995b, *Sound Intensity (2nd Ed.)*, E & FN Spon, London.

Fahy, F.J., 2001, *Foundations of Engineering Acoustics*, Academic Press, London.

Fahy, F.J., 2004, Some applications of the reciprocity principle in experimental vibroacoustics, *Acoustical Physics* 49, 217.

Fahy, F.J. and R.S. Langley, 2004, High-frequency Structural Vibration, In *Advanced Applications in Acoustics, Noise and Vibration*, Eds. F.J. Fahy and J.G. Walker, Spon Press, London.

Fahy, F.J. and D.J. Thompson, 2004, The effect of perforation on the radiation efficiency of vibrating plates, In Proceedings of the Institute of Acoustics Spring Meeting, Southampton, UK, Vol.26 (CD ROM).

Fahy, F.J. and J.G. Walker (Eds.), 1998, Fundamentals of noise and vibration, E & FN Spon, London.

Filippi, P.J.T., 1977, Layers potentials and acoustic diffraction, *J. Sound Vib.* 54, 473.

Finnveden, S., 1997, Simplified equations of motion for the radial-axial vibrations of fluid filled pipes, *J. Sound Vib.* 208, 685.

Fuller, C.R., 1981, The effects of wall discontinuities on the propagation of flexural waves in cylindrical shells, *J. Sound Vib.* 75, 207.

Fuller, C.R., 1983, The input mobility of an infinite circular cylindrical elastic shell filled with fluid, *J. Sound Vib.* 87, 409.

Fuller, C.R. and F.J. Fahy, 1981, Characteristics of wave propagation and energy distributions in cylindrical elastic shells filled with fluid, *J. Sound. Vib.* 81, 501.

Fuller, C.R., S.J. Elliott and P.A. Nelson, 1996, *Active Control of Vibration*, Academic Press, London.

Gardonio, P., 2002, A review of active techniques for aerospace vibration and noise control, *Journal of Aircraft* 39, 206.

Gardonio, P., and S.J. Elliott, 1998, Driving Point and Transfer Mobility Matrices for Thin Plates Excited in Flexure, ISVR Technical Report No. 277, ISVR, University of Southampton.

Gardonio, P. and M.J. Brennan, 2004, Mobility and impedance methods in structural dynamics, Chapter 9 of *Advanced Applications in Noise, Acoustics and Vibration*, Eds. F.J. Fahy and J.G. Walker, E & FN Spon., London.

Gardonio, P. and S.J. Elliott, 2004, Smart panels for active structural acoustic control, *Journal of Smart Materials and Structures* 13, 1314.

Gardonio, P. and S.J. Elliott, 2005a, Modal response of a beam with a sensor–actuator pair for the implementation of velocity feedback control, *J. Sound Vib.* 284, 1.

Gardonio, P. and S.J. Elliott, 2005b, Smart panels with velocity feedback control systems using triangularly shaped strain actuators, *J. Acoust. Soc. Am.* 117, 2046.

Gardonio, P., N.S. Ferguson and F. J. Fahy, 2001a, A modal expansion analysis of noise transmission through circular cylindrical shell structures with blocking masses, *J. Sound Vib.* 244, 259.

Gardonio, P., Y.-S. Lee, S.J. Elliott and S. Debost, 2001b, Analysis and measurement of a matched volume velocity sensor and uniform force actuator for active structural acoustic control, *J. Acoust. Soc. Am.* 110, 3025.

Gardonio, P., E. Bianchi and S.J. Elliott, 2004, Smart panel with multiple decentralized units for the control of sound transmission. Part I: theoretical predictions; Part II: design of the decentralised control units; Part III: control system implementation, *J. Sound Vib.* 274, 163.

Gladwell, G.M.L., 1966, A variational formulation of damped acousto-structural vibration problems, *J. Sound Vib.* 4, 172.

Gladwell, G.M.L. and G. Zimmermann, 1966, On energy and complementary energy formulations of acoustic and structural vibration problems, *J. Sound Vib.* 3, 233.

Godano, Ph., S. Simonnin and B. Semeniuk, 2000, Rieter Ultra Light: Une avanceé majeure dans l'allègement de l'insonorisation automobile, Proceedings of the Conference 'Confort automobile et ferroviaire', Le Mans, France, (CD ROM from Rieter Automotive Systems, Winterthur, Switzerland).

Gomperts, M.C., 1977, Sound radiation from baffled, thin rectangular plates, *Acustica* 37, 93.

Guigou-Carter, G. and M. Villot, 2003, Modelling of sound transmission through lightweight elements with stiffeners, *Building Acoustics* 10, 193.

Hansen, C.H. and S.D. Snyder, 1997, Active control of noise and vibration, E & FN Spon., London.

Harrari, A. and B.E. Sandman, 1990, Radiation and vibrational properties of submerged stiffened cylindrical shells, *J. Acoust. Soc. Am.* 88, 1817.

Heckl, M., 1960, Untersuschungen an orthotropen Platten, *Acustica* 10, 109.

Heckl, M., 1962a, Measurements of absorption coefficients of plates, *J. Acoust. Soc. Am.* 34, 803.

Heckl, M., 1962b, Vibration of point-driven cylindrical shells, *J. Acoust. Soc. Am.* 34, 1553.

Heron, K.H., 1979, Acoustic radiation from honeycomb panels, Proceedings of the Fifth European Rotorcraft and Powered Lift Aircraft Forum, Amsterdam, The Netherlands.

Hixon, E.L., 1976, Mechanical impedance, Chapter 10 in *Shock and Vibration Handbook*, Eds. C.M. Harris and C.E. Crede, McGraw-Hill Inc., New York.

Hodges, C.H. and J. Woodhouse, 1989, Vibration isolation from irregularity in a nearly periodic structure: theory and experiments, *J. Acoust. Soc. Am.* 74, 894.

Hodgson, D.C. and M.M. Sadek, 1983, A technique for the prediction of the noise field from an arbitrary vibrating machine, Proceedings Inst. Mech. Eng., Part C 197, 189.

Holmer, C.I. and F.J. Heymann, 1980, Transmission of sound through pipe walls in the presence of flow, *J. Sound. Vib.* 70, 275.

Hongisto, V., M. Lindgren and R. Helenius, 2002, Sound insulation of double walls – experimental parametric study, *Acta Acustica united with Acustica* 88, 904.

Hopkins, C. and P. Turner, 2005, Field measurement of airborne sound insulation between rooms with non-diffuse sound fields at low frequencies, *Applied Acoustics* 66, 1339.

Hoskins, R.F., 1979, *Generalized Functions*, Horwood, West Sussex.

Ingemansson, S. and T. Kihlman, 1959, Sound Insulation of Frame Walls, Trans. Chalmers. Univ. Technol., Gothenburg; No. 2.

Jayachandran, V. and J.Q. Sun, 1997, Unconditional stability domains of structural control systems using dual actuator-sensor pairs, *J. Sound Vib.* 208, 159.

Johnson, M.E. and S.J. Elliott, 1995, Active control of sound radiation using volume velocity cancellation, *J. Acoust. Soc. Am.* 98, 2176.

Jones, R.E., 1981, Field sound insulation of load-bearing sandwich panels for housing, *Noise Control Eng.* 16, 90.

Joseph, P., C.L. Morfey and C.R. Lowis, 2003, Multi-mode sound transmission in ducts with flow, *J. Sound Vib.* 264, 523.

Josse, R. and C. Lamure, 1964, Transmission du son par une paroi simple, *Acustica* 14, 266.

Junger, M.C. and D. Feit, 1986, *Sound, Structures, and their Interaction* (2nd Ed.), The MIT Press, Cambridge, Massachusetts.

Keane, A.J. and W.G. Price (Eds.), 1997, *Statistical Energy Analysis: An Overview with Applications in Structural Dynamics*, Cambridge University Press, Cambridge, England.

Kennard, E.H., 1953, The new approach to shell theory: circular cylinders, *J. Appl. Mech.* 20, 33.

Kerschen, E.J. and J.P. Johnston, 1980, Mode selective transfer of energy from sound propagating in circular pipes to pipe wall vibration, *J. Acoust. Soc. Am.* 67, 1931.

Kihlman, T., 1967a, Transmission of Structure-borne Sound in Buildings, Rep. 9. National Swedish Institute for Building Research, Stockholm, Sweden.

Kihlman, T., 1967b, Sound transmission in building structures of concrete, *J. Sound Vib.* 11, 435.

Kim, J.-W. and J.S. Bolton, 2004, Free wave propagation in sandwich-panels with poro-elastic cores, Proceedings of Noise-Conference 2004, Baltimore, CD ROM.

Kinsler, L.E., A.R. Frey, A.B. Coppens and J.V. Sanders, 1999, *Fundamentals of Acoustics* (4th Ed.), John Wiley and Sons, Inc., New York.

Koopmann, G.H. and H. Benner, 1982, Method for computing the sound power of machines based on the Helmholtz integral, *J. Acoust. Soc. Am.* 71, 78.

Koopmann, G.H. and J.B. Fahnline, 1997, *Designing Quiet Structures*, Academic Press, London.

Kreyszig, E., 1999, *Advanced Engineering Mathematics*, John Wiley & Sons, Inc., New York.

Kropp, W. and E. Rebillard, 1999, On the air-borne sound insulation of double wall constructions, *Acustica with Acta Acustica* 85, 707.

Kropp, W., A. Pietrzyk and T. Kihlman, 1991, On the meaning of the sound reduction index at low frequencies, *Acustica with Acta Acustica* 2, 379.

Kuo, S.M. and D.R. Morgan, 1996, *Active Noise Control Systems: Algorithms and DSP Implementations*, John Wiley & Sons, Inc., New York.

Kurra, S. and D. Arditi, 2001, Determination of sound transmission loss of multilayered elements. Part 2: experimental study, *Acta Acustica with Acustica* 87, 592.

Kurtze, C. and B.G. Watters, 1959, New wall design for high transmission loss or high damping, *J. Acoust. Soc. Am.* 31, 739.

Lagrange, J.L., 1788, Mechanique Analytique.

Lai, H.-Y., 2001, Numerical optimisation based on genetic algorithm for multi-layered acoustic liner design, In Proceedings of Noise-Conference 01, Portland, Maine, USA.

Lam, Y.K., 1995, Noise transmission through profiled metal cladding: Part III: double skin SRI prediction, *Building Acoustics* 2, 403.

Lam, Y.K. and R.M. Windle, 1995a, Noise transmission through profiled metal cladding: Part I: single skin measurements, *Building Acoustics* 2, 341.

Lam, Y.K. and R.M. Windle, 1995b, Noise transmission through profiled metal cladding: Part II: single SRI prediction, *Building Acoustics* 2, 357.

Langley, R.S., 1994a, The modal density and mode count of thin cylinders and curved panels, *J. Sound Vib.* 169, 43.

Langley, R.S., 1994b, Spatially-averaged frequency response envelopes for one- and two-dimensional structures, *J. Sound Vib.* 178, 483.

Langley, R.S., 1994c, Wave motion and energy flow in cylindrical shells, *J. Sound Vib.* 169, 29.

Langley, R.S., 1999, Some perspectives on wave-mode duality, In *IUTAM Symposium on SEA*, Eds. F.J. Fahy and W.G. Price, Kluwer Academic Publishers, Dordrecht, The Netherlands.

Langley, R.S. and A.N. Bercin, 1994, Wave intensity analysis of high frequency structural vibrations, *Phil. Trans. Roy. Soc. Series* A 346, 489.

Laulagnet, B., 1998, Sound radiation by a simply-supported unbaffled plate, *J. Acoust. Soc. Am.* 103, 2451.

Laulagnet, B. and J.-L. Guyader, 1989, Modal analysis of a shell's acoustic radiation in light and heavy fluids, *J. Sound Vib.* 131, 397.

Lee, C.K., 1990, Theory of laminated piezoelectric plates for the design of distributed sensor/actuators. Part I: governing equations and reciprocal relationships, *J. Acoust. Soc. Am.* 87, 1144.

Leissa, A.W., 1993a, Vibration of plates, Acoustical Society of America through the American Institute of Physics, New York.

Leissa, A.W., 1993b, Vibrations of Shells, Acoustical Society of America through the American Institute of Physics, New York.

Leppington, F.G., E.G. Broadbent, K.H. Heron and S.M. Mead, 1986, Resonant and non-resonant acoustic properties of elastic panels. I. The radiation problem, Proceedings Roy. Soc. Lond. A 406, 139.

Leppington, F.G., K.H. Heron, E.G. Broadbent and S.M. Mead, 1987, Resonant and non-resonant acoustic properties of elastic panels. II. The transmission problem, Proceedings Roy. Soc. Lond. A 412, 309.

Li, W.L., 2001, An analytical solution for the self- and mutual radiation resistances of a rectangular plate, *J. Sound Vib.* 245, 1.

Liamshev, L.M., 1958, Scattering of sound by a thin bounded rod, *Sov. Phys. Acoust.* 4, 50.

Liamshev, L.M., 1960, Theory of sound radiation by thin elastic plates and shells, *Sov. Phys. Acoust.* 5, 431.

Liew, K.M., C.M. Wang, Y. Xiang and S. Kitipornchai, 1998, *Vibration of Mindlin Plates*, Elsevier Science, Oxford, England.

Lighthill, M.J., 1964, *Fourier Analysis and Generalised Functions*, Cambridge University Press, Cambridge, England.

Lin, G.F. and J.M. Garrelick, 1977, Sound transmission through periodically framed parallel plates, *J. Acoust. Soc. Am.* 61, 1014.

Ljunggren, S., 1983, Generation of waves in an elastic plate by a vertical force and by a moment in the vertical plane, *J. Sound Vib.*, 90, 559.

Ljunggren, S., 1984, Generation of waves in an elastic plate by a torsional moment and a horizontal force, *J. Sound Vib.* 93, 161.

Ljunggren, S., 1987, Sound reduction of a double-wall cylinder at discrete angles of incidence, *Acustica* 63, 229.

Lomas, N.S. and S.I. Hayek, 1977, Vibration and acoustic radiation of elastically supported rectangular plates, *J. Sound Vib.* 52, 1.

Lueg, P., 1936, Process of silencing sound oscillations, U.S. Patent, No. 2,043,416.

Lyon, R.H. and R.G. DeJong, 1995, *Theory and Application of Statistical Energy Analysis* (2nd Ed.) Butterworth-Heinemann, Newton, Massachusetts.

Macadam, J.A., 1976, The measurement of sound radiation from room surfaces in lightweight buildings, *Appl. Acoust.* 9, 103.

Maidanik, G., 1962, Response of ribbed panels to reverberant acoustic fields, *J. Acoust. Soc. Am.* 34, 809.

Maidanik, G., 1966, The influence of fluid loading on the radiation from orthotropic plates, *J. Sound Vib.* 3, 288.

Manning, J. and G. Maidanik, 1964, Radiation properties of cylindrical shells, *J. Acoust. Soc. Am.* 36, 1691.

Mead, D.J., 1975, Wave propagation and natural modes in periodic systems: 1. mono-coupled systems, *J. Sound Vib.* 40, 1.

Mead, D.J., 1990, Plates with regular stiffening in acoustic media: vibration and radiation, *J. Acoust. Soc. Am.* 88, 391.

Mead, D.J., 1999, *Passive Vibration Control*, John Wiley & Sons, Inc., New York.

Mead, D.J. and A.K. Mallik, 1978, An approximate theory for the sound radiated from a periodic line-supported plate, *J. Sound Vib.* 61, 315.

Meirovitch, L., 1967, *Analytical Methods in Vibrations*, Macmillan, New York.

Meirovitch, L., 1990, Dynamics and Control of Structures, John Wiley & Sons, Inc., New York.

Mindlin, R.D., 1951, Influence of rotary inertia and shear on flexural motions of isotropic elastic plates, *J. Appl. Mech.* 18, 31.

Mkhitarov, R.A., 1972, Interaction of the vibrational modes of a thin bounded plate in a liquid, *Sov. Phys. Acoust.* 18, 123.

Mollo, C.G. and R.J. Bernhard, 1989, Generalised method of predicting optimal performance of active noise controllers, *AIAA J.* 27, 1473.

Morfey, C.L., 2001, *Dictionary of Acoustics*, Academic Press, London.

Morse, P.M., 1948, *Vibration and Sound* (2nd Ed.), McGraw-Hill, New York (reprinted by the Acoustical Society of America, 1981).

Morse, P.M. and H. Feshbach, 1953, *Methods of Theoretical Physics*, McGraw-Hill, New York.

Morse, P.M. and K.U. Ingard, 1968, Theoretical acoustics, Princeton University Press, Princeton, New Jersey.

Muggleton, J.M., M.J. Brennan and R.J. Pinnington, 2002, Wavenumber prediction of waves in buried pipes for water leak detection, *J. Sound Vib.* 249, 939.

Muller, P., 1983, Torsional-flexural waves in thin-walled open beams, *J. Sound Vib.* 87, 115.

Naghshineh, K. and G.H. Koopmann, 1993, Active control of sound power using acoustic basis functions as surface velocity filters, *J. Acoust. Soc. Am.* 93, 2740.

Naghshineh, K., G.H. Koopmann and A.D. Belegundu, 1992, Material tailoring of structures to achieve a minimum radiation condition, *J. Acoust. Soc. Am.* 92, 841.

Nelson, P.A. and S.J. Elliott, 1992, *Active Control of Sound*, Academic Press, London.

Newland, D.E., 1989, *Mechanical Vibration Analysis and Computation*, Longman Scientific and Technical, Harlow, England.

Nikiforov, A.S., 1981, Acoustic interaction of the radiating edges of a plate, *Sov. Phys. Acoust.* 27, 87.

Northwood, T.D., 1970, Transmission Loss of Plasterboard Walls, Build. Res. Note No. 66, National research Council of Canada, Ottawa.

Norton, M. and D.G. Karczub, 2003, Fundamentals of noise and vibration analysis for engineers (2nd *Ed.*), Cambridge University Press, Cambridge, England.

Ochmann, M. and F.P. Mechel, 2002, Analytical and numerical methods in acoustics, Chapter 0 in *Formulas of Acoustics*, Ed. F.P. Mechel, Springer Verlag, Berlin.

Oden, J.T., 1972, *Finite Elements of Nonlinear Continua*, McGraw-Hill, New York.

O'Hara, G.J., 1966, Mechanical impedance and mobility concepts, *J. Acoust. Soc. Am.* 41, 1180.

Olson, H.F. and E.G. May, 1953, Electronic sound absorber, *J. Acoust. Soc. Am.* 25, 1130.

Oppenheimer, C.H. and S. Dubowsky, 1997, A radiation efficiency for unbaffled plates with experimental verification, *J. Sound Vib.* 199, 473.

Pallett, D.S., 1972, Applications of Statistical Methods to the Vibrations and Acoustic Radiation of Fluid-loaded Cylindrical Shells, PhD Thesis, Pennsylvania State University, State College, Pennsylvania.

Pankoke, J., 1993, Measurement of Structure-borne Interior Noise using Acoustic Reciprocity, MSc Dissertation, University of Southampton, England.

Paulitsch, C.P., P. Gardonio and S.J. Elliott, 2006, Active vibration control using an inertial actuator with variable internal damping, *J. Acoust. Soc. Am.* 119, 2136.

Pavic, G., 1992, Vibroacoustical energy flow through straight fluid filled pipes, *J. Sound Vib.* 154, 411.

Petitjean, B. and I. Legrain, 1996, Feedback controllers for active vibration suppression, *Journal of Structural Control* 3, 111.

Petyt, M., 1998, *Introduction to Finite Element Vibration Analysis*, Cambridge University Press, Cambridge, England.

Petyt, M.P., J. Lea and G.H. Koopmann, 1976, A finite element method for determining the acoustic modes of irregular shaped cavities, *J. Sound Vib.* 45, 495.

Pierce, A.D., 1989, *Acoustics: An Introduction to its Physical Principles and Applications*, Acoustical Society of America, New York.

Pierri, R.A., 1977, Study of a Dynamic Absorber for Reducing the Vibrations and Sound Radiation of Plate-like Structures, MSc. Dissertation, University of Southampton, England.

Pilkey, W.D., 1964, Clebsch's method for beam deflections, *Journal of Engineering Education* 54, 170.

Pinnington, R. and A. Briscoe, 1994, Externally applied sensor for axisymmetric waves in a fluid-filled pipe, *J. Sound Vib.* 173, 503.

Pretlove, A.J., 1965, Free vibrations of a rectangular panel backed by a closed rectangular cavity, *J. Sound Vib.* 2, 197.

Preumont, A., 2002, *Vibration Control of Active Structures*, Kluwer Academic Publisher, The Netherlands.

Quirt, J.D., 1982, Sound transmission through windows: I. Single and double glazing, *J. Acoust. Soc. Am.* 72, 834.

Quirt, J.D., 1983, Sound transmission through windows: II. Double and triple glazing, *J. Acoust. Soc. Am.* 74, 534.

Randall, R.B., 1977, *Frequency Analysis*, Brüel and Kjaer, Naerum, Denmark.

Rao, S.S., 1995, *Mechanical Vibrations*, Addison-Wesley Publishing Company, New York.

Rayleigh, Lord, 1896, *The Theory of Sound* (2nd Ed.), (reprinted by Dover, New York, 1945).

Reddy, J.N., 1984a, *An Introduction to the Finite Element Method*, McGraw-Hill, New York.

Reddy, J.N., 1984b, Energy and Variational Methods in Applied Mechanics, John Wiley & Sons, Inc., New York.

Reddy, J.N., 1997, *Mechanics of Laminated Composite Plates*, CRC Press, New York.

Renji, K. and P.S. Nair, 1996, Modal density of composite honeycomb sandwich panels, *J. Sound Vib.* 195, 687.

Rennison, D.C., 1977, The Vibrational Response of, and the Acoustic Radiation from, Thin-walled Pipes, PhD Thesis, University of Adelaide, Australia.

Rindel, J.H. and D. Hoffmeyer, 1991, Influence of the stud distance on sound insulation of gypsum board walls, Proceedings of Inter-noise 91, 279.

Roure, A., 1985, Self adaptive broadband active noise control systems, *J. Sound Vib.* 101, 429.

Rschevkin, S.N., 1963, *Lectures on the Theory of Sound*, Pergamon Press, London.

Rubin, S., 1967, Mechanical immitance- and transmission-matrix concepts, *J. Acoust. Soc. Am.* 41, 1171.

Rumerman, M.L. 2002, The effect of fluid loading on radiation efficiency, *J. Acoust. Soc. Am.* 111, 75.

Sayhi, M.N., Y. Ousset and G. Verchery, 1981, Solution of radiation problems by collocation of integral formulations in terms of single and double layer potentials, *J. Sound Vib.* 74, 187.

Scharton, T.D. and R.H. Lyon, 1968, Power flow and energy sharing in random vibration, *J. Acoust. Soc. Am.* 43, 1332.

Schenck, H.A., 1967, Improved integral formulation for acoustic radiation problems, *J. Acoust. Soc. Am.* 44, 418.

Schenck, H.A., 1968, Improved integral formulation for acoustic radiation problems, *J. Acoust. Soc. Am.* 44, 41.

Schroter, V. and F.J. Fahy, 1981, Radiation from modes of a rectangular panel into a coupled fluid layer, *J. Sound Vib.* 74, 575.

Serati, P.M. and S.E. Marshall, 1989, Comparison of experimental and analytical estimations for the modal density of a ring-stiffened cylinder, Proceedings of Inter-Noise 89, Newport Beach, California, 1199.

Sewell, E.C., 1970, Transmission of reverberant sound through a single leaf partition surrounded by an infinite rigid baffle, *J. Sound Vib.* 12, 21.

Seybert, A.F., D. Hamilton and P.A. Hayes, 1998, Prediction of radiated noise from machine components using the BEM and the Rayleigh integral, *Noise Control Eng.* 46, 77.

Sharp, B., 1978, Prediction methods for the sound transmission of building elements, *Noise Control Eng.* 11, 53.

Shuku, T. and K. Ishihara, 1973, The analysis of the acoustic field in irregularly shaped rooms by the finite element method, *J. Sound Vib.* 29, 67.

Skudrzyk, E., 1968, *Simple and Complex Vibratory Systems*, The Pennsylvania State University Press, Pennsylvania, USA.

Skudrzyk, E., 1971, *The Foundations of Acoustics*, Springer-Verlag, New York.

Smith, P.W., 1962, Response and radiation of structural modes excited by sound, *J. Acoust. Soc. Am.* 34, 640.

Smolenski, C.P. and E.M. Krokosky, 1973, Dilatational-mode sound transmission in sandwich panels, *J. Acoust. Soc. Am.* 54, 1449.

Snowdon, J.C., 1968, *Vibration and Shock in Damped Mechanical Systems*, John Wiley & Sons, Inc., New York.

Snyder, S.D., and N. Tanaka, 1995, Calculating total acoustic power output using modal radiation efficiencies, *J. Acoust. Soc. Am.* 97, 1702.

Soedel, W., 2004, *Vibrations of Shells and Plates* (3rd Ed.), Marcel Dekker Publishing Co., New York.

Sommerfeld, A., 1949, *Partial Differential Equations in Physics*, Academic Press, New York.

Stepanishen, P., 1978, Radiated power and radiation loading of cylindrical surfaces with non-uniform velocity distribution, *J. Acoust. Soc. Am.* 63, 328.

Stepanishen, P., 1982, Modal coupling in the vibration of fluid-loaded cylindrical shells, *J. Acoust. Soc. Am.* 71, 813.

Strawderman, W.A., S.H. Ko and A.N. Nuttall, 1979, The real roots of the fluid-loaded plate, *J. Acoust. Soc. Am.* 66, 579.

Sullivan, J.M., J.E. Hubbard Jr. and S.E. Burke, 1996, Modeling approach for two-dimensional distributed transducers of arbitrary spatial distribution, *J. Acoust. Soc. Am.* 99, 2965.

Sun, J.Q., 1996, Some observations on physical duality and collocation of structural control sensors and actuators, *J. Sound Vib.* 194, 765.

Szechenyi, E., 1971a, Modal densities and radiation efficiencies of unstiffened cylinders using statistical methods, *J. Sound Vib.* 19, 65.

Szechenyi, E., 1971b, Sound transmission through cylinder walls using statistical considerations, *J. Sound Vib.* 19, 83.

Thompson, L.L., 2006, A review of finite-element methods for time-harmonic acoustics, *J. Acoust. Soc. Am.* 119, 1315.

Timoshenko, S., D.H. Young and W. Jr. Weaver, 1992, *Vibration Problems in Engineering* (*4th Ed.*) John Wiley & Sons, Inc., New York.

Tseo, G.G, 1972, The zoning of radiation modes of a finite plate on a wavenumber diagram, *J. Acoust. Soc. Am.* 52, 1564.

Twee, L.W. and D.R. Tree, 1978, Three methods for predicting the insertion loss of close-fitting acoustical enclosures, *Noise Control Eng.* 10, 74.

Variyart, W. and M.J. Brennan, 2002, Simplified dispersion relationships for *in-vacuo* pipes, *J. Sound Vib.* 256, 995.

VDI., 1996, Noise at pipes, Guideline VDI 3733, Verein Deutche Ingenieure, VDI, Düsseldorf, Germany.

Vér, I.L., 1992, Interaction of sound waves with solid structures in noise and vibration control engineering, Eds. L.L. Beranek and I. Vér, John Wiley and Sons, Inc., New York.

Villot, M., G. Chavériat and J. Roland, 1992, Phonoscopy: an acoustical holography technique for plane structures radiating into enclosed spaces, *J. Acoust. Soc. Am.* 91, 187.

Villot, M., C. Guigou and L. Gagliardini, 2001, Predicting the acoustical radiation of finite size multi-layered structures by applying spatial windowing on infinite structures, *J. Sound Vib.* 245, 433.

Vitiello, P., P.A. Nelson and M. Petyt, 1989, Numerical Studies of the Active Control of Sound Transmission Through Double Partitions, ISVR Technical Report No. 183, University of Southampton.

Von Estorff, O., 2000, *Boundary Elements in Acoustics*, WIT Press, Southampton, England.

Von Venzke, G., P. Dämmig and H.W. Fischer, 1973, Der Einfluss von Versteifungen auf die Schallabstrhlung und Schalldämmung von Metallwänden, *Acustica* 29, 29.

Wallace, C.E., 1972, Radiation resisistance of a rectangular panel, *J. Acoust. Soc. Am.* 51, 946.

Warburton, G.B., 1951, The vibration of rectangular plates, Proceedings of the Institute of Mechanical Engineering 168, 371.

Warburton, G.B., 1976, *The Dynamical Behaviour of Structures* (*2nd Ed.*), Pergamon, Oxford, England.

Warburton, G.B., 1978, Comments on "Natural frequencies of cylindrical shells and panels in vacuum and in a fluid", *J. Sound Vib.* 60, 465.

Watson, G.N., 1966, *A Treatise on the Theory of Bessel Functions* (*2nd Ed.*), Cambridge University Press, Cambridge.

Weinreich, G., 2005, Interactions of a fluid with an almost surrounding shell, *Acta Acustica with Acustica* 91, 197.

Wilkinson, J.P.D., 1967, Modal densities of certain shallow structural elements, *J. Acoust. Soc. Am.* 43, 245.

Williams, E.G., 1983a, A series expansion of the acoustic power radiated from plane sources, *J. Acoust. Soc. Am.* 73, 1520.

Williams, E.G., 1983b, Numerical evaluation of the radiation from unbaffled, finite plates using the FFT, *J. Acoust. Soc. Am.* 74, 343.

Wolde, T.T., J.W. Verheij and H.F. Steenhoek, 1975, Reciprocity methods for the measurement of mechano-acoustical transfer functions, *J. Sound Vib.* 42, 49.

Wu, T.W., 2000, *Boundary Element Acoustics, Fundamentals and Computer Codes*, WIT Press, Southampton.

Xu, M.B. and W.H. Zhang, 2000, Vibrational power input and transmission in a circular cylindrical shell filled with fluid, *J. Sound Vib.* 234, 387.

Zienkiewicz, O.C. and R.L. Taylor, 2000, *The Finite Element Method*, (5th Ed.), Butterworth, London.

Index

Printed and bound by CPI Group (UK) Ltd, Croydon, CR0 4YY

03/10/2024

01040413-0018